Fertilizers and Pesticides: Assessment and Applications

Fertilizers and Pesticides: Assessment and Applications

Edited by Edwin Tan

New York

Published by Syrawood Publishing House,
750 Third Avenue, 9th Floor,
New York, NY 10017, USA
www.syrawoodpublishinghouse.com

Fertilizers and Pesticides: Assessment and Applications
Edited by Edwin Tan

International Standard Book Number: 978-1-68286-707-5 (Hardback)

Cataloging-in-Publication Data

Fertilizers and pesticides : assessment and applications / edited by Edwin Tan.
p. cm.
Includes bibliographical references and index.
ISBN 978-1-68286-707-5
1. Fertilizers. 2. Pesticides. 3. Agricultural chemicals. 4. Soil fertility. I. Tan, Edwin.
S633 .F47 2019
631.8--dc23

TABLE OF CONTENTS

PREFACE

This book has been a concerted effort by a group of academicians, researchers and scientists, who have contributed their research works for the realization of the book. This book has materialized in the wake of emerging advancements and innovations in this field. Therefore, the need of the hour was to compile all the required researches and disseminate the knowledge to a broad spectrum of people comprising of students, researchers and specialists of the field.

Fertilizers are natural or artificial substances that are applied to the soil or plant tissues to facilitate the growth of the plant by providing essential nutrients. They also enhance the water retention capacity of the soil. These can be naturally prepared or industrially produced. Fertilizers provide varied macronutrients such as nitrogen, potassium, sulfur, phosphorous, calcium, etc. Pesticides are the substances that are used to control pests and weeds. They are classified as herbicides, insecticides, bactericides, fungicides, etc. depending on the target organism. Pesticides help in improving crop yield and quality. This book is compiled in such a manner, that it will provide in-depth knowledge about the varied aspects of pesticides and fertilizers. Different approaches, evaluations and advanced studies have been included herein. This book is a vital tool for all researching or studying agricultural science, crop science and soil science as it gives incredible insights into emerging trends and concepts.

At the end of the preface, I would like to thank the authors for their brilliant chapters and the publisher for guiding us all-through the making of the book till its final stage. Also, I would like to thank my family for providing the support and encouragement throughout my academic career and research projects.

Editor

A Chemical-Physical Procedure to Reduce Levels of Potentially Toxic Elements (PTEs) in Municipal Sewage Sludges

Baffi C[1]*, Cella F[2], Fumi I[2] and Trevisan M[1]

[1]*Centro di Ricerca BIOMASS, Facoltà di Scienze Agrarie, Alimentari e Ambientali, Università Cattolica del Sacro Cuore, Piacenza, Italy*
[2]*Syngen, Agrosistemi, Piacenza, Italy*

Abstract

The current waste policy in EU aims at reducing the amounts of sludges to be disposed in landfill, encouraging their energy reuse (e.g. energy source, fertilizers, etc.). In this perspective a reduction of potentially toxic elements (PTEs) and water is welcome, in view of a more sustainable reutilization of sludges, e.g. in agriculture. Aim of this work was to set up a chemical-physical procedure, named modified Fenton process (patent n° PC2012A000008 of 22/03/2012), which was able to improve the reduction of some PTEs (Zn, Ni, Cu, Pb, Cd and Cr) and water in municipal sewage sludges from anaerobic digestion, better than a conventional Fenton process, avoiding significant decreases of total organic carbon (TOC).

Keywords: Municipal sewage sludge; Potentially toxic elements; Fenton reaction; Sludge dewatering

Introduction

During the last 20 years sewage sludge production has increased from 6.5 M tons of dry solids (DS) in 1992 (EU-15) to 9.8 M tons of DS in 2006 (EU-27-All member States) (EC 2006 [1] with forecasted values exceeding 13 M tons of DS up to 2020 (Milieu Ltd., WRc and RPA) [2]. On average, the preferred final destination of sludges remains landfill (38%), followed by recycling (24-25%), incineration (20-22%) and composting (15-18%). The waste policy (Directive 2008/98/EC) [3] has pointed out that, in order to minimise the negative effects of the generation and management of wastes on human health and the environment, all member States should reduce the use of resources and favour the practical application of the waste hierarchy a) prevention b) preparing for re-use c) recycling d) other recovery (e.g. energy recovery) e) disposal. The following actions should be improved: removing metals by chemical or microbiological leaching, dewatering, to reduce sludge volumes and relative costs, preserving acceptable levels of organic matter in treated sludges for their profitable future reuse. Municipal sludges originate from processes of treatment of waste-waters in which physical-chemical processes are involved, so they tend to concentrate PTEs (Milieu Ltd., WRc and RPA) [2] with increased potential risks for the environment and human health [4]. In order to efficiently reduce PTEs in sewage sludges, the suggested approaches are the control of the point or the diffuse sources and the extractive removal [5]. For the latter, the chemical leaching with use of inorganic or organic salts is employed, which consists of four steps: solubilisation of the PTEs; separation of the water-phase containing the mobilized PTEs and the sludge particles; removal of sludge particles from the leachate containing PTEs and precipitation of PTEs present in the leachate and their subsequent removal [4]. The extraction is frequently preceded by an oxidation treatment, known as classic Fenton reaction, where H_2O_2, activated by iron salts at pH of 2.5-3.0, has proved to be more effective, than H_2O_2 alone, in the destruction of toxic organics in wastewaters [6]. Moreover Fenton's reagent was observed to improve sludge dewaterability (Neyens and Baeyens) [7]; this is an important feature in that it allows high saving-powers and a more sustainable reuse of sludges. Numerous are the studies concerning the effects of Fenton's oxidation process on the reduction of organic contaminants in wastewaters and landfill leachates [8-10] but few concern PTEs reduction and dewatering.

The aim of this work was to set up a chemical-physical procedure, called modified Fenton treatment (patent n° PC2012A000008 of 22/03/2012), performed on municipal sludges from anaerobic digestion, to improve the reduction of some PTEs levels, to increase the dewatering efficacy, with a low TOC decrease, respect to a classic Fenton treatment.

Materials and Methods

Reagents and instrumentation

The following reagents, pure for analysis (Carlo Erba, Milano, Italy), were used: HCl 37% (m/v), HNO_3 65% (m/v), H_2SO_4 96% (m/v), H_2O_2 (40% m/v), $Fe_2SO_4 \times 7H_20$. Multi element standard solutions were prepared by dilution with distilled water, containing the same quantities of acids as the samples. The distilled water used had a conductivity ≤ 0.1 μS cm^{-1} (residual at evaporation <1 ppm). The following instruments were used: pH-meter CRISON mod. Basic 20 meter (Barcelona, Spain); stove mod. PID system (MPM Instruments, Bernareggio, MB, Italy); CN elemental analyser (mod. Vario Max CN Macro Elementar Analyzer, Hanau, Germany); open digester system (Digiprep mod. Jr, SCP SCIENCE, Baie d'Urfè, QC H9X 4B6, Canada) (time=120 minutes; temperature=95°C); inductively coupled plasma optical emission spectrometer PerkinElmer Optima 2100 DV simultaneous ICP-OES (PerkinElmer Life and Analytical Sciences, Shelton, CT, USA), equipped with an AS-93 plus autosampler, a concentric glass (Meinhard) with Cyclonic Spray Chamber (Table 1).

Sample preparation and methods of analysis

The municipal sewage sludges was obtained from an anaerobic

***Corresponding author:** Baffi C, Centro di Ricerca BIOMASS, Facoltà di Scienze Agrarie, Alimentari e Ambientali, Università Cattolica del Sacro Cuore, Piacenza Italy, E-mail: claudio.baffi@unicatt.it

	Perkin Elmer Optima 2100 DV Simultaneous ICP
Frequency	40 MHz, free-running
Incident power	1400 W
Reflected power	<5 W
	Ar gas flow rate:
Plasma	15.0 L min^{-1}
Auxiliary	0.20 L min^{-1}
Nebulizer	0.70 L min^{-1}
Peristaltic pump	1.20 mL min^{-1}
Nebulizer	Concentric glass (Meinhard) with cyclonic spray chamber
Autosampler	AS 93plus

Table 1: ICP-OES Instrumental operating parameters.

wastewater treatment plant (full capacity of 163,000 equivalent inhabitants) sited in Piacenza, Italy. Samples were collected in polypropylene bottles, shipped cold, and kept at 4°C before use. Sampling, pH and Total Solids (TS) determination were carried out according to the National Official Methods for sewage sludges (IRSA CNR, 1985) [11]. Before analysis of PTEs and carbon, the sample was dried at 105°C overnight, homogenized and ground to 0.2 mm for PTE analysis and to 0.5 mm for total C analysis. For PTE determination (Cd, Cr, Cu, Ni, Pb and Zn) an aqua regia digestion (1:3=HNO_3:HCl) was performed in an open digestion system. 0.50 g of sample was weighed and put in contact with 14 mL of aqua regia, at T=95°C for approx. 120 minutes; the digests were brought to a final volume of 50 mL, filtered by paper (Schleicher and Schuell n° 589 Blue ribbon ashless) and analyzed for metal contents by inductively coupled plasma emission spectrometry. Total Carbon (TC) was determined by sample combustion on a CN elemental analyser on about 0.250 g of sludge sample; Inorganic Carbon (IC) was determined according to the Official Methods of Soil Analysis (MIPAF G.U.R.I) [12]. Total Organic Carbon content (TOC) was calculated by the subtraction of inorganic carbon from total carbon: TOC=TC-IC. The values of the main chemical-physical parameters of the sludge are shown in Table 2.

Experimental procedures

To compare the efficacy of the proposed procedure vs. the conventional Fenton method, a RBD (Randomized Block Design) experiment was set up in laboratory; three treatments plus a control were arranged, with five replicates.

A first treatment, called P, was performed with use of H_2O_2 alone; it consisted in putting in contact 80 g of H_2O_2 to an aliquot of 1.6 kg of sludge for different times (60 min and 180 min). The mixtures were brought to a value of 7% of TS in order to facilitate subsequent operations such as filtering etc. Then the samples were submitted to vacuum filtration on a Buchner funnel (diameter 90 mm) on extra-rapid filter paper Whatman 41 for 8 minutes, followed by 5 minutes on a press. Determination of TS, PTEs and TOC were carried out on the residues of the filtration, called samples P_1 and P_3; the former refers to a contact time of 60 min, the latter to a contact time of 180 min with H_2O_2.

A second treatment, called R, concerned the use of the classic Fenton's reaction. 80 g of H_2O_2 and 22.85 g of $FeSO_4 \times 7H_2O$ (Fe^{2+} concentration equal to 3.58 mg g^{-1} DS and H_2O_2 equal to 62.5 mg g^{-1} DS with a ratio Fe^{2+}/ H_2O_2 of 0.057) were added to an aliquot of 1.6 kg of sludge. Aliquots of H_2SO_4 (50% v/v) were also added to reach pH 3.0 (classic Fenton oxidation); this mixture was brought to a value of 7% TS, as in the case of the first treatment. pH determinations were made on samples, named R_1 and R_3, in virtue of different contact times with H_2O_2 (the former refers to 60 min; the latter to 180 min). These samples were submitted to vacuum filtration on a Buchner funnel (diameter 90 mm) on extra rapid filter paper Whatman 41 for 8 minutes, followed by 5 minutes on the press under a pressure ranging between 0 and 15 bars for 5 minutes. Determinations of TS, PTEs and TOC were carried out on the residues of the filtration (called R_1 and R_3).

A third treatment, called Q, concerned the procedure with use of modified Fenton's reaction. 80 g of H_2O_2 and 22.85 g of $FeSO_4 \times 7H_2O$ (Fe^{2+} concentration equal to 3.58 mg g^{-1} DS and H_2O_2 equal to 62.5 mg g^{-1} DS with a ratio Fe^{2+}/H_2O_2 of 0.057) were added to an aliquot of 1.6 kg of sludge. Aliquots of H_2SO_4 (50% v/v) were also added to reach pH 3.0 (classic Fenton reaction), then others were added to reach pH 2.0; this mixture was brought to a value of 7% of TS, as seen before. pH determinations were carried out on Q_1 and Q_3 samples, referring to contact time with H_2O_2 of 60 min and 180 min., respectively. Then Q_1 and Q_3 samples were submitted to vacuum filtration on a Buchner funnel (diameter 90 mm) on extra-rapid filter paper Whatman 41 for 8 minutes, followed by 5 minutes on the press (pressure of 0-15 bar for 5 minutes). Determinations of TS, PTEs and TOC were carried out on the residues of the filtration. A treatment, called O, was prepared as a control and it had no treatment with H_2O_2, no press, no filtration and it was brought to a value of 7% of TS. All information concerning experimental data design is shown in Table 3.

Statistical data analysis

IBM SPSS vs. 19.0 software (2010) was used to carry out statistical data processing.

parameter	Measure units	value
pH		8.10
Total moisture	%	80.29
Total solids (TS)	%	19.71
Residue at 550°C	%	48.28
Lost to calcination	%	51.72
Total organic carbon (TOC)	% on DS	28.86
Total Cd	mg kg^{-1} DS	3.05
Total Cr	mg kg^{-1} DS	194
Total Cu	mg kg^{-1} DS	612
Total Ni	mg kg^{-1} DS	125
Total Pb	mg kg^{-1} DS	82.0
Total Zn	mg kg^{-1} DS	1031

Table 2: Mean values of chemical-physical parameters of the sludge.

Treatment	sample	Contact time with H_2O_2 min	pH	H_2O_2 added (% of sludge weight)
O*	O	---	8.04	0
P	P_1 P_3	60 180	7.97 8.09	5 5
R	R_1 R_3	60 180	3.34 3.32	5 5
Q	Q_1 Q_3	60 180	2.02 2.06	5 5

*O: control (no treatment); P: addition of H_2O_2 alone; R: classic Fenton's treatment; Q: proposed modified Fenton's treatment.

Table 3: Experimental design with treatments and values of pH and added H_2O_2.

Results

Yields of leached PTEs

Table 4 shows the data relative to the leached aliquots of PTEs from sludge, which are calculated as percentage variation (ΔPTE%) with respect to the values observed for the control treatment (O). The outcomes of a statistical comparison (one-way ANOVA + LSD test) are also reported.

Very low or no decreases are observed for the treatment with H_2O_2 alone (P_1 and P_2); the pH of this treatment is 7.97 and 8.09 respectively, very close to that of the control (8.04).

When pH levels decreased, following Fenton's treatment, the values of leached PTEs increased, for both R_1 and R_3, which show similar pH values (3.34 and 3.32 respectively). With regard to this, a more prolonged contact time for H_2O_2 (R_3) proves better than a shorter time (R_1) for obtaining higher values of leached PTEs: -68.8 vs. -66.9 for Zn; -52.1 vs. -51.6 for Ni; -12.9 vs. -8.73 for Cu; -3.45 vs. -3.20 for Cd. This treatment did not prove suitable for Pb and Cr, in that no decrease was observed for the leached aliquots.

When the modified Fenton treatment was applied, further decreases were observed with respect to the classic Fenton reaction for all investigated PTEs but, in some cases (Zn, Cu and Cr) the highest values were observed for the lower contact time (Q_1 at pH=2.02) respect to the longest contact time (Q_3 at pH 2.06). In fact, for these three PTEs the leached values were: for Zn (-76.9 for Q_1; -73.1 for Q_3); for Cu (-33.4 for Q_1; -19.0 for Q_3); for Cr (-4.14 for Q_1; -3.04 for Q_3). On the contrary, Ni, Cd and Pb showed higher leached values for the longer contact time with H_2O_2 (Q_3) than with the shorter time (Q_1), with the following values: for Ni (-62.2 for Q_1; -62.5 for Q_3); for Cd (-17.0 for Q_1; -21.2 for Q_3); for Pb (-3.44 for Q_1; -3.51 for Q_3).

Dewatering efficiency

Table 5 shows the mean percentage values of Total Solids (TS %) and dewatering (%) which were much affected (P<0.01) by treatments.

The dewatering values were calculated as the percentage difference with respect to the control (treatment O). The addition of H_2O_2 alone gives rise to TS values (range 27-29%) similar to the control (26.8%) with very low dewatering values (-0.65%; -2.96%). Fenton's reactions proved to be better than H_2O_2 alone, with values of TS ranging from 37.8% to 44.2% and dewatering values from -15.0% to -23.8%; a contact time of 60 min proved to be as good as 180 min. The modified Fenton treatment proved to be better than classic Fenton for obtaining better TS values (range 44.0-44.2% vs. 37.8-38.1%) and for better dewatering percentages (range: -23.5; -23.8% vs. -15.0 ;-15.5%).

Total organic carbon (TOC)

Table 6 shows data concerning TOC contents in sludge (expressed as g kg^{-1} DS). When H_2O_2 alone is used, no appreciable variation of TOC with respect to the control is observed (28.05 g kg^{-1}DS-28.19 g kg^{-1}DS vs. to 28.24 g kg^{-1} DS). In presence of Fenton's reactions a reduction of TOC is observed (range 25.64-26.57 g kg^{-1}DS) compared to the control (28.24 g kg^{-1} DS). The modified Fenton treatment gives higher values of TOC with respect to the classic Fenton (range 26.23 g kg^{-1} DS-26.57 g kg^{-1}DS with respect to 25.64 g kg^{-1}DS-26.07 g kg^{-1}DS) with a value of 26.57 g kg^{-1}DS for Q_3 (180 min of contact time), significantly higher (P<0.01) than the values observed for the other treatments (R_1, R_3 and Q_1) which employed Fenton's reaction.

Samples	ΔZn %	ΔNi %	ΔCu %	ΔCd %	ΔPb %	ΔCr %
P1	0.51 A*	-0.41 A	-0.02 B	0.39 A	-1.76 AB	1.18 B
P3	2.66 A	1.74 A	5.41 A	-0.16 A	0.26 AB	3.52 AB
R1	-66.9 B	-51.6 B	-8.73 C	-3.20 A	1.33 AB	3.21 AB
R3	-68.8 B	-52.1 B	-12.9 D	-3.45 A	1.94 A	5.31 A
Q1	-76.9 C	-62.2 C	-33.4 F	-17.0 B	-3.44 B	-4.14 C
Q3	-73.1 BC	-62.5 C	-19.0 E	-21.2 B	-3.51 B	-3.04 C

* In this and the following tables statistical comparisons are intended for columns; capital letters point out significant (P<0.01) differences among means

Table 4: Mean values of percentage variations of PTEs leached; statistical comparisons following One-way ANOVA + LSD test.

samples	Total Solids %	Dewatering %
O	26.8 A	----
P1	29.0 B	-2.96 B
P3	27.3 AB	-0.65 A
R1	37.8 C	-15.0 C
R3	38.1 C	-15.5 C
Q1	44.0 D	-23.5 D
Q3	44.2 D	-23.8 D

Table 5: Mean values of percentages of Total Solids (TS) and dewatering; statistical comparisons following One-way ANOVA + LSD test.

samples	TOC (g kg^{-1} DS)
O	28.24 E
P1	28.19 ED
P3	28.05 D
R1	26.07 B
R3	25.64 A
Q1	26.23 B
Q3	26.57 C

Table 6: Mean values of Total Organic Carbon (TOC) content (g kg^{-1} DS); statistical comparisons following One-way ANOVA + LSD test.

Discussion

Yields of leached PTEs

Working conditions for pH, ferrous ion and hydrogen peroxide concentrations were established when Advanced Oxidation Processes (AOP), such as Fenton or Fenton-like systems, were successfully employed to remove organic contaminants from wastewaters; a recent review, dealing with iron-free Fenton-like systems for activating H_2O_2, was published [13] which concerned the removal of organic pollutants with high chemical stability and/or biodegradability. The setting up of the above mentioned parameters is still valid when Fenton or Fenton-like systems are applied before the extraction of heavy metals and/or PTEs from wastewaters [4]; here other chemical-physical processes (e.g. speciation) have to be held into consideration, together with the type of extraction (sequential, single) and the type of extractant (single acid, mixture of acids, chelating agents).

The reduction of pH alone, although sometimes allows instant metal solubilisation, as observed for Zn [14], failed for the majority of PTEs unless supported by an increase in the sludge Oxygen Reduction Potential (ORP); this could be caused by biological and chemical oxidation, obtained by H_2O_2 addition [15]. The H_2O_2 addition alone fails to reduce metal contents in a neutral environment and needs more acidic conditions through the use of different acids (H_2SO_4, HCl, HNO_3, etc.).

The pH is considered a key factor for the efficiency of Fenton and Fenton-like reactions, which consist in the activation of H_2O_2 by ferrous (Fe^{2+}) ions to generate hydroxyl radicals (HO•) via a complex reaction sequence (Fenton) [16-18]. The need for strictly acidic conditions is necessary for all AOP, such as Fenton's systems, because they are essential for preventing iron precipitation. The efficacy of the Fenton and Fenton-like processes to degrade organic compounds in wastewaters reaches its optimum at a pH of about 3 [7] and it is reduced at higher and lower pH levels.

At higher pH levels, the oxidation efficiency of Fenton's reagent may decrease because ferric ions could form $Fe(OH)_3$ [19], which will not react with H_2O_2, because of its low activity [20]. The ferric ions in the solution that can react with H_2O_2 are reduced with the reaction as follows: $Fe^{3+} + H_2O_2 \leftrightarrow Fe\text{-}OOH^{2+} + H^+$. This reaction is the rate-limiting step. Moreover, at high pH levels, the auto decomposition of H_2O_2 is accelerated.

At very low pH levels, in presence of high H_2O_2 concentrations, the formation of $FeOOH^{2+}$ will slow down which consequently causes the production rates of ferrous ions and hydroxyl radicals to decrease as well. The earlier reaction may delay the Fenton reaction. Moreover at very low pH levels, an iron complex species $[Fe\ (H_2O)_6]^{2+}$ exists, which reacts more slowly with hydrogen peroxide than the other species [21]. This phenomen was also influenced by the concentration of ferrous ion present. In addition, the peroxide becomes solvated in the presence of high H^+ ion concentrations to form stable oxonium ions $[H_3O_2]^+$. Oxonium ions make hydrogen peroxide more stable and reduce its reactivity with ferrous ions [22,23].

If Fenton or Fenton-like reactions are used in procedures with metal or PTE extraction, the above mechanisms are still valid, together with considerations concerning metal speciation and kinetics, due to different extraction procedures and types of extractant.

Our results strengthen the assumption of very low efficacy of H_2O_2 use alone (pH about 8.0) for all investigated PTEs (Zn, Ni, Cu, Cd, Pb and Cr) both with 60 min and 180 min of contact time with H_2O_2. The use of classic Fenton's reaction improved the leached quotas much more than H_2O_2 alone, by virtue of lower pH levels (3.34 for R_1 and 3.32 for R_3). The most leached PTEs were Zn (about 67%) and Ni (about 51%); at a lesser extent Cu (8-12%) and Cd (3.2-3.4%); no leaching was observed for Pb and Cr. The leached quota increased with increased contact times for H_2O_2, from R_1 to R_3, for all PTEs.

The modified Fenton procedure, carried out at a further lower pH (pH ≈ 2.0), improved all quotas of the leached PTEs, with respect to the classic Fenton procedure. This agrees with results of other authors (Lake) [24,25] About an increased extractability of metals following progressive acidification steps.

These results are consistent with those observed in trials made on extractability of metals (Zn, Cu, Cd and Pb), carried out on an anaerobically digested sludge with the use of three different acids: nitric acid, citric acid and oxalic acid [4]. For some PTEs (Ni, Cd and Pb) a longer contact time with H_2O_2 allowed higher leached quotas, as occurred in the classic Fenton procedure. The amounts of Zn and Cr were slightly higher for lower times. The behaviour of Cu for which a sharp decrease of leached quota was observed for longer contact times was unusual. This could be explained, as observed by other authors (Marchioretto) [4], by the fact that Cu presents high affinity for organic complexes and a more prolonged contact time would have promoted the formation of Cu soluble complexes, more extractable with organic acids (e.g. citric or oxalic acid) than with HNO_3. These authors observed a percentage of 40.5% of Cu extracted by HNO_3 at a pH=1.0, higher than our result (33.4% at pH=2.2). Low values of leached Pb and Cr were observed in this trial for the modified Fenton procedure in comparison with no values for the classic Fenton. A chemical leaching with HCl at a pH level of 1 should probably be adopted; this would allow a solubilisation of almost 100% of Pb and also provide the best extraction yield for Cr (72%) [4]. The low values of leached Pb were probably due to the formation of insoluble $PbSO_4$ following the treatment with H_2SO_4; a further acidification step with the use of HCl is probably needed to allow a complete Pb dissolution. For Cr, in order to increase the leached quota, the use of a closer digestion system (e.g. microwave oven) would be preferable instead of the open system used.

Dewatering efficiency

The temperature, hydrogen peroxide concentration, pH and reaction time affect the dewaterability of the sludges [26]. Positive effects of Fenton's peroxidation on dewaterability of sludges were observed [7,27,28]. Fenton's reagent is known to have different treatment functions, depending on the $H_2O_2/FeSO_4$ ratio [29]. When the amount of Fe^{2+} employed exceeds that of H_2O_2, the treatment tends to have an effect of chemical coagulation; when the two amounts are reversed, the treatment tends to have the effect of chemical oxidation [7,30,31]. In this study classic and modified Fenton reactions (respectively at pH 3.0 and 2.0) were performed in the presence of 3.58 g Fe^{2+} kg^{-1} DS and with 62.5 g H_2O_2 kg^{-1} DS (ratio $Fe^{2+}/\ H_2O_2$ equal to 0.057) which gave better dewatering results in preliminary trials than other tested ratios (0.028, 0.114 and 0.228) obtained by varying the amounts of H_2O_2 added. The employed ratio (0.057) allowed us to obtain satisfactory results (about 23% of water reduction for modified Fenton and approx. 15% for classic Fenton). Good results were obtained by adding 1.67 g Fe^{2+} kg^{-1} DS and 25 g H_2O_2 kg^{-1} DS (ratio $Fe^{2+}/\ H_2O_2$ equal to 0.067) after proving other ratios with additions of quantities of H_2O_2 variables in the range 5-50 g H_2O_2 kg^{-1} DS [26]. Good dewater ability results were also obtained with a lower ratio $Fe^{2+}/\ H_2O_2$ (0.016) by using 20 g Fe^{2+} kg^{-1} DS and 125 g H_2O_2 kg^{-1} DS [32]. In our study a time of 60 min proved to be good enough for dewatering, similar to times of 80 min [32], 60-90 min [26] and 30 min [9].

Total organic carbon (TOC)

When H_2O_2 alone is used no appreciable loss of TOC was observed (less than 1%) 0); this agrees with observations of Authors (Neyens and Bayens) [7] in which the use of H_2O_2 alone was not effective for decreasing high concentrations of certain refractory contaminants. The reduction of TOC was more pronounced (-8.44 on average) for classic Fenton than for modified Fenton (-6.51% on average), both values being low, less than 10%. In general higher reduction values of organics are observed with the application classic Fenton to wastewater treatment. The reason for this can be found considering that classic Fenton process includes two steps: a 1st step where oxidation takes place at low pH (2.5-3.0) and a 2nd step where coagulation is observed at higher pH values; this last process (here not present) is retained to have a primary role in the selective removal of organics [7]. Moreover the modified Fenton treatment revealed more efficient than classic Fenton in maintaining high TOC values, in particular after 180 m of contact (Q_3) time respect to a lower time (60 min) (Q_1).

Conclusions

This laboratory study highlights the importance of more acidic conditions (pH=2.0) to improve the efficacy of the classic Fenton reaction applied as a pre-treatment to the chemical leaching concerning PTEs contained in a municipal anaerobic sewage sludge. In this study a modified Fenton treatment, carried out at pH 2.0, proved to be more reliable than classic Fenton (approx. pH 3.0) in decreasing the contents of Zn, Ni, Cu and Cd, in increasing dry solids with a higher dewater ability percentage and in preserving higher amounts of TOC. Further research is needed to improve Pb and Cr removal aliquots.

Acknowledgements

The Authors thank Mrs. Marina Modenesi (IREN, Municipal Purification Plant of Piacenza) for her kindness in supplying basic materials.

References

1. http://epp.eurostat.ec.europa.eu/portal/page/portal/environment/data/main_tables.
2. Milieu Ltd, WRc, Risk and Policy Analysts Ltd. (RPA) (2010) Environmental, economic and social impacts of the use of sewage sludge on land. Final report, Part III: Project Interim Reports, DG ENV.G.4/ETU/2008/0076r, 10.2.2010.
3. CEC (Council of the European Communities) (2008) Directive 2008/98/EC of the European Parliament and of the Council of 19 November 2008 on waste and repealing certain Directives. Official Journal of the European Union N°.L 312/3-30.
4. Marchioretto MM (2003) Heavy metals removal from anaerobically digested sludge. PhD. Thesis Wageningen University, the Netherlands.
5. Rulkens WH, van Voorneburg F, Joziasse J (1989) Removal of heavy metals from sewage sludges. In: Sewage Sludge Treatment and Use. Elsevier Applied Science. The Netherlands.
6. Huang CP, Dong C, Tang Z (1993) Advanced chemical oxidation: its present role and potential future in hazardous waste treatment. Waste Management 13: 361-377.
7. Neyens E, Baeyens J (2003) A review of classic Fenton's peroxidation as an advanced oxidation technique. Journal of Hazardous Materials B98: 33-50.
8. Christensen TH, Kjeldsen P, Bjerg PL, Jensen DJ, Christensen JB, et al. (2001) Biogeochemistry of landfill leachate plumes. Applied Geochemistry 16: 659-718.
9. Zhang H, Choi HJ, Huang CP (2005) Optimization of Fenton process for the treatment of landfill leachate. Journal of Hazardous Materials B125: 166-174.
10. Renou S, Givaudan JC, Poulain S, Dirassouyan F, Moulin P (2008) Landfill leachate treatment: Review and opportunity. Journal of Hazardous Materials 150: 468-493.
11. IRSA Istituto di Ricerca sulle Acque, CNR. (1985) Analytical methods for sewage sludges (Vol 3) Chemical-physical parameters. 64 (10): 1-5.
12. MIPAF (Ministero Delle Politiche Agricole E Forestali) (1999) Official Methods of soil chemical analysis. DM 13 sept. 1999. Ordinary Supplement to the Official Italian Gazette GURI n° 248 of 21 Oct. 1999.
13. Bokare AD, Choi W (2014) Review of iron-free Fenton-like systems for activating H_2O_2 in advanced oxidation processes. Journal of Hazardous Materials 275: 121-135.
14. Tyagi RD, Couillard D, Tran F (1988) Heavy metals removal from anaerobically digested sludge by chemical and microbiological methods. Environmental Pollution 50: 295 - 316.
15. Hayes TD, Jewell WJ, Kabrick RM (1980) Heavy metal removal from sludges using combined biological/chemical treatment. In: The 34th Industrial Waste Conference, Purdue University, West Lafayette, Indiana.
16. Fenton HJH (1894) Oxidation of tartaric acid in presence of iron. J Chem Soc Trans 65: 899-910.
17. Walling C, Goosen A (1973) Mechanism of the ferric ion catalysed decomposition of hydrogen peroxide: effects of organic substrate. J Am Chem Soc 95: 2987-2991.
18. Walling C (1975) Fenton's reagent revisited. Accounts of Chemical Research 8: 125-131.
19. Snoeyink VL, Jenkins D (1982) Water Chemistry, Wiley, New York.
20. Lu MC, Lin CJ, Liao CH, Ting WP, Huang RY (2001) Influence of pH on the dewatering of activated sludge by Fenton's reagent. Water Science and Technology 44: 327 - 332.
21. Xu XR, Li XY, Li XZ, Li HB (2009) Degradation of melatonin by UV, UV/H_2O_2, Fe^{2+}/H_2O_2 and UV/Fe^{2+}/H_2O_2 processes. Sep Purif Technol 68: 261-266.
22. Kavitha V, Palanivelu K (2005) Destruction of cresols by Fenton oxidation process. Water Res 39: 3062-3072.
23. Kwon BG, Lee DS, Kang N, Yoon J (1999) Characteristics of p-chlorophenol oxidation by Fenton's reagent. Wat Res 33: 2110-2118.
24. Lake DL (1987) Chemical speciation of heavy metals in sewage sludge and related matrices. In: Heavy metals in wastewater and sludge treatment processes-Vol I Ed. By Lester J.N., CRC Press Inc., Boca Raton, Florida, USA.
25. Marchioretto MM, Bruning H, Loan NTP, Rulkens WH (2002) Removal of heavy metals from anaerobically digested sludge. Water Science and Techn 46: 1-8.
26. Neyens E, Baeyens J, Weemaes M, De Heyder B (2002) Environmental Engineering Science 19: 27-35.
27. DewilR, Baeyens J, Neyens E (2005) Fenton peroxidation improves the drying performance of waste activated sludges: an electron microscopic study on resin-embedded samples. Water Resources 35: 3018-3024.
28. Pere J, Alen R, Viikari L, Eriksson L (1993) Characterization and dewatering of activated sludge from the pulp and paper industry. Water Science and Technology 28: 193-201.
29. Yoon J, Lee Y, Kim S (2001) Investigation of the reaction pathway of OH radicals produced by Fenton oxidation in the conditions of wastewater treatment. Water Science and Technology 44: 15 - 21.
30. Kang YW, Hwang KY (2000) Effects of reaction conditions on the oxidation efficiency in the Fenton process. Water Resources 34: 2786-2790.
31. Yoon J, Kim Y, Huh J, Lee Y, Lee D (2002) Roles of oxidation and coagulation in Fenton process of the removal of organics in landfill leachate. Journal of Industrial and Engineering Chemistry 8: 410-418.
32. Liu H, Yang J, Shi Y, Li Y, He S, et al. (2012) Conditioning of sewage sludge by Fenton's reagent combined with skeleton builders. Chemosphere 88: 235-239.

2

Evaluation of Biofertilizers in Cultured Rice

Nino Paul[1], Pompe C Cruz[2], Edna A Aguilar[2], Rodrigo B Badayos[3] and Stephan Hafele[1]*

[1]International Rice Research Institute, Los Baños, Laguna, Philippines
[2]Crop Science Cluster, College of Agriculture, University of the Philippines at Los Baños, Laguna, Philippines
[3]Agricultural Systems Cluster, College of Agriculture (CA), University of the Philippines at Los Baños, Laguna, Philippines

Abstract

Biofertilizers are becoming increasingly popular in many countries and for many crops, but very few studies on their effect on grain yield have been conducted in rice. Therefore, we evaluated three different biofertilizers (based on *Azospirillum*, *Trichoderma*, or unidentified rhizobacteria) in the Philippines during four cropping seasons between 2009 and 2011, using four different fertilizer rates (100% of the recommended rate [RR], 50% RR, 25% RR, and no fertilizer as Control). The experiments were conducted under fully irrigated conditions in a typical lowland rice environment. Significant yield increases due to biofertilizer use were observed in all experimental seasons with the exception of the 2008/09 DS. However, the effect on rice grain yield varied between biofertilizers, seasons, and fertilizer treatments. In relative terms, the seasonal yield increase across fertilizer treatments was between 5% and 18% for the best biofertilizer (*Azospirillum*-based), but went up to 24% in individual treatments. Absolute grain yield increases due to biofertilizer were usually below 0.5 $t \cdot ha^{-1}$, corresponding to an estimated additional N uptake of less than 7.5 kg N ha^{-1}.

The biofertilizer effect on yield did not significantly interact with the inorganic fertilizer rate used but the best effects on grain yield were achieved at low to medium fertilizer rates. Nevertheless, positive effects of the biofertilizers even occurred at grain yields up to 5 $t \cdot ha^{-1}$. However, the trends in our results seem to indicate that biofertilizers might be most helpful in rainfed environments with limited inorganic fertilizer input. However, for use in these target environments, biofertilizers need to be evaluated under conditions with abiotic stresses typical of such systems such as drought, soil acidity, or low soil fertility.

Keywords: *Azospirillum*; biofertilizer; grain yield; inorganic fertilizer; PGPR; plant growth-promoting rhizobacteria; rice; *Trichoderma*

Introduction

Biofertilizers are becoming increasingly popular in many countries and for many crops. They are defined as products containing active or latent strains of soil microorganisms, either bacteria alone or in combination with algae or fungi that increase the plant availability and uptake of mineral nutrients [1]. In general, they contain free-living organisms associated with root surfaces but they may also include endophytes, microorganisms that are able to colonize the intercellular or even intracellular spaces of plant tissues without causing apparent damage to the host plant. The concept of biofertilizers was developed based on the observation that these microorganisms can have a beneficial effect on plant and crop growth (e.g.) [2]. Consequently, a range of plant growth-promoting rhizobacteria (PGPR) has been identified and well characterized. Direct beneficial effects can occur when the microorganisms provide the plants with useful products.

The best known case of this are microorganisms that can directly obtain N from the atmosphere and convert this into organic forms usable by plants. Such biological nitrogen fixers (BNF) include members of the genus *Rhizobium*, *Azospirillum*, and blue-green algae. Rhizobia are symbiotically associated with legumes and nitrogen fixation occurs within root or stem nodules where the bacterium resides [3]. The genus *Azospirillum* also has several N-fixing species, which are rhizobacteria associated with monocots and dicots such as grasses, wheat, maize and *Brassica chinensis* L. [4,5]. *Azospirillum* strains have been isolated from rice repeatedly, and recently the strain *Azospirillum* sp. B510 has been sequenced [6,7].

Considerable N fixation by *Azotobacter* spp. and *Azospirillum* spp. in the rice crop rhizosphere was reported repeatedly [6,8], but others [9] questioned such high amounts of non-symbiotic N fixation in agriculture. Instead, it was hypothesized that the beneficial effect of *Azospirillum* inoculums may not derive from its N-fixing properties but from its stimulating effect on root development [2], probably often triggered by phytohormones [10]. This view was confirmed by Okon and Labandera-Gonzales [11], who concluded that the main effect of *Azospirillum* spp. is the stimulation of the density and length of root hairs, the rate of appearance of lateral roots, and the root surface area. Phytohormone production and a beneficial effect on plant growth were also shown for a range of other microorganisms [12,13].

Another important genus for biofertilizer producers is *Trichoderma*, a fungus present in nearly all soils. *Trichoderma* spp. thrives in the rhizosphere and can also attack and parasitize other fungi. *Trichoderma* spp. have been known for decades to increase plant growth and crop yield [14-16], to improve crop nutrition and fertilizer uptake [16,17], to speed up plant growth and enhance plant greenness [18], as well as to control numerous plant pathogens [19-21]. A part of these effects may also be related to the fact that some *Trichoderma* spp. seem to hasten the mineralization of organic materials [22], thus probably releasing nutrients from soil organic matter. Positive effects on plant nutrition were also described for other organisms, and many soil bacteria may enhance the mineral uptake of the plant, as for example by the increased solubility of phosphate in the soil solution [23].

***Corresponding author:** Stephan Hafele, International Rice Research Institute, Los Baños, Laguna, Philippines, E-mail: s.hafele@cgar.org

There is a wide range of reports on the effect of biofertilizer application in crops grown in non-flooded soils (unlike lowland rice), and the technology for *Rhizobium* inoculation of leguminous plants is well established. A review on results from *Azospirillum* inoculation experiments across the world and covering 20 years was conducted by Okon and Labandera-Gonzales [11]. They found a success rate of 60–70% with statistically significant yield increases on the order of 5–30%. However, the vast majority of these trials were on wheat, maize, sorghum, or millet, and only one of the experiments included in the analysis was on rice. Consequently, results from biofertilizer use in rice are still rare. Some reports from groups promoting the use of biofertilizers indicated considerable yield increases upon their use. *Trichoderma harzianum*, used as a coating agent for rice seed, was reported to result in a 15–20% yield increase compared with rice plants receiving full inorganic fertilizer rates only [22]. As already mentioned above Sison [8], reported enhanced growth and development of rice and maize after the use of biofertilizer containing *Azospirillum spp*, and asserted the biofertilizer would provide 30–50% of the crop's N requirement. Similarly, Razie F and Anas I [6] claimed that the inoculation of rice seedlings with *Azotobacter* spp. and *Azospirillum* spp. was able to substitute for the application of inorganic N fertilizer, and that this technology enabled rice yields of 3.9 to 6.4 $t{\cdot}ha^{-1}$ (yield increases in comparison with the control were about 2.0–3.0 $t{\cdot}ha^{-1}$). Another study tested the effect of rice root inoculation with *Azospirillum* spp. under different N fertility levels, and found a more pronounced yield response at lower levels of inorganic N fertilization [24]. Generally, rice yield increases in this study were lower, and ranged around 0.5 $t{\cdot}ha^{-1}$. A yield-increasing effect on rice by inoculation with *Azospirillum* sp. strain B510 was also shown by Isawa et al. [25] but the experiment was conducted in pots only.

Based on these reports, it can be assumed that biofertilizers could offer an opportunity for rice farmers to increase yields, productivity, and resource use efficiency. And, the increasing availability of biofertilizers in many countries and regions and the sometimes aggressive marketing brings ever more farmers into contact with this technology. However, rice farmers get little advice on biofertilizers and their use from research or extension because so little is known on their usefulness in rice. Necessary would be recommendations describing under which conditions biofertilizers are effective, what their effect on the crop is, and how they should best be used.

To start addressing these issues, we conducted this study, testing different biofertilizers in an irrigated lowland rice system in the Philippines during four seasons. The objectives of the study were (1) to evaluate the effects of different biofertilizers on irrigated rice grain yield, (2) to investigate possible interactions of the effect of these biofertilizers with different inorganic fertilizer rates, and (3) to determine, based on the results, whether biofertilizers are a possible option to improve the productivity of rice production and under which conditions they give good results.

Materials and Methods

Site Description

The experiments were conducted during two dry seasons (DS) and two wet seasons (WS). In the 2008/09 DS and the 2009 WS, an experimental site at the Central Experimental Station of the University of the Philippines at Los Baños (CES-UPLB) was used, whereas the experiment in the 2010 WS and the 2010/11 DS was conducted at the Experimental Station of the International Rice Research Institute (IRRI) in Los Baños (ES-IRRI). Both experimental sites were located in close vicinity (about 1 km apart) in Laguna Province, Philippines (14°11' North, 121°15' East, 21 masl), in a typical lowland rice production area with the dominant soil type "anthraquic Gleysols" [26].

Detailed soil characteristics were analyzed only for the field at ES-IRRI (Table 1) but the soil at CES-UPLB was similar. The soil at both sites had a fine texture (clayey loam) and a high cation exchange capacity (CEC). Topsoil pH values at CES-UPLB in the 2009 DS and WS were 6.9 and 6.8, respectively, while pH values of 6.9 (2010 WS) and 6.5 (2011 DS) were observed at the ES-IRRI site. The soil organic carbon concentrations at both farms were relatively high, ranging between 1.5% and 1.9%. Related to this, organic N concentrations were also high at both farms (0.15–0.27%). The high soil organic matter content also caused high P availability as indicated by high Olsen P values, which were far above the critical low level of 10–15 $mg{\cdot}kg^{-1}$ [27]. Similarly, the exchangeable K was adequate for both experimental sites at the start of the cropping seasons [27].

Experimental Treatments and Design

In all four seasons, the experiment was a two-factor experiment arranged in a randomized complete block design (RCBD) with three replications. Main plots were assigned to four different fertilizer levels: i) the full recommended rate (100% RR) of inorganic fertilizer; ii) 50% RR, 25% of RR, and the Control treatment in which no inorganic fertilizer was applied. However, the recommended rate changed between seasons and was 120 kg N ha^{-1}, 60 kg P_2O_5 ha^{-1}, and 60 kg K_2O ha^{-1} in the DS, and 90 kg N ha^{-1}, and 30 kg P_2O_5 ha^{-1}, 30 kg K_2O ha^{-1} in the WS. The exact N, P, and K amounts applied are given in Table 2.

Subplots (30 m^2 each) were assigned to the different biofertilizers tested in the experiment. Three different biofertilizers available in the Philippines were used, and an overview of their characteristics is given in Table 3. The products were Bio-N® (BN), BioGroe® (BG), and BioSpark® (BS; the same product was called BioCon in 2009). In addition, a Control treatment was used in which no biofertilizer was applied. Thus, the total number of treatment combinations tested was 16.

BN was developed in the early 1980s by Paredes JC [28]. According to the distributor (BIOTECH, UPLB), it contains *Azospirillum lipoferum* and *A. brasilense*, isolated from *Saccharum spontaneum* (local name is

Site Soil type		UPLB Anthraquic Gleysols		IRRI Anthraquic Gleysols	
		2008/2009 DS	2009 WS	2010 WS	2010/2011 DS
pH (1:1)	-	6.9	6.8	6.9	6.5
Total organic C	g kg^{-1}	18.6	15.9	16.2	15.0
Total soil N	g kg^{-1}	2.7	1.6	1.5	1.5
Olsen P	mg kg^{-1}	55	40	35	30
Avail K	cmol kg^{-1}	-	-	1.26	1.32
Exch K	cmol kg^{-1}	1.50	1.06	1.50	1.50
Exch Ca	cmol kg^{-1}	-	-	18.9	18.1
Exch Mg	cmol kg^{-1}	-	-	13.5	13.3
Exch Na	cmol kg^{-1}	-	-	1.01	1.00
CEC	cmol kg^{-1}	-	-	33.6	33.0
Clay	g kg^{-1}	-	-	441	445
Silt	g kg^{-1}	-	-	332	355
Sand	g kg^{-1}	-	-	227	200

Table 1: Average top-soil characteristics (0–15 cm depth) for all experimental seasons and both experimental sites.

Fertilizer Rate	Unit	Dry Season	Wet Season
0% RR	$N-P_2O_5-K_2O$ in kg·ha^{-1}	0-0-0	0-0-0
25% RR	$N-P_2O_5-K_2O$ in kg·ha^{-1}	30-15-15	22.5-7.5-7.5
50% RR	$N-P_2O_5-K_2O$ in kg·ha^{-1}	60-30-30	45-15-15
100% RR	$N-P_2O_5-K_2O$ in kg·ha^{-1}	120-60-60	90-30-30
or			
0% RR	N-P-K in kg·ha^{-1}	0-0-0	0-0-0
25% RR	N-P-K in kg·ha^{-1}	30-7-13	22.5-3-6
50% RR	N-P-K in kg·ha^{-1}	60-13-25	45-7-13
100% RR	N-P-K in kg·ha^{-1}	120-26-50	90-13-25

Table 2: Inorganic fertilizer treatments in all four experimental seasons as ratio of the Recommended Rate (RR) and as actual nutrients applied in the dry and wet season.

Product ID	BN	BG	BS
Product name	Bio-N®	BioGroe®	BioSpark®
Active ingredient	*Azospirillum lipoferum, A. brasilense*	Plant growth-promoting rhizobacteria (not defined)	*Trichoderma parceramosum, T. pseudokoningii*, and UV-irradiated strain of *T. harzianum*
Active organism	Bacteria	Bacteria	Fungi
Product type	Dry powder in 200-g pack	Dry powder in 100-g pack	Dry powder in 250-g pack
Carrier medium	Sterile charcoal/ soil mixture	Sterile charcoal/soil mixture	Dry organic medium (rice hull)
Producer declared cell number	10^8 cfu g^{-1}	-	10^9 cfu g^{-1}
Shelf life	3 months	6 months	24 months
Product amount recommended and used	1000 g 40 kg^{-1} seed	400 g 40 kg^{-1} seed	200 g 40 kg^{-1} seed
(for 1 ha) 2011 biofertilizer costs	US$6.82	US$3.64	US$6.36
needed for 40 kg seed			
Elemental contents*			
N %	0.13	0.34	1.27
P %	0.091	0.063	0.687
K %	0.22	0.24	0.72
Supplier	BioTech UPLB	BioTech UPLB	BioSpark Corp.

*Source: Analytical Service Laboratory, GQNPC, IRRI.

Table 3: Characteristics of the three biofertilizer used and tested.

Talahib). BN is available in dry powder form in a 200-gram package, which can be used for seed inoculation, direct broadcasting on seeds, or mixed with water as a root dip. The BN product has a shelf-life of 3 months and the package we used was well before its expiry date. BN is specifically targeted at rice and corn.

The second product tested was BG, developed by Paterno of BIOTECH at UPLB. It contains unknown plant growth-promoting bacteria (rhizobacteria) that influence root growth by producing plant hormones and providing nutrients in soluble form [28].

The last product tested was BS, developed by Cuevas [29]. According to personal information from her, it contains three different species of *Trichoderma* isolated from Philippine forest soils (including *Trichoderma harzianum*), and is mass-produced using a pure organic carrier [29]. The product can be used for seed coating or for soil application in the seedbed.

Crop Establishment and Management

In all experiments, rice variety PSB Rc18, a modern-type variety with 120 days duration, was used. Seed for the BN and Control treatments was soaked for 24 h, incubated for another 24 h, and sown using the modified *dapog* (mat) method. BN was prepared in a slurry solution and applied by dipping the roots of the seedlings into the slurry, 1 h before transplanting in the field. For the BG and BS treatments, seeds were initially also soaked for 24 h. The biofertilizers BG and BS were then applied by mixing the seeds with the biofertilizer product, thus coating the seeds. BG and BS were applied at 400 g 40 kg^{-1} seed and 200 g 40 kg^{-1} seed, respectively. The seed-biofertilizer mixture was then incubated for 10 hours in an open jute sac to allow cooling, followed by 14 hours incubation in the closed sack like the control. Seeds were sown using the modified *dapog* method. In all treatments, 14-day-old seedlings were transplanted at 2–3 seedlings per hill with a planting distance of 20 cm. Missing hills were replanted within 7 days after transplanting (DAT).

Inorganic fertilizers used for the fertilizer treatments were urea (46-0-0 $N-P_2O_5-K_2O$) and compound (14-14-14 $N-P_2O_5-K_2O$) fertilizer. Compound fertilizer was applied basal just before transplanting according to the treatment. The remaining N was applied in equal splits at 10 DAT and at 55 DAT. A water depth of 3–5 cm was aimed for at every irrigation from early tillering until 1–2 weeks before physiological maturity. To control insect pests and diseases in the 2010 WS and 2010/11 DS, granular Furadan was applied 20 DAT at a rate of 33 kg·ha^{-1} and Hopcin was applied at a rate of 0.8 L·ha^{-1} at flowering. Molluscicide was applied right after transplanting to control golden apple snails in the field. Post-emergence herbicide was applied once at the 2-3-leaf stage of emerging weeds. Hand-weeding was done thereafter as needed. Application rates were based on the recommended rate of the specific pesticides that were used.

Sampling and Statistical Analysis

Grain yields were determined in the study for a 5-m^2 (2.5 m-2.0 m) designated sampling area, which was strategically located at the center of each subplot, leaving at least two border rows. Grain moisture content was determined immediately after threshing (Riceter grain moisture meter, Kett Electric Laboratory, Tokyo, Japan) and all grain yields are reported at 14% moisture content. The data gathered in the study were statistically analyzed using the procedures described by Gomez and Gomez [30]. Analysis of variance was conducted using SAS (Version 9.0) and treatment means were compared by the least significant difference (LSD) and were considered significant at $p \leq 0.05$.

Results and Discussion

In all four seasons and across the biofertilizer treatments, grain yield increased with increasing amounts of applied fertilizer (Table 4). However, this increase was not always statistically significant and the yield increase varied considerably between seasons. Overall, the lowest grain yields occurred in the 2009 WS, ranging only from 1.9 to 2.7 t·ha^{-1}. Generally, low yields in that season were due to a typhoon that caused considerable damage through flooding of the experimental field and lodging of the crop. For this reason, the crop was harvested prematurely by about 1 week, which further reduced attainable yields.

Grain yields in the other three experimental seasons were similar and ranged from 4.0 to 5.2 t·ha^{-1} in the 2009 DS, from 3.4 to 5.1 t·ha^{-1} in the 2010 WS, and from 3.8 to 5.6 t·ha^{-1} in the 2010/11 DS. These ranges already indicate a relatively low yield increase due to fertilizer application in the 2008/09 DS (up to 1.2 t·ha^{-1} for the full fertilizer rate of 120-60-60 kg N-P_2O_5-K_2O ha^{-1}) and the 2009 WS (up to 0.8 t·ha^{-1} for the full fertilizer rate of 90-30-30 kg N-P_2O_5-K_2O ha^{-1}). A higher response to inorganic fertilizer was achieved in the 2010 WS (up to 1.7 t·ha^{-1} for the full fertilizer rate of 90-30-30 kg N-P_2O_5-K_2O ha^{-1}) and the 2010/11 DS (up to 1.8 t·ha^{-1} for the full fertilizer rate of 120-60-60 kg N-P_2O_5-K_2O ha^{-1}).

The effects of biofertilizer treatments on grain yield, depending on the inorganic fertilizer treatment, are shown in Table 4. Significant yield increases due to biofertilizer use were observed in all experimental seasons with the exception of the 2008/09 DS. In the 2010/11 DS, no significant difference between the three biofertilizers tested was detected, but all three achieved better yields than the Control. The biofertilizer achieving the highest average grain yields across all four inorganic fertilizer treatments and in all four seasons was BN. Statistically significant interactions between biofertilizer treatment and inorganic fertilizer treatment could not be detected in any season (at $p \leq 0.05$), suggesting that the effect of the biofertilizer was independent of the inorganic fertilizer rate. However, there was a trend of higher yield increases due to biofertilizer use at low to medium inorganic fertilizer rates (Table 4). This trend was most obvious for the BN biofertilizer whereas the performance of the BS and BG biofertilizers was less consistent.

The grain yield increase due to biofertilizer only (0% RR inorganic fertilizer treatment) usually ranged from 200 to 300 kg grain ha^{-1} for the best biofertilizers with the exception of the 2010 WS, when the BN treatment had an almost 800 kg·ha^{-1} better grain yield than the Control. In relative terms (Table 5), the seasonal yield increase across fertilizer treatments was between 5% and 18% for the BN biofertilizer (up to 24% for individual treatment combinations), between 3% and 13% for the BS biofertilizer (up to 24% for individual treatment combinations), and between 1% and 9% for the BG biofertilizer (up to 28% for individual treatment combinations). For the calculation of the relative yield increase, only average values could be compared and no statistical analysis could be conducted.

Season	Bio-fertilizer treatment***	Inorganic fertilizer treatment**				
		0% RR	25% RR	50% RR	100% RR	Mean*
		Grain yield (kg·ha^{-1})				
2008/09 DS	BG	4016	4421	4569	5134	4508 a
	BN	4163	4753	4900	5081	4683 a
	BS	4351	4569	4375	5173	4610 a
	Control	4062	4440	4630	4799	4534 a
	Mean*	**4158 c**	**4548 b**	**4617 b**	**5034 a**	
2009 WS	BG	1963	1975	2502	2383	2206 bc
	BN	2149	2417	2420	2604	2398 a
	BS	2005	2179	2287	2674	2286 ab
	Control	1902	2000	2038	2165	2026 c
	Mean*	**2005 c**	**2143 bc**	**2038 ab**	**2456 a**	
2010 WS	BG	4326	4303	4670	4596	4482 ab
	BN	4197	4529	5131	4794	4663 a
	BS	3952	4336	4578	4732	4399 bc
	Control	3389	4245	4274	4716	4219 c
	Mean*	**3965 c**	**4353 b**	**4659 a**	**4710 a**	
2010/11 DS	BG	4145	4665	4926	5556	4825 a
	BN	4009	5049	5262	5519	4960 a
	BS	3955	4876	5175	5492	4861 a
	Control	3801	4420	4707	5265	4548 b
	Mean*	**3977 c**	**4751 b**	**5014 b**	**5458 a**	

*In each season, mean values in a column or row followed by the same letter are not significantly different at 5% of significance.

Table 4: Grain yield of the variety PSB Rc18 as affected by inorganic fertilizer level and biofertilizer treatments in all four experimental seasons and both sites.

The effect of biofertilizer on the agronomic efficiency of N fertilizer (AEN) is shown in Table 6. For these calculations, the yield of each treatment was compared with the grain yield baseline (the Control treatment in which no biofertilizer and no inorganic fertilizer were used) and the yield increase was divided by the N rate applied. Again, only average values could be compared and no statistical analysis was possible. The results (Table 6) indicate considerably higher overall AEN values in the 2010 WS and the 2010/11 DS. Also, the AEN values are generally higher at low N rates and decrease with higher N application rates. The biggest AEN increase caused by biofertilizer occurred at the lowest N fertilizer rate (25% RR treatment), and, among the different biofertilizers tested, the BN biofertilizer resulted in the highest and most consistent AENs.

In our experiments, the selected biofertilizers were used as recommended by the producers but we could not check the viability or the contents of the products. Thus, we did not verify whether the biofertilizers contained the declared organisms (Table 6; the contents of BG remained unidentified) or the required number of living cells in the inoculate. The importance of quality control and regulation for biofertilizer production was emphasized by Reddy and Giller [31], who also pointed out that the frequent absence of such mechanisms can cause non-functional products. Maintenance of high standards for *Azospirillum* inoculants with proven efficient strains and cell numbers on the order of $1\cdot10^9$ to $1\cdot10^{10}$ colony-forming units (cfu) g^{-1} or mL^{-1} was also requested by Okon and Labandera-Gonzales [11].

But, the fact that the products in our study caused a significant effect on grain yield in three out of four seasons (only two out of four seasons for BG) indicated that the biofertilizers tested had sufficient active ingredients and that the producers maintained a good quality over the four seasons (or 2.5 years). Theoretically, the effect of the biofertilizers could also have been caused by non-living ingredients but the applied amount was so small that even micronutrients could not explain the observed effects. Also, no micronutrient deficiencies are known from either of the two experimental sites.

The general effect of inorganic fertilizer was as expected, and grain yields increased continuously with increasing fertilizer rates (Table 4). However, the response to inorganic fertilizer was low in the 2008/09 DS and the 2009 WS, as also indicated by the low AEN (Table 5). Good and economic values for AEN are usually 15–20 kg grain yield per kg N applied, and, at AEN < 10, inorganic fertilizer use may give negative economic returns depending on the input and output prices [32,33]. Low response in the 2009 WS can be explained by the negative effects of a typhoon and the early harvest. The low response in the 2008/09 DS could be due to the combination of a very fertile soil (high grain yield in the 0% RR treatment) and a limited yield potential in that season (low maximum yields in the 100% RR treatment).

The tested biofertilizers did increase grain yield significantly, and especially the BN biofertilizer did so consistently. Even in seasons in which no significant effect could be detected due to the yield variability

Season	Biofertilizer treatment***	Inorganic fertilizer treatment**				
		0% RR	25% RR	50% RR	100% RR	Mean
			Relative yield increase (%)*			
2008/09 DS	BG	-1	0	-1	7	1
	BN	2	7	6	6	5
	BS	7	3	-6	8	3
	Control	-	-	-	-	-
2009 WS	BG	3	-1	23	10	9
	BN	13	21	19	20	18
	BS	5	9	12	24	13
	Control	-	-	-	-	-
2010 WS	BG	28	1	9	-3	8
	BN	24	7	20	2	12
	BS	17	2	7	0	6
	Control	-	-	-	-	-
2010/11 DS	BG	9	6	5	6	6
	BN	5	14	12	5	9
	BS	4	10	10	4	7
	Control	-	-	-	-	-

* The relative yield increase was calculated for treatment means and in comparison to the control without biofertilizer use but within the same inorganic fertilizer treatment; ** RR: Recommended rate: 120-60-60 kg N-P_2O_5-K_2O ha^{-1} in the DS; 90-30-30 kg N-P_2O_5-K_2O ha^{-1} in the WS;

*** Biofertilizer treatments are described in detail in the text.

Table 5: Relative yield increase over the Control treatments with the same inorganic fertilizer rate for all biofertilizers tested, in all seasons and at both experimental sites.

between plots, the grain yield with biofertilizer was usually better than without. The seasonal yield increase across fertilizer treatments was between 5% and 18% for the BN biofertilizer (up to 24% for individual treatments; Table 5), which is within the 5–30% range reported for *Azospirillum* inoculums and non-rice crops by Bashan and Levanony and Okon and Labandera-Gonzales [4,11]. Similarly, the here-observed yield increase for the *Trichoderma*-based BS (3–13%) was close to the 15–20% rice yield increase described by Cuevas [22].

The trend of yield increases between the different inorganic fertilizer treatments was not so clear across seasons but yield increases were often lower at higher inorganic fertilizer rates, which were also reported by Rajabamamohan et al. [24]. Absolute grain yield increases due to biofertilizer were usually below 0.5 t·ha^{-1} (Table 1), corresponding to an estimated additional N uptake of less than 7.5 kg N ha^{-1} (based on 0.5% N in straw, 1.0% N in grain, and harvest index 0.5). Both values are far below grain yield increases and additional N uptake reported by Razie and Anas and Sison [6,8], but similar to the rice grain yield increases reported by Rajabamamohan et al. [24].

The calculated AEN values (Table 6) suggested higher N use efficiency for treatments with biofertilizer use. Increased nutrient uptake and fertilizer use efficiency were also reported for *Trichoderma* spp. [16,17,34] and for *Azospirillum* spp. [11]. But, the results could be explained in several ways. One possibility is that the biofertilizer stimulated root growth and thereby increased the uptake of indigenous N from the soil (the higher AEN would then be only an artifact of the calculation method). Second, the increased root growth could reduce N fertilizer losses, and the third option could be biological N fixation (which could explain the superior performance of the BN biofertilizer, supposedly containing organisms capable of biological N fixation). But, our experiment cannot answer the question of which process or combination of processes is at work here, if that is possible at all under field conditions [9].

Summary and Conclusions

The study was conducted to evaluate the effect of different biofertilizers on the grain yield of lowland rice, and investigate possible interaction effects with different inorganic fertilizer amounts. The results showed significant yield increases for all products tested in some seasons but the most consistent results were achieved by the *Azospirillum*-based biofertilizer. In most cases, the observed grain yield increases were not huge (0.2 to 0.5 t·ha^{-1}) but could provide substantial income gains given the relatively low costs of all biofertilizers tested. The positive effect of the tested biofertilizers was not limited to low rates of inorganic fertilizers and some effect was still observed at grain yields up to 5 t·ha^{-1}. However, the trends in our results seem to indicate that the use of biofertilizers might be most helpful in low- to medium-input systems. The results achieved can already be used to develop better advice for farmers on biofertilizer use in lowland rice, but several important questions remain. In particular, biofertilizers need to be evaluated under conditions with abiotic stresses typical for most low- to medium-input systems (e.g., under drought or low soil fertility) and with a range of germplasm because their effect might depend also on the variety used. More upstream-oriented research would be needed to better understand the actual mechanisms involved, which in turn could also contribute to making the best use of biofertilizers in rice-based systems.

Season	Bio-fertilizer treatment**	0% RR Reference grain yield (kg·ha^{-1})	Inorganic fertilizer treatment *		
			25% RR	50% RR AEN***	100% RR
			(kg grain yield increase kg^{-1} N applied)		
2008/09 DS	BG		12	8	9
	BN		23	14	8
	BS		17	5	9
	Control	4062	13	9	6
2009 WS	BG		3	13	5
	BN		23	12	8
	BS		12	9	9
	Control	1902	4	3	3
2010 WS	BG		41	28	13
	BN		51	39	16
	BS		42	26	15
	Control	3389	38	20	15
2010/11 DS	BG		29	19	15
	BN		42	24	14
	BS		36	23	14
	Control	3801	21	15	12

* RR: Recommended rate: 120-60-60 kg N-P_2O_5-K_2O ha^{-1} in the DS; 90-30-30 kg N-P_2O_5-K_2O ha^{-1} in the WS; ** Biofertilizer treatments are described in detail in the text; *** For the estimation of AEN in each experimental season, the grain yield of the treatment without inorganic fertilizer and biofertilizer (0% RR and Control) was used as reference.

Table 6: Estimated agronomic efficiency (AEN) of applied N depending on the inorganic fertilizer treatment and the biofertilizer used.

References

1. Vessey JK (2003) Plant growth promoting rhizobacteria as biofertilizers. Plant Soil 255: 571–586.
2. Davison J (1988) Plant beneficial bacteria. Biotechnol 6: 282–286.
3. Dela Cruz, RE State of the art in biotechnology: Crop production. In: Biotechnology for Agriculture, Forestry and the Environment; PCARRD: Los Baños, Laguna, Philippines, 1993.
4. Bashan Y, Levanony H (1990) Current status of Azospirillum inoculation technology: Azospirillum as a challenge for agriculture. Can. J. Microbiol 36: 591–608.
5. Singh S, Rekha PD, Arun AB, Hameed A, Singh S, et al. (2011) Glutamate wastewater as a culture medium for Azospirillum rugosum production and its impact on plant growth. Biol. Fert. Soils 47: 419–426.
6. Razie F, Anas I (2008) Effect of Azotobacter and Azospirillum on growth and yield of rice grown on tidal swamp rice fields in south Kalimantan. Jurnal Tanah dan Lingkungan 10: 41–45.
7. Kaneko T, Minamisawa K, Isawa T, Nakatsukasa H, Mitsui H, et al. (2010) Complete genomic structure of the cultivated rice endophyte Azospirillum sp. B510. DNA Res 17: 37-50.
8. Sison, MLQ Available biotechnologies and products. Presented at the workshop on promoting popular awareness and appreciation of biotechnology, Cagayan de Oro City, Philippines, 16th February 1999.
9. Giller KE, Merckx R (2003) Exploring the boundaries of N_2-fixation in cereals and grasses: A hypothetical and experimental framework. Symbiosis 35: 3–17.
10. Ereful NC, Paterno ES (2007) Assessment of cytokinin production in some plant growth-promoting bacteria. Asia Life Sci. 16: 137–152.
11. Okon Y, Labandera-Gonzales CA (1994) Agronomic applications of Azospirillum: An evaluation of 20 years worldwide field inoculation. Soil Biol. Biochem 26: 1591–1601.
12. Fernando LM, Merca FE, Paterno ES (2010) Isolation and partial structure elucidation of gibberellin produced by plant growth promoting bacteria (PGPB) and its effect on the growth of hybrid rice (Oryza sativa L.). Philipp. J. Crop Sci 35: 12–22.
13. Difuntorum-Tamabalo D, Paterno ES, Barraquio W, Duka IM (2006) Identification of an indole-3-acetic acid-producing plant growth-promoting bacterium (PGPB) isolated from the roots of Centrosema pubescens Benth. Philipp. Agr. Sci 89: 149–156.
14. Lindsey DL, Baker R (1967) Effect of certain fungi on dwarf tomatoes grown under gnotobiotic conditions. Phytopathology 57: 1262–1263.
15. Chang Y-C, Chang Y-C, Baker R, Kleifeld O, Chet I (1986) Increased growth of plants in the presence of the biological control agent Trichoderma harzianum. Plant Diseases 70: 145–148.
16. Harman GE (2000) Myths and dogmas of biocontrol. Changes in perceptions derived from research on Trichoderma harzianum T-22. Plant Diseases 84: 377–393.
17. Yedidia I, Srivastva AK, Kapulnik Y, Chet I (2001) Effect of Trichoderma harzianum on microelement concentrations and increased growth of cucumber plants. Plant Soil 235: 235–242.
18. Harman GE (2006) Overview of mechanisms and uses of Trichoderma spp. Phytopathology 96: 190–194.
19. Weindling R (1932) Trichoderma lignorum as a parasite of other soil fungi. Phytopathology 22: 837–845.
20. Shoresh M, Harman GE (2008) The molecular basis of shoot responses of maize seedlings to Trichoderma harzianum T22 inoculation of the root: a proteomic approach. Plant Physiol 147: 2147-2163.
21. Cuevas VC, Sinohin AM, Orajay JI (2005) Performance of selected Philippine species of Trichoderma as biocontrol agents of damping off pathogens and as growth enhancer of vegetables in farmer's field. Philipp. Agr. Sci 88: 63–71.
22. Cuevas VC (1991) Rapid composting for intensive rice land use. In Innovation for Rural Development; SEAMEO-SEARCA: Los Baños, Philippines 5–10.
23. Goldstein AH, Liu ST (1987) Molecular cloning and regulation of a mineral phosphate solubilizing gene from Erwinia herbicola. Nat. Biotech 5: 72–74.
24. Rajabamamohan RV, Nayak DN, Charyulu PBBN,Adhy TK (1983) Yield response of rice to root inoculation with Azospirillum. J. Agr. Sci 100: 689–691.
25. Isawa T, Yasuda M, Awazaki H, Minamisawa K, Shinozaki S, et al. (2010) Azospirillum sp. strain B510 enhances rice growth and yield. Microbes Environ 25: 58-61.
26. World Reference Base for Soil Resources (2006) Food and Agriculture Organization of the United Nations: Rome, Italy 128.
27. Fairhurst TH, Witt C, Buresh RJ, Dobermann A (2007) A Practical Guide to Nutrient Management, 2nd ed. International Potash Institute: Singapore 89.
28. Paredes JC, Go I (2008) Putting biofertilizers to good use. BioLife 5: 4–10.
29. Cuevas VC (2006) Soil inoculation with Trichoderma pseudokoningii rifai enhances yield of rice. Philip. J. Sci. 135: 31–38.
30. Gomez KA, Gomez AA (1984) Statistical Procedures for Agricultural Research, 2nd ed.; John Wiley & Sons: New York, NY, USA 680.
31. Reddy LN, Giller KE (2008) How effective are effective micro-organisms? LEISA Magazine 24: 18–19.
32. Haefele S M, Sipaseuth N, Phengsouvanna V, Dounphady K, Vongsouthi S (2010) Agro-economic evaluation of fertilizer recommendations for rainfed lowland rice. Field Crop. Res 119: 215–224.
33. Witt C, Buresh RJ, Peng S, Balasubramanian V, Dobermann A (2007) Nutrient management. In Rice: A Practical Guide to Nutrient Management; Fairhurst, T.H., Witt, C., Buresh, R.J., Dobermann, A., Eds.; International Rice Research Institute, International Plant Nutrition Institute, and the International Potash Institute: Singapore, 2007; pp. 1–45.
34. Harman GE (2001) Microbial tools to improve crop performance and profitability and to control plant diseases. In Proceedings of International Symposium on Biological Control of Plant Diseases for the New Century—Mode of Action and Application Technology, Taichung City.

3

Early Summer Slender Aster Control in Bermudagrass using Bioherbicide *Phoma macrostoma*

Smith J[1,2], Wherley B[1]*, Baumann P[1], Senseman S[1], White R[1] and Falk S[2]

[1]*Department of Soil and Crop Sciences, Texas A&M University, College Station, TX 77843, USA*
[2]*The Scotts Miracle-Gro Company, Marysville, OH 43041, USA*

Abstract

Phoma macrostoma is a fungus being developed as a natural herbicide (bioherbicide) for selective weed control in turfgrass. Previous research with this product is limited to cool-season turfgrass, and information is limited on appropriate application rates or efficacy at higher temperatures and weeds associated with warm-season turf. Field studies were conducted in College Station, TX to evaluate efficacy of the bioherbicidefor slender aster (*Aster subulatus*var. *ligulatus* Shinners) control in common bermudagrass. In 2011, applications of 128 g m^{-2}, split-applied between days 0 and 28 resulted in good control (88%); however, 64 and 32 g m^{-2} rates failed to provide adequate control. In 2012, single applications of 128 gave excellent control (94%), while the 32 and 64 g m^{-2} rates gave poor control (54 and 68%, respectively) relative to untreated plots.No injury to common bermudagrass occurred in either study.

Keywords: Natural weed control; Organic weed control; Bioherbicide; Photobleaching

Introduction

In recent decades, synthetic herbicides have come under increased scrutiny around the world. While some legislation has targeted the removal of herbicides for cosmetic use, others have banned the use of all weed and feed products, only allowing the use of herbicides for spot treatment applications [1]. With growing pressure to ban synthetic herbicides, the need for alternative weed control options has increased.

Currently, there are few effective natural options for weed control in turfgrass systems. Some alternative herbicides available for use include vinegar, essential oils (clove and cinnamon oil), citric acid, fatty acids (pelargonic acid), and combinations of these different products. These products are primarily used as nonselective herbicides, with effective weed control being dependent upon product concentrations with vinegar, citric acid and clove oils providing better control at higher use rates [2-4]. Use of higher product rates, however, results in greater potential for crop injury [5].

Corn gluten meal (CGM) is a granular-applied herbicide that is a byproduct of commercial corn milling, containing approximately 10% nitrogen by weight. It is primarily used in turf as a crabgrass pre-emergence herbicide, but may inhibit other broadleaf and grassy weeds [6]. One disadvantage to CGM is that it is effective only as a pre-emergence product, and has minimal post-emergent activity on established weeds [7]. As a granular-applied product, challenges associated with CGM include relatively high use rates (60 to 120 g m^{-2}) and inconsistent reports of weed control [8-10]. While natural herbicides do exist, finding products that are safe to turfgrass and provide consistent, effective weed control has been a challenge.

The bioherbicide *Phoma macrostoma* is a natural herbicide being developed by the Scotts-Miracle Gro Company, Marysville, OH. It is produced from the solid fermentation of the fungus *Phoma macrostoma* on grain. *Phoma macrostoma* was discovered in Canada, when field isolates were collected from infected Canada thistle exhibiting symptoms of bleaching and chlorosis [11]. To date, this bioherbicide has been evaluated primarily in northern climates on weeds, including dandelion (*Taraxacum officinale* Weber ex F. H. Wigg.), Canada thistle (*Cirsium arvense* (L.) Scop.), chickweed (*Stellaria media* (L.) Vill.) and English daisy (*Bellis perennis* L.), with maximal reported efficacy at temperatures ranging from 15 to 25°C [12]. Currently, limited research has been conducted to determine if this product could provide weed control at higher temperatures associated with southern climates in which warm-season turfgrass is managed.

Slender aster (*Aster subulatus*var. *ligulatus* Shinners) is a troublesome summer annual weed that thrives under high temperatures in many areas of the southern U.S. This plant's ability to grow in a prostrate growth pattern allows it to survive mowing, making it problematic in southern turfgrass. Slender aster can be difficult to control if not treated early on in its growth cycle, because it becomes woody as the season progresses, necessitating repeated herbicide applications. Early summer applications of synthetic herbicide formulations, 2, 4-D+MCPP+Dicamba, do provide good control of slender aster (Paul Baumann, personal communication), but the focus of this study was to evaluate the activity of the bioherbicide on slender aster.

The objectives of this study were to 1) evaluate efficacy of the bioherbicide *Phoma macrostoma* at elevated temperatures following early summer applications, 2) determine effective application rates for slender aster control and 3) evaluate potential phytotoxicity on common bermudagrass (*Cynodon dactylon* L. Pers.).

Materials and Methods

Field studies were conducted during late spring/early summer of 2011 and 2012 at the Texas A&M University Turfgrass Research Field Laboratory, College Station, TX. In 2011, trials were initiated on 1 June 2011, and carried out until 12 August 2011. The 2012 trials were initiated

***Corresponding author:** Wherley B, Department of Soil and Crop Sciences, Texas A&M University, College Station, TX 77843, USA, E-mail: b-wherley@tamu.edu

on 2 May 2012, and carried out until 26 June 2012. The studies were conducted on an established stand of common bermudagrass (*Cynodon dactylon* L. Pers.), intermixed with slender aster (*Aster subulatus* var. *ligulatus* Shinners) at the 3 to 4-leaf stage, and approximately 7 cm tall at treatment. Soil at the site was a Lufkin fine sandy loam soil with a pH of 9.8. Field plots were arranged as a randomized complete block design (RCBD) with four replications. Individual plots measured 0.91 m×0.91 m with a 0.3 m buffer between plots. Turf was mowed weekly to a height of 6.4 cm. Just prior to initiation, the study area was fertilized at a rate of 49 g N ha^{-1} using 46-0-0 (N: P: K) urea fertilizer.

Granular bioherbicide was produced under contract for The Scotts Miracle-Gro Company, at a pilot scale manufacturing facility on grain using solid state fermentation. The experimental bioherbicide product supplied for this experiment had half the potency that will be delivered in the commercially produced batches, as such higher application rates were used to compensate in this study. To initiate the studies, plots were irrigated to dampen weed and turf foliage, and treatments were applied to the dampened foliage *via* shaker jar at application rates of 32, 64 or 128 g m^{-2}. In 2011, treatments were split-applied, with half of the herbicide applied at trial initiation and half applied 28 days after treatment (DAT). For 2012, the same overall rate of herbicide was applied, but treatments were applied once at trial initiation. Granules were left on weed foliage for 24 hours, at which time granules were washed off of plant foliage and into the soil by irrigation.

Temperature and rainfall data were recorded by an onsite weather station during the studies. Mean daily air temperature for 2011 was 31°C, with an absolute maximum of 41°C and minimum of 20°C during the study period. For 2012, mean daily temperature was 27°C, with maximum of 41°C and minimum of 17°C. During the study period, plots were irrigated 4 to 5 times weekly receiving a total of 25 mm of water per week. Additionally, rainfall of 81 mm and 93 mm were received throughout the course of the 2011 and 2012 study periods, respectively.

Slender aster weed counts were made using a 0.91 m^{-2} grid rating system consisting of thirty-six 12.7 cm×12.7 cm squares. Squares which contained green slender aster plants were totaled (0-36) and used to

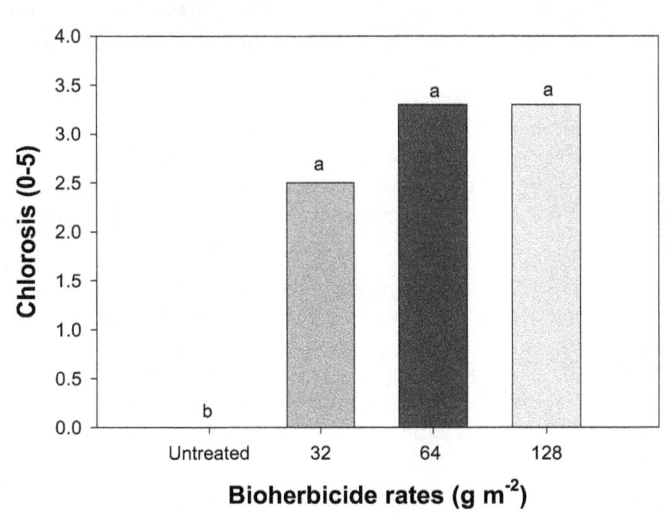

Figure 1: Bleaching and chlorosis of slender aster (0=no bleaching or chlorosis, 5=complete necrosis) 13 days after treatment (DAT). Trial was initiated on May 2, 2012, in College Station, TX, with single bioherbicide applications made at the beginning of the trial. Means followed by the same letter are not significantly different according to Tukey's Test ($P \leq 0.05$).

calculate percent weed control based on the Henderson-Tilton Method [13]. Weed counts were recorded prior to treatment applications and biweekly, thereafter for the duration of the trials. Weed chlorosis/necrosis was also evaluated biweekly using a scale of 0 to 5, with 0=no chlorosis injury, and 5=complete necrosis. Phytotoxicity of common bermudagrass in plots was monitored using a scale of 0=no injury to 5=complete chlorosis.

Data were subjected to analysis of variance using the general linear model, univariate test procedure using SPSS ver. 21.0 (IBM Corp, Armonk, NY) to determine statistical significance of the results. Means separation procedures were performed using Tukey's test at the $P \leq 0.05$ level.

Results and Discussion

Initial slender aster populations in the selected test areas were very high. Grid counts were taken prior to study initiation and showed test plots contained slender aster in ~ 34 out of 36 grids in 2011, and ~ 33 out of 36 grids in 2012. By the conclusion of the study, significant reductions in weed populations were seen in both 2011 and 2012 with the higher two rates of the bioherbicide. The final grid counts for the 128 g m^{-2} rate were 3.5 and 1.8 out of 36 grids in 2011 and 2012, respectively.The 64 g m^{-2} ended the study with slender aster in 5.8 and 9.8 out of 36 grids in 2011 and 2012, respectively.Though the 32 g m^{-2} rate did not significantly reduce final slender aster populations in 2011 (17 grids out of 36), weed populations were significantly reduced in 2012, ending the study with slender aster in 10.5 out of 36 grids. While the same total application amounts were applied in both years, the bioherbicide treatments were split-applied between day 0 and day 28 in 2011, and applied entirely at day 0 in 2012. No injury to bermudagrass was observed following application in either year. Ongoing research with this product using other warm-season species has shown no injury in other major warm-season grasses [14].

Chlorosis and Bleaching

A potential concern of natural products may be a slow or delayed efficacy, relative to synthetic products. Turfgrass managers and home owners prefer rapid control, indicating that the treatments are working. Therefore, rapid visual weed injury and decline following application is an important characteristic of an effective natural consumer product. During both years, initial foliar chlorosis and bleaching of slender aster became evident at all rates within 3 to 4 DAT. By 13 DAT in the 2012 study, moderate (3.25/5) foliar bleaching of weeds was observed in plots receiving both the 64 and 128 g m^{-2} application rates, with slightly less (2.5/5) chlorosis noted in 32 g m^{-2} rate plots (Figure 1). Photobleaching of susceptible weeds was followed by necrosis and gradual decomposition of weeds in plots over the course of 2-4 weeks.

Weed Control

While equivalent total rates of the bioherbicide were applied in both studies, the split applied applications of 2011 resulted in considerably delayed control relative to the single application of 2012 (Figure 2). Herbicide-induced chlorosis was quickly evident in treated plots within the first two weeks of both years, but these did not result in significant differences in control, until weeds had fully decomposed. In 2011, by 28 DAT, the 64 and 128 g m^{-2} bioherbicide rates exhibited significantly improved control (18 and 11% control, respectively) compared to untreated plots (Figure 2). By the end of the 2011 study (72 DAT), the highest rate of bioherbicide (128 g m^{-2}) provided significantly improved slender aster control (88%), when compared to the untreated. In

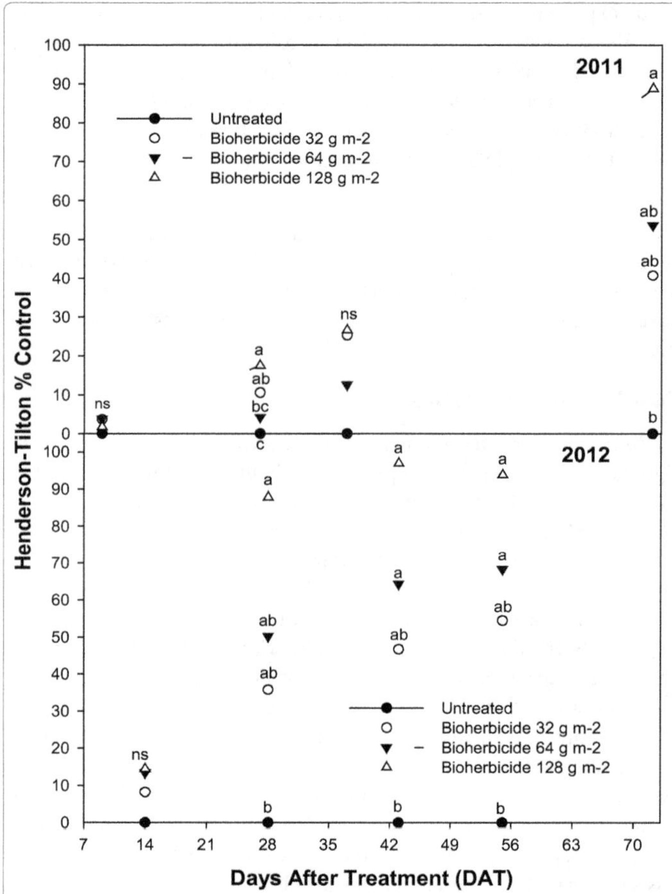

Figure 2: Henderson-Tilton % control of slender aster.Percentages were calculated using the Henderson-Tilton formula based on treated vs. untreated plots. Trials were initiated on June, 1 2011, and May 2, 2012, in College Station, TX. Bioherbicide was split applied in 2011, and single applications were made in 2012 at trial initiation. Means followed by the same letter are not significantly different according to Tukey's Test ($P \leq 0.05$).

addition, the lower two rates of bioherbicide (32 and 64 g m^{-2}) provided marginal control (41 and 54%, respectively) when compared to the untreated.

Onset of weed injury and subsequent control following application occurred much more rapidly in 2012 compared with 2011, likely due to bioherbicide treatments being applied as a single application. Another factor which could have contributed to the more rapid weed injury in the 2012 trial is that treatments were applied in May, on slightly younger weeds. Herbicide activity has been shown to occur much more rapidly when applied early in the weed life cycle to younger weeds when herbicide uptake and translocation are favored [15]. As in 2011, no significant differences in weed control were observed in any treatment until 28 DAT, at which time the 128 g m^{-2} bioherbicide treatment provided 88% control. Rates of 32 g m^{-2} and 64 g m^{-2} again provided marginal (~ 50%) control; however, these were not statistically different from untreated plots, due to a naturally occurring decline in weed population in untreated plots. By the end of the 2012 study (55 DAT), both the 64 and 128 g m^{-2} bioherbicide rates provided significantly improved control relative to untreated plots, (68 and 94%, respectively). Final levels of slender aster control were similar between both years, with the 128 g m^{-2} application rate providing 88 to 94% control, the 64 g m^{-2} application providing 55-65% control, and the 32 g m^{-2} application rate providing 40 to 50% control.

Temperature Effects

Previous research has shown this bioherbicide effectively controlling weeds under mild temperatures ranging from 15°C to 25°C [12]. However, prior to this research, it was not known what levels of control could be expected under higher temperatures. Despite the high temperatures around 41°C (with mean temperatures of 31°C in 2011 and 27°C in 2012) in both years, the bioherbicide provided effective control of slender aster in this study. Based on these results, the bioherbicide appears to retain good efficacy as a natural weed control product during summer months in areas receiving high temperatures, following application. It should be noted, however, that irrigation was provided frequently (4 to 5 times weekly) during this study, and may have also contributed to success of the bioherbicide under these conditions. Daily irrigation may not be agronomically appropriate or feasible in some situations, especially where municipal water restrictions limit the frequency of irrigation allowable on a landscape.

Potential Carryover

Another area of interest is the potential of this product to persist and carry over into subsequent seasons in the soil. Although no analysis of microbial fractions were attempted in this study, field observations indicate limited to no carryover into the following year, as successive weed seeds germinated in plots and produced green, healthy slender aster plants. However, research has shown that clay soils will retain the bioherbicide product longer than sandy loam soils [16], similar to those used in this study. Our observations are consistent with the findings of Zhou et al. [17], who found that residual activity of *Phoma macrostoma* begins to decline after 4 months, with no negative effects seen on susceptible plants in the year following application.

Implications on Use in Lawns

The results from this study demonstrate that the bioherbicide is effective at controlling slender aster, with a final level of control dependent on rate applied. Though significant control was not observed either year at the lowest application rate (32 g m^{-2}), limited activity was observed. Again, it should be noted that the material tested only contained half of the potency of the target commercial product intended for consumer use. Therefore, the amount of product used in these studies was twice that of the anticipated final commercial product. While a 128 g m^{-2} rate may be required for lawns with high weed pressure, lower rates (32 and 64 g m^{-2}) may be adequate for light weed infestations. Furthermore, unlike CGM, which must be broadcast applied at high rates over the entire lawn for preemergence control, the bioherbicide could be used as a remedial postemergence weed spot-treatment product, thereby reducing the total application amount substantially.

Effective weed control using the bioherbicide has been observed under a range of temperatures and appears to be limited to the season of application and site of placement, with no apparent phytotoxicity to desirable warm-season turfgrass. While future research is needed to more clearly define the spectrum of weed species controlled by this bioherbicide, it appears to be a suitable candidate for use as a natural broadleaf weed control option for lawns.

References

1. Anonymous (2010) Alberta environment and sustainable resource development. Fertilizer/herbicide Combination Product Fact Sheet.
2. Abouziena HFH, Omar AAM, Sharma SD, Singh M (2009) Efficacy comparison of some new natural product herbicides for weed control at two growth stages. Weed Technol 23: 431-437.

3. Boyd NS, Brennan EB (2006) Burning nettle, common purslane, and rye responseto a clove oil herbicide. Weed Technol 20: 646-650.

4. Evans GJ, Bellinder RR, Goffinet MC (2009) Herbicidal effects of vinegar and a clove oil product on redroot pigweed (*Amaranthus retroflexus*) and velvetleaf (*Abutilon theophrasti*). Weed Technol 23: 292-299.

5. Evans GJ, Bellinder RR (2009) The potential use of vinegar and a clove oil herbicide for weed control in sweet corn, potato, and onion. Weed Technol 23: 120-128.

6. Bingaman BR, Christians NE (1995) Greenhouse screening of corn gluten meal as a natural control product for broadleaf and grass weeds. Hort Sci 30: 1256-1259.

7. Christians N, Liu D, Unruh JB (2010) The use of protein hydrolysates for weed controls: Protein hydrolysates in biotechnology. Springer 127-133.

8. Christians NE (1993) The use of corn gluten meal as a natural preemergence weed control in turf. Int Turfgrass Soc Res J 7: 284-290.

9. Patton A, Weisenberger D (2012) Evaluation of crab¬grass control with various dimension formulations and corn gluten meal. 2011 Annual Report, Purdue University Turfgrass Scientific Program 31-32.

10. Stier JC (1999) Corn gluten meal and other natural products for weed control in turfgrass. University of Wisconsin-Department of Horticulture and University of Wisconsin-Extension Turfgrass Bulletin.

11. Graupner PR, Carr A, Clancy E, Gilbert J, Bailey KL, et al. (2003) The Macrocidins: Novel cyclic tetramic acids with herbicidal activity produced by *Phoma macrostoma*. J Nat Prod 66: 1558-1561.

12. Bailey KL, Pitt WM, Falk S, Derby J (2011) The effects of *Phoma macrostoma* on nontarget plant and target weed species. Biol Control 58: 379-386.

13. Henderson CF, Tilton EW (1955) Tests with acaricides against brown wheat mite. J Econ Entomol 48: 157-161.

14. Smith J, Wherley B, Baumann P, Falk S (2012) Evaluation of Phoma macrostoma toxicity on warm-season turfgrasses.

15. McCarty LB, Murphy TR (1994) Control of turfgrass weeds: Turf weeds and their control. Turgeon AJ, Kral DM, Viney MK (Eds), American Society of Agronomy and Crop Science Society of America, Madison, Wisconsin, USA 209-248.

16. Bailey KL, Pitt WM, Derby J, Walter S, Taylor W, et al. (2010) Efficacy of *Phoma macrostoma*, a bioherbicide, for control of dandelion (*Taraxacum officinale*) following simulated rainfall conditions. Am J Plant Sci Biotechnol 4: 35-42.

17. Zhou L, Bailey KL, Derby J (2004) Plant colonization and environmental fate of the biocontrol fungus *Phoma macrostoma*. Biol Control 30: 634-644.

4

Effects of Inoculation of *Sinorhizobium ciceri* and Phosphate Solubilizing Bacteria on Nodulation, Yield and Nitrogen and Phosphorus Uptake of Chickpea (*Cicer arietinum* L.) in Shoa Robit Area

Birhanu Messele* and L.M. Pant

Department ofAgro-ecology, Menschen fur Menschen,Agro-Technical andTechnology College, P. O. Box 322, Harar, Ethiopia

Abstract

A field experiment was conducted during the 2006/07 growing season to assess the effects of inoculation of *Sinorhizobium ciceri* and phosphate solubilizing bacteria on the performance of chickpea variety" DZ-10-11" in Shoa Robit area, Ethiopia. Three levels of NP fertilizer and four levels of inoculants were used for the experiment. Treatments were laid down in a Randomized Complete Block Design (RCBD) in a factorial combination with three replications. The result of this study revealed that inoculation of *Sinorhizobium ciceri* alone increased dry matter yield by 156.58% and nodule number by 117.96% over the control whilst the addition of 18/20 kg NP ha^{-1} as urea and DCB resulted in 149.6% increase of dry matter yield and 143.6% increase in nodule number per plant over the uninoculated control. There was also a marked increase in nodule dry weight (200%), as a result of *Sinorhizobium ciceri*+ 18/20 kg NP ha-1 as urea and DCB, indicating the importance of phosphorus for nodule tissue development. Similarly inoculation of *Pseudomonas sp.*+ 18/20 kg NP ha^{-1} as urea and DCB also increased nodule dry weight, nodule number, nodule volume and seed yield by 240%, 188.52%, 151.81% 142.95% respectively over the control, indicating the efficiency of the bacteria in solubilizing phosphate in DCB. On the other hand inoculation of *Sinorhizobium ciceri*+ *Pseudomonas sp.* with 18/20 kg NP ha^{-1} as urea and DCB increased nodules number per plant by 208.8% and nodule dry weight by 220% and nodule volume by 221.24%, dry matter by 172.09% over uninoculated control at mid flowering stage of chickpea. Similarly inoculation of *Sinorhizobium ciceri*+*Pseudomonas sp.* With 18/20 kg NP ha-1 as urea and DAP increased nodule number, nodule dry weight, nodule volume and dry matter by 271.59%, 220%, 241.97%, 181.40% respectively over uninoculated control at mid flowering stage.

Introduction

Chickpea (*Cicer arietinum* L) belongs to the family *Fabaceae* (earlier Leguminoseae) and sub family papilonaceae [1]. It is most probably originated in an area of present day south-eastern Turkey and adjoining Syria where three mild annual species of *Cicer* viz *C. bijigum*, *C. aerhinosperum, and C. reticulatum* are found [2].

Chickpea is one of the most important cool season food legumes in Ethiopia and grown on heavy black soils (Vertisols). It is mainly cultivated between 1400-2300 m.a.s.l where annual rainfall ranges from 700-1200 mm. Chickpea, being a legume, can be used to restore fertility in crop rotation [3]. Hence, the farmers in Ethiopia commonly rotate chickpea with cereals such as wheat, barley and teff. Despite the above fact, chickpea yield in the country is extremely low. The national average yield is 0.8-0.9 t ha^{-1} [4], whereas at farmer's field the average yield is 0.6 t ha^{-1}.

Being a legume crop, chickpea can obtain a significant portion of its nitrogen requirement through symbiotic N_2 fixation when grown in association with effective and compatible *Rhizobium* strains [5]. Most Ethiopian soils, similar to the agricultural soils of other countries in the tropics, are generally low in nitrogen (N) and phosphorus (P). These two nutrients are often limiting the crop production in Ethiopia [6]. For pulse production, P is the major limiting nutrient followed by N. This is because P not only affects legume growth, but also nodule formation and development [7-9]. The phosphate solubilizing microorganisms have the capacity to dissolve the insoluble phosphatic compounds present in the soil and also solubilize rock phosphate, bone meal and basic slag [10,11]. The field experiments done on inoculation of P-solubilizing bacteria in various crops have shown 10-15% increases in crop yields in 10 out of 37 experiments [12].

Various authors reported increased yield responses of pulses to seed inoculation of *Rhizobium* [13-15] and phosphate-solubilizing bacteria (PSB) [16]. When inoculated, these organisms colonize the rhizosphere and enhance plant growth by providing it with N and P [17].

In Ethiopia, there is very little information on combined or dual inoculation of *Rhizobium* and PSB on crop productivity [18]. Hence, it is of great practical importance to study the combined effect of these organisms on nodulation, plant growth and nutrition and legume crop yields. Adoption of such technologies by farmers will help in minimizing production costs and at the same time, avoid the environmental hazards [19]. Therefore, in view of this, a field study on chickpea was carried out at farmers' field in Shoa Robit with the specific objective to study the effects of inoculation of *Sinorhizobium ciceri* and phosphate solubilizing bacteria and their interaction on nodulation, growth, yield, and nitrogen and phosphorus uptake of chickpea.

Material and Methods

Description of the study area: The study was conducted on a

***Corresponding author:** Birhanu Messele, Department of Agro-Ecology, Menschen fur Menschen, Agro-Technical and Technology College, P. O. Box 322, Harar, Ethiopia, E-mail: birhanmartha@gmail.com

farmer's field under Merye Peasant Association around Shoa Robit, the capital of Kewet woreda located 11°55'N and 37°20'E at 1300 m.a.s.l, in North Shoa of the Amhara National Regional Sate. It is located at a distance of about 225 km to the North east of Addis Ababa on the way to Dessie. The Kewet woreda is classified under hot to warm moist agro-ecological zone. The annual rainfall, from meteorological records of last 14 years, is 1023.8 mm and the temperature ranges from 8°C to 37°C with a mean daily minimum and maximum temperatures of 16.6°C and 31°C [20]. The soil of the area is typically dark grey when dry and very dark grey-brown when moist with clay texture. The clay is montmorillonite, so the soils have high shrinkage capacity when dry and high swellings when wet for long time. The farmer's field, where the experiment was conducted, is located around 7 km away from Shoa Robit town.

There are two distinct growing seasons in the area viz., 'Belg' (March-July) and 'Meher' (August-December). The study was conducted during the Meher 2006/07 cropping season under rainfed condition with supplemental irrigation when required.

Experimental procedures

Soil sampling, preparation and analysis: Pre-sowing surface soil samples were collected at 0-30 cm diagonally from five spots in the experimental field, then were composited and processed for soil analysis before sowing. Composite soil samples were analysed for organic matter, total N by Kjeldahl digestion and distillation method; available soil P was extracted and analysed; soil texture, cation exchange capacity and soil pH were measured using standard laboratory procedures. At harvest, soil samples were collected at 0-30 cm from each plot and composited treatment-wise for determination of available P.

Treatments: Urea as a source of nitrogen and two sources of P namely, dried and crushed bone (DCB) and diammonium phosphate (DAP) were used for the field experiment. They were used in combination with inoculants as per the treatment. Lignite -based inoculants of *Sinorhizobium ciceri* (strain EAL 001) and local isolate of phosphate-solubilizing bacteria, *Pseudomonas sp.* singly or in combination and with or without N and P sources (urea, DCB and DAP) were used to get the treatments.

Experimental design: The experiment consisted of 12 treatments with a factorial combination of 3 levels of NP fertilizer (0/0, 18/20 as urea and DAP and 18/20 as urea and DCB kg ha^{-1}), and with four levels of inoculants (Uninoculated, *Sinorhizobium ciceri* (EAL 001), *Pseudomonas sp.* and *Sinorhizobium ciceri* (EAL 001)+*Pseudomonas sp.*). Treatments were laid down in a Randomized Complete Block Design (RCBD) with three replications making a total number of 36 plots. The full dose of N and P fertilizers were applied using row methods of application at planting time.

As per design of the experiment, field layout was prepared and each treatment was assigned randomly to experimental units within a block. The size of each plot was 2.8×4 m (11.2 m^2). The spacing between blocks and plots was 1.5 m and 0.5 m, respectively. The chickpea variety 'DZ-10-11' was used for planting. This variety was chosen on the basis of its resistance for ascochyta blight disease and better performance for many years in the mid altitudes areas of Ethiopia.

Seed inoculation: Seeds were inoculated with lignite-based inoculants of *Sinorhizobium ciceri* (EAL 001, Mojo isolate) and/or phosphate solubilizing bacteria, *Pseudomonas sp.* (Jimma isolate) at the rate of 7 g/kg of seeds as per the treatment. Carrier based cultures were mixed with small amount of 10% solution of sucrose in cool and clean water to form a thick slurry. The slurry was poured over the dry seeds so as to uniformly coat the seeds with the inoculant. For combined inoculation, inoculants of *Sinorhizobium ciceri* and phosphate solubilizers were mixed in equal proportions (7 g/kg seed) and applied to the seeds in the similar manner. All inoculations were done just before planting under shade to maintain the viability of microbial cells.

Sowing: The experiment was planted on September 28, 2006. Chickpea seeds of variety DZ-10-11 were sown in seven rows per plot at 40 cm row to row and 10 cm plant to plant distance.

Agronomic practices: The experimental field was weeded two times during the growing season. The first weeding was done 15-days after planting of chickpea to avoid competition during early stage of crop growth. The second weeding was undertaken one month later. Weeds, in general were not a serious constraints to chickpea. At about the poding stage of the crop, the insecticide Selectron was sprayed at the rate of 1.04 liter ha^{-1} to control ball worms.

Data collected

Nodulation: The data on nodulation parameters were taken at mid flowering stage of chickpea. Five competitive plants were randomly taken from second border rows from each side of the plot for nodulation parameters (number of nodules, nodule volume and nodule dry weight per plant) and dry weight of plants at mid flowering. In each plot, whole root system of a plant was completely exposed and carefully uprooted for nodulation parameters. The roots were gently washed under running tap water over a sieve to avoid loss of detached roots. The nodules from all the plants were removed and separately spread on the sieve for some minutes until the water drained off. The total number of nodules was counted and the mean value of five plants was recorded as the average number of nodules per plant. The color on inside of nodules was observed by cutting with the help of a sharp blade.

The collected nodules were immersed in a previously measured volume of water in a measuring cylinder. The volume of water displaced by nodules from 5 plants was considered as nodule volume and converted to average nodule volume per plant. After determination of nodule volume, the nodules were dried in an oven at 70°C to constant weight to determine nodule dry weight per plant. The average of five plants was taken as nodule dry weight per plant.

Dry matter: Dry matter of plants was determined at mid flowering stage of the crop from plants sampled for nodulation. The sampled plants were placed in labeled perforated paper bags and oven-dried at 70°C to a constant weight. The average dry weight of five plants was measured to determine dry weight per plant.

Yield and yield components: At physiological maturity, five competitive plants from net area were sampled for the determination of number of pods per plant. Number of seeds per pod was determined for 20 randomly sampled pods from the same five plants.

One hundred seed weight was also determined by counting 100 seeds and weighing on a sensitive electronic balance. Harvest index was determined as the ratio of grain yield with above ground dry biomass per plot. Yield per plot was determined by harvesting the chickpea from the central three rows of a net size of 1.2×3 m (3.6 m^2) leaving the boarder rows and 0.5 m row length on every end of each row. Chickpea plants were harvested from each plot at physiological maturity, the harvested plants was sun-dried in the open air for 3-4 weeks, weighed to determine above ground biomass yield and then threshed and weighed to determine the grain yield of each plot. Finally, yield per plot was converted to per ha basis. Straw yield was calculated by subtracting grain yield from the corresponding above ground biomass yield.

Plant tissue sampling and analysis for N and P: At physiological maturity, five non-border plants were harvested and partitioned into grain and straw. The grain and straw sample materials was separately air-dried, oven dried at 70°C to a constant weight, ground to pass a 1 mm sieve and saved for laboratory analysis of grain and straw N and P concentration.

Phosphorus in grain and straw sub-samples was determined using metavanadate method. Samples were calcinated in the furnace overnight at 450°C and the ash was dissolved in 20% nitric acid (HNO_3) to liberate organic P. The phosphorus in the solution was determined colorimetrically using molybdate and metavanadate for color development. The reading of phosphorus was made at 460 nm in spectrophotometer. Total N in the grain and straw sub-samples were quantitatively determined by a Kjeldahl procedure that included a salicylic acid predigest ion step to convert nitrate to ammonium.

Phosphorus uptake by grain and straw were determined from the phosphorus content of the respective parts after multiplying with the grain yield and straw yield, respectively. Total phosphorus uptake was then calculated as the summation of grain and straw uptake. Similarly, N uptake in the grain was determined after multiplying nitrogen content of the grain by grain yield, and straw nitrogen uptake was determined by multiplying nitrogen content in the straw by straw yield. Total nitrogen was recorded as the sum of grain N uptake and straw N uptake.

Statistical analysis

All data collected were subjected to the analysis of variance (ANOVA) appropriate to factorial experiments in Randomized complete Block Design using SAS software (SAS Inistitute, 1989 [21]).

Results

Some selected physical and chemical properties of the soil of experimental site before planting

Results of the laboratory analysis (Table 1) for the soil sample taken before planting indicated that the textural class of the experimental soil was loam. It had high N, 1.26% while the available P content was 7.64 ppm, the P level of the soil can be rated as low.

The percent organic carbon content of the soil sample is low (1.74%). The organic matter content of the soil sample is computed by multiplying the organic carbon content with a conversion factor 1.724 and the result showed that medium organic matter content of the soil (2.99%). The pH value of soil is 8.3 (in 1:2.5 soil: water suspension) which according to Landon (1984) was rated as high and shows the cation exchange capacity of the soil was medium (29.57 cmol (+) kg^{-1}.

Particle size distribution (%)			Textural class	pH (1:2.5 H_2O)	OC (%)	TN (%)	AP (ppm)	CEC (Cmol (+)/kg)	PBS
Sand	Silt	Clay							
26	42	32	Loam	8.3	1.74	1.26	7.64	29.57	89

OC: Organic carbon; TN: Total nitrogen; AP: Available phosphorous; CEC: Cation exchange capacity; PBS: Percent base saturation

Table 1: Selected physical and chemical properties of the soil of experimental site before planting.

Number of nodules per plant

As presented in table 2, the analysis of variance test showed a significant statistical difference among treatments in relation to number of nodules per plant. The highest number of nodules (140.60 nodules $plant^{-1}$) were recorded with inoculation of *Sinorhizobium ciceri+ Pseudomonas sp.* along with 18/20 kg NP ha^{-1} as urea and DAP, followed by *Pseudomonas* sp.+18/20 kg NP ha^{-1} as urea and DAP (108.10 nodules $plant^{-1}$), *Sinorhizobium ciceri+Pseudomonas sp.*+18/20 kg NP ha^{-1} as urea and DCB (108.07 nodules $plant^{-1}$) and *Pseudomonas sp.* alone (104.93 nodules $plant^{-1}$) which were significantly higher than uninoculated control. Compared to the control all the treatments recorded higher nodule number per plant, however, inoculation of *Sinorhizobium* without urea+DCB/DAP and uninoculated treatment with urea and DCB/DAP gave nodules at par with the control.

Nodule dry weight

F-test indicated significant difference ($P \leq 0.01$) among treatments with respect to nodule dry weight per plant (Table 2). Maximum nodule dry weight (1.20g $plant^{-1}$) was recorded with inoculation of *Pseudomonas sp.* along with 18/20 kg NP ha^{-1} applied as urea and DCB and *Sinohizobium ciceri*+18/20 kg NP ha^{-1} as urea and DAP (1.20 g $plant^{-1}$) followed by *Pseudomonas sp.*+18/20 kg NP ha^{-1} as urea and DAP (1.1 g $plant^{-1}$) and (*Sinorhizobium ciceri* + *Pseudomonas sp.*)+18/20 kg NP ha^{-1} as urea and DAP/DCB.

Although inoculation of *Sinohizobium ciceri* alone and combined inoculation of *Sinorhizobium ciceri* and *Pseudomonas sp.* in the absence of P source resulted only in marginal increase in nodule dry weight over the untreated control, but inoculation of *Pseudomonas sp.* alone in the presence or absence of P source showed a considerable increase in nodule dry weight over uninoculated control. In general, the treatments having P solubilizers alone or with *Sinorhizobium ciceri* with phosphorous source recorded higher nodule dry weight as compared to treatments without P solubilizers (Table 2). Therefore, the results have indicated that due to increased availability of phosphorus from soil or DAP/DCB which resulted due to solubilization of P by inoculated phosphate solubilizers, the infection of roots by inoculated rhizobia increased and resulted in higher number and mass of nodules. Application of 18/20 kg NP ha^{-1} as urea and DAP/DCB, inoculation of *Sinorhizobium* cicer without N and P source and *Sinorhizobium ciceri* and phosphorous solublizing bacteria without N and P source were at par with the uninoculated control.

Nodule volume

Nodule volume was significantly ($P < 0.01$) affected by application of treatments (Table 2). The highest nodule volume (4.67 ml) was recorded with inoculation of (*Sinorhizobium ciceri+Pseudomonas sp.*)+18/20 kg N P ha^{-1} as urea and DAP followed by (*Sinorhizobium ciceri* + Pseudomonas)+18/20 kg N P ha^{-1} as urea and DCB (4.27ml) and *Pseudomonas sp.*+18/20 kg N P ha^{-1} as urea and DCB (2.93ml). However, all other treatments gave nodule volume at par with the control.

Effect on dry matter yield per plant at mid flowering

The dry matter production at mid flowering stage of chickpea was significantly ($P \leq 0.05$) influenced by different treatments (Table 2). Among the treatments, inoculation of (*Sinorhizobium ciceri*, EAL 001+*Pseudomonas sp.*)+18/20 kg N P ha^{-1} as urea and DAP (7.8 g $plant^{-1}$) and (*Sinorhizobium ciceri*, EAL 001+*Pseudomonas sp.*)+18/20 kg N P ha^{-1} as urea and DCB (7.4 g $plant^{-1}$) were significantly ($P \leq 0.05$)

Treatments	Number of nodule plant^{-1}	Nodule dry wt. plant^{-1}(g)	Nodulevol.plant^{-1} (ml)	Dry matter plant-1(g)
Uninoculated	51.77f	0.5d	1.93bc	4.30d
18/20 kg NP ha^{-1} as urea and DAP	64.67def	0.7dc	2.067bc	4.27d
18/20 kg NP ha^{-1} as urea and DCB	57.23f	0.7dc	2.43bc	5.93abcd
Sinorhizobium	61.07ef	0.8bcd	1.97bc	6.73abc
Sinohizobium+18/20 kg N P ha^{-1} as urea and DAP	90.13bcd	1.2a	2.60bc	5.33bcd
Sinorhizobium +18/20 kg N P ha^{-1} as urea and DCB	74.33cdef	1.0abc	1.53c	6.43abcd
Pseudomonas sp.	104.93b	1.0abc	2.47bc	5.07cd
Pseudomonas sp.+18/20 kg N P ha^{-1} as urea+DAP	108.10b	1.1ab	2.7bc	5.17bcd
Pseudomonas sp.+18/20kg NP ha^{-1} as urea and DCB	97.60bc	1.2a	2.93b	5.57abcd
Sinorhizobium +Pseudomonas sp.	84.33bcde	0.8bcd	2.53bc	4.20d
Sinorhizobium +Pseudomonas sp.+18/20 kg NP ha^{-1}as urea and DAP	140.60a	1.1ab	467a	7.80a
Sinorhizobium +Pseudomonas sp.+18/20 kg NP ha^{-1}as urea and DCB	108.07b	1.1ab	4.27a	7.40ab
LSD (5%)	27.02	0.3387	1.25	2.31
CV (%)	18.36	21.43	27.53	23.99

CV: Coefficient of variance, DCB: Dried and crushed bone, DAP: Diammonium phosphate, N: Nitrogen and P: Phosphorous. Means within a column followed by the same letter(s) are not significantly different.

Table 2: Nodulation of chickpea and dry matter yield plant^{-1} at mid flowering stage as affected by inoculation of *Sinorhizobium ciceri+Pseudomonas sp.*

Treatments	SY (kg/ha)	HSW (g/p)	Harvest index	Percent yield
Uninoculated	790.83f	12.67bc	0.4c	100
18/20 kg NP ha^{-1} as urea and DAP	1176.47bcd	12.83abc	0.5b	148.76
18/20 kgNP ha^{-1} as urea and DCB	1081.00d	13.40abc	0.5b	136.69
Sinorhizobium ciceri, EAL 001	879.63ef	14.73a	0.47b	111.22
Sinohizobium ciceri, EAL 001 +18/20 kg NP ha-1 as urea and DAP	790.73f	14.37abc	0.47b	99.98
Sinorhizobium ciceri, EAL 001 +18/20kg NP ha^{-1} as urea and DCB	941.10e	13.47abc	0.47b	119.00
Pseudomonas sp.	1113.00cd	14.60ab	0.5b	140.73
Pseudomonas sp.+18/20 kg NP ha^{-1} as urea+DAP	1209.43abc	12.90abc	0.5b	152.93
Pseudomonas sp.+18/20 kg NP ha^{-1} as urea and DCB	1130..47bcd	12.90abc	0.47b	142.94
Sinorhizobium ciceri, EAL 001+*Pseudomonas sp.*	856.63ef	12.40c	0.5b	108.32
Sinorhizobium ciceri, EAL 001+*Pseudomonas sp.*+18/20kg NP ha^{-1}as urea and DAP	1318.53a	13.03abc	0.57a	166.72
Sinorhizobium ciceri, EAL 001+*Pseudomonas sp.*+18/20 kg NP ha^{-1} as urea and DCB	1241.73ab	13.47abc	0.6a	157.02
LSD (5%)	118.81	1.98	0.07	-
CV (%)	6.72	8.75	7.80	-

Means within a column followed by the same letter(s) are not significantly different. NPP: Number of pods per plant; NSP: Number of seeds per pod; SY: Seed yield; HSW: Hundred seed weight; Nr: Number; AGW: Above ground weight; DCB: Dried and crushed bone; DAP: Diammonium phosphate; N: Nitrogen and P: Phosphorous

Table 3: Effects of inoculation of *Sinorhizobium ciceri*, EAL 001+*Pseudomonas sp.* on number of pods per plant, number of seeds per pod, seed yield, percent yield increase and hundred-seed weight of chickpea.

0.05) increased dry matter yield per plant over the uninoculated control. The increase in dry matter yield due to combined inoculation of *Sinorhizobium* and phosphate solubilizing bacteria might be due to synergetic effect, which enhanced nitrogen and phosphorus availability to the plant. In agreement with the present finding, Application of 18/20 kg NP ha^{-1} as a source urea and DAP gave dry matter yield in par with the control.

Inoculation of *Sinorhizobium ciceri*, EAL 001 alone (6.73 g plant^{-1}) increased dry matter yield significantly over the control (4.3 g plant^{-1}). The increased dry matter yield due to inoculation of *Sinorhizobium ciceri*, EAL 001 alone could be the result of increased nitrogen fixation and its supply to chickpea, which enhanced crop growth.

Effect on number of pods per plant and seed per pod

The number of pods per plant was significantly different ($P \leq 0.01$) among the treatments (Table 3). All the treatments gave higher number of pods per plant over the control. The highest number of pods per plant (122.63) was recorded with inoculation of (*Sinorhizobium ciceri*, EAL 001+*Pseudomonas sp.*)+18/20 kg NP ha^{-1} as urea and DAP followed by (*Sinorhizobium ciceri*, EAL 001+*Pseudomonas sp.*)+18/20 kg NP ha^{-1} as urea and DCB (119.87) and *Pseudomonas sp.*+18/20 kg NP ha^{-1} as urea and DAP (106.80) as compared to the control.

Uninoculated treatment with 18/20 kg N P ha^{-1} as urea and DAP or DCB, *Sinorhizobium ciceri*, EAL 001+18/20 kg NP ha^{-1} as urea and DCB and *Pseudomonas sp* alone also gave a marginal increase in number of pods per plant over the control. However, other treatments were at par with the control.

Seeds per pod were not significantly ($P \geq 0.05$) affected by inoculation of *Sinorhizobium ciceri*, EAL 001 and *Pseudomonas sp.* with or without the P source.

Effect on seed yield and hundred-seed weight

Results (Table 3) showed that seed yield was significantly affected by application of different treatments. Inoculation of *Sinorhizobium ciceri*, EAL 001+*Pseudomonas sp.* in the presence of 18/20 kg NP ha^{-1} as urea and DAP (1318.53 kg ha^{-1}) was superior to all other treatments followed by inoculation of *Pseudomonas sp.* and *Sinorhizobium ciceri*, EAL 001 in the presence of 18/20 kg NP ha^{-1} as urea and DCB (1241.73 kg ha^{-1}) and *Pseudomonas sp.* with 18/20 kg NP ha^{-1} as urea and DAP (1209.43 kg ha^{-1}). These three treatments showed significant ($P < 0.01$) increase in seed yield over the uninoculated control. Inoculation of phosphate solubilizing bacterial isolates alone with or without DCB/DAP as P source and uninoculated+18/20 kg NP ha^{-1} as urea and DAP

Treatments	Total N uptake (kg ha^{-1})	Total P uptake (kg P ha^{-1})	Olsen P (ppm)
Uninoculated	5.7	4.1	8.6
18/20 kg NP ha^{-1} as urea and DAP	7.2	4.8	9.5
18/20 kg NP ha^{-1} as urea and DCB	7.9	5.8	10.4
Sinorhizobium ciceri, EAL 001	6.5	4.2	10
Sinorhizobium ciceri, EAL 001+18/20 kg NP ha^{-1} as urea and DAP	9.1	6.3	9.2
Sinorhizobium ciceri, EAL 001+18/20 kg NP ha^{-1} as urea and DCB	6.0	6.3	10.4
Pseudomonas sp.	9.2	6.2	8.4
Pseudomonas sp.+18/20 kg NP ha^{-1} as urea andDAP	10.1	6.8	9.9
Pseudomonas sp.+18/20 kg NP ha^{-1} as urea and DCB	7.7	5.2	8.5
Sinorhizobium ciceri, EAL 001+*Pseudomonas sp.*	6.7	4.8	9.0
Sinorhizobium ciceri, EAL 001+*Pseudomonas sp.*+18/20 kg NP ha^{-1} as urea and DAP	7.9	5.4	11.5
Sinorhizobium ciceri, EAL 001+*Pseudomonas sp.*+18/20 kg NP ha^{-1} as urea and DCB	8.8	5.7	9.0

Table 4: Uptake of nitrogen, phosphorus and available phosphorus content of the soil after harvest as influenced by inoculation of Sino*Rhizobium* ciceri, EAL 001 and phosphate solubilizing bacteria.

or DCB also gave a marginal increase over the control, however, the remaining treatments are statistically at par each other and with the control.

As presented in table 3, inoculation of (*Sinorhizobium ciceri*, EAL 001+*Pseudomonas sp*).+18/20 kg NP ha^{-1} as urea and with the cheap source of phosphorous as dried and crushed bone resulted 57.02% yield increase over the control. In general, the response to inoculation of phosphate solubilizing bacteria singly or in combination with *Sinorhizobium ciceri*, EAL 001 was found to be greater in the presence or absence of DCB and DAP as phosphorus source compared to inoculation of *Sinorhizobium ciceri*, EAL 001 with or without phosphorous source . Hundred-seed weight on the other hand was not significantly ($P \geq 0.05$) affected by any of the treatments (Table 3).

Harvest index

Harvest index was observed to be significantly ($P \leq 0.01$) affected by the treatments (Table 3). The highest value of harvest index (0.60) was recorded with inoculation of *Sinorhizobium ciceri*, EAL 001+*Pseudomonas sp.* in the presence of 18/20 kg NP ha^{-1} as urea and DCB followed by *Sinorhizobium ciceri*, EAL 001+*Pseudomonas sp.* in the presence of 18/20 kg NP ha^{-1} as urea and DAP (0.57). The other treatments were at par with each other, however, significantly different from the control.

Total nitrogen uptake

As presented in table 4, inoculation of *Pseudomonas sp.*+18/20 kg NP ha^{-1} as urea and DAP (10.1 kg N ha^{-1}) resulted the highest total nitrogen uptake of chickpea compared to the control and followed by inoculation of *Pseudomonas sp.* alone (9.2 kg N ha^{-1}) and inoculation of. *Sinorhizobium ciceri*, EAL 001 *Pseudomonas sp.*+18/20 kg NP ha^{-1} as urea and DCB (8.8 kg N ha^{-1}), and *Sinorhizobium ciceri*, EAL 001+*Pseudomonas sp.*+18/20 kg NP ha^{-1} as urea and DAP (7.9 kg N ha^{-1}). Application of 18/20 kg NP ha^{-1} as urea and DCB also resulted in the total nitrogen uptake of Chickpea in similar manner with *Sinorhizobium ciceri*, EAL 001+*Pseudomonas sp.*+18/20 kg NP ha^{-1} as urea and DAP. However, inoculation of *Sinorhizobium ciceri*, EAL 001 alone with 18/20 kg NP ha^{-1} as urea and DCB resulted in the total nitrogen uptake of chickpea at par with the control.

Total phosphorus uptake

Total phosphorus uptake was markedly affected due to various treatments (Table 4). The maximum total P uptake was observed due to inoculation of *Pseudomonas sp.*+18/20 kg N P ha^{-1} as urea and DAP (6.8 kg P ha^{-1})followed by *Sinorhizobium ciceri*, EAL 001+18/20 kg NP ha^{-1} as urea and DAP (6.3 kg P ha^{-1}) and *Pseudomonas sp.* alone (6.2 kg P ha^{-1}). Inoculation of *Sinorhizobium ciceri*, EAL 001+*Pseudomonas sp*+18/20 kg NP ha^{-1} as urea and DAP or urea and DCB also resulted in a higher total phosphorus uptake as compared to the control. The increased total P uptake as a result of inoculation of *Sinorhizobium ciceri*, EAL 001 and *Pseudomonas sp.* could be due to increased availability of N and P which enhances crop growth.

Effect on available phosphorus at harvest

As presented in table 4, available phosphorus immediately after harvesting chickpea was found to be highest with inoculation of *Sinorhizobium ciceri*, EAL 001+*Pseudomonas sp.*+18/20 kg NP ha^{-1} as urea and DAP (11.5 ppm P) followed by *Sinorhizobium ciceri*, EAL 001+18/20 kg NP ha^{-1} as urea and DCB and application of 18/20 kg NP ha^{-1} as urea and DCB. (10.4 ppm P each).

Application of 18/20 kg N P ha^{-1}as urea and DAP (9.5 ppm P) resulted higher available phosphorous immediately after harvest as compared to control.

In all treatments the available phosphorus at harvest was higher than the soil P status (7.64 ppm) before planting. Inoculation of *Pseudomonas sp.* either singly or in combination with *Sinorhizobium ciceri*, EAL 001 in the presence or absence phosphorus source increased the available phosphorus immediately after crop harvest.

Discussion

Microbial processes such as biological nitrogen fixation, phosphate solubilization and cellulose degradation etc., could supplement the nutrient requirements of crops. The contributions of these microbial processes are enhanced by introducing efficient microbes in the rhizosphere. For instance, symbiotic nitrogen fixation rates could be markedly increased by introducing highly efficient, competitive and persistent strains of Rhizobia [22]. Similarly, besides solubilizing the native phosphorus sources in the soil, phosphate solubilizing microorganisms could increase the effectiveness of mineral P fertilization [23]. Inoculation of seeds or soil with efficient nitrogen fixing and phosphate solubilizing microorganisms change the rhizosphere population, consequently affecting plant growth.

The results of field study on the effects of inoculation of *Sinorhizobium ciceri*, EAL 001 and *Pseudomonas sp.* on chickpea in presence or absence of N P sources showed positive response for most of the parameters. The nodulation parameters such as nodule number per plant, nodule volume per plant and nodule dry weight were significantly affected by single or combined inoculation of *Sinorhizobium ciceri*, EAL 001 and *Pseudomonas sp.* on chickpea. It also increased seed yield (166.7%) compared to the control. This implies that interaction between the two organisms, PSB and *Sinorhizobium ciceri*, EAL 001, benefited the crop in terms of growth and yield.

The response to inoculation was also more when DAP used in combination with *Pseudomonas sp.* and *Sinorhizobium ciceri*, EAL 001. The results indicated that the integrated use of chemical fertilizers and inoculants increase the growth, yield and yield parameters due to solubilization of fixed P in the soil. Therefore, the results of the study indicated that single or combined inoculation of *Sinorhizobium ciceri*, EAL 001 and *Pseudomonas sp.* on chickpea was beneficial for initiating formation of effective nodules under Shoa Robit soil conditions. The dry matter production at mid flowering, yield and yield components were significantly affected by single or combined inoculation of *Sinorhizobium ciceri*, EAL 001 and *Pseudomonas sp.* in the presence of N P sources on chickpea but, hundred seed weight, and seed per pod, were not affected by the treatments. Therefore, it can be concluded that use of effective inoculants is promising under Shoa Robit conditions in relation to growth and yield of chickpea. Total nitrogen uptake, total phosphorous uptake and available phosphorous at harvest increased due to the treatments.

In all treatments the available phosphorus at harvest was higher than the soil P status before planting (7.64 ppm). The increased in available phosphorus because of inoculation with phosphate solubilizing bacteria could be explained by solubilization of native phosphate by these organisms. Therefore the uses of inoculants made from effective strains will not only increase crop yield where fertilizer uses is negligible but it will also help in maintaining and enhancing soil fertility. However, the response to microbial inoculants is dependent on several soil factors including organic matter, temperature, moisture, aeration and nutrient status of the soils. Therefore the results of present study need to be evaluated and re-confirmed by conducting extensive field trials under varying soil fertility conditions with different sources and rates of N P fertilizers in combination with *Sinorhizobium ciceri*, EAL 001 and *Pseudomonas sp.*

Acknowledgement

The funding for this study was provided by the Haramaya University, Biofertilizer project. We thank all those who gave us constructive ideas in shaping this manuscript. However, only the authors did participate in the conduction of the study, data collection and interpretation and article preparation.

References

1. Kupich FK (1977) The Delimitation of the Tribe *Viceae* (Leguminosae) and the Relationship of *Cicer arietinum* L. Botanical Journal of the Linnaean Society 74: 131-162.
2. Saxena MC, Singh KB (1987) The Chickpea CAB International, the International Center for Agricultural Research in the Dry Areas. Aleppo Syria.
3. Baldev B (1988) Cropping patterns in pulse crops (Eds) Oxford and IBH Publishing Co., Pvt. Ltd., New Delhi 513-517.
4. CSA (Centeral Statisical Authority) (2000) Time Series Data on Production and Yield of Major Crops. Statistical Bulletin, Addis Ababa 56: 11.
5. Kyei-Boahen ES, Slinkard AE, Walley FL (2002) Evaluation of Rhizobial Inoculation Methods for Chickpea. Agronomy Journal 94 : 851-859
6. Asgelil Dibabe (2000) Effect of fertilizer on the yield and nodulation pattern of Faba bean on a nitosol of Adet Northwestern Ethiopia. Ethiopian Journal of Natural Resources 2: 237-244.
7. Andrew CS, Jones RK (1978) The Phosphorus Nutrition of Tropical Forage Legumes. E.J. (Eds.). Mineral nutrition of legumes in tropical and subtropical soils, CSIRO, Brisbane 295-311.
8. Haque I, Nnadi LA, Saleem MMA (1986) Phosphorus Management with Special Reference to Forage Legumes in Sub-Saharan Africa (Eds.) Potentials of Forage Legumes in Farming Systems of Sub-Saharan Africa. ILCA, Addis Ababa, Ethiopia 100-119.
9. Uloro Y, Richter C (1999) Phosphorus Efficiency of Different Varieties of *Phaseolus vulgaris* L. and *Sorghum bicolor* (L.) moench on an Alfisol in the Eastern Ethiopian Highlands. Ethiopian Journal of Natural Resources 1: 187-200.
10. Kucey RMN, Tanzen HH, ME Legett (1989) Microbiologically Mediated Increases in Plant Available Phosphorus. Advances in Agronomy 42: 199-221.
11. Asfaw Hailemariam (1997) The Effect of Phosphorus Solubilizing Fungus on the Growth and Yield of Teff. Ethiopian Science and Technology Research Report. Addis Ababa, Ethiopia 16-21.
12. Tandon HL (1987) Phosphorous Research and Production in India. Fertilizer Development and Consultation Organization, New Delhi, 160.
13. Dorosinksy LM, Kadyrov AA (1975) Effect of inoculation on nitrogen fixation by chickpea, its crop and content of protein. Mikrobiologiya 44: 1103-1106.
14. Moawad H, Badr El-Din SMS, Khalfallah MA (1988) Field Performance of Rhizobial Inoculants for Some Important Legumes in Egypt. In: Beck, D.P. and L.A. Materon (Eds. 89). Proceedings of a Workshop on Biological Nitrogen Fixation on Mediterranean-Type Agriculture, April 14-17, 1986, ICARDA, Syria.
15. Beyene D (1988) Biological Nitrogen Fixation Research on Grain Legumes in Ethiopia: An Overview. Pp 73-78. In: Beck, D.P. and Materon, L.A. (Eds.). Proceedings of a Workshop on Biological Nitrogen Fixation on Mediterranean-Type Agriculture, April 14-17, 1986, ICARDA, Syria.
16. Gaur AC (1985) Phosphate-Solubilizing Microorganisms and Their Role in Plant Growth and Crop Yields. Proceedings of Soil Biology Symposium, Hissa-R, India 125-135.
17. Kundu BS, Gaur AC (1980) Establishment of Nitrogen Fixing and Phosphate Solubilizing Bacteria in the Rhizosphere and Their Effect on Yield and Nutrient Uptake of Wheat Crop. Plant and Soil 57: 223-230.
18. Haile W (1999) Isolation and Characterization of Phosphate solublizing bacteria from Some Ethiopian Soils and their Effect on the Growth of Faba Bean. M.Sc. Thesis, Addis Ababa University, Addis Ababa 65.
19. Galal YGM, El-Ghandour IA, El-Akel EA (2002) Stimulation of Wheat Growth and N Fixation through *Azospirillum* and *Rhizobium* Inoculation. A Field Trial with ^{15}N Techniques. Plant Nutrition 92: 666-667.
20. Kewet Wereda Bureau of Agriculture (1995) Annual Report on Agriculture and Rural Development 15.
21. SAS Institute (1989) SAS/STAT User's Guide. SAS Publ., Cary, NC.
22. Materon LA, Keatinge JDH, Beck DP, Yurtsever N, Altuntas S (1995) The Role of Rhizobial Biodiversity in Legume Crop Productivity in the West Asian Highlands. II. *Rhizobium leguminosarum*. Journal of Experimental Agriculture 31: 485-491
23. Mikanova O, Novakova J (2002) Evaluation of the P-Solubilizing Activity of Soil Microorganisms and its Sensitivity to Soluble Phosphate. Rostlinna Vyroba 48: 397-400.

5

Efficacy of *Cymbopogon Schoenanthus* L. Spreng (Poaceae) Extracts on Diamondback Moth Damaging Cabbage

Bakouma Laba[2], Amen Y Nenonéné[2], Yao Adjrah[1*], Koffi Koba[2], Wiyao Poutouli[3], Komlan Sanda[2]

[1]*Laboratoire de Microbiologie et de Contrôle de qualité des Denrées Alimentaires, Ecole Supérieure des Techniques Biologiques et Alimentaires (ESTBA), Université de Lomé BP 12281, Lomé - Togo*
[2]*Unité de recherche sur les Agroressources et la Santé Environnementale, Ecole Supérieure d'Agronomie, Université de Lomé, BP 20131, Lomé - Togo*
[3]*Laboratoire de Biologie Animale et de Zoologie, Faculté des Sciences, Université de Lomé, BP 1515, Lomé - Togo*

Abstract

This study aims to examine the insecticidal properties of the aerial part of *Cymbopogon schoenanthus*. Cabbage plants were sprayed with the aqueous extracts of *C. schoenanthus* leaves as treatment, and the damage levels of *Plutella xylostella* was assessed. *In vitro*, the emulsified essential oil concentrations were used in a contact test on the larvae in order to assess the mortality effects. The larvae survival time was only 22 seconds with *C. schoenanthus* emulsified oil treatment (2 g/l), whilst it exceeded 44,100 seconds (over 12 hours) for the dimethoate. The nutrition test showed that at 48 h period, a significant effectiveness against larvae was observed with emulsified oil treatment 2 g/l (60% mortality) versus 10% of mortality for dimethoate. *Cymbopogon schoenanthus* can validly be used as alternative in *P. xylostella* management. The results of the field experiments showed no significant difference between the treatments and the control in terms of marketable cabbages harvested.

Keywords: Cabbage; *Plutella xylostella*; *Cymbopogon schoenanthus*; Lomé

Introduction

Apart from being consumed as a current vegetable, cabbage has been valued for medicinal purposes in treating headaches, gout, and diarrhea, inflammatory and gastrointestinal disorders [1-4]. Some researchers have focused their works on the capacity of cabbage to reduce the risks of some cancers [5,6], especially due to the content of glucosinolates and derived products, flavonoids and other phenolics in cabbage [7-9]. The antioxidant activity of these compounds has been shown to correlate with vitamin C and phenolic phytochemicals content [4,6,10,11].

The intensified growing of cabbage has led to a common problem of high pest infestation, which is caused mostly by the Diamondback Moth (DBM) *Plutella xylostella* [12-14]. DBM larvae are very difficult pests to control [15] and therefore, are the greatest threat to crucifer production in many parts of the world. The losses can reach the 90% [13,16]. This explains the large use of insecticides in crucifer production. Owing to its polyvoltine characteristics and serious overlap of generations, this pest can easily develop resistance to various kinds of insecticides [17-19], including biological one such as *Bacillus thuringiensis* [20-22], and particularly in sub-tropical and tropical countries [23]. To overcome resistance, farmers resort to increasing frequency and rates of pesticide applications and to mixed cocktails of pesticides [16]. In addition, the information about the chemical composition of these pesticides is not always publicly available [24]. In Togo, vegetable producers currently apply seven different chemical insecticides (fipronil, chlorpyrifos ethyl, cypermethrin, dimethoate, endosulfan, *Bacillus thurgensis* and acephate) on cabbage and in over 11 applications within 3 months of crop growth prior to harvest. Indiscriminate use of pesticides constitutes one of the main environmental and public health problems in developing countries leading to harmful effects on the ecosystems, the health of both farmers and consumers [25-29].

Biological options in an integrated pest management (IPM) approach offer a solution to sustainable control of DBM. In West Africa, farmers use botanical pesticides, such as plants extracts of *Azadirachta indica* A. Juss, *Melia azedarach* L. against DBM [30]. *Cymbopogon schoenanthus* is efficient in the biological control against pests [31,32]. *Cymbopogon schoenanthus* L. Spreng. (Poaceae) or lemon grass, originally from India, is a warm climate aromatic plant that grows in Togo [31]. It's essential oils are very rich in piperitone [31,33,34] which is responsible for the insecticidal activity of this plant [31,35]. This present work aims at using aqueous leaf extracts and low concentrations of essential oil of *C. schoenanthus* as an integrated management approach to the control populations of *P. xylostella* and their larvae *in vitro* and the field.

Materials and Methods

Experimental site

Experiments were conducted from May 2009 to August 2010 at Agricultural Teaching Experimental Station (ATES) of University of Lomé (UL), Togo. Plant extractions, nutrition and contact tests on *P. xylostella* were carried out respectively in Chemistry and Plant Biology laboratories of the *Ecole Supérieured'Agronomie* (ESA) of UL.

Plant and materials

Cabbage cultivar "KK cross" purchased in a seeds shop in the outskirts of Lomé was used for the assay. Leaves of *C. schoenanthus* were collected at the Agricultural Teaching Experiment Station of UL where this plant was cultivated just for experimental purposes.

***Corresponding author:** Yao Adjrah, Laboratoire de Microbiologie et de Contrôle de qualité des Denrées Alimentaires, Ecole, Supérieure des Techniques Biologiques et Alimentaires (ESTBA), Université de Lomé BP 12281, Lomé-Togo, E mail: neladjrah@gmail.com

Source of insects

Fourth instars larvae of *P. xylotella* were collected in cabbage growing field and from a small rearing unit of *P. xylostella* of ATES.

Chemical insecticide dimethoate was purchased from a local supplier as 'Calidim' 400 EC (Caliope Chemical Industries Ltd, France).

P. xyllostella larvae control with emulsified oil of C. *schoenanthus in vitro*

Contact tests: Ten fourth instars larvae were introduced into a Petri dish, and were sprayed (using ULV apparatus) with the five formulations: DW (Distilled water); THE1 (essential oil 2 g, hand soap containing soda 2 g, distilled water 96 g); THE2 (essential oil 1 g, hand soap containing soda 2 g, distilled water 97 g); SW (hand soap containing soda 2 g, distilled water 98 g); and DT (40 µl aqueous solution containing 0.25 mg of active ingredient). The soap is an adjuvant used to support oils on the leaf surface and during larvae contact. The test was replicated four times with each formulation. Larval behavior was observed with a magnifying glass, and a chronometer was used to determine the duration of larval survival. The parameter measured was mortality, and larvae were considered dead if they did not move either their head or their thorax when touched.

Nutrition test: A randomized complete block design five treatments each replicated 4 times was used. Fresh cabbage leaves discs (diameter 5 cm) were sprayed with the formulations DW, THE1, THE2, SW and DT and then placed in Petri dishes. Ten fourth instar larvae were introduced into each Petri dish. The development of the larvae feeding on sprayed leaves discs was observed at 24, 48, 72 and 96 hours. Three parameters were recorded: mortality, larval stage and adult emergence.

Aqueous extracts preparation: Aqueous extracts of *Cymbopogon schoenanthus* leaves were obtained according to the following methodology: 1) three years old leaves were harvested, chopped to fine particles size and shade dried at room temperature for four days; 2) three formulations, TS50, TS100, TS150, were obtained with 50 g, 100 g and 150 g respectively of chopped dried leaves infused in one liter of water for 24 hours.

Effects of *C. schoenanthus* leaves aqueous extracts on the field: In our investigation, different concentrations of aqueous extracts of *C. schoenanthus* leaves were used as an integrated management approach to the control of *P. xylostella* under field conditions. An untreated small plot of cabbage was placed about 15 m from the test plots to serve as a control. After one month in seedbed, the seedlings were transplanted to the plots. Plot size was 3.50 m × 1.20 m each, with a spacing of 75 cm between plants and 60 cm between rows, to create two rows of 5 plants each. Spacing between plots and replications were 1.5 m and 1.5 m, respectively. Before the planting, 30 kg of organic manure collected from extensive poultry farming was applied to each plot. The seedlings were sprinkled with water two times per day. No insecticide was used in the seedbed, but after the planting, the organophosphorus insecticide dimethoate was used weekly at 400 g of active ingredient per hectare (the dose indicated by manufacturer) to control DBM on every plot until the pre-heading stage. After this phase, three doses of *C. schoenanthus* aqueous extracts (TS50: 50 g/l, TS100: 100 g/L and TS150: 150 g/L) were applied weekly at a rate of 6 liters for the 4 plots during 4 weeks. One liter of dimethoate prepared solution was used for 4 plots, and the treatment was stopped 15 days before harvest to respect persistence time. Sprays were applied with a manually operated knapsack sprayer at 1.5 liter per treatment. A randomized complete block design (RCBD) with four treatments replicated four times was used. At harvest, variables recorded for analysis were the yield, the circumference, the weight of the heads. For damage analysis, the cabbage leaves were classified according to the level of the perforations caused by the larvae (Table 1).

Statistical analysis

Significant differences among data concerning cabbage head circumference, weight, yield, level of damage, and *P. xylostella* larval mortality and adult eclosion were determined with analysis of variance using Systat 5.0 software. Pairwise comparisons were done using the Fisher LSD at $p < 0.05$. All data are presented as means ± standard deviation.

Results

Toxicity of *C. schoenanthus* oil to *P. xylostella* larvae (contact test)

The formulations THE1 and THE2 of *C. schoenanthus* emulsified oil caused a faster mortality of the larvae than dimethoate suspension. The mean survival times in a group of 10 larvae were 22, 33 and 44100 seconds respectively for THE1, THE2 and dimethoate (Table 2). Larvae mortality time was significantly lower in the *C. schoenanthus* emulsified oil treatment compared to the control dimethoate (ANOVA: $F_{0.05(4)}=333.73$; $P<0.001$). Thus *C. schoenanthus* emulsified oil acts more quickly than synthetic insecticide used in the present study. Larvae treated with the soap preparation (SW) survived for 330 seconds.

Nutrition Test

The results in Table 3 showed that different concentration of pesticide tested affected differently the instars of *P. xylostella* larvae. Feeding damage was observed on the leaf discs after 24 hours indicating that the larvae fed on the leaves. Generally, THE2, SW and DT treatments caused similar death rates, which ranged from 5-12.5%. This value turned around 10%. These formulations did not differ significantly from each other. However, larval mortalities were 25% and 60%, respectively, at 24 h and over 48 h for THE1. A significant effectiveness against larvae of *P. xylostella* was observed when cabbage leaves were treated with THE1 dose (ANOVA: $F_{0.05(4)}=26.25$; $P<0.001$).

Scale	Perforations caused by the larvae (damage)
Degree 0	No damage
Degree 1	The first external leaves of the head: 1% of the leaves perforated
Degree 2	2 - 5% of the leaves perforated
Degree 3	6 - 10% of the leaves perforated, but the head is not damaged
Degree 4	11 - 30% of the leaves perforated with minor damage on the head
Degree 5	More than 30% of the leaves perforated and head damaged with presence of holes and waste resulting from the metabolism of the larvae.

Table 1: Classification of cabbage leaves damaged.

Insecticides	THE1	THE2	SW	DT	DW
Mean Survival	22.5	33.2	330.0	44100	86400
Time (sec)	±4.9*	± 6.3*	± 77.4	± 9467.8	± 0.0

Larvae were sprayed with DT, SW, DW and the essential oil of *C. schoenanthus* at 2 doses (THE1 and THE2). Data are Means ± SD. *P < 0.05 (THE1 and THE2 vs DT). DW: distilled water; THE1: essential oil 2g + hand soap containing soda 2 g + distilled water 96 g; THE2: essential oil 1 g + hand soap containing soda 2 g + distilled water 97 g SW: hand soap containing soda 2 g + distilled water 98 g; DT: 40 µl aqueous solution containing 0.25 mg of active ingredient.

Table 2: Mean survival time of the larvae of *P. xylostella* after treatment with emulsified oil of *C. schoenanthus* (100% mortality).

As in the contact or feeding tests, the formulation THE1 appeared more effective. Figure 1 reports percentage of *P. xylostella* adult eclosion at 96 hours. Concerning the absolute control by the DW and THE2 treatments, 95% and 65% adults, respectively, have evolved. THE1 (7.5%) significantly (ANOVA: $F_{0.05(4)}=9.5$; $P<0.005$) prevented the adult evolution than DW, SW and THE2. The adult eclosion rate recorded with dimethoate was 37.5%.

Harvest Data

Overall, insect populations were low. Generally, it was observed that the cabbage plants thrived well, and all the harvested heads of cabbage from each treatment plot were marketable. First attacks began at pre-heading stage, and an ave rage of 3-5 larvae were recorded per 10 plants. By comparing the various treatments at the harvest period, a strong presence of larvae (more than 4 larvae per 10 plants) was observed on certain heads harvested from the dimethoate-treated plot. However, the heads treated with dimethoate were slightly larger than those treated with the botanical aqueous extracts: 61.70 ± 3.71 cm versus 59.85 ± 5.86 cm; 59.53 ± 1.96 cm and 60.32 ± 2.86 cm for TS50, TS100 and TS150 respectively. The head weight also showed a slight difference between the dimethoate treatment and the botanical extracts: 1.69 ± 0.32 kg versus 1.41 ± 0.36; 1.57 ± 0.31 kg and 1.56 ± 0.19 kg for TS50, TS100 and TS150, respectively (Table 4). There were no statistically significant differences between the dimethoate and *C. schoenanthus* aqueous extract treatments (ANOVA: $F_{0.05(3)}=0.38$; $P>0.7$). The yield was slightly higher and the level of damage concerning the heads harvested from control plot (dimethoate treatment) was relatively of low quality compared to the treatments the aqueous extracts the botanical aqueous extracts (4.22 ± 0.34 kg/m^2 and 2.30 ± 0.14 versus 3.45 ± 0.40 to 3.93 ± 0.28 kg/m^2 and 1.80 ± 0.72 to 2.22 ± 0.73 respectively). The ANOVA test for yield and damage parameters showed also no significant difference between the insecticide and the botanical treatment (ANOVA: $F_{0.05(3)}=0.84$; $P>0.4$).

Larvae mortality induced by insecticides				
Insecticides	24 h	48 h	72 h	96 h
THE1	25.0±6*	60.0±5**	60.0±5**	60.0±5**
THE2	10.0±6	12.5±2	12.5±2	12.5±2
SW	5.0±2	10.0±4	10.0±4	10.0±4
DT	10.0±4	10.0±4	10.0±4	10.0±4
DW	0.0±0	7.5±4	7.5±4	10.0±4

Larvae were nourished with fresh cabbage leaves treated with formulations DT, SW, DW and the essential oil of *C. schoenanthus* at 2 doses (THE1 and THE2). Data in percent are Means ± SD. *P < 0.05 (THE1 vs DT); ***P* < 0.05 (THE1 vs DT). DW: distilled water; THE1: essential oil 2g + hand soap containing soda 2 g + distilled water 96 g; THE2: essential oil 1 g + hand soap containing soda 2 g + distilled water 97 g SW: hand soap containing soda 2 g + distilled water 98 g; DT: 40 µl aqueous solution containing 0.25 mg of active ingredient.

Table 3: Mean percentage of larvae mortalities induced by the insecticides and larvae developing to pupal stage.

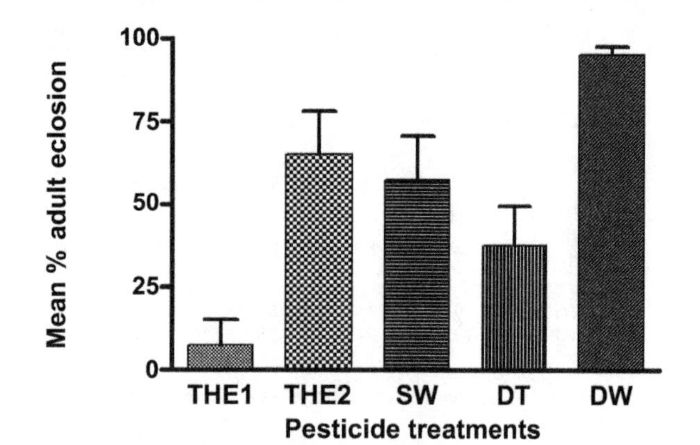

Larvae were nourished with fresh cabbage leaves treated with formulations DT, SW, DW and the essential oil of C. schoenanthus at 2 doses (THE1 and THE2). Data in percent are Means ± SD. *P < 0.05 (SW vs DW). **P < 0.05 (THE1 and DT vs DW).
DW: distilled water; THE1: essential oil 2g + hand soap containing soda 2 g + distilled water 96 g; THE2: essential oil 1 g + hand soap containing soda 2 g + distilled water 97 g;
SW: hand soap containing soda 2 g + distilled water 98 g; DT: 40 µl aqueous solution containing 0.25 mg of active ingredient.

Figure1: Evolution of the larvae of P. xylostella towards the adult forms at 96 hours.

Insecticide	Head circum- ference (cm)	Head weight (Kg)	Yield (Kg/m²)	Level of damage
DT	61.70±3.71	1.69±0.32	4.22±0.34	2.30±0.14
TS50	59.85±5.86	1.41±0.36	3.45±0.40	1.80±0.72
TS100	60.32±2.86	1.57±0.21	3.93±0.28	2.22±0.73
TS150	59.53±1.96	1.56±0.19	3.75±0.23	2.15±0.29

Cabbage plants were sprayed weekly with DT and *C. schoenanthus* aqueous extracts at 3 doses (TS50, TS100 and TS150). Measurement was made on ten samples in each case. Data are Means ± SD and non significant (*P*>0.05). DT: Dimethoate treatment (400 g of active ingredient per hectare); TS50: *C. schoenanthus* aqueous extracts treatment (50g/l); TS100: *C. schoenanthus* aqueous extracts treatment (100g/l); TS150: *C. schoenanthus* aqueous extracts treatment (150g/l).

Table 4: Effects of aqueous extracts of *C. schoenanthus* and dimethoate on cabbage Yield and quality.

Discussion

The objective of this study was to investigate the activities of aqueous leaves extracts and low concentrations of essential oil of *C. schoenanthus* to control populations of *P. xylostella* and their larvae. The aqueous leaf extracts of this plant were quite effective against *P. xylostella*, and achieved the similar protection of cabbage to dimethoate. The results of our investigation confirm the reports of Idrissou [36] which revealed that both *Azadirachta indica* and *C. schoenanthus* extracts had equivalent efficiency against DBM of cruciferous plants. The observations showed that the essential oils of *C. schoenanthus* can cause effective mortality of *P. xylostella* larvae. The essential oil extracted from *C. schoenanthus* contained a high percentage of monoterpenes. Studies carried out in Togo showed that the major component of *C. schoenanthus* extract was piperitone with a value bordering 70% [37,33], and was responsible for the insecticidal activity [35]. This component, isolated and purified by Author's name [38], had strong insecticidal activity against eggs, neonate larvae and adults of *Callosobruchus maculatus* at very low concentrations [31]. These observations corroborates the low concentrations of the formulation THE1 (2 g/l) effective effect on larvae of *P. xylostella* in our study. This formulation would ensure a good protection of cabbage against DBM. It can validly used as alternative in *P. xylostella* management.

References

1. de Sousa EMBD, Câmara APC, Costa WA, Costa ACJ, Oliveira HNM et al. (2005) Evaluation of the Extraction Process of the Essential Oil from *Cymbopogon schoenanthus* with Pressurized Carbon Dioxide. Braz. Arch. Biol. Technol. 48: 231-236.

2. Singh J, Upadhyay AK, Bahadur A, Singh B, Singh KP et al. (2006) Antioxidant phytochemicals in cabbage (*Brassica oleracea* L. var. *capitata*). Scientia Horticulturae 108: 233-237.

3. McDougall GJ, Fyffe S, Dobson P, Stewart D (2007) Anthocyanins from red

cabbage-stability to simulated gastrointestinal digestion. Phytochemistry 68: 1285-1294.

4. Samec D, Piljac-Zegaraca J, Bogovic M, Habjanic K, Gruz J (2011) Antioxidant potency of white (*Brassica oleracea* L. var. *capitata*) and Chinese (*Brassica rapa* L. var. *pekinensis* (Lour.)) cabbage: The influence of development stage, cultivar choice and seed selection. Scientia Horticulturae 128: 78-83.

5. Fowke JH, Longcope C, Hebert JR (2000) Brassica Vegetable Consumption Shifts Estrogen Metabolism in Healthy Postmenopausal Women. Cancer Epidemiology, Biomarkers & Prevention 9: 773-779.

6. Tanongkankit Y, Chiewchan N, Devahastin S (2011) Evolution of anticarcinogenic substance in dietary fibre powder from cabbage outer leaves during drying. Food Chemistry 127: 67-73.

7. Vrchovska V, Sousa C, Valentao P, Ferreres F, Pereira JA et al. (2006) Antioxidative properties of tronchuda cabbage (*Brassica oleracea* L. var. costata DC) external leaves against DPPH, superoxide radical, hydroxyl radical and hypochlorous acid. Food Chemistry 98: 416-425.

8. Heo HJ and Lee CY (2006) Phenolic phytochemicals in cabbage inhibit amyloid β protein induced neurotoxicity. LWT - Food Science and Technology 39: 331-337.

9. Volden J, Borge GIA, Bengtsson GB, Hansen M, Thygesen IE et al. (2008) Effect of thermal treatment on glucosinolates and antioxidant-related parameters in red cabbage (*Brassica oleracea* L. ssp. *capitata* f. *rubra*). Food Chemistry 109: 595-605.

10. Nilnakara S, Chiewchan N, Devahastin S (2009) Production of antioxidant dietary fibre powder from cabbage outer leaves. Food and Bioproducts Processing 87: 301-307.

11. Hounsome N, Hounsome B, Tomosc D, Edwards-Jones G (2009) Changes in antioxidant compounds in white cabbage during winter storage. Postharvest Biology and Technology 52: 173-179.

12. Ester A, de Putter H, van Bilsen JGPM (2003) Filmcoating the seed of cabbage (*Brassica oleracea* L. convar.*capitata* L.) and cauliflower (*Brassica oleracea* L.var. *botrytis* L.) with imidacloprid and spinosad to control insect pests. Crop Protection 22: 761-768.

13. Mazlan N, Mumford J (2005) Insecticide use in cabbage pest management in the Cameron Highlands, Malaysia. Crop Protection 24: 31-39

14. Ayalew G (2006) Comparison of yield loss on cabbage from Diamondback moth, *Plutella xylostella* L. (Lepidoptera: Plutellidae) using two insecticides. Crop Protection 25: 915-919.

15. Sayyed AH, Saeed S, Noor-Ul-Ane M, Crickmore N (2008) Genetic, Biochemical, and Physiological Characterization of Spinosad Resistance in *Plutella xylostella* (Lepidoptera: Plutellidae). J Econ Entomol 101: 1658-1666.

16. Macharia I, Löhr B, De Groote H (2005) Assessing the potential impact of biological control of *Plutella xylostella* (diamondback moth) in cabbage production in Kenya. Crop Protection 24: 981-989

17. Diaz-Gomez O, Rodriguez JC, Shelton AM, Lagunes-T A, Bujanos MR (2000) Susceptibility of *Plutella xylostella* (L.) (Lepidoptera: Plutellidae) Populations in Mexico to Commercial Formulations of *Bacillus Thuringiensis*. J Econ Entomol 93: 963-970.

18. Xu J, Shelton AM, Cheng X (2001) Comparison of Diadegma insulare (Hymenoptera: Ichneumonidae) and *Microplitis plutellae* (Hymenoptera: Braconidae) as Biological Control Agents of *Plutella xylostella* (Lepidoptera: Plutellidae): Field Parasitism, Insecticide Susceptibility, and Host-Searching. Journal of Economic Entomology 94:14-20.

19. Qian L, Cao G, Song J, Yin Q, Han Z (2008) Biochemical mechanisms conferring cross-resistance between tebufenozide and abamectin in *Plutella xylostella*. Pesticide Biochemistry and Physiology 91: 175-179.

20. Ferré J, Real MD, Rie JV, Jansens S, Peferoen M (1991) Resistance to the *Bacillus thuringiensis* bioinsecticide in a field population of *Plutella xylostella* is due to a change in a midgut membrane receptor. Proc Natl Acad Sci USA 88: 5119-5123.

21. Masson L, Mazza A, Brousseau R, Tabashinik B. (1995) Kinetics of *Bacillus thuringiensis* Toxin Binding with Brush Border Membrane Vesicles from Susceptible and Resistant Larvae of *Plutella xylostella*. J Biol Chem 270: 11887-11896.

22. Soufbaf M, Fathipour Y, Karimzadeh J, Zalucki MP (2010) Bottom-Up Effect of Different Host Plants on *Plutella xylostella* (Lepidoptera: Plutellidae): A Life-Table Study on Canola. Journal of Economic Entomology 103: 2019-2027.

23. Schroer S, Ehlers RU (2005) Foliar application of the entomopathogenic nematode Steinernema carpocapsae for biological control of diamondback moth larvae (*Plutella xylostella*). Biological Control 33: 81-86.

24. Ng PJ, Fleet GH, Heard GM (2005) Pesticides as a source of microbial contamination of salad vegetables. International Journal of Food Microbiology 101: 237-250.

25. Ciglasch H, Busche J, Amelung W, Totrakool S, Kaupenjohann M (2008) Field Aging of Insecticides after Repeated Application to a Northern Thailand Ultisol. J Agric Food Chem 56: 9555-9562.

26. van der Werf HMG (1996) Assessing the impact of pesticides on the environment. Agriculture Ecosystems and Environment 60: 81-96.

27. Barnard C, Daberkow S, Padgitt M, Smith ME, Uri ND (1997) Alternative measures of pesticide use. The Science of the Total Environment 203: 229-244

28. Isin S, Yildirim I (2007) Fruit-growers' perceptions on the harmful effects of pesticides and their reflection on practices: The case of Kemalpasa, Turkey. Crop Protection 26: 917-922.

29. Waichman AV, Eve E, da Silva Nina NC (2007) Do farmers understand the information displayed on pesticide product labels? A key question to reduce pesticides exposure and risk of poisoning in the Brazilian Amazon. Crop Protection 26: 576-583.

30. Godonou I, James B, Atcha-Ahowé C, Vodouhè S, Kooyman C et al. (2009) Potential of Beauveria bassiana and Metarhizium anisopliae isolates from Benin to control *Plutella xylostella* L. (Lepidoptera: Plutellidae). Crop Protection 28: 220-224.

31. Ketoh GK, Koumaglo HK, Glitho IA (2005) Inhibition of *Callosobruchus maculatus* (F.) (Coleoptera: Bruchidae) development with essential oil extracted from *Cymbopogon schoenanthus* L. Spreng. (Poaceae), and the wasp *Dinarmus basalis* (Rondani) (Hymenoptera: Pteromalidae). Journal of Stored Products Research 41: 363-371.

32. Ketoh GK, Koumaglo HK, Glitho IA, Huignard J (2006) Comparative effects of *Cymbopogon schoenanthus* essential oil and piperitone on *Callosobruchus maculatus* development. Fitoterapia 77: 506-510.

33. Koba K, Sanda K, Raynaud C, Nenonene YA, Millet J, et al. (2004) Activités antimicrobiennes d'huiles essentielles de trois *Cymbopogon* sp. africains vis-à-vis degermes pathogènes d'animaux de compagnie. Ann Méd Vét 148: 202-206.

34. Khadri A, Serralheiro MLM, Nogueira JMF, Neffati M, Smiti S, et al. (2008) Antioxidant and antiacetylcholinesterase activities of essential oils from *Cymbopogon schoenanthus* L. Spreng. Determination of chemical composition by GC–mass spectrometry and ^{13}C NMR. Food Chemistry 109: 630-637.

35. Kullenberg J (1951) Traité de zoologie des insectes. Tome 10 Fascicule 2: 31- 40.

36. Idrissou N (2003) Etude de l'efficacité de l'extrait aqueux de *Cymbopogon schoenanthus* Treng. Contre la teigne des crucifères *Plutella xylostella* L. (Lepideptora: Hyponomeutidae). Mémoire d'Ingénieur Agronome, Ecole Supérieure d'Agronomie, Université de Lomé. 63p.

37. Ketoh GK, Glitho IA, Huignard J (2002) Susceptibility of the bruchid *Callosobruchus maculatus* (Coleoptera: Bruchidae) and its parasitoid *Dinarmus basilis* (Hemenoptera: Pteromalidae) to three essential oils. Journal of Economic Entomology 95: 174-182.

38. Ketoh GK, Glitho IA, Koumaglo KH and Garneau FX (2000) Evaluation of essential oils from six aromatic plants in Togo for *Callosobruchus maculatus* F. pest Control Insect Science and its Application 20: 45-49.

Biopesticidal Formulation of *Beauveria Bassiana* Effective against Larvae of *Helicoverpa Armigera*

Agarwal Ritu[1]*, Choudhary Anjali[1], Tripathi Nidhi[1], Patil Sheetal[2] and Bharti Deepak[1]

[1]*Assistant Professor, Department of Biotechnology and Biochemistry, Career College, Bhopal 462 023, M.P, India*
[2]*Trainee student, Department of Biotechnology and Biochemistry, Career College, Bhopal, 462 023, M.P, India*

Abstract

The present study was to emphasize entomopathogens in pest management of cash crops over chemical pesticides, optimization of media for growth of *Beauveria bassiana* and bioassay of different formulations for their efficacy as marketable and easily applicable biopesticide. *Beauveria bassiana* (Order: Hypocreales, Family: Cordycipitaceae) popularly known as white muscardine entomogenous fungi was isolated from soyabean fields of Misrod, Bhopal. Media for optimal growth of fungus were standardized. Its different formulations viz. carrier based powder formulation, oil based formulation and bentonite oil based formulation using homogenizers were prepared. These formulations were bioassayed against *Helicoverpa armigera* (Order: Lepidoptera, Family: Noctuidae), the most destructive pest in soyabean. Bentonite based liquid formulation was observed to be most effective as determined by measuring larval mortality as well as viability of fungal spores and ease of applicability.

Keywords: Biopesticide; *Beauveria bassiana; Helicoverpa armigera*; Formulations; Bentonite

Abbreviations

BOBLF: Bentonite Oil-Based liquid Formulation; OBLF: Oil-Based Liquid Formulation; CBPF: Carrier Based Powder Formulation

Introduction

Over reliance on broad spectrum pesticides has been severely condemned in different parts of the world after International Conference on Chemicals Management. Since then, an alternative eco friendly strategy for the management of noxious insect pests has been explored to trim down the harmful effects of chemical insecticides on humanity. Studies of biodiversity in agro ecosystems and the delivery of ecosystem services to agricultural products have usually ignored the contribution of entomopathogens in the regulation of pest populations (Tscharntke et al.).

In recent years, crop protection based on biological control of crop pests with microbial pathogens like virus, bacteria, fungi and nematodes has been recognized as a valuable tool in pest management [1,2]. The appropriate use of eco friendly microbial biopesticide can be engaged in recreation of sustainable organic crop production by providing a stable pest management program. In light of this understanding, extensive work has been done on various species of bacteria and fungi (especially entomopathogenic that parasitize insects) implicated as effective biocontrol agents.

Till date, various entomopathogenic fungi such as *Lecanicillium* sp. [3,4], *Beauveria bassiana* [5-7] and *Metarhizium anisopliae* [8-11] have been effectively used to control aphids, lepidopteron larvae and other pests.

Sixteen different mycotoxins have been analysed of which the cyclodepsipeptidic mycotoxin, beauvericin, produced by *Beauveria bassiana* has been documented to be most effective for its larvicidal properties. Sowjanya Sree and Padmaja [10] reported the ultrastructural effects of crude beauvericin on the salivary glands of 9-day-old *S. litura* larva after 24 h of treatment with the mycotoxin at a dosage of 0.147 μg/g body wt. (LD_{50}). *Helicoverpa armigera* (Hubner) is commonly known as the gram pod borer as it is a serious pest on pulses (Nahar et al., David and Ananthakrishnan). It has been reported that *M.anisopliae* is effective against *H. armigera* [12].

The present investigation was carried out to highlight the significance of entomopathogens in pest management of cash crops, optimization of media for growth of *Beauveria bassiana* and bioassay of different formulations for their efficacy as marketable and easily applicable biopesticide. The investigation established the role of *Beauveria bassiana* bentonite formulation as an effective biocontrol agent against the most destructive pest *Helicoverpa armigera.*

Material and Methods

Chemicals

Sabouraud-Dextrose media, Czapek Dox media, Yeast Extract and Bentonite were purchased from HiMedia Biosciences Mumbai, India. Corn oil, gum and glycerine were purchased from local market. Maize flour was prepared by grinding maize grains in laboratory. Ponds talcum powder was used as one of the carriers.

Culture

Beauveria bassiana employed in this study was isolated from soil collected from soyabean field of Misrod, Bhopal (Rice technology bulletin, 2003 [13,14]. *Beauveria bassiana* was grown on Czapek Dox media modified with maize flour as per findings of Ramle et al. [15] and on Sabouraud-dextrose media with yeast extract [16] for growth optimization of the entomopathogenic fungus. Once the fungus was identified [17], pure cultures were obtained followed by flask cultures to develop the fungal mat.

***Corresponding author:** Ritu Agarwal, Department of Biotechnology, Career College, Bhopal 462 023, M.P, India, E-mail: ritu.agarwal@rediffmail.com

Formulations

For preparing the formulations viz. carrier based powder formulation (CBPF) using talcum powder, glycerine and gum, oil-based liquid formulation (OBLF) using corn oil, gum and glycerin, bentonite oil-based liquid formulation (BOBLF) using corn oil, gum, glycerin and bentonite [6] using a knife mill grindomix (Remi, India). Fungal spore suspensions in different types of formulations were prepared from the fungal mats.

Viability of spores was assessed before and after preparation of formulations by serial dilution test in modified Czapek-Dox broth and after six months, viability of spores was again carried out in the prepared formulations as per Maheshwari [14]. Spore count of each formulation was done using Neubaeur chamber [18]. A range of concentration from 10% to 100% of each formulation was also prepared using distilled water and assayed at laboratory scale in triplicate and the optimal concentration for effective infection was screened for assay at the levels of pot culture and field assay. Reproducibility of each assay was established in triplicate.

Laboratory Bioassay

The fungal formulations of *Beauveria bassiana* were assayed against larvae of *Helicoverpa armigera* in the laboratory. Eggs of *Helicoverpa armigera* collected from the fields and were kept in optimum condition maintained at 23 ± 2°C and 35 ± 5% relative humidity under a 16:8 (L:D) cycle in a growth chamber. The larvae obtained from these eggs were fed with soya beans and sugar beet roots and leaves. The three days old larvae were divided into four groups of control and treated with different formulations and subsequently examined for symptoms of infection and mortality at regular intervals of 24 hours (Table 1).

Pot culture assay

In pot culture experiments, *Beauveria bassiana* formulations were evaluated against first and third instar larvae at 60% concentration of formulations with dosage equivalent to 2.1×10^4 spores ml^{-1}, calibrated on the surface area of the pot. As a control, contents of formulation were sprayed excluding *Beauveria bassiana*. The larvae were released in pots pre planted with sugar beet plants and allowed to settle for 24 hrs. In the test against larvae, different formulations carrying the fungal spores were applied to pots in replicates of four with seven larvae each. The pots were watered immediately. Larvae were examined for infection due to *Beauveria bassiana* at weekly intervals [19]. The experiments were repeated three times.

Field evaluation

Beauveria bassiana was tested at 2.1×10^4 spores ml^{-1} in a randomized block design (RBD) and plot size of 24 sq. m. in triplicate in soyabean fields of Misrod, Bhopal having four month old stalks. Spray method was used in the toxicity assays and as a control, contents of formulation was sprayed excluding *Metarhizium anisopliae*. First instar larvae were collected from the experimental plots about a month later by completely uprooting the plants or digging deep on either side of the rows. These larvae were brought to the laboratory in individual boxes filled with moist soil and reared on sugar beet and soya bean roots and leaves, which were changed every week. The proportion of larvae that showed symptoms of fungal attack after treatments with optimal concentration of all three formulations, were recorded as per protocol of Samuel et al. [20].

Statistical analysis

Each assay was established in triplicate for each parameter. The statistical analysis was performed using mean as a base for central tendency followed by calculation of deviation using standard error. Statistical significance was drawn by comparing the p value from students "*t*" test table. Significantly different from the control if $p<0.05$, significant if $p<0.01$, highly significantly if $p<0.001$.

Results and Discussion

The fungus, *Beauveria bassiana* showed fast growth on a medium consisting of yeast extract i.e Sabouraud-dextrose-yeast extract media which exhibited a shorter fungal mat development period of 15 days as compared to 20 days on Czapek Dox media.

Among the three different formulations, it was found that the bentonite based liquid formulation exhibited the highest efficacy of infection against Helicoverpa armigera. A range of concentration from 10% to 100% of each formulation was assayed in the laboratory bioassay and it was found that a 60% concentration of each formulation showed optimum results (Table 1 and Figure 1).

Thereafter, for pot culture assay and field trial, the concentration of formulations was assayed at 60% as in laboratory assay the result obtained from 60% to 100% were nearby. Preparation of varied formulations did not hamper the viability of spores as was assessed before and after, by serial dilution test in modified Czapek-Dox broth and provided an easy applicability in the liquid bentonite formulation. After six months, the viability of spores was again assessed in the formulations and it was found that the spores were still viable, although better result was seen in bentonite formulation as compared to the other two.

Beauveria bassiana is cited to be highly active against more than hundreds of insect pests and highly selective in its parasitisation [21]. The fungus, *Beauveria bassiana* was cultured with excellent results on a medium consisting of yeast extract i.e Sabouraud-dextrose-yeast extract [15]. Among the three different formulations, it was found that the bentonite based liquid formulation had the highest efficacy.

S. No	Types of formulation used	Required days for infection on larvae
1.	BOBLF	05
2.	OBF	12
3.	CBPF	09

(BOBLF): bentonite oil-based liquid formulation
(OBLF): oil-based liquid formulation
(CBPF): carrier based powder formulation

Table 1: Infection rate of different formulations of *Beauveria bassiana* without dilution on larvae of *Helicoverpa armigera*.

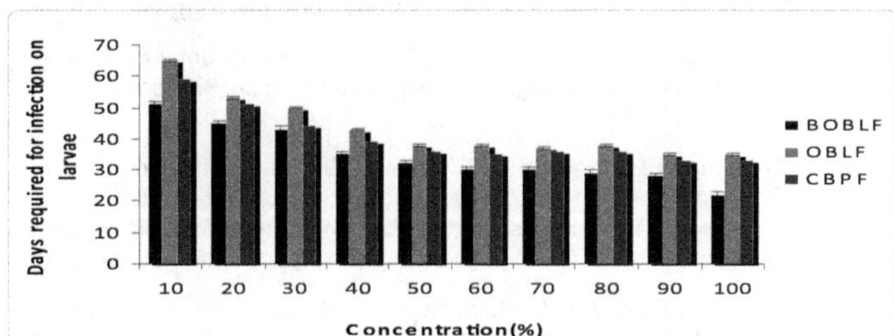

Figure 1: Showing number of days required by different concentration of formulations of *Beauveria bassiana* to infect larvae of *Helicoverpa armigera* in laboratory assay.

Anand et al. [1] has reported that the fungal pathogen *Beauveria bassiana* infects pupae of *Spodoptera litura* in a dose–dependent manner for each of the formulations investigated which corroborates the findings of the present investigation. They also reported that *Beauveria bassiana* was more infective and resulted in maximum average percent mortality among the three species viz. *Metarhizium anisopliae*, *Beauveria bassiana* and *Lecanicillium* sp. under investigation.

In the lab assay, 52% of larvae treated with 2.7×10^9 spores per ml of *Beauveria bassiana* developed infection in 7 days as against no infection in control, a finding also published by Chen et al. [22] during their work on *Metarhizium anisopliae*. Laboratory bioassay by Chandler and Davidson [8], documented higher mortality in *M. anisopliae* (ARSEF 7487) than *L. muscarium* (ARSEF 7037) in all soil based studies.

In pot culture assay, *Beauveria bassiana* caused low level of infection in larvae at the dosage equivalent to 2.1×10^4 spores ml^{-1} hectare since the fungus required 30-35 days incubation period to produce disease symptoms. In test, with a higher dosage range (1×10^6 - 10^8 spores per ml) observations made at monthly intervals, showed similar dosage dependent infection rates.

In the field trials with *Beauveria bassiana*, infection rates in larvae collected from the experimental plot were remarkably higher at the higher dosage of 2.1×10^4 spores ml^{-1} than at the lower dosage. The larvae collected 25 days after treatment developed infection in a maximum period of 32 days after collection from the field. This was in accordance with the field bioassay of Anand et al. [1].

Conclusion

Formulations of *Beauveria bassiana* can thus serve as an effective broad spectrum biocontrol agent for soyabean and various other cash crops. As also evident from the study, Bentonite is an effective carrier of the *Beauveria bassiana* in terms of being economical, maintaining the biological activity and increasing the ease of application. This carrier can also serve to be good medium for other fungal biopesticides for instance *Beauveria bassiana*, further studies on which are in progress.

Acknowledgement

The authors are grateful to their academic councils for support in the form of infrastructural facilities made available for undertaking the present study.

References

1. Anand R, Prasad B, Tiwary B N (2009) Relative susceptibility of Helicoverpa armigera pupae to selected entomopathogenic fungi. Biological Control 54: 85-92.
2. Rao C, Devi KU, Khan PAA (2006) Effect of combination treatment with entomopathogenic fungi *Beauveria bassiana* and Nomuraea rileyi (Hypocreales) on Helicoverpa armigera (Lepidoptera: Noctuidae). Biocontrol Science and Technology 16:221-232.
3. Jung HS, Lee HB, Kim K, Lee EY (2006) Selection of Lecancillium strains for aphid (Myzus persicae) control. The Korean Journal of Mycology 34: 112-118.
4. Ownley BH, Gwinn KD, Vega FE (2010) Endophytic fungal entomopathogens with activity against plant pathogens: ecology and evolution. BioControl 55:113-128.
5. Quesada-Moraga E, Maranhao EAA, Valverde-Garcia P, Santiago-Alvarez C (2006) Selection of *Beauveria bassiana* isolates for control of the whiteflies Bemisia tabaci and Trialeurodes vaporariorum on the basis of their virulence, thermal requirements and toxicogenic activity. Biological Control 36: 247-287.
6. Sivasundaram V, Rajendran L, Muthumeena K, Suresh S, Raguchander T, et al. (2007) Effect of talc-formulated entomopathogenic fungus Beauveria against leaffolder (Cnaphalocrosis medinalis) in rice. World J Microbiol Biotechnol 24:1123-1132.
7. Chandler D, Davidson G (2005) Evaluation of entomopathogenic fungus *Metarhizium anisopliae* against soil-dwelling stages of cabbage maggot (Diptera: Anthomyiidae) in glasshouse and field experiments and effect of fungicides on fungal activity. J Econ Entomol 98: 1856-1862.
8. Dong C, Zhang J, Chen W, Huang H, Hu Y (2007) Characterization of a newly discovered China variety of *Metarhizium anisopliae* (M. anisopliae var. dcjhyium) for virulence to termites, isoenzyme, and phylogenic analysis. Microbiological Research 162: 53-61.
9. Sowjanya SK, Padmaja V, Murthy YL (2008) Insecticidal activity of destruxin, a mycotoxin from *Beauveria bassiana*(Hypocreales), against Helicoverpa armigera (Lepidoptera: Noctuidae) larval stages. Management Science. 64: 119-125.
10. Tullu B, Anthonieke M, Constantianus JMK, Willem T, Bart GJK (2010) Factors affecting fungus-induced larval mortality in Anopheles gambiae and Anopheles stephensi. Malar J 9: 22.
11. Sahayaraj K, Francis Borgio J (2010) Virulence of entomopathogenic fungus *Beauveria bassiana* (Metsch.) Sorokin on seven insect pests. Indian Journal of Agricultural Research 44: 195-200.
12. Management of armyworms and cutworms (2003) Rice technology bulletin, Philippines.
13. Maheshwari DK, Dubey RC (2004) Practical Microbiology. S Chand group, India.
14. Ramle M, Hisham H, Wahid MB, Kamarudin N, Ahmad ASR (2005). Mass production of *Beauveria bassiana*using solid fermentation and wet harvesting methods. Proc. Of the PIPCO 2005 International palm oil congress-agriculture conference. 928-943.
15. Luz C, Tai MHH, Santos AH, Rocha LFN, Albernaz DAS, et al. (2007) Ovicidal activity of entomopathogenic Hyphomycetes on Aedes aegypti (Diptera: Culicidae) under laboratory conditions. J Med Entomol 44:799-804.
16. Aneja KR (2005) Identification of an unknown microorganism. In: Experiments in Microbiology, Plant Pathology and Biotechnology. New Age international publishers, New Delhi, India.
17. Nugrohol, Ibrahim YB (2004) Laboratory bioassay of some entomopathogenic fungi against broad mite (Polyphagotarsonemus latus Bank). Int J Agric Biol 6: 223-225.
18. Suganya T, Selvanarayanan V (2009) Varietal variation in tomato vis-a-vis pathogenicity of *Beauveria bassiana* (Bals.) Vuill. to leaf eating caterpillar, Spodoptera litura Fab. Karnataka Journal of Agricultural Science 22: 572-574.
19. Samuel P, Raquel A, Marcela-Ines S, Ana-Mabel M (2007) Pathogenicity of two entomopathogenic fungi on Trialeurodes vaporariorum and field evaluation of a Paecilomyces fumosoroseus isolate. Southwest Entomology 32: 43-52.
20. Zimmermann G (2006) The entomopathogenic fungus *Beauveria bassiana*and its potential as a biocontrol agent. Pest Management Science 37: 375-379.
21. Chen Bin, Yue Z, FuRong G (2006) Virulence of *Beauveria bassiana*to Spodoptera mauritia and Mythimna separata and its control efficacy in fields. Environmental Science 9: 685-692.

7

Evaluation of Certain Botanical Preparations against African Bollworm, *Helicoverpa armigera* Hubner (Lepidoptera: noctuidae) and Non Target Organisms in Chickpea, *Cicer arietinum* L

Nigussie Lulie and Nagappan Raja*

Department of Biology, Faculty of Natural and Computational Sciences, Post Box-196, University of Gondar, Gondar, Ethiopia

Abstract

Aqueous extract of individual and mixed form of *Azadirachta indica* A. Juss seeds kernel and leaves of *Milletia ferruginea*, Hochst and *Croton macrostachyus* Hochst was tested against African bollworm, *Helicoverpa armigera* Hubner. Antifeedant activity of selected plant extract was tested at 1%, 2.5%, 5% and 10% concentration against 4th instar larvae of *H. armigera* in the laboratory and 5% concentration was tested under field condition. All the tested plant extract showed 100% protection at 5% and 10% concentration. Among the various botanical treatment Neem Seed Kernel Extract (NSKE) and NSKE+BLE (Birbira Leaves Extract) was effective at 2.5% concentration with minimum pod damage. In the field observation among the botanicals, reduction of larval population in the treatment of NSKE, BLE and NSKE+BLE was statistically not significant ($p>0.05$; LSD). The lowest percentage pod damage (0.45%) was observed in Diazinon 60% EC treated plot followed by NSKE treated plot (3.90%) after second spray.

The highest mean yield was obtained from NSKE treated plot (781 g) followed by Diazinon 60% EC treated plot (719.33 g), NSKE+BLE (656.67 g) and BLE treated plot (653.33 g). Five days after second treatment there was a significant difference in the mean number of ants between NSKE+BLE, control and other treatments. The highest mean number of spiders was observed in control plot (3.6) and lowest (0.3) was in plot treated with Diazinon 60% EC. The reduction of lady bird beetle population among the botanicals treated plots was statistically not significant ($p>0.05$; LSD). The mean number of wasp population was the highest in control plot (3.3) and there was no wasp observed in Diazinon 60% EC treated plot. In conclusion, even though Diazinon 60% EC was found to be effective by considering the interaction of beneficial in the field botanical preparations are much better particularly NSKE and also suitable to spray under rain fed condition to protect the crop by small farming communities.

Keywords: Botanicals; *Helicoverpa armigera; Azadirachta indica; Milletia ferruginea; Croton macrostachyus*; Non-target organisms; Aqueous extract

Introduction

Chickpea is an important source of protein, carbohydrate, fibres, oil, calcium, phosphorus, magnesium, iron, zinc, β-carotene, unsaturated fatty acids. In addition, improves soil fertility by fixing atmospheric nitrogen in to soil [1,2]. Ethiopia is a largest producer of chickpea in Africa accounting for about 46% of production during 1994-2006 [3] and seventh largest producer in worldwide [4]. One of the constraints to reduce yield loss in chickpea is African bollworm, *Helicoverpa armigera* which causes 80% pod damage in early sown chickpea [5]; 21 to 36% pod damage in central highlands [6] and 100% pod damage in some localities of Yilmana Densa and Achefer areas in Gojjam in the 1990s [7]. Globally botanical pest management is gaining appreciation because of multiple mode of action such as antifeedant which inhibit normal development of insects, repellent, antijuvenile hormone activity, oviposition/ hatching deterrence, antifertility or growth disrupters and chemosterilants [8]. According to Purohit and Vyas [9] about 2121 plant species are reported to use in pest management programs. In Ethiopia, even though with rich floral diversity, about 30 plant species are recorded and most of them are used traditionally for the management of storage pests [10].

Azadirachta indica (Neem) and their products are considered as effective botanical pesticides due to controlling wide variety of insect pest including *H. armigera* [11]. Tebkew et al. [12] reported that crude neem extracts prepared from neem seeds collected from Melka Woreda has significantly reduced the percentage pod damage; similarly pod damage on treated chickpea plot was lower than untreated plots [13]. The predatory lady bird beetle (*Mallada signatus*) pupation was delayed when they fed with neem product treated larvae of *H. armigera* thereby increasing the individual predatory activity [14,15]. In East Africa, predatory ants and anthocorids are most important natural enemies of H. armgiera on corn, sorghum and sunflower [16]. Among the predators, ants kill up to 90% of *H. armigera* pupae in the soil [17]. Some other natural enemies such as wasps feed on egg and larvae; ants feed on egg, larvae and pupae; preying mantids feed on egg; spiders feed on egg, larva and adults of *H. armigera* [18]. However, indiscriminate use of chemical pesticides and their continuous application create intolerable environment to natural enemies and also to prevent re-entry in treated areas [16,19]. The neem products and their utilization are increasing world wide but in Ethiopia their use was not well explored in farming communities.

***Corresponding author:** Nagappan Raja, Department of Biology, Faculty of Natural and Computational Sciences, Post Box-196, University of Gondar, Gondar, Ethiopia, E-mail: nagappanraja@yahoo.com

Milletia ferruginea (Birbira) belongs to the family Leguminaceae and subfamily Papilionoideae is a multipurpose tree widely distributed in Ethiopia [20,21]. The seeds and roots containing the toxic principle rotenone was widely used as insecticides and piscicides (to kill fishes) [22,23]. The crude extracts from the seed was toxic to stored grain pest *Sitophilus zeamais* [24]; aqueous and organic solvent extracts of seeds have killing effects on three species of aphids by contact [25] and aqueous seed extracts proved to show 100% mortality on termites under laboratory conditions [26].

Croton has economic importance due to its essential oil content and various biologically active substances such as terpenoid, flavonoids and alkaloids [27-31]. Several species of Croton are reported to have wide range of insect controlling properties. According to Kubo et al. [32] diterpenes from *C. cajucara* inhibit the growth of *Heliothis virescens*. The dichloromethane and ethyl acetate fraction of *C. urucurana* causes 65% mortality in the larvae of *Anagsta kuehniella* due to the action of phenolic compounds catechin and gallocatechin [33]. The adult mortality of *Dystercus maurus* was significantly higher in *C. urucurana* treated insects [34]. Even though, biopotential of Croton species are explored well in worldwide but there is no much work in Ethiopia. Therefore, present study was aimed to evaluate aqueous extract of *Azadirachta indica* seeds kernel, leaves of *Milletia ferruginea* and *Croton macrostachyus* against African bollworm *Helicoverpa armigera* and some non-target organism's interaction in chickpea field.

Materials and Methods

Plant materials collection and extraction

Mature and healthy *Azadirachta indica* seeds were collected from Metema area; located at 175 km west of Gondar town, Ethiopia. The pulp of the seed was removed initially and shade dried; later seed coat was removed from the dried seeds and the kernel was powdered by using mortal and pestle. Sixteen hours before spraying, botanical extract was prepared from the powder by following the procedure of Jain and Bhargava [35]. According to this procedure to prepare 5% concentration, 5 kg of neem seed kernel powder was dissolved in 100 liters of water and used to spray on one hectare of crop land. Since the experimental plot size in the present study for one treatment was 21.6 m2, for this reason 10.8 g of neem seed kernel powder was soaked in 21.6 ml of water and stirred periodically to mix the contents well. After 16 h, contents were filtered by shema cloth and the volume was made up to 216 ml by adding freshwater. Finally, 30 ml of detergent soap solution was added as an emulsifier and sprayed. Leaves of *Milletia ferruginea* and *Croton macrostachyus* were collected in and around Gondar town, shade dried and powdered separately by using mortar and pestle. Three days before spray schedule, powdered leaves were soaked in water and stirred well periodically to facilitate thorough mixing. According to the procedure of Dodia et al. [36], 5% leaf extract was prepared and 30 ml of detergent soap was added as an emulsifier and sprayed.

Antifeedant activity

Antifeedant activity of the plant extract was tested at 1%, 2.5%, 5% and 10% concentration against IVth instar larvae of *H. armigera* in the laboratory by following the procedure of Jain and Bhargava [35] and Dodia et al. [36]. Healthy immature chickpea pods were collected from field and sprayed with respective concentration of plant extract individually. In a clean glass container 5 sprayed pods were kept and single healthy 4th instar larva of *H. armigera* was added to avoid cannibalism. The chickpea pods treated with soap solution was considered as untreated control and field recommended dose (1 L/hectare) of Diazinon 60% EC was considered as positive control. The experiment was replicated three times and number of pod damaged by the larva of *H. armigera* was recorded up to 48 h and percentage of pod damage was calculated.

Field spraying of botanical preparations

Field study was conducted at Maksegnit, Gondar Zuria Woreda of North Gondar administrative zone in Amhara regional state, Ethiopia. The study area was located at an altitude ranging from 1912 -2848 masl, latitude 12°11′ 24″ N – 12°39′ 40″ N and longitude 37°24′ 48″ E -37°36′ 00″ E. The mean annual rain fall in the study area was 992.5mm and annual temperature was ranged from 13.5°C to 28.5°C. The study area was characterized by wet season from June to September and dry season from October to May. The botanical treatment was arranged in Randomized Complete Block Design (RCBD) with three replications [37]. The study plot (3 m×2.4 m) was divided into three blocks and each block further divided in to eight subplots. There was 1.5 m space between blocks and 1m space between subplots to facilitate easy movement for spraying and data collection and also to avoid mixing of botanical spray from one plot to another. Local desi variety of chickpea seeds were sown (<5 cm depth) in eight rows in each plot in early September 2011. The distance between the rows was 30cm and between the plants was 10 cm. Among the eight plots in each block six was treated with botanical preparations, one with chemical pesticide (Diazinon 60% EC) and one was untreated control. The details of the treatments given in the field was as follows; 5% Neem Seed Kernel Extract (NSKE); 5% Croton Leaves Extract (CLE); 5% Birbira Leaves Extract (BLE); 2.5% NSKE+2.5% BLE; 2.5% CLE+2.5% BLE; 2.5% NSKE+2.5% CLE; positive control Diazinon 60% EC and untreated Control. The field spraying was done in the evening time after sun shed to increase the residual action of botanicals since they degrade rapidly by UV light. A total of two round sprays were undertaken before harvesting.

Estimation of pods damage

The number of chickpea pods damaged by the larvae of bollworm was counted before and after each spray schedule on predetermined and tagged five chick pea plants/plot. After the completion of botanical treatment, total number of damaged and undamaged pods was counted and percentage of pod damage was calculated by using the following formula.

$$\%\ \text{pod damage} = \frac{\text{Total number of pods damaged}}{\text{Total number of pods}} \times 100$$

Estimation of non-target organisms

The number of non-target organism such as ants, wasps, lady bird beetles and spiders were counted and recorded on each plot. Data were taken for three times before treatment, after first and second treatment. Mean number of non-target organisms per plot was calculated.

Estimation of African bollworm larvae

The number of bollworm larvae was counted by visual observation on ten chickpea plant/plot and recorded. The data was collected before treatment, after first and second treatments. Mean number of *H. armigera* larvae per 10 plants was calculated.

Estimation of yield and yield loss

Chickpea crops were harvested from each plot separately, thrashed and weighed by using balancer. Mean yield weight was calculated for each treated plots and projected yield for hectare was calculated. The percentage yield loss was calculated by using the following formula suggested by Judenko [38] and Walker [39].

$$\% \text{ Yield loss} = \frac{\text{MCYTP} - \text{MCYUP}}{\text{MCYTP}} \times 100$$

Whereas MCYTP=Mean Chickpea Yield of Treated Plot

MCYUP=Mean Chickpea Yield of Untreated Plot

Statistical analysis of data

The data collected from antifeedant activity of plant extracts in the laboratory and field trial was subjected to ANOVA followed by Least Significant Difference (LSD) test to separate individual mean significant difference at 5% level (p<0.05) by using to SPSS version 16.

Results

Antifeedant activity of the plant extracts against IVth instar larvae of *H. armigera*

Antifeedant activity of tested plant extracts results demonstrates that at 1% concentration minimum pod damage of 13.33% was observed in NSKE and NSKE+BLE which was statistically not significant (p>0.05; LSD). However, compared to other treatments and untreated control the result was statistically significant. At 2.5% concentration, NSKE+BLE combination showed minimum pod damage (3.33%) compared to other treatments. Among the different botanicals treatment except CLE, remaining results were statistically not significant (p>0.05; LSD). The chickpea pods treated at 5%, 10% concentration of the plant extracts and positive control Diazinon 60% EC showed 100% protection against *H. armigera* larval infestation. All the control groups were observed with 100% pod damage (Table 1).

Effect of botanical extracts on *H. armigera* larvae in the field

Pre and post spray count for mean number of *H. armigera* larvae was recorded and presented in table 2. The distribution of larval population in the plot was not uniform before treatment. The plot assigned for different botanical treatment results showed significant difference within the plot (p<0.05; LSD). The minimum number of larvae (9.0) was observed in plot allotted for the treatment of NSKE+BLE. Five days after first treatment with botanical application larval population was reduced significantly at 5% level (p<0.05) compared to control. Among the various botanical treatment, reduction of larval population in the treatment of NSKE, BLE and NSKE+BLE was statistically not significant (p>0.05; LSD). Five days after the second application mean number of *H. armigera* larvae was further reduced in each treatment.

Treatments	Concentration tested			
	1%	2.5%	5%	10%
NSKE	13. 33 ± 5.77^{c}	10.0 ± 10.0^{c}	0.0 ± 0.00	0.0 ± 0.00
CLE	36.6 ± 5.77^{b}	30.0 ± 10.00^{b}	0.0 ± 0.00	0.0 ± 0.00
BLE	36.6 ± 5.77^{b}	13.3 ± 5.77^{c}	0.0 ± 0.00	0.0 ± 0.00
NSKE + BLE	13.3 ± 5.77^{c}	3.33 ± 5.77^{c}	0.0 ± 0.00	0.0 ± 0.00
CLE + BLE	36.6 ± 5.77^{b}	13.3 ± 5.77^{c}	0.0 ± 0.00	0.0 ± 0.00
NSKE + CLE	36.6 ± 5.77^{b}	13.3 ± 5.77^{c}	0.0 ± 0.00	0.0 ± 0.00
Diazinon 60% EC	0.0 ± 0.00^{d}	0.0 ± 0.00c^{d}	0.0 ± 0.00	0.0 ± 0.00
Control	100.0 ± 0.00a	100.0 ± 0.00a	100.0 ± 0.00	100.0 ± 0.00

Values are mean percentage pod damage ± standard deviation of three replications. Within the column similar alphabets are statistically not significant by LSD (p>0.05).

Table 1: Mean percentage of pod damaged by 4th instar larva of *H. armigera* exposed to different concentration of aqueous extracts.

Treatment	Before spraying	After first spray	After second spray	% reduction
NSKE	13.33 ± 2.51^{bc}	6.66 ±2.08^{c}	1.33 ± 1.15^{e}	90.02
CLE	13.0 ± 4.00^{a}	10.66 ± 3.51^{b}	6.66 ± 2.51^{b}	48.77
BLE	10.14 ± 1.52^{cd}	7.14 ± 1.00^{c}	3.28 ± 2.31^{d}	67.65
NSKE + BLE	9.0 ± 4.00^{d}	4.33 ± 1.52^{cd}	3.33 ± 1.52^{d}	63.00
CLE + BLE	12.33 ± 7.02^{ab}	8.0 ±4.58^{c}	4.33 ± 1.15^{cd}	64.88
NSKE + CLE	13.3 ± 6.02^{a}	7.0 ± 2.64^{c}	5.3 ± 2.08^{bc}	60.15
Diazinon 60% EC	10.6 ± 2.51^{c}	2.6 ± 2.51^{d}	0.0 ± 0.00^{f}	100
Control	10.1 ± 2.00^{cd}	17.3 ± 3.57^{a}	14.0 ± 2.64^{a}	-38.66

Values are mean ± standard deviation of three replications. Within the column similar alphabets are statistically not significant by LSD (p>0.05).

Table 2: Mean number of H. armigera larvae recorded in experimental plot before and after spraying.

The minimum number of larva (1.33) was observed in NSKE treated plot followed by BLE (3.28) and NSKE+BLE (3.33). In positive control Diazinon 60% EC treated plot, there was no *H. armigera* larva. The percentage reduction of larval population was 100% in Diazinon 60% EC treated plot followed by NSKE (90.02%) when compared to before spraying. However, in control plot the larval population was increased compared to before spraying.

Chick pea pod damaged by the larvae of *H. armigera*

Mean number and the percentage of pods damaged by *H. armigera* before treatment, after first and second treatments were presented in table 3. Five days after first treatment, mean number of pods damaged by the larvae of *H. armigera* was significantly (p<0.05) decreased in treatments compared to control. The lowest percentage pod damage (0.45%) was observed in Diazinon 60% EC treated plot followed by NSKE treated plot (3.90%) after second spray. There was no statistically significant pod damage between NSKE and BLE treatment. The highest percentage of pod damage was observed in untreated plot (22.22%) followed by CLE+BLE (14.52%), CLE (14.33%), NSKE+CLE (9.41%), NSKE+BLE (8.00%) and BLE (5.22%) treated plot. Among the six botanical treatments NSKE was proved to be more effective in reducing pod damage in the field.

Values are mean ± standard deviation of pods damaged in 5 plants/ plot. Within the column similar alphabets are statistically not significant by LSD (p>0.05).

Effect of botanicals on chickpea yield

The yield of processed chickpea at the end of cropping season from each treatment was recorded (Table 4). The highest mean yield was obtained from NSKE treated plot (781 g) followed by Diazinon 60% EC treated plot (719.33 g), NSKE + BLE (656.67 g) and BLE treated plot (653.33 g). Where as lowest mean chickpea yield was obtained in control plot (419.33 g) followed by CLE (522 g) and NSKE + CLE (525 g) and CLE + BLE treated plots (556.67 g). Overall yield was significantly higher in treated plots compared to untreated plot. From the result it is clear that highest yield was recorded from plots treated with NSKE and positive control Diazinon 60% EC. The yield obtained from NSKE treated plot and Diazinon 60% EC treated plot was statistically not significant (p>0.05; LSD value 89.19). The overall percentage of yield loss was 46.31% and 41.70% if the filed was not treated with NSKE and Diazinon 60% EC respectively.

Effect of botanicals and chemical pesticides on non target organisms

The mean number of non target organisms in the field before

Treatment	Before spraying	After first spray	After second spray	Total number of pods
NSKE	16.6 ± 3.57^{c} (9.18%)	8.8 ± 3.41^{f} (4.87%)	7.06 ± 2.8^{d} (3.9%)	180.67
CLE	22.3 ± 5.16^{b} (15.07%)	25.6 ± 6.25^{b} (17.3%)	21.2 ± 4.81^{b} (14.33%)	147.93
BLE	18.7 ± 2.93^{c} (11.36%)	12.1 ± 2.57^{e} (7.35%)	8.6 ± 2.92^{d} (5.22%)	164.6
NSKE + BLE	20.8 ± 3.91^{b} (13.21%)	16.8 ± 2.5^{d} (10.67%)	12.6 ± 2.91^{c} (8.00%)	157.47
CLE + BLE	28.8 ± 6.34^{a} (19.36%)	29.8 ± 5.78^{a} (20.03%)	21.6 ± 6.00^{b} (14.52%)	148.73
NSKE + CLE	28.8 ± 5.47^{a} (19.08%)	21.2 ± 4.93^{c} (14.05%)	14.2 ± 4.97^{c} (9.41%)	150.87
Diazinon 60% EC	25.9 ± 5.78^{b} (14.64%)	2.26 ± 2.08^{g} (1.27%)	0.8 ± 1.47^{e} (0.45%)	176.87
Control	20.5 ± 7.14^{bc} (13.13)	32.6 ± 9.75^{a} (20.88%)	34.7 ± 15.12^{a} (22.22%)	156.13

Table 3: Mean number of damaged pods recorded in control and experimental plot.

Treatment	Yield/g/plot	Projected yield per/ ha in kg	% yield loss / hectare compared to control if not treated
NSKE	781 ± 16.28^{a}	1446.3	46.31%
CLE	522 ± 49.11^{cd}	966.67	19.67%
BLE	653.33 ± 15.27^{b}	1209.87	35.82%
NSKE + BLE	656.67 ± 70.23^{b}	1216.05	36.14%
CLE + BLE	556.67 ± 18.58^{c}	1030.87	24.59%
NSKE + CLE	525 ± 39.23^{cd}	972.22	20.13%
Diazinon 60% EC	719.33 ± 49.00^{ab}	1332.09	41.70%
Control	419.33 ± 95.44^{d}	776.54	-

Values are mean ± standard deviation of three replications. Similar alphabets within the column was statistically not significant by LSD (P>0.05).

Table 4: Chickpea yield per plot in grams.

spraying, after first and second spraying was recorded. When consider the population of ants before treatment there was a significant difference between the plots (p<0.05, LSD). Five days after first treatment there was no significant difference among the botanical treatments. However, there was a significant difference between botanical treatments and Diazinon 60% EC and untreated control. The highest number of ants population was recorded in control (5.33) and lowest was in Diazinon 60% EC treated plot (0). The mean number of ants in six botanical treatments ranged from 1.3 to 2.0 which was not statistically significant (p>0.05, LSD). Five days after second treatment there was a significant difference in the mean number of ants between NSKE+BLE, and control. There was no significant difference among NSKE, CLE, BLE, CLE+BLE and NSKE+CLE treated plots (Table 5).

The mean number of spider population in pre treatment count was statistically not significant (p>0.05; LSD) between the plots assigned for botanical treatment. Five days after first treatment there was a significant difference between positive control, control and other treatments. The highest mean number of spiders was observed in control plot (3.6) and lowest (0.3) was in plot treated with Diazinon 60% EC. Five days after second treatment, spider population was significantly (p<0.05; LSD) decreased in the treated plots compared control (Table 6).

In the case of lady bird beetles before treatment in all the plots there was no significant difference (p>0.05; LSD). Five days after first treatment among the botanicals treated plots the result was statistically not significant (p>0.05; LSD). However, compared to chemical pesticide treated plot and control plot result was statistically significant (p<0.05; LSD). The highest mean number of lady bird beetles was observed in control plot (3.66) followed by NSKE treated plot (2.6) and the difference was statistically not significant at 5% level by LSD (Table 7).

When consider the wasp population, LSD test showed no significant difference before treatment. Five days after first treatment, in general mean number of wasps was decreased in the experimental plot compared to control plot. The highest mean number of wasp population was recorded in control plot (3.3) and there was no wasp observed in Diazinon 60% EC treated plot. Five days after second treatment, there was a slight decrease in the mean number of wasps compared to first treatment. The LSD test showed that wasp population in the plot treated with NSK, CLE and BLE was significantly increased compared to Diazinon 60% EC treated plot. Among the botanicals treatment, mean number of wasp population was statistically not significant (p>0.05; LSD) except NSKE + CLE treated plot (Table 8).

Discussion

Natural products in insect pest management programs are gaining recognition in recent years due to environmental pollution, pest resistance and resurgence caused by indiscriminate use of synthetic chemical pesticides. In Ethiopia, marginal farmers cannot afford the cost of chemical pesticides and moreover chemical pesticides are not advisable for crops that are mainly grown under rain fed condition particularly in highlands of Ethiopia. Therefore, an attempt was made to find out ecofriendly pest management strategies by utilizing locally available plant materials. In the laboratory findings higher percentage of pod damage was observed at lower concentration of botanicals than at higher concentration. Among the botanicals tested, NSKE was found to be most effective as compared to other botanical extracts even at lower concentration. The better results of NSKE may be due to antifeedant or repellent property and this is in line with the observation of Gilani [40] who has reported that neem plant extracts deter insects from feeding. Redferen et al. [41] also reported that neem compound azadirachtin

Treatments	Before treatment	After first treatment	After second treatment
NSKE	2.0 ± 1.00^{a}	1.6 ± 0.57^{b}	1.33 ± 1.15^{bc}
CLE	2.0 ± 1.00^{a}	2.0 ± 1.00^{b}	1.66 ± 1.15^{bc}
BLE	2.0 ± 1.00^{a}	1.3 ± 0.57^{b}	1.66 ± 0.57^{bc}
NSKE+BLE	1.6 ± 1.15^{a}	1.3 ± 0.57^{b}	0.6 ± 0.57^{c}
CLE + BLE	2.6 ± 0.57^{a}	2.0 ± 0.00^{b}	2.33 ± 0.57^{b}
NSKE +CLE	1.6 ± 0.57^{a}	1.3 ± 0.57^{b}	1.0 ± 1.00^{bc}
Diazinon 60% EC	2.0 ± 1.00^{a}	0.0 ± 0.00^{c}	0.0 ± 0.00^{c}
Control	2.33 ± 1.15^{a}	5.6 ± 0.57^{a}	5.33 ± 1.15^{a}

Values are mean ± standard deviation of three replications. Similar alphabets within the column was statistically not significant (p>0.05; LSD).

Table 5: Mean number of ants recorded in experimental and control plots.

Treatments	Before treatment	After first treatment	After second treatment
NSKE	2.6 ± 1.15^{ab}	2.0 ± 1.73^{b}	1.66 ± 2.08^{b}
CLE	1.6 ± 0.57^{b}	1.3 ± 0.57^{b}	1.0 ± 1.00^{b}
BLE	1.3 ± 0.57^{b}	1.3 ± 0.57^{b}	1.0 ± 0.00^{b}
NSKE+BLE	1.3 ± 0.57^{b}	1.0 ± 1.00^{b}	0.6 ± 0.57^{b}
CLE + BLE	2.0 ± 1.00^{b}	1.6 ± 0.57^{b}	1.33 ± 0.57^{b}
NSKE +CLE	2.0 ± 1.00^{b}	1.0 ± 0.00^{b}	0.66 ± 0.57^{b}
Diazinon 60% EC	2.33 ± 0.57^{a}	0.3 ± 0.57^{c}	0.33 ± 0.57^{b}
Control	2.66 ± 0.57^{a}	3.6 ± 0.57^{a}	3.33 ± 0.57^{a}

Values are mean ± standard deviation of three replications. Similar alphabets within the column was statistically not significant (p>0.05; LSD).

Table 6: Mean number of spiders recorded in experimental and control plots.

Treatments	Before treatment	After first treatment	After second treatment
NSKE	2.3 ± 1.15^{a}	2.6 ± 0.57^{ab}	2.33 ± 0.57^{ab}
CLE	2.0 ± 1.00^{a}	2.0 ± 1.00^{b}	2.0 ± 1.00^{b}
BLE	2.33 ± 0.57^{a}	2.0 ± 1.00^{b}	1.33± 1.15^{b}
NSKE+BLE	1.6 ± 0.57^{a}	1.3 ± 0.57^{b}	1.33 ± 0.57^{b}
CLE + BLE	2.33 ± 0.57^{a}	2.0 ± 1.00^{b}	2.0 ± 1.00^{b}
NSKE +CLE	1.6 ± 0.57^{a}	1.6 ± 0.57^{b}	1.33 ± 0.57^{b}
Diazinon 60% EC	2.3 ± 1.15^{a}	0.6 ± 0.57^{c}	1.0 ± 0.00^{b}
Control	2.0 ± 1.00^{a}	3.6 ± 0.57^{a}	3.66 ± 0.50^{a}

Values are mean ± standard deviation of three replications. Similar alphabets within the column was statistically not significant (p>0.05; LSD).

Table 7: Mean number of predatory lady bird beetles recorded in experimental and control plots.

Treatments	Before treatment	After first treatment	After second treatment
NSKE	1.6 ± 0.57^{a}	1.33 ± 1.15^{bc}	1.0 ± 1.00^{b}
CLE	1.6 ± 0.57^{a}	1.3 ± 0.57^{bc}	1.0 ± 0.00^{b}
BLE	2.33 ± 0.57^{a}	1.6 ± 0.57^{b}	1.66 ± 0.57^{b}
NSKE+BLE	2.0 ± 0.00^{a}	1.6 ± 0.57^{b}	1.33 ± 1.15^{bc}
CLE + BLE	1.3 ± 0.57^{a}	1.0 ± 1.00^{bc}	0.66 ± 1.15^{bc}
NSKE +CLE	2.0 ± 1.00^{a}	1.3 ± 0.57^{bc}	0.33 ± 0.57^{c}
Diazinon 60% EC	1.6 ± 0.57^{a}	0.0 ± 0.00^{c}	0.0 ± 0.00^{c}
Control	1.66 ± 0.57^{a}	3.3 ± 0.57^{a}	3.0 ± 1.00^{a}

Values are mean ± standard deviation of three replications. Similar alphabets within the column was statistically not significant (p>0.05; LSD).

Table 8: Mean number of wasps recorded in experimental and control plots.

has antifeedant effect on insects.

In the field study, reduction in mean number of *H. armigera* larval population was higher in positive control Diazinon 60%EC followed by NSKE. The second best botanical treatment was BLE. The reduction in mean number of *H. armigera* on chickpea plot treated with botanicals may be associated with individual or combined properties of either antifeedant or repellent or oviposition deterrent or antifertility. Among the botanicals, NSKE was found to be superior and it is agreed with the report of Rajput et al. [42]. They have reported that neem products are superior in reducing pest population due to the repellent activity against the larvae of different instars on the treated plots as a result lower the number of *H. armigera* compared to control plot and /or due to the antifeedant effect on the larvae of H. arimigera. The different neem extracts (aqueous, ethanolic and hexane) have shown ovicidal properties against the eggs of *H. armigera* [43]. In addition to NSKE, CLE also showed about 48.77% reduction in the mean number of *H. armigera*. This is in accordance with the findings of Barbozasilva et al. [44]; they have reported that cis-dehydrocrotonin extracted from *Croton cajucar* bark inhibits the growth of *Heliothis virescens*. The percentage reduction of *H. armigera* in BLE treated plots may be due to the antifeedant and/or toxic effects. It is known that rotenone have been used as insecticides since 1848 when they were applied to plants to control leaf eating caterpillars. Bekele [24] also observed that crude extracts from the seeds of Birbira was toxic to *Sitophilus zeamais*. Ishaaya et al. [45] suggested that Birbira products have both contact and stomach poison to insects and kill insects slowly but causes them to stop feeding almost immediately.

The overall pod damage was significantly lower in treated plots as compared with the control. Among the treatments, the lowest pod damage was observed in Diazinon 60% EC and NSKE treated plots. The second best botanical treatment with higher protection was BLE. The biopotency of NSKE and BLE may be due to antifeedant, repellency and oviposition deterrence properties of these extracts, consequently lowering the number of *H. armigera* larvae from the treated plots. This is in line with the findings of Sehgal and Ujagir [46]; they have reported that using neem seed extracts in chickpea field, the pod damage was lowered compared to untreated plots. Sadawarte and Sarode [47] also indicated similar results that neem was effective in reducing pod damage at 5 or 6% concentration. Jeyakumar and Gupta [48] reported that neem product was superior in anti-ovipostion activity on *H. armigera*. The overall insect populations and pod damage was lower in treated plots than untreated plots resulting in higher yields. Chickpea yield was significantly higher in plot treated with NSKE and positive control (Diazinon 60% EC). Sadaworte and Sarode [47] also suggested that NSKE can be used in place of the highly toxic insecticides because of its safety to beneficial insects and lowest cost.

During the pod formation stage of chickpea plants, spiders, ants, lady bird beetles and wasps were observed in all plots due to availability of prey. These non target organisms are generalized predator and also natural enemies of *H. armigera*; they can feed either in egg or larva or pupa or adult moth. After treatment, mean number of each non target organism was reduced in treated plots. The highest reduction was observed in Diazinon 60% EC treated plots. It is known that chemical pesticides are not safer to non-target organism due to its contact toxicity. The reduction in the number of natural enemies in botanicals treated plots may be due to less availability of prey. Therefore, it is possible that the natural enemies may migrate to the control plot or nearby plots in search for their prey. In general, botanical extracts, particularly NSKE proved to have significant reduction in insect population thereby chickpea yield was increased. The cost of commercially available Diazinon 60% EC was 200 Ethiopian Birr ($10.899). However, if the farmers prepare botanical pesticides by themselves the cost of input becomes nil since the materials are locally available. Therefore, these widely and freely available eco-friendly botanical biopesticides are suitable for resource poor farming community to protect their chickpea crop particularly under rain fed condition against African bollworm *H. armigera*.

Acknowledgements

First author thank the University of Gondar for providing financial assistance under teaching and learning program to conduct post graduate research.

References

1. Muehlbaver FJ, Abebe T (1997) *Cicer arietinum* L. New Crop Fact Sheet.
2. Guar PM, Tripathi S, Gowda CLL, Ranga Rao GV, Sharma HC, et al. (2010) Chickpea seed production manual. International Crops Institute for Semi arid tropics.
3. EARI (2006) (Ethiopian Agricultural Research Institute) Crop protection department progress report for the period 1994-2006.
4. Jones R, Audi P, Shiferaw B, Gwata E (2006) Production and marketing of kabuli or chickpea seeds in Ethiopia; experience from Ada district. International Crops Research Institute for the Semi-Arid Tropics, Kenya.
5. ICRISAT (1991) Growing chickpea in Eastern Africa. India.
6. Gelatu B, Million E (1996) Chickpea in Ethiopia. In: Adaptation of chickpea in the West Asia and North African region.1-6.
7. Melaku W, Melkamu A, Birhane A, Fentahun M (1998) Research on insect pests and disease of field crops in North Western Ethiopia.12-14.
8. Metcalf RL, Metcalf ER (1992) Plant kairomones in insects ecology and control. 1st edn. Chapman and Hall, UK.
9. Purohit SS, Vyas SP (2004) Medicinal plant cultivation: A scientific approach. 165-168.

10. Tesfahun G, Bayu W, Teafaye A (2000) Indigenous techniques of crop pest control in Wollo. Pest Manag J Ethiopia 6: 64-68.

11. Roopa PK, Gouda B, Mandihalli (2003) Ovipositional effects of neem extracts on different insects. Karnataka J Agri Sci 16: 251.

12. Tebkew D, Adane T, Asmare D (2002) Potentials for botanicals in controlling the African bollworm. In: proceedings of the national workshop on African bollworm management in Ethiopia: status and need. 106-114.

13. Gossa MW (2007) Effects of neem extracts on the feeding, survival, longevity and fecundity of African bollworm, *Helicoverpa armigera* (Hubner) (Lepidoptera: Noctuidae) on cotton.

14. http://www.nap.edu/openbook.php?record_id=1924&page=1

15. Ma DL, Gordh G, Zalucki MP (2000) Toxicity of biorational insecticides to Helicoverpa spp. (Lepidoptera: Noctuidae) and predators in cotton field. Inter J Pest Manag 46: 237-240.

16. Van den Berg H (1993) Natural control of *H. armigera* in smallholder crops in East Africa. PhD Thesis, Wageningen University, Wageningen, Netherlands.

17. Greathead DJ, Grling DJ (1985) Distribution and economic importance of Heliothis and their natural enemies and host plants in Southern and Eastern Africa. IOBC Heliothis Work Group, India.

18. Desta G, Ermias S, Ridwan M (2004) African bollworm management in Ethiopia status and needs. 24.

19. Emden HF Van (1989) Pest control. 2nd Edition, Cambridge University Press. 27-37.

20. Bekele J, Daniel G, Meried N, Emiru S (2002) Toxicity of Birbira (*Milletia ferruginea*) seed crude extract to some insect pests as compared to other botanical and synthetic insecticides. 11th NAPRECA Symposium Book of Proceedings, Antananarivo, Madagascar 88-96.

21. Karunamoorthi K, Bishaw D, Mulat T (2009) Toxic effects of traditional Ethiopian fish poisoning plant *Milletia ferruginea* (Hochst) seed extracts on aquatic macro invertebrates. Eur Rev Med Pharmacol Sci 13:179-185.

22. Holden MJ, Raitt DFS (1974) Manuals of Fisheries Science Part 2 - Methods of resource investigation and their application. Food and Agriculture Organization of the United Nations, Italy.

23. Bekele J (1998) Investigation of flavonoids from Birbira. 167.

24. Bekele J (2002) Evaluation of the toxicity potential of *Milletia ferruginea* (Hochst) Baker against *Sitophilus zeamais*. Inter J Pest Manag 48: 29-32.

25. Mulatu B (2007) Contact bioassay of an endemic plant to Ethiopia on three aphid species. Ethiopian J Biol Sci 1: 51-62.

26. Getahun D, Jembere B (2006) Evaluation of toxicity of crude extracts of some botanicals on different castes of macrotermes termites. Pest Management Journal of Ethiopia 10: 15-24.

27. Peres MT, Delle Monache F, Cruz AB, Pizzolatti MG, Yunes RA (1997) Chemical composition and antimicrobial activity of Croton urucurana Baillon (Euphorbiaceae). J Ethnopharmacol 56: 223-226.

28. Peres MT, Pizzolatti MG, Yunes RA, Monache MD (1998) Clerodane diterpenes of Croton urucurana. Phytochem 49: 171-174.

29. Suárez AI, Compagnone RS, Salazar-Bookaman MM, Tillett S, Delle Monache F, et al. (2003) Antinociceptive and anti-inflammatory effects of Croton malambo bark aqueous extract. J Ethnopharmacol 88: 11-14.

30. Anazetti MC, Melo PS, Durán N, Haun M (2004) Dehydrocrotonin and its derivative, dimethylamide-crotonin induce apoptosis with lipid peroxidation and activation of caspases-2, -6 and -9 in human leukemic cells HL60. Toxicology 203: 123-137.

31. Fischer H, Machen TE, Widdicombe JH, Carlson TJ, King SR, et al. (2004) A novel extract SB-300 from the stem bark latex of Croton lechleri inhibits CFTR-mediated chloride secretion in human colonic epithelial cells. J Ethnopharmacol 93: 351-357.

32. Kubo I, Asaka Y, Shibata K (1991) Insect growth inhibitory nor-diterpenes, cis-dehydrocrotonin and trans-dehydrocrotonin, from *Croton cajucara*. Phytochem 30: 2545-2546.

33. Silva LB, Silva W, Macedo MLR, Peres MTLP (2009) Effects of Croton urucurana extracts and crude resin on Anagasta kuehniella (Lepidoptera: Pyralidae). Braz Arch Biol Technol 52: 653-664.

34. Silva LB, Xavier ZF, Silva CB, Faccenda O, Candido ACS, et al. (2012) Insecticidal effects of Croton urucurana extracts and crude resin on Dysdercus maurus (Hemiptera: Pyrrocoridae). J Entomol 9: 98-106.

35. Jain PC, Bhargava MC (2007) Entomology novel approaches. 165-309.

36. Dodia DA, Patel IS, Patel GM (2010) Botanical pesticides for pest management.

37. Gomez KA, Gomez AA (1984) Statistical procedures for agricultural research, (2ndedn). Wiley, USA.

38. Judenko E (1973) Analytical method for assessing yield losses caused by pests on cereal crops with and without pesticides. Tropical Pest Bulletin 31.

39. Walker PT (1997) The assessment of crop losses in cereals. Insect Sci Appl 4: 97-104.

40. Gilani G (2001) Neem the wonder tree. 3-7.

41. Redferen RE, Warthen JD, Liebel EL, Mills GD (1980) The antifeedant and growth disrupting effect of Azadirachtin on Sopdoptera ferugiperda and Onceopeltus faseiatus. 87-91.

42. Rajput AA, Sarwar M, Bux M, Tofique M (2003) Evaluation of synthetic and some plant origin insecticides against *Helicoverpa armigera* (Hubner) on chickpea. Pak J Biol Sci 6: 496-499.

43. Jhansi BR, Singh RP (1993) Biological effects of neem on insect pests. 12: 17-19.

44. Silva LB, Silva W, Macedo MLR, Peres MTLP (2009) Effects of Croton urucurana extracts and crude resin on Anagasta kaehniella. Braz Arch Biol Technol 52.

45. Perry AS, Yamamoto I, Ishaaya I, Perry RY (1998) Insecticides in agriculture and environment: Retrospects and prospects (Applied agriculture). (1stedn) Springer, USA.

46. Sehgal VK, Ujagir R (1990) Effect of synthetic pyrethroids, neem extracts and other insecticides for the control of pod damage by *Helicoverpa armigera* (Hübner) on chickpea and pod damage-yield relationship at Pantnagar in northern India. Crop Protection 9: 29-32.

47. Sadawarte AK, Sarode SV (1997) Effect of neem seed extract, cow dung and cow urine alone and in combination against the pod borer complex on pigeon pea. International Chickpea and Pigeon pea 4: 37.

48. Jeyakumar P, Gupta GP (1999) Effect of neem seed kernel extract (NSKE) on *Helicoverpa armigera*. Pesticide Res J 11: 32-36.

8

Effects of Synthetic Insecticides and Crude Botanicals Extracts on Cabbage aphid, *Brevicoryne brassicae* (L.) (Hemiptera: Aphididae) on Cabbage

Tadele Shiberu* **and Mulugeta Negeri**

Department of Plant Sciences, College of Agriculture and Veterinary Sciences, Ambo University, Ethiopia

Abstract

The study was conducted to evaluate the insecticidal action of two new chemical insecticides and three locally available botanicals. These materials were evaluated for their efficacy against cabbage aphids, *Brevicoryne brassicae* (L.), under laboratory and field conditions. The laboratory result revealed that Lambda Cyhalothrin (Triger 5 E.C™) and Emamectin benzoate (Cutter 112 E.C™) were non-significant difference from the standard check within 1 day after application. The field results showed between the new insecticides, Triger 5 E.C™ and Cutter 112 E.C™ were non-significant difference from that of the standard check (Diazinon 60% E.C™) at both locations (Guder and Mutulu) within three days after application. Extracts of neem (*Azadirachta indica*) seed, Hop bush (*Dodonae angustifolia*) fresh leaf and Lemon grass (*Cymbopogon citrates*) gave positive performance under laboratory while the efficacy percent were decline at both field locations on field. *Azadirachta indica* seed, *Dodonae angustifolia* and *Cymbopogon citrates* were given 59.48, 57.16 and 52.50 at Mutulu and 53.92, 37.26 and 62.72 at Guder location mortality percent, respectively, after 3rd day observation in first spray. The obtained results revealed that all tested materials were exhibited mortality rate action against cabbage aphids. Powdered neem seed, *Dodonae angustifolia* fresh leaf and leaves of *Cymbopogon citrates* caused moderately mortality percent against cabbage aphid within 3 days after applications. Higher damage and low yield were observed in control and botanicals application. This study indicated that the two insecticides Cutter 112 E.C™ and Triger 5 E.C™ and botanicals could be recommended as an alternative management option of *Brevicoryne brassicae* (L.) in Ethiopia.

Keywords: Cabbage; Botanicals; Chemical insecticides; Cabbage aphid

Introduction

Cabbage (*Brassica oleracea* L. var. *capitata*) is cultivated in about 31,783.54 ha [1]. Its annual production is 43,483.94 ton with 3, 230,180.49 ton head cabbage, estimated yield per hectare is 9.1 ton [1]. Ethiopia has the favorable agro-climatic conditions for the production of cabbage head for fresh market. However, cabbage yield and quality has been shown to be influenced by several factors, among of these factors cabbage aphids are very serious in cabbage growing areas of Ethiopia especially in dry season. A number of insect species including cabbage aphid *Brevicoryne brassicae* inflict damage on brassica crops in Ethiopia [2]. The cabbage aphid, *Brevicoryne brassicae* L. (Homoptera: Aphididae), is found on brassica crops with worldwide distribution and severe damage and outbreaks [3]. Cabbage aphid cause significant yield losses in many crops in the family Brassicaceae, which includes the mustards and crucifers. Continued feeding by aphids causes yellowing, wilting and stunting of plants [4]. Severely infested plants become covered with a mass of small sticky aphids, which can eventually lead to leaf death and decay [5]. Cabbage aphids feed on the underside of the leaves and on the center of the cabbage head [6]. They prefer feeding on young leaves and flowers and often go deep into the heads of brussels sprouts and cabbage [7]. Colonies of aphids are found on upper and lower leaf surfaces, in leaf folds, along the leafstalk, and near leaf axils. It also secretes honey dew while feeding on leaves that reduces the quality of cabbage. With development of sooty mold on this honey dew, the produce have became sooty and unmarketable [8-11]. Controlling cabbage aphid is not an easy practice although synthetic chemicals are apparently available for use. Effective pest control is no longer a matter of heavy application of limited insecticides, because continuous use of chemicals promotes development of pesticide resistance in the target pests, pest resurgence, emergence of secondary pests, affects non-target insects species, affects the environment and human health. Therefore use of alternatives including botanicals, biopesticides and new generation synthetic insecticides is essential to grow health crops of cabbage. The use of different botanical insecticides to protect plants from pests is very promising because of several distinct advantages [12]. Pesticidal plants are generally much safer than conventionally used synthetic pesticides. Pesticidal plants have been in nature as its component for millions of years without any adverse effect on the ecosystem [12]. Plant product pesticides can be undertaken into practical applications in natural crop protection, which can help the small-scale farmers [13]. The use of natural and easily biodegradable crop protection inputs like *Azadirachtin* can be a useful component of an IPM strategy since the compound is known for its low toxicity against beneficial insects [14]. Therefore, two newly introduced insecticides Emamectin benzoate (Cutter 112 E.C.™) and Lambda Cyhalothrin (Triger 5 E.C.™) and three locally available plants (*Azadirachta indica* seed, *Dodonae angustifolia* fresh leaf and *Cymbopogon citrates*) leaf were evaluated to determine their efficacy percent on mortality of *cabbage aphids* under laboratory as well as on the fields. *Dodonaea angustifolia*, commonly known as hop bush, called "Etecha" in afan oromo and "Kitkita" in Amharic is a perennial shrub belonging to Sapindaceae family. This plant commonly found in West shawa of Ethiopia, the branches used for teeth brush because of its interesting aroma and the leaves used for cultural medicinal value.

***Corresponding author:** Tadele Shiberu, Department of Plant Sciences, College of Agriculture and Veterinary Sciences, Ambo University, Ethiopia, E-mail: tshiberu@yahoo.com

Materials and Methods

Description of the study area

The experiment was carried out under irrigation on two locations of farmers' field at Toke Kutaye district West Shawa Zone, Oromia Regional state, Ethiopia. The district is 125 km far away from Addis Ababa and 12 km from Ambo town having an altitude of 1990 meter above sea level, latitude of 08° 59" 01.1" north and longitude of 37° 46" 27.6" east. The average annual rainfall is 1028.7 mm and maximum and minimum temperatures of the area 26.2°C and 12.3°C, respectively.

Preparation of botanicals and insecticides

Fresh leaves of *Dodonae angustifolia* and *Cymbopogon citrates* were collected from the surrounding areas of Ambo town, cut into small pieces, and dried under shade. Dried the leaves and seeds of *Azadirachta indica* were ground and mixed with water at 5% concentration level (w/v) i.e., 10g of powder in 100ml of water and filtered through cheese cloth. The extract suspensions were sprayed on the cabbage aphid populations using hand sprayer at the rate of 150 liters per hectare. There was two times spray at seven days interval. The two insecticides Emamectin benzoate (Cutter 112 E.C™) and Lambda Cyhalothrin (Triger 5 E.C™) mixed separately with 100 liter of water and sprayed at the rates of 1 and 0.4 liter per hectare, respectively.

Treatments and experimental design

The laboratory study was conducted at Ambo University Plant science department, entomology laboratory using complete randomized design with three replications. The leaf powder extracts were diluted with distilled water to obtain 10% test solutions, which were used in the bioassay studies. The nymphs and adults of cabbage aphids used for the study were collected from the host cabbage crops in the cabbage fields.

Laboratory treatments application

Studies were conducted for a period of 24hrs in the laboratory 100 nymph and adults were inserted in a Petridis and provided with fresh leaves of cabbage that was collected from the field to serve as food source, botanical crude extracts and insecticides were applied on against aphid insects in the Petridis at defined rate. All the experiments were conducted at 24°C ± 2 and70% humidity of laboratory condition. Twenty four hour latter larval mortality was observed.

Field treatment application

The fields' experiments were scouted every week for the signs and symptoms of aphids' damage and occurrence till reach economic threshold level. Botanicals were sprayed in the afternoon at 4 p.m. Aphid population pre-spray counts were made before 24 hrs of each spray. Hand sprayers were used and suspension performed at the plant stage of 6-8 true leaf and when aphid population is reached its economic threshold level 2% infestation [15]. Opfer and McGrath [4] also stated that when more than 20% of leaves are infested with aphids, then an insecticide application is recommended. However, we used at 2% infestation level. The field study was laid out in Randomized Complete Block Design with three replications. Plant to plant and row-to-row distances were kept at 40 × 80 cm. The single plots size was 6 m^2. The Copenhagen variety of cabbage was brought from local market. Seedlings were raised on a seedbed. Ready seedlings were transplanted. All agronomic practices and management treatments were applied as recommended. Head cabbages were harvested at maturity stage in an extended harvesting period.

Data collection

The experimental fields were scouted every week for the signs and symptoms of aphids' damage and occurrence till reach economic threshold level. Numbers of aphid cabbage per cabbage head were counted and their economic threshold levels were determined. After treatment application numbers of live cabbage aphids' post-treatment data were recorded after 1 and 3 days of each spray. Finally, the cabbage head yield was measured and expressed as kg/ha.

Data analysis

Data were analyzed with SAS version 9.1 [16] and the effects of the treatments were compared. Least significant difference (LSD) test was used to determine the differences among the means for different treatments and yield parameters. Efficacy percentages were calculated by using the following formula:

$$\text{Efficacy (\%)} = \frac{Pr\,SC - Posc}{Pr\,SC} X100$$

Where: PrSC=Pre Spray Count

PoSC=Post Spray Count

Result and Discussion

Effect of botanicals against Cabbage aphid

The result presented in (Table 1a) indicted there are significant (P<0.01) differences among the treatments. The presented data are

Treatments	Efficacy percentage (%)		
	Under laboratory condition	Under field condition	
		Mutulu site	
		1 day ADTA	3 days ADTA
Lambda Cyhalothrin (Triger 5% E.C™)	100(90.00)^a	85.04(67.21)^a	95.05(77.08)^a
Emamectin benzoate (Cutter 112 E.C™)	100(90.00)^a	83.36(68.87)^a	94.2275.85)^a
Azadirachta indica seed	75.67(60.67)^c	50.81(45.57)^b	59.48(50.19)^b
Dodonae angustifolia fresh leaf	87.33(68.87)^b	49.09(42.43)^b	57.16(49.02)^b
Cymbopogon citrates	72.06(58.05)^c	36.34(36.87)^b	52.50(46.72)^b
Diazinon 60 E.C™	100(90.00)^a	97.67(81.87)^a	99.22(84.26)^a
Water/control	14.33(21.97)^d	12.64(21.13)^c	11.57(20.26)^c
MSE	3.12	7.78	6.11
LSD at 0.01	7.77	19.4	15.25
CV (%)	3.97	13.12	9.12

Note: Means with the same letter are not significantly different ADTA (After Day of Treatment Application) Figures in parentheses are Arcsin *percent* transformed value.

Table 1a: First spray mean efficacy of insecticides and botanicals on cabbage aphids, *Brevicoryne brassicae* under laboratory and field conditions.

pertaining to mean percent reduction of aphid population reveals that, among all the treatments the new chemical insecticides and the standard check (Diazinon 60% E.C.™), Emamectin benzoate (Cutter 112 E.C™) and Lambda Cyhalothrin (Triger 5 E.C™) were recorded to be significantly superior in efficacy against cabbage aphids of first sprayed (97.67, 96.29), (95.05, 98.48,) and (94.22, 98.15) within three days of application, respectively. All insecticides gave the highest mean percent reduction of aphids (Table 1b). However, among the botanicals extracts *Cymbopogon citrates* and *Azadirachta indica* gave moderate mortality rate in both locations and *Dodonae angustifolia* (57.16, 37.26) showed the lowest values but significantly higher than control treatment. The results were similar with those of Eileen and Sydney [17]. They reported that the treated insect with *Azadirachta indica* extract usually cannot molt to its next life stage and dies within a few days and acts primarily as a repellent when applied to a plant, and may kill an insect within 24 hrs (Table 2). The obtained results were in agreement with those of Ahmad and Akhtar [18]. They reported that aphids developed resistance against chemical insecticides including pyrethroids (cypermethrin, lambdacyhalothrin, bifenthrin and deltamethrin) and neonicotinoids (imidacloprid, acetamiprid, and thiamethoxam). Aphids resistance level increased progressively in concurrence with regular use of vegetables. A field experiment was undertaken to evaluate

Treatments	Efficacy percentage (%)	
	Under field condition	
	Guder Site	
	1 day ADTA	3 days ADTA
Lambda Cyhalothrin (Triger 5% E.C™)	92.86(74.66)^{a}	98.48(81.87)^{a}
Emamectin benzoate (Cutter 112 E.C™)	91.97(73.579^{a}	98.15(81.84)^{a}
Azadirachta indica seed	44.18(41.55)^{b}	53.92(47.29)^{b}
Dodonae angustifolia fresh leaf	36.31(36.87)^{c}	37.26(37.46)^{c}
Cymbopogon citrates	48.39(43.85)^{b}	62.72(52.53)^{b}
Diazinon 60 E.C™	94.32(75.82)^{a}	96.29(78.46)^{a}
Water/control	20.06(26.56)^{c}	16.57(24.35)^{d}
MSE	9.33	4.35
LSD at 0.01	23.27	10.86
CV (%)	15.26	6.58

Note: Means with the same letter are not significantly different
ADTA (After Day of Treatment Application)
Figures in parentheses are Arcsin *percent* transformed value.

Table 1b: First spray mean efficacy of insecticides and botanicals on cabbage aphids, *Brevicoryne brassicae* under field conditions.

the efficacy of fresh leaves powder extracts of *D. angustifolia* against cabbage aphids during the dry season in irrigation and it gave promising results. Similarly, a field experiment was conducted in Sorapet, India, to evaluate the efficacy of crude extracts of *D. angustifolia* against *Earias vitella* on rain fed cotton. It was reported that the extracts of the product was also drastically reduced the number of larvae similar to a neem product. The extracts drastically reduced the number of larvae *Earias vitella* [19]. Naqvi [20] stated that *Azadirachtin* based neem pesticides having diverse pest control properties affect insect growth, disturb adult fertility and different negative physiological processes in insects such as metamorphosis in addition to direct toxicity and anti-feedant and oviposition deterrent effects. Gandhi et al. [21] tested the performance of neem oil as seed treatment against aphid on Okra crop and found excellent results up to 45 days after treatment and provided better yield compared to control. Muhammad et al. [22] reported that *Azadirachtin* based biosal performed well with 59.77 % reduction against green peach aphid, *Myzus persicae* in Pakistan. Ahmed et al. [23] also evaluated the toxicity of *Azadirachtin* based formulations Neem-Azal T/S and Neemix against mature and immature stages of bean aphid (*Aphis fabae Scop.*) both products had remarkable effects on adult aphid when used as systemic insecticides (Tables 3a-3d). In another comparative study with neem oil 2% and neem seed water extract at 3% caused significant reduction in the population of jassid, whitefly and thrips on cotton up to 168 hrs but lost their efficacy at 336 hrs [24] (Table 4). Dodia et al. [25] reported that *Cymbopogon citrates* had an effect on the onion thrips management. Shiberu et al. [26] also mentioned that *Cymbopogon citrates* was reduced the population number of onion thrips at the rate of 10 g/liter of water. An overall performance of two consecutive sprays in two different locations against the population of cabbage aphid by insecticides gave as an average effect after third day of application. Triger 5%^{MT} (97.26) and Cutter 112 E.C^{MT} (97.24%) were revealed. The most effective control on cabbage aphids. However, *Azadirachta indica* (56.49), *Cymbopogon citrates* (55.20), and *Dodonae angustifolia* (48.96) gave less effective control on cabbage aphid as compared with tested insecticides.

Frequency of application

The frequency of application made two times within 15 days intervals at the recommended application rates. After 15 days of the first application the aphid infestation was reach economic threshold level. Therefore, the second sprays were made. Application timing is very important to keep aphids under control [5,6].

Treatments	Efficacy percentage (%)			
	On field			
	Mutulu site		Guder Site	
	1 day ADTA	3 days ADTA	1 day ADTA	3 days ADTA
Lambda Cyhalothrin (Triger 5% E.C™)	92.97(74.65)^{a}	96.03(78.46)^{a}	98.03(81.87)^{a}	99.50(84.26)^{a}
Emamectin benzoate (Cutter 112 E.C™)	96.53(80.02)^{a}	97.10(80.01)^{a}	99.03(84.26)^{a}	99.50(84.26)^{a}
Azadirachta indica seed	48.8(44.43)^{c}	49.10(44.42)^{c}	46.67(43.28)^{b}	63.47(52.53)^{b}
Dodonae angustifolia fresh leaf	56.20(48.44)^{b}	56.67(49.02)^{b}	34.17(35.67)^{c}	44.76(42.13)^{c}
Cymbopogon citrates	36.87(37.46)^{d}	42.67(40.97)^{c}	39.83(39.23)^{bc}	62.90(52.53)^{b}
Diazinon 60 E.C™	95.93(78.46)^{a}	98.70(84.26)^{a}	98.67(84.26)^{a}	99.27(84.26)^{a}
Water/control	9.73(18.43)^{e}	12.50(21.13)^{d}	14.70(22.79)^{d}	16.17(23.58)^{d}
MSE	2.51	2.65	3.95	2.41
LSD at 0.01	6.29	6.59	9-86	6.02
CV (%)	4.01	4.09	6.42	3.48

Note: Means with the same letter are not significantly different
ADTA (After Days Treatment Application)
Figures in parentheses are Arcsin *percent* transformed value

Table 2: Second spray mean efficacy of insecticides and botanicals on cabbage aphids, *Brevicoryne brassicae* under field conditions.

Mutulu site							
1Day ATA							
Spray	T_1	T_2	T_3	T_4	T_5	T_6	T_7
1st Spray	92.97(74.66)^a	83.36(65.65)^a	50.81(45.579)^a	49.09(44.43)^a	36.34(36.37)^a	97.67(81.87)^a	12.60(21.13)^a
2nd Spray	96.53(80.02)^a	84.9(67.21)^a	49.80(45.0)^a	56.2(48.459)^a	36.87(37.46)^a	95.93(78.46)^a	14.83(22.79)^a
MSE	2.25	3.19	7.58	3.4	7.74	3.89	2.14
CV (%)	2.53	3.55	15.07	6.46	21.16	4.02	15.61

Note: Means with the same letter are not significantly different
ATA (After Treatment Application)
Figures in parentheses are Arcsin percent transformed value
T_1(Lambda Cyhalothrin (Triger 5 E.C™)), T2(Emamectin benzoate (Cutter 112 E.C™)), T_3(*Azadirachta indica*), T_4(*Dodonae angustifolia*), T_5(*Cymbopogon citrates*), T_6(Diazinon 60 E.C™), and T_7(Water/control)

Table 3a: Effect of spray and time on efficacy of treatments at Mutulu Site After Days Treatment Application.

Mutulu site							
	3 Days ATA						
Frequency of spray	T_1	T_2	T_3	T_4	T_5	T_6	T_7
1st Spray	95.05(77.08)^a	94.22(75.82)^a	59.10(50.18)^a	54.12(47.29)^a	52.50(46.71)^a	99.22(84.26)^a	11.51(20.26)^a
2nd Spray	96.03(78.46)^a	97.0(80.03)^a	63.47(52.53)^a	48.38(43.85)^a	62.90(52.53)^a	99.27(84.26)^a	16.35(23.57)^a
MSE	1.96	5.15	4.64	10.46	8.67	1.46	2.38
CV (%)	2.05	5.39	7.56	20.41	15.02	1.47	17.07

Note: Means with the same letter are not significantly different
ATA (After Treatment Application)
Figures in parentheses are Arcsin *percent* transformed value
T_1(Lambda Cyhalothrin (Triger 5 E.C™)), T2(Emamectin benzoate (Cutter 112 E.C™), T_3(*Azadirachta indica*), T_4(*Dodonae angustifolia*), T_5(*Cymbopogon citrates*), T_6(Diazinon 60 E.C™), and T_7(Water/control)

Table 3b: Effect of spray and time on efficacy of treatments at Mutulu Site after treatment application.

Guder site							
1Day ATA							
Frequency of spray	T_1	T_2	T_3	T_4	T_5	T_6	T_7
1st Spray	92.86(74.66)^a	91.67(73.57)^a	44.18(41.55)^a	28.24(31.95)^a	48.39(43.85)^a	94.32(75.82)^a	20.06(26.56)^a
2nd Spray	98.03(81.87)^a	99.03(84.26)^a	50.17(45)^a	34.17(35.67)^a	46.50(43.28)^a	98.67(84.26)^a	14.70(22.78)^b
MSE	5.59	7,08	4.56	7.18	8.84	3.29	0.91
CV (%)	5.53	7.37	9.66	19.83	16.94	3.42	5.26

Note: Means with the same letter are not significantly different
ATA (After Treatment Application)
Figures in parentheses are Arcsin percent transformed value
T_1(Lambda Cyhalothrin (Triger 5 E.C™)), T2(Emamectin benzoate (Cutter 112 E.C™), T_3(*Azadirachta indica*), T_4(*Dodonae angustifolia*), T_5(*Cymbopogon citrates*), T_6(Diazinon 60 E.C™), and T_7(Water/control)

Table 3c: Effect of spray and time on efficacy of treatments at Guder Site after treatment application.

Guder site							
	3 Days ATA						
Frequency of spray	T_1	T_2	T_3	T_4	T_5	T_6	T_7
1st Spray	98.48(81.87)^a	98.14(81.87)^a	53.92(47.29)^a	37.30(37.46)^a	64.92(53.73)^a	96.30(78.46)^a	16.47(23.58)^a
2nd Spray	99.50(84.26)^a	99.50(84.26)^a	63.43(52.53)^a	44.67(42.13)^a	62.90(52.53)^a	99.27(84.26)^a	16.23(23.58)^a
MSE	1.25	1.67	8.15	2.54	3.01	1.83	2.37
CV (%)	1.26	1.69	13.09	6.19	4.72	1.87	14.46

Note: Means with the same letter are not significantly different
ATA (After Treatment Application)
Figures in parentheses are Arcsin percent transformed value
T_1(Lambda Cyhalothrin (Triger 5 E.C™)), T2(Emamectin benzoate (Cutter 112 E.C™), T_3(*Azadirachta indica*), T_4(*Dodonae angustifolia*), T_5(*Cymbopogon citrates*), T_6(Diazinon 60 E.C™), and T_7(Water/control)

Table 3d: Effect of spray and time on efficacy of treatments at Guder Site after treatment application.

Effect of treatments on yield

There was a significant difference among treatments and untreated check but no significant ($P<0.001$) difference among the insecticides when compared to the standard check (Diazinon 60 E.C) and also no significant difference among insecticides and botanicals except *Cymbopogon citrates* where it gave less yield compared to other treatments but better than water treated check. However, cabbage head yields in the standard check (Diazinon 60 E.C^{MT}) gave high yield percentage in all locations compared to untreated plot.

Conclusion and Recommendation

The results in all locations indicated that the mortality rate

Treatments	Yield/plot in kg	
	Mutulu site	Guder site
Lambda Cyhalothrin (Triger 5% E.C™)	9.50^{a}	7.67^{ab}
Emamectin benzoate (Cutter 112 E.C™)	8.67^{a}	7.00^{ab}
Azadirachta indica seed	7.83^{ab}	6.00^{ab}
Dodonae angustifolia fresh leaf	7.33^{ab}	6.00^{ab}
Cymbopogon citrates	7.17^{ab}	5.67^{b}
Diazinon 60 E.C™	8.83^{a}	8.00^{a}
Water/control	6.17^{b}	5.50^{b}
MSE	0.58	0.87
LSD at 0.01	2.42	2.18
CV (%)	12.31	12.34

Note: Means with the same letter are not significantly different.

Table 4: Effect of insecticides and botanicals on cabbage yield against cabbage aphids, *Brevicoryne brassicae.*

percentage of the two newly introduced insecticides Emamectin benzoate (Cutter 112 E.C™) and Lambda Cyhalothrin (Triger 5 E.C™) were comparable and effective when compared to the standard check (Diazinon 60 E.C™) in reducing the number of cabbage aphid population. The yield of all treatments except lemon grass extract and the untreated check were similar. Therefore, the newly introduced insecticides Emamectin benzoate (Cutter 112 E.C™) and Lambda Cyhalothrin (Triger 5 E.C™) and botanicals could be recommended to be considered as alternative insecticides for the management of cabbage aphid, *Brevicoryne brassicae* (L.) under Ethiopian condition.

References

1. CSA (2012) Report of Federal Democratic Republic of Ethiopia, Statistical Report on Socio-Economic Characteristics of the Population in Agricultural Households, Land Use, Area and Production of Crops. Addis Ababa, Ethiopia, pp. 17-20.
2. Tesdeke A, Gashawbeza A (1994) Progress in vegetable management research: 1985-1992. Pp.187-193. In: E. Hearth and Lemma, D.(eds.). Proceedings of the Second National Horticultural Workshop, Addis Ababa, Ethiopia, 1-3 December 1992, IAR/FAO.
3. Kessing JLM, Mau RFL (1991) Cabbage aphid, Brevicoryne brassicae (Linnaeus). Crop Knowledge Master. Department of Entomology, Honolulu, Hawaii.
4. Opfer P, McGrath D (2013) Oregon vegetables, cabbage aphid and green peach aphid. Department of horticulture. Oregon State University, Corvallis.
5. Griffin RP, Williamson J (2012) Cabbage, Broccoli and other cole crop insect pests HGIC 2203, Home and Garden information center. Clemson cooperative extension, Clemson University.
6. Hines RL, Hutchison WD (2013) Cabbage aphids on Vegetable IPM resource for the Midwest. University of Minnesota, Minneapolis, MN.
7. Natwick ET (2009) Cole crops: cabbage aphid UC Pest Management Guidelines. University of California Agriculture & Natural Resources.
8. Von-Dohlen CD, Rowe CA, Heie OE (2006) A test of morphological hypotheses for tribal and subtribal relationships of Aphidinae (Insecta: Hemiptera: Aphididae) using DNA sequences. Molec Phylogenet Evol 38: 316-329.
9. Farag NA, Gesraha MA (2007) Impact of four insecticides on the parasitoid wasp, Diaertiella rapaeand its host aphid, Brevicoryne brassicae under laboratory conditions. Res J Agric Biol Sci 3(5): 529-533.
10. Lu WN, Wu YT, Kuo MH (2008) Development of species-specific primers for the identification of aphids in Taiwan. Appl Entomol Zool 43: 91-96.
11. Toper-Kaygin A, Çota F, Gorur G (2008) Contribution to the aphid (Homoptera: Aphididae) species damaging on woody plants in dresser for managing Homopterous sucking pests of Okra (Abelmoschus esculentus (L.) (Moench). J Pest Sci 79: 103-111
12. Maribet LP, Aurea CR (2008) Insecticidal action of five plants against maize weevil, Sitophilus zeamais motsch. (Coleoptera: Curculionidae). KMITL Sci Tech J Vol 8: 24-34.
13. Binggeli P (1999) Lantana camara L (Verbenaceae). Retrieved October 12, 2005 from http//www.memberslycos.co.uk/woodyPlant Ecology/docs/web spd. htm.
14. Koona P, Njoya J (2004) Effectiveness of Soybean Oil and Powder from Leaves of Lantana camara Linn. (Verbenaceae) as Protectants of Stored Maize against Infestation by Sitophilus zeamais Motsch. (Coleoptera: Curculionidae). Asia Network for Scientific Information.
15. Saskatchewan Agriculture, 2012. Economic threshold level of pest activity when control action is suggested to prevent economic injury. Threshold Guide Vegetable Crops.
16. SAS Institute, 2003. The SAS system for windows, version 9.0. SAS, Institute, Cary, NC.
17. Eileen AB, Sydney G, Park B (2013) Natural Products for Managing Landscape and Garden pests in Florida. ENY-350, University of Florida.
18. Ahmad M, Akhtar S (2013) Development of insecticide resistance in field populations of Brevicoryne brassicae (Hemiptera: Aphididae) in Pakistan. J Econ Entomol 106: 954-958.
19. Malarvannan S, Subashini HD (2007) Efficacy of Dodonaea angustifolia Crude Extracts against Spotted Bollworm, Earias vitella (Fab.) (Lepidoptera: Noctuidae). J of Entomol 4: 243.
20. Naqvi SNH. 1996. Prospects and development of a neem based pesticide in Pakistan. Proceed. 16th Congr. Zool., Islamabad, 16: 325-338.
21. Gandhi I, Gunasekaran PK, Tongmin SA (2006) Neem oil as a potential seed dresser for managing Homopterous sucking pests of Okra (Abelmoschus esculentus (L.) (Moench). J Pest Sci 79: 103-111.
22. Muhammad FA, Muhammad AH, Farzana P, Nikhat Y, Muhammad FUK (2010) Comparative management of cabbage aphid (Myzus persicae (Sulzer) (Aphididae: Hemiptera) through bio and synthetic insecticides. Entomol 32: 1.
23. Ahmed AAI, Gesraha MA, Zebitz CPW (2007). Bioactivity of two neem products on Aphis fabae. J Appl Sci Res 3: 392-398.
24. Khattak MK, Mamoon-Ur-Rashid, Hussain AS, Islam T (2006) Comparative effect of neem (Azadirachtin indica A. Juss) oil, neem seed water extract and Baythroid TM against whitefly, jassid and thrips on cotton. Pak Enomol 28: 31-37.
25. Dodia DA, Patel IS, Patel GM (2008) Botanical pesticides for Pest Management, pp. 354.
26. Shiberu T, Negeri M, Selvaraj T (2013) Evaluation of Some Botanicals and Entomopathogenic Fungi for the Control of Onion Thrips (Thrips tabaci L.) in West Showa, Ethiopia. J Plant Pathol Microb 4: 1-7.

Exploration of Different *Azospirillum* Strains from Various Crop Soils of Srivilliputtur Taluk

G. Pandiarajan[1], N. Tenzing Balaiah[2*] and B. Makesh Kumar[1]

[1]*Department of Plant Biology and Plant Biotechnology, G.Venkataswamy Naidu College, Kovilpatti, Tamilnadu, India*
[2]*Department of Plant Biology and Plant Biotechnology, Ayya Nadar Janaki Ammal College Sivakasi, Tamilnadu, India*

Abstract

Exploration of different *Azospirillum* strains from various crop soils of Srivlliputtur Taluk was analysed. There are three different species of strains were found and analysed with different parameters like their morphology, motility, catalase and biotin content. However, the biology was also been studied which includes the vitamin and carbon utilization of the *Azospirillum*. In this results showed that there are two different species of *Azospirillum* were identified in all the soils namely *A.lipoferum*, and *A.brasilense*. The morphology, biochemical parameters of these two species of *Azospirillum* are varied according to the soil in which they are identified. It is concluded that the strains of *Azospirillum* will help to the plants for better growing by means of utilization of various parameters from the soil to the plants and these strains are used as very efficient biofertilizers in the crop plants from all over the world.

Keywords: Crop soils; Srivilliputtur Taluk; Exploration; *Azospirillum*

Introduction

Biofertilizer is a wide term which includes a diverse category of bioinoculants such as nitrogen fixers, phosphate solubilizers, phosphate mobilizers and plant growth promoting rhizobacteria. Biofertilizers are the organisms which are naturally present in all types of soils. Applications of these biofertilizers are environment friendly, means to supplement nutrient to the plants. Important biofertilizers are nitrogen fixers, phosphate solubilizers and phosphate mobilizers [1]. However, the concept of biofertilizers was developed with the discovery of nitrogen fixing *Azospirillum*. *Azospirillum* was first reported by *Beijerinck* [2] and it was named as *Spirillum lipoferum* by *Schroeder* [3]. It gained the reputation of being the most studied plant associative bacterium only after it was rediscovered by *Dobereiner J* [4] in the roots of *Digitaria decumbens*. The meaning of *Azospirillum* was from 'azite' a French word meaning nitrogen and 'spira' a Greek word meaning spiral. *Azospirillum* means a small nitrogen spiral. The name *lipoferum* means fat bearing; *brasilense* from brasiliensis pertaining to the country of Brazil, South America. They also explored the high potentiality of *Azospirillum* as microbial inoculant with tropical grasses and other crop plants. It has been enlarged to encompass other possible bacterial associations by adopting the terminology "Diazotropic biocoenosis".

Azospirillum is a gram negative, motile, curved rod of variable size, ranging from 0.5 – 1μm in length, exhibits spirillar movement and polymorphism, containing poly-β- hydroxy butyrate (PHB) granules and fat droplets [4,5] They contain peritrichous flagella and polar flagellum used for swarming [6]. It was associated with the root system of plants making use of the nutrients exuded by them. *Azospirillum* is colonized the root region of crop plants in large numbers and fixes substantial amount of nitrogen [7] and they exerted beneficial effects on plant growth and yield many crops of economic importance [8].It is used extensively in rice and other cereal crops as biofertilizers [8]. Tarrand et al. [9] and Magalhaes et al. [10] proposed as *Azospirillum*, the genus distinguished into two species based on physiological and morphological differences between various strains on DNA homology experiments like *Azospirillum brasilense* and *A. lipoferum* [9,10] Later, four additional *Azospirillum* species were described, *A. amazonense*. Isolated from many grasses in the Amazonian area of Brazil [11], the salt tolerant species *A. halopraeferans*, associated exclusively with roots of kallar grass [12], *A. irakense* [13] and *A. dodereinerae* [14].

Azospirillum is grown in N-free medium, it behaves as microaerophilic, fixes nitrogen and when supplemented with nitrogen it grows as an aerobe [4]. The population levels of *Azospirillum* were reported from 10^4 to 10^6 cells per gram of dry soil or root by Magalhaes et al. [10] and 7 x 10^4 per gram of fresh roots in *Stenocerus pruinosus* and 1.1 x 104 in *Opuntia ficus - indica* by Mascarua-Esparza et al [15]. Maximum *Azospirillum* numbers were detected in a laterite soil and the minimum in an extremely acid sulphate saline kari soil [16]. The present study aims to isolate, identify, screen and study the biology of *Azospirillum* spp from various soils of Srivilliputtur Taluk, Virudhunagar District, Tamil Nadu, India.

Methodology

Study area

Tamil Nadu is situated in Southern end of India, towards east of Kerala north of Andhra Pradesh and Karnataka. Several folds or parts of Western Ghats separate the states of Tamil Nadu and Kerala. The area of investigation, Srivilliputtur is located in the Southern side of Virudhunagar District; this area is a boundary of Theni and Madurai Districts in North, Tirunelveli District in South and Kerala in Southwest.

Collection of soil and root samples

Soil and Root samples were collected from different crop soils in various places of Srivilliputtur Taluk, Virudhunagar District. The soil

***Corresponding author:** N. Tenzing Balaiah, Department of Plant Biology and Plant Biotechnology, ANJAC Sivakasi, Tamilnadu, India, E-mail: tens007@rediffmail.com

samples (Rhizosphere soil) were air dried under shade and used for isolation and enumeration of *Azospirillum*.

Isolation and enumeration *Azospirillum*

For the isolation of *Azospirillum*, the rhizosphere soil samples were serially diluted from 104 to 106 using sterile water. One mille liter of the soil diluents from each dilution was transferred to the tubes containing 10 mille liter of nitrogen free malic acid semi solid medium and kept it in incubation for three days at 37±2°C.

Enumeration of *Azospirillum* in soil samples were carried out by Most Probable number method (MPN). One mille liter successive dilutions of 10^4, 10^5 and 10^6 soil samples were transferred to test tubes containing nitrogen free malic acid semi solid medium. Then the tubes were incubated at room temperature for 3 days. The positive tubes were counted and the population was calculated and expressed as number of *Azospirillum* per gram dry weight of soil samples.

$$\textit{Azospirillum population} = \frac{\text{MPN value x middle dilution x middle dilution used}}{\text{Dry weight of the soil sample}}$$

Identification and screening of *Azospirillum*

The selected bacterial strains were identified using standard biochemical tests as listed in the Bergey's Manual of Determinative Bacteriology. The following tests/methods were used for isolation and identification of *Azospirillum* from the soils.

Cell morphology: To study the cell morphology, log phase cultures of *Azospirillum* were properly diluted with sterile distilled water. Smear was prepared on clean glass slides by using a loopful of culture dilution. The smears were air-dried, heat fixed and stained with crystal violet. The cells were observed under light microscope with oil immersion objective to see the size and shape.

Motility: Motility was tested by hanging drop method. Slides were prepared with cultures and motility was observed under oil immersion.

Biotin requirement: The biotin requirements of the bacterial isolates are tested using semisolid nitrogen free malic acid medium prepared in two sets of tubes, one set of medium prepared with the addition of biotin (100 µg l^{-1}) and other without biotin. The growth was observed by the change in colour from yellowish green to blue.

Catalase test: Selected strains were inoculated on LB agar plates and incubated at 28°C for 24 h. About 3 to 4 drops of 3 percent H_2O_2 solution was allowed to flow over the culture. Formation of bubbles or effervescence indicated a positive result.

Biology of *Azospirillum*

Utilization of carbon, nitrogen, amino acid and vitamin sources

The utilization of different carbon, nitrogen, amino acid and vitamin sources of different *Azospirillum* isolates were estimated in LB broth. Filter sterilized carbon, nitrogen, amino acid and vitamin sources were inoculated aseptically into the sterile medium at 1 percent level. The *Azospirillum* cultures were inoculated at the rate of 1.0 mille liter and incubated at room temperature. The growth was observed by the turbidity of the broth read at 560 nm.

Results and Discussion

Totally fifteen *Azospirillum* strains were isolated from the soil samples collected from different crop plants at various places of Srivilliputtur Taluk, Virudhunagar district. The isolated strains were brought to pure culture by several subcultures. The purified *Azospirillum* strains were maintained in nutrient agar slants and stored at 4°C for future use (Table 1). The result showed that the population level of *Azospirillum* was higher in cotton followed by tomato. Among different crop plants, *Azospirillum* population was least in soil samples collected from bhendi (Table 1). According to Haahtela et al. [17] and Eckert et al. [14] Azospirilla were isolated from a wide variety of plants including many grasses and cereals from all over the world, in tropical, temperate and cold climates. In addition to isolation of *Azospirillum*, enumerates the population level in soil samples were also collected from different crop plants. *Azospirillum* population was higher in cotton and least in soil samples collected from bhendi. The association of these organisms with the roots of non-Graminae family plants was also reported by [18]. Seasonal variation in MPN counts, which showed a similar pattern of decrease or increase with the variable climatic conditions, confirm that all types of microorganisms were influenced by the temperature fluctuations in a similar fashion in tea soils Govindan and Purushothaman [19]. Plantation crops like areca

S.No	Crop Plants	Code Number	Number of Strains Isolated	Population level (x 10^5/g soil dry wt.)
1	Tomato	T1 T2 T3	3	5.7
2	Bendi	BD1 BD2 BD3	3	0.6
3	Bringal	B1 B2 B3	3	2.0
4	Chillies	C1 C2 C3	3	3.6
5	Cotton	CT1 CT2 CT3	3	16.6

Table 1: Isolation of Total Number of *Azospirillum* Strains and their Population level in different Crop Plants.

S. No	COLOUR CHANGE				pH Change	Gram staining
	Strains	1st day	2nd day	3rd day	pH	Gram stain
1.	T1	Light blue	Blue	Dark blue	9.6	-ve
2.	T2	Green	Light blue	Dark blue	9.3	-ve
3.	T3	Light blue	Blue	Dark blue	9.8	-ve
4.	BD1	Light blue	Light blue	Dark blue	9.6	-ve
5.	BD2	Light blue	Light blue	Dark blue	9.5	-ve
6.	BD3	Light blue	Light blue	Dark blue	9.4	-ve
7.	B1	Light blue	Blue	Dark blue	9.4	-ve
8.	B2	Green	Light blue	Dark blue	9.6	-ve
9.	B3	Light blue	Blue	Dark blue	9.7	-ve
10.	C1	Light blue	Blue	Dark blue	9.8	-ve
11.	C2	Light blue	Light blue	Dark blue	9.7	-ve
12.	C3	Light blue	Light blue	Dark blue	9.7	-ve
13.	CT1	Light blue	Blue	Dark blue	9.4	-ve
14.	CT2	Green	Light blue	Dark blue	9.6	-ve
15.	CT3	Light blue	Blue	Dark blue	9.8	-ve
16.	Control	Green	Green	Green	7.0	--

Table 2: Results of colour changes, pH and Gram staining in the medium used for the growth of various *Azospirillum* strains.

nut, cashew, cocoa, rubber, cardamom and sapota grown in acid soils colonized by *Azospirillum* in their root system [20].

Selection of Efficient Strains

The results indicated that the *Azospirillum* strains were differed in their ability to change colour intensity of the medium. The difference between the regions in nitrogen fixing ability was related to geographic variables, including soil type [21]. In this study, the isolated strains were screened under *in vitro* condition. The isolated *Azospirillum* strains were screened by noting the changes of colour and pH in the *Azospirillum* growing medium. After selection the *Azospirillum* strains were screened by noting the colour and pH change of the medium. These two observations were used for the screening of *Azospirillum* in the laboratory conditions. In the nitrogen free malate broth, *Azospirillum* strains were able to change colour. The initial green colour changed into blue colour. The ability of colour change was differed between the strains. The result revealed that all the strains were able to change the colour and pH of the medium on 3rd day. But some strains started to change the colour even in the first day after inoculation. Among fifteen strains, the T1, T3, B1, B3, C1, CT1 and CT3 strains were superior in colour change than others (Table 2), however, among fifteen strains, the ability to pH changes is differed between the strains. In strains T3, C1 and CT3 were increased in pH than other strains (Table 2).

Identification of *azospirillum*

Based on the present study, the *in vitro* experiment shows that the totals of 15 *Azospirillum* strains were screened for their efficiency using various biochemical tests. In Gram stain all the fifteen strains were Gram negative because all the strains were stained with safranin rather than crystal violet. Based on the staining, fifteen strains were revealed Gram negative (Table 2), but the size of the selected strains were ranged from 1-2.0µm in diameter and 2.0-3.5µm in length. Microscopic examination of the isolates revealed that they were vibroid in shape and in the motility levels of all the fifteen strains were showed spiral movement. In catalase and biotin content of *Azospirillum*, all the strains except T1, T2 were negative to catalase test. (Table 3). The selected fifteen strains were cultured in nitrogen free semisolid medium with or without biotin. Biotin test is one of the important test to differentiate the *Azospirillum* species mainly *A. lipoferum* and *A. brasilense* (Table 3). According to Okon and Itzigsohn R [22], *Azospirilla* was gram negative,

S.No	Strains	Identification	Catalase	Biotin
1	T1	*A.brasilense*	+	-
2	T2	*A.brasilense*	+	-
3	T3	*A.lipoferum*	-	+
4	BD1	*A.lipoferum*	-	+
5	BD2	*A.lipoferum*	-	+
6	BD3	*A.lipoferum*	-	+
7	C1	*A.lipoferum*	-	+
8	C2	*A.lipoferum*	-	+
9	C3	*A.lipoferum*	-	+
10	CT1	*A.lipoferum*	-	+
11	CT2	*A.lipoferum*	-	+
12	CT3	*A.lipoferum*	-	+

Table 3: Identification, catalase and biotin contents of various Azospirillum strains in different crop plants.

S. No	Strains	Carbon Source				
		Glucose	Lactose	Sucrose	Fructose	Maltose
1.	T1	0.357 +	0.357 +	0.349 +	1.191 +++	0.831 ++
2.	T2	0.562 ++	0.765 ++	0.760 ++	1.463 +++	0.760 ++
3.	T3	0.636 ++	0.670 ++	0.914 ++	1.437 +++	0.815 ++
4.	BD1	1.094 +++	1.003 +++	0.931 ++	0.932 ++	1.063 +++
5.	BD2	1.169 +++	0.799 ++	1.066 +++	1.413 +++	0.972 ++
6.	BD3	1.146 +++	0.683 ++	0.947 ++	1.451 +++	0.914 ++
7.	B1	0.845 ++	0.604 ++	1.025 +++	1.467 +++	0.713 ++
8.	B2	0.930 ++	0.683 ++	0.811 ++	1.543 +++	0.741 ++
9.	B3	1.050 +++	0.512 ++	1.106 +++	1.512 +++	0.834 ++
10.	C1	0.891 ++	0.865 ++	1.141 +++	1.455 +++	1.136 +++
11.	C2	0.830 ++	0.637 ++	1.473 +++	1.551 +++	1.212 +++
12.	C3	1.126 +++	0.820 ++	0.712 ++	1.543 +++	0.971 ++
13.	CT1	0.956 ++	0.998 ++	1.029 +++	1.521 +++	0.973 ++
14.	CT2	0.760 ++	0.851 ++	1.360 +++	1.462 +++	1.070 +++
15.	CT3	0.890 ++	0.712 ++	0.814 ++	1.568 +++	1.114 +++

+ Fair ; ++ Good; +++ Excellent

Table 4: Carbon utilization by *Azospirillum* strains.

S. No	Strains	Vitamin source			
		(Vitamin B) Nicotinic acid	(Vitamin B1) Thiamine	(Vitamin B6) Pyridoxine	(VitaminB12) (Myoinositol
1.	T1	0.713 ++	1.298 +++	0.678 ++	1.234 +++
2.	T2	0.443 +	0.978 ++	0.896 ++	1.340 +++
3.	T3	0.772 ++	1.236 +++	0.834 ++	0.978 ++
4.	BD1	0.725 ++	0.631 ++	0.428 +	1.456 +++
5.	BD2	1.273 +++	0.748 ++	0.853 ++	1.389 +++
6.	BD3	0.813 ++	0.734 ++	0.896 ++	1.467 +++
7.	B1	0.461 +	0.967 ++	0.438 +	1.389 +++
8.	B2	0.678 ++	0.834 ++	0.942 ++	1.271 +++
9.	B3	0.616 ++	0.616 ++	0.836 ++	0.945 ++
10.	C1	0.483 +	0.976 ++	1.134 +++	1.432 +++
11.	C2	0.893 ++	1.343 +++	1.224 +++	1.446 +++
12.	C3	0.842 ++	0.941 ++	1.036 +++	1.258 +++
13.	CT1	1.282 +++	1.134 +++	1.454 +++	1.468 +++
14.	CT2	0.841 ++	0.964 +++	0.916 ++	1.396 +++
15.	CT3	1.125 +++	0.978 ++	0.822 ++	1.464 +++

+ Fair ; ++ Good; +++ Excellent

Table 5: Utilization of Vitamins by *Azospirillum* strains.

the cured rods of variable size exhibits spirillar movement and contain PHB as reserve food material.

Utilization of carbon sources

The *Azospirillum* strains utilized different types of chemicals as carbon sources. The utilization of these different types of carbon sources varied from strain to strain. There was a marked difference between the *Azospirillum* species in the pattern of carbohydrate utilization. The preferential carbon sources varied from strain to strain. Most of the strains preferred fructose and maltose as carbon source but sucrose was moderately utilized while lactose was found to be a poor source (Table 4). The disaccharides except lactose, starch and mannitol supported little growth of isolates belonging to both species, while trisaccharides, polysaccharides did not support their growth. This suggests that sugars are supposed to be very poor substrates of carbon and energy for A. brasilense but better source for *A. lipoferum* [23].

Utilization of vitamin source

All *Azospirillum* strains were utilized for various vitamin sources. The preferential vitamin sources varied from strain to strain. Most of the strains preferred myoinosital, as vitamin source but nicotinic acid was also moderately utilized while, thiamine, pyridoxine was found to be a poor source (Table 5).

Conclusion

Our results demonstrate remarkably different strains of *Azospirillum* from various crop soils in Srivilliputtur Taluk.

The isolated strains were contains different features in their morphology, carbon utilization and their vitamin contents. This type of results which would be very useful in the field of agriculture to develop organic/bio fertilizer.

Acknowledgement

Authors are thankful to the University Grants Commission for funding the financial support by minor research project to Dr. N.T. Balaiah.

References

1. Baby UI (2002) Biofertilizers in tea. Planters' Chronicle 98: 395.
2. Beijerinck MW (1925) Parasitenkd. Infektionskr HygAbt 2: 353-359.
3. Schroeder M (1932) Dis Assimilation des Luffstickstoffs durcheininge Bakterien Zentrab Bakteriol. Prasaitenkd, Infesktoinskr Hyg Abt 2: 178-212.
4. Dobereiner J, Day JM (1976) Associative symbiosis in tropical grasses: characterization of microorganisms and nitrogen-fixing sites. Proceedings *of the First International Symposium on Nitrogen Fixation.* (eds.) WE Newton and CJN Pullman. Washington State University Press 518-536.
5. Okon Y (1985) *Azospirillum* as a potential inoculant for agriculture. TIBTECH 3: 223-228.
6. Okon Y, Albrecht S.L, Burris R.H (1976) Factors affecting growth and nitrogen fixation of *Spirillum lipoferum.* J Bacteriol 127: 1248-1254.
7. Okon Y, Vanderleyden J (1997) Root associated *Azospirillum* species can stimulate Plants. ASM News 63: 366-370.
8. Jayaraj J, Muthukrishnan S, Liang G.H (2004) Transfer of a plant chitinase gene into a nitrogen fixing *Azospirillum* and study of its expression. Can J Microbiol 50: 509-513.
9. Tarrand J, Kreig NR, Dobereiner J (1978) A taxonomic study of the *Spirillum lipoferum* group, with descriptions of a new genus *Azospirillum* gen nov and two species, *Azospirillum lipoferum* Beijerin (h) comb and *Azospirillum* sp. Can J Microbiol 24: 967-980.
10. Magalhaes FM, Baldani J, Santo J, Kuykendall JR, Dobereiner J (1983) A new acid tolerant *Azospirillum* species. Ann Acad Brar Clien 55: 417-430.
11. Falk EC, Dobereinur J, Jojnson JL, Kreig R (1985) Deonyribo nucleic acid homology of *Azospirillum amazomense.* Int J Syst Bacteriol 35: 117-121.
12. Reinhold B, Hurek T, Fendrik I, Pot B, Gillis M, et al. (1987) *Azospirillum halopraeferens* sp. nov., a nitrogen¬ fixing organism associated with roots of kallar grass (Leptochola fusca) (L.) kunth.) Int J Syst Bacteriol 37: 43-51.
13. Khammas KM, Ageron E, Grimont PAD, Kaiser P (1989) *Azospirillum* irakense p. Nov., a nitrogen fixing bacterium associated with rice roots and rhizosphere soils. Res Microbiol 140: 679-693.
14. Eckert B, Weber OB, Kirchhof G, Halbritter A, Stoffels M, et al. (2001) *Azospirillum* doebereinerae sp. nov., a nitrogen fixing bacterium associated with the C4-grass Miscanthus. Int J Syst Evol Microbiol 51: 17-26.
15. Mascarua-Esparza MA, Villa-Gonzalex R, Carbellro-Melled J (1987) Acetylene reduction and indoleacetic acid production by *Azospirillum* isolates from coctaccous plants. Plant Soil 91-95.
16. Charyulu PBBN, Rajaramamohan Rao VR (1980) Influence of various soil factors on nitrogen fixation by *Azospirillum* spp. Soil Boil Biochem 12: 343-346.
17. Haahtela K, Wartiovaara T, Sundman V, Skujins J (1981) Root associated N2 fixation (acetylene reduction) by Enterobacteriaceae and *Azospirillum* strains in cold Climate spodosolo. Appl Environ Microbiol 41: 203-206.
18. Kumari ML, Kavimandan SK, Subba Rao NS (1976) Occurrence of nitrogen fixing Spirillum in roots of rice, sorghum, maize and other plants. Indian J Exptl Biol 14: 638-639.
19. Bezbaruah B (1999) Microbial dynamics in tea (*Camellia Sinensis* (L) O Kuntze) plantation soils. Global Advances in Tea Science 563-574.
20. Govindan M, Purushothaman D (1985) Association of nitrogen fixing bacteria with certain plantation crops. Natl Acad Sci Lett 8: 1631-1665.
21. Han SO, New PB (1998) Variation in nitrogen fixing ability among natural isolates of *Azospirillum.* Microbiol Ecol 36: 193-201.
22. Okon Y, Itzigsohn R (1992) Poly-β-hydroxy butyrate metabolism in *Azospirillum brasilense* and ecological role of PHB in the rhizosphere biodegradable plastic production. FEMS Microbiol Rev 103: 131-139.
23. Konde BK (1984) Utilization of carbon and nitrogenous compounds by dinitrogen fixing *Azospirillum* strains. Indian J Microbiol 24: 44-47.

10

Effect of Vermicompost Prepared from Aquatic Weeds on Growth and Yield of Eggplant (*Solanum melongena* L.)

A. Gandhi* and U. Sivagama Sundari

Department of Botany, Annamalai University, Annamalai Nagar-608 002, Tamil Nadu, India

Abstract

The aim of the present investigation was to study the effect of vermicompost prepared from two different aquatic weeds on eggplant (*Solanum melongena* L.) growth and yield under greenhouse conditions. The experiment was conducted at the botanical garden of Annamalai University during December, 2011 to June, 2012. Vermicompost was prepared from cow dung and aquatic weeds i.e., *Azolla* and *Eichhornia* by using earthworms (*Eudrilus eugeniae*). The pot experiment was conducted with four treatments via T_1–(Control), T_2 (Cow dung), T_3 (*Azolla*), and T_4 (*Eichhornia*). The experimental results showed significant variations in plant growth and yield on par with the physico-chemical properties of different vermicomposts. The growth characters of brinjal such as plant height, number of leaves per plant were observed at 20th day, 40th day and 80th day from the date of planting. There was maximum value of growth parameters observed in egg plant treated with *Azolla*-vermicompost followed by *Eichhornia*-vermicompost and cow dung-vermicompost. The yield parameters such as number of days for flowering, number of fruits per plant and fruit length and width also showed similar trend of growth parameters. The investigation clearly reveals that the biochemical properties of vermicompost play a major role in the growth and development of egg plant.

Keywords: Vermicompost; Earthworms; *Azolla*; *Eichhornia*; Cow dung; Eggplant; Aquatic weeds

Introduction

Increasing population of the world has doubled the food demands and inundated the available land sources [1]. The need of increased food production in most developing countries becomes an ultimate goal, to meet the dramatic expansion of their population [2]. Among the major food crops, vegetables are the most important one by cultivation and consumption. The nutritional content of vegetables varies considerably as they contain a great variety of other phytochemicals and other antioxidant properties. Generally, vegetables are cultivated in all part of the world by using different inputs like chemical fertilizers and pesticides, organic fertilizers, biofertilizers and biopesticides, etc. In recent days, the use of different organic fertilizers, biofertilizers and biopesticides are being recommended not only to minimize the use of hazardous chemical inputs but also for sustainable crop production particularly in vegetables' cultivation.

Among the immature vegetables, brinjal not only occupies a major area in cultivation but also by consumption in Tamil Nadu. Egg plant, *Solanum melongena* L. also known as Aubergine in Europe, Brinjal in India, is one of the non-tuberous species of the night shade family *Solanaceae* [3]. The varieties of *Solanum melongena* L. show a wide range of fruit shapes and colours, ranging from white, yellow, green through degrees of purple pigmentation to almost black [4]. It is an economically important crop in Asia, Africa and the Subtropics (India and Central America) and it is also cultivated in some warm temperate regions of the Mediterranean and South America [5]. The fruits are known for being low in calories and having a mineral composition beneficial for human health. They are also rich source in Potassium, Magnesium, Calcium and Iron [6]. Unripe fruit of egg plant is primarily used as cooking vegetable [7]. Egg plant is perennial but grown commercially as an annual crop. Asia has the largest egg plant production which comprises more than 90% of the world production and 299,770 ha in area of cultivation. It has many medicinal values and its fruit helps to lower the blood cholesterol levels, and is suitable as a part of a diet to help regulate high blood pressure [8].

Fertilizers provide plants with the nutrients necessary for healthy growth. Apart from the macronutrients, there is a known suite of micronutrients that play important roles in the plants metabolism. Fertilizers can be applied as either organic or inorganic. Inorganic fertilizers, compost or manure prepared from vegetative matter or animal excreta has been utilized due to its high value of physical and chemical properties. But in modern agriculture, the chemical fertilizers and pesticides are being applied indiscriminately with desire of getting higher yield which deteriorate the soil fertility as well as crop quality. But in recent years, the chemical fertilizers have produced undesirable effects on the soil [9]. The foliar application of humic acid on vegetables particularly in brinjal increases growth and yield parameters when compared to chemical nitrogen fertilizers [10]. Using of organic fertilizers serves as a good and suitable source to supply soil food elements. Among the organic manure, vermicompost is one of the best organic manure in increasing the crop yield. It contains growth regulators like growth hormones which increase the growth and yield of crops [11]. Compost plays an important role for improving soil physical properties and contains higher levels of relatively available nutrient elements, which are essential for plant growth [12].

Vermicomposting involves the bio-oxidation and stabilization of organic material by the joint action of earthworms and microorganisms. Although it is the microorganism, that biochemically degrade the organic matter, earthworms are the crucial drivers of the process, as they aerate and fragment the substrate there by drastically altering the

***Corresponding author:** A. Gandhi, Department of Botany, Annamalai University, Annamalai Nagar-608 002, Tamil Nadu, India, E-mail: drgbot@gmail.com.

microbial activity and increasing the surface area thus making much more microbial activity and further decomposition [13]. Vermicompost is being a stable fine granular organic matter, when added to soil, it loosens the soil and improves the passage to the entry of air. The mucus associated with the cast being hydroscopic absorbs water and prevents water logging and improves water holding capacity. The organic carbon in vermicompost releases the nutrients slowly and steadily into the system and enables the plant to absorb nutrients. The soil enriched with vermicompost provides additional substances that are not found in chemical fertilizers [14]. Aquatic weed plants grow very luxuriously in lotic and lentic type of water bodies, they have a devastating effect on water quality. They can bring rivers and lakes to a standstill and destroy the livelihoods of communities that depend on them. Nowadays, the aquatic weeds are obnoxious to eradicate from natural environment which create pollution. So the present study was carried out to examine the effect of vermicompost prepared from different aquatic weeds such as *Azolla spp. and Eichorrnia spp.* on growth and yield of egg plant.

Materials and Methods

Epigeic species, *Eudrilus eugeniae* was obtained from M/s Vishal vermifarm, Nellore, Andhra Pradesh, India and maintained in a rearing box by feeding cow dung for further studies. Common weeds such *Azolla* sp. and *Eichhornia* sp. were collected from local ponds located around Annamalai University, Annamalai Nagar, Tamil Nadu, India. The fresh aquatic weed biomass were washed with tap water and chopped into small pieces. The chopped weed biomass were made as a heap individually under shady conditions and decomposing bacterial culture was inoculated (Bacillus sp MTTC No.: 297), and moisture was maintained up to 60% by spraying water regularly. The heaps were turned up 7 days gap to accelerate decomposition and after 30 days, the pre-composted aquatic weed biomass were collected and fed to the earthworms during vermicomposting.

Preparation of vermicompost

Vermicompost of different aquatic weeds were prepared on clay pots, sized 12 inch height and 9 inch width. The clay pots were filled with sandy soil followed by dried coconut epicarp upto 1/4th of pot height for providing shelter to earthworms. The pre-composted aquatic weed biomass of *Azolla* and *Eichhornia* were mixed with 30 days old cow dung at 4:1 ratio and filled in the pots up to top individually with uniform biomass weight. Simultaneously, only cow dung also filled in pots as control. Moisture was adjusted to 60% and 50 numbers of adult earthworms (*Eudrilus eugeniae*) from rearing box were transferred to each vermipots and covered with jute gunny sheets, and kept under complete shade. Moisture of the earthworm feed mixture was maintained between 50-60% by spraying water regularly. The formation of vermicasting was observed after one week from the date of introducing earthworms. The number of days for 100% conversion of filled feed material into vermicastings was recorded. The vermicastings were harvested, and stored for further studies. The harvested vermicomposts were analyzed for physical and chemical properties such as pH, electric conductivity, organic carbon, nitrogen, phosphorous, potassium, calcium, magnesium, sodium, chloride, sulphate and carbon and nitrogen ratio at Department of Soil Science, Tamil Nadu Agriculture University, Coimbatore, India.

A pot experiment was conducted at Botanical Garden, Annamalai University, Annamalai nagar Tamil Nadu during December, 2011 to June, 2012. Experiment was laid out in randomized design with three replications. Altogether there were 12 pots, three replicates in each for control, cow dung vermicompost (CV), *Azolla* vermicompost (AV), Eichorrnia vermicompost (EV). Egg plant (*Solanum melongena* L.) was grown as test crop. 20 days old eggplant seedlings of local variety planted in pots and applied different vermicompost as uniform dosage by soil application. Treatments consisted of T_1-control with 100% recommended dose of inorganic NPK; T_2-cowdung vermicompost supplemented with 50% NPK; T_3-*Azolla* vermicompost supplemented with 50% NPK; T_4–Eichorrnia vermicompost supplemented with 50% NPK. Inorganic NPK was applied through urea, single super phosphate (SSP) and muriate of potash (MOP). Inorganic NPK and vermicompost were applied to egg plant by soil stench method at the time of planting and 40th day, and 80th day from the date of planting.

Analysis of physico-chemical and biological properties

Soil samples were collected from each pot from 0-15 cm depth in two different periods: Initial pot soil mixture before planting of eggplant seedlings and pot soil after final harvesting of fruits were analyzed for soil pH, electrical conductivity, organic carbon, available nitrogen, phosphorus and potassium at soil testing laboratory, Department of agriculture, Cuddalore, Tamil Nadu, India.

Growth and yield parameters of egg plant

The plant height, number of leaves per plant was recorded at 20th day, 40th day and 80th day from the date of planting. The number of days for flowering, number of fruits per plant, fruit length and width were recorded. The results were statistically analyzed.

Results and Discussions

Table 1 shows the physical and chemical properties of different vermicompost. There were significant differences in each physical and chemical properties of the prepared aquatic weeds vermicompost. The chemical analysis of experimental soil has presented in table 2. *Azolla* vermicompost shows high value in both pH (6.9) and electrical

S.NO.	PARAMETERS	CV	AV	EV
1.	pH	6.6	6.9	6.8
2.	Electric conductivity	1.68	2.85	2.24
3.	Organic carbon (%)	12.40	18.50	16.40
4.	Nitrogen (%)	0.62	1.12	0.96
5.	Phosphorous (%)	0.50	0.65	0.32
6.	Potassium (%)	0.54	0.62	0.74
7.	Calcium (ppm)	295	385	410
8.	Magnesium (ppm)	113	102	202
9.	Sodium (ppm	45	85	73
10.	Chloride(ppm)	32	66	48
11.	Sulphate (ppm)	10	12	15
12.	C/N ratio	20:23	26:32	27:26

Data represents mean value of three determinations
CV=Cow dung vermicompost; AV=*Azolla* vermicompost; EV=*Eichornnia* vermicompost

Table 1: Physical and chemical properties of different vermicomposts.

S.No	Parameters	Before planting	After harvest			
			T_1-control	T_2-CV	T_3-AV	T_4-EV
1.	pH	7.1	6.8	7.2	6.9	7.1
2.	EC(mmhos/cm/25°C)	1.34	1.21	1.31	1.15	1.13
3.	Organic carbon (%)	0.65	0.52	0.78	0.72	0.68
4.	Available nitrogen(mg/100 gm soil)	124.23	87.35	112.43	120.12	118.23
5.	Available phosphorus(mg/100 gm soil)	8.23	7.65	7.85	7.21	7.24
6.	Available potassium(mg/100 gm soil)	0.78	1.87	1.56	1.34	1.25

Data represents mean value of three determinations.

Table 2: Chemical analysis of the experimental soil.

S. No	Treatments	Plant height (cm)			Number of leaves per plant		
		20th day	40th day	80th day	20th day	40th day	80th day
1.	T_1-Control	6.42 ± 0.031	13.45 ± 0.040	24.26 ± 0.097	5.31 ± 0.021	12.35 ± 0.061	20.34 ± 0.081
2.	T_2-CV	8.01 ± 0.032	17.9 6± 0.071	29.12 ± 0.087	5.72 ± 0.017	15.45 ± 0.046	24.55 ± 0.073
3.	T_3-AV	9.12 ± 0.027	20.24 ± 0.101	32.56 ± 0.162	6.05 ± 0.030	18.12 ± 0.054	28.32 ± 0.141
4.	T_4-EV	8.75 ± 0.043	19.22 ± 0.057	30.34 ± 0.151	5.82 ± 0.023	16.78 ± 0.083	26.05 ± 0.078

Values are mean ± SD; sample size (n)=6

Table 3: The effect of different vermicomposting on plant height and number of leaves per plant at 20th day, 40th day and 80th from the date of planting.

S.No	Control	Number of days for flowering	Number of fruits per plant	Fruit length (cm)	Fruit width (cm)
1.	T_1-Control	72.65 ± 0.363	8.34 ± 0.033	6.21 ± 0.031	3.85 ± 0.011
2.	T_2-CV	65.86 ± 0.263	10.35 ± 0.031	7.85 ± 0.030	4.17 ± 0.016
3.	T_3-AV	63.25 ± 0.189	14.12 ± 0.070	9.05 ± 0.027	5.25 ± 0.026
4.	T4-EV	64.56 ± 0.322	12.85 ± 0.064	8.12 ± 0.032	4.75 ± 0.023

Values are mean ± SD; sample size (n)=6

Table 4: The effect of different vermicomposting on number of days for flowering, number of fruits per plant, fruit length and width.

conductivity (2.85). The maximum percentage of organic carbon was observed in AV (18.50%) followed by EV (16.40%) compared with control (12.40%). The major macronutrients of nitrogen and phosphorous were high in AV (1.12 % and 0.65%) followed by EV (0.96 % and 0.32%). But the potassium content of 0.74% was observed as high in EV than AV (0.62%).

The application of different vermicompost such as CV, AV and EV showed significant difference in vegetative parameters of brinjal, observed at 20th, 40th and 80th day from date of planting. Table 3 shows the effect of different vermicompost on plant height and number of leaves per plant as 20th day, 40th day and 80th day. The maximum value of plant height and number of leaves per plant were observed in plants treated with AV followed by EV and CV. The present reports is an agreement with the reports of Abdullah Adil Ansari and Kumar Sukhraj [15] who found that the availability of macronutrients and micronutrients in vermicompost and vermiwash enhanced plant growth and yield in Okra.

Table 4 exhibits the effect of different vermicompost on number of days for flowering, number of fruits per plant, fruit length and fruit width. There was significant difference in number of days of flowering. It was observed that the number of days for flowering reduced to 63.25 days in AV when compared to EV (64.56) and CV (65.86). But the maximum number of days (72.65) for flowering was recorded in untreated plants (control). Gorakh Nath et al. [16] reported that the application of vermicompost of different animal and agro waste along with neem oil/garlic/custard apple reduced the number of days for flowering in brinjal. The highest fruit yield per plant was observed in AV (14.12) followed by EV (12.85), CV (10.35) and Control (8.34). The maximum value of fruit length was recorded in AV treated plants (9.05cm) compared to plants treated with EV, CV and untreated control. The similar trend of fruit length was observed in fruit width of brinjal treated with different vermicomposts. Moranditochaee et al. [4] reported that the application of vermicomposting increases growth and yield parameters of eggplant in general. Several workers were reported that application of vermicomposting because of supplying optimum nourishment condition caused to improve growth, yield and yield components in crops [17-19]. Similarly, Nuruzzaman et al. [20] observed significant increase in growth and yield parameters of Okra when applied biofertilizers with cowdung.

Agriculture in modern times is getting more and more dependent upon the steady supply of artificial fertilizer with the introduction of green revolution technologies. [13]. Vermicompost is one of the best organic manure in increasing the crop yield; they aerate and fragment the substrate there by drastically altering the microbial activity. But the nutrient status of produced vermicompost differs on the type of biodegradable waste usage during vermicomposting. It results variations in plant response such as growth and yield parameters when it is applied. The application of organic fertilizers has an emphatic effect on plant growth and production [21]. The soil enrich with vermicompost provides additional substances that are not found in chemical fertilizers [14]. Nowadays, it is difficult to manage the aquatic weeds in lotic and lentic types of water bodies. So the present investigation proves that the conversion of aquatic weed biomass into vermicompost is an effective eco-friendly technology for not only managing the rapid growth of aquatic weeds but also can fertilize the crops for sustainable production, particularly vegetable crops.

References

1. Hussain J, Rehman NU, Khan AL, Hussain H, Al-harrasi A, et al. (2011) Determination of Macro and Micronutrients and Nutritional Prospects of Six Vegetables Species of Mardan. Pak J Bot 43: 2829-2833.

2. El-Shaikh KAA, Mohammed MS (2009) Enhancing Fresh and Seed Yield of Okra and reducing chemical phosphorus Fertilizer Via using VA Mycorrhizal inoculants. World Journal of Agricultural Sciences 5: 810-818.

3. Kantharajah AS, Golegaonkar PG (2004) Somatic embryogenesis in eggplant. Scientia Horticulturae 99: 107-117.

4. Maral moraditochaee, Hamid Reza Bozorgi, Nesa halajisani (2011) Effects of Vermicompost Application and Nitrogen Fertilizer Rates on Fruit Yield and Several Attributes on Eggplant (*Solanum melongena* L.) in Iran. World Applied Sciences Journal 15: 174-178.

5. Sihachkr D, Chaput MH, Serraf L, Ducreux (1993) Regeneration of plants from protoplasts of eggplant (*Solanum melongena* L.). In: Y.P.S. Bajai, (Edn), Biotechnology in Agriculture and Forestry, Plant Protoplast and Genetic Engineering. Springer, Berlin 108-122.

6. Zenia M, Halina B (2008) Content of macroelements in Egg plant fruits depending on Nitrogen fertilization and plant training method. J Elementol 13: 269-275.

7. Gargi Chakravarty, Kalita MC (2011) Comparative evaluation of organic formulation of P. flourescens based biopesticides and their application in the management of bacterial wilt of brinjal (*Solanum melongena* L.) African Journal of Biotechnology 10: 7174-7182.

8. Thomas TH (1998) Handbook of Vegetable Science and Technology: production, composition, storage and Processing. Plant Growth Regulation 26: 141-142.

9. Ntanos DA, Koutroubas SD (2002) Dry matter and N accumulation and translocation for Indica and Japonica rice under Mediterranean conditions. Field Crops Research 74: 93-101.

10. Azarpour E, Motamed MK, Moranditochaee M, Bozorgi HR (2012) Effects of bio, mineral nitrogen management, under humic acid foliar spraying on fruit yield and several traits of Egg plant (*Solanum melongena* L.). Afr J Agric Res 7: 1104-1109.

11. Canellas LP, Oliveres FL, Olorovola AL, Facanda AR (2002) Humic acid isolated from Earthworm compost enhance root elongation; aerial root emergence and plasma membrane H+ ATP activity Maize root. Plant Physiol 130: 1951-1957.

12. Abdel-Mouty MM, Mahmoud AR, EL-Desuki M, Rizk FA (2011) Yield and fruit quality of Egg plant as Affected by organic and Mineral Fertilizers Application. Research Journal of Agriculture and Biological Sciences. 7: 196-202.

13. Dominguez J, Edwards CA, Subler S (1997) A comparison of vermicomposting and composting. Biocycle, 38: 57-59.

14. Kale RD, Mallesh BC, Kubra B, Bagyaraj DJ (1992) Influence of vermicompost application on the available macronutrients and selected microbial populations in a paddy field. Soil Biol Biochem 24: 1317-1320.

15. Ansari AA, Sukhraj K (2010) Effect of vermiwash and vermicompost on soil parameters and productivity of Okra (*Abelmoschus esculentus*) in Guyana. Afr J Agric Res 5: 1794-1798.

16. Nath G, Singh DK, Singh K (2011) Productivity enhancement and nematode management through vermicompost and biopesticides in Brinjal (*Solanum melongena* L). World Applied Sciences Journal 12: 404-412.

17. Federico A, Miceli G, Santiago-Borraz J, Molina JAM, et al. (2007) Vermicompost as a soil supplement to improve growth, yield and fruit quality of tomato (Lycopersicum esculentum). Bioresour Technol 98: 2781-2786.

18. Vijaya D, Padmadevi SN, Vasandha S, Meerabhai RS, P Chellapandi (2008) Effect of vermicomposted coir pith on the growth of *Andrographis paniculatum*. J Organic Systems 3: 51-56.

19. Hernandez A, Castillo H, Ojeda D, Arras A, Lopez J, Sanchez E (2010) Effect of vermicompost and compost on lettuce production. Chil J Agr Res 70: 583-589.

20. Nuruzzaman M, Ashrafuzzaman M, Islam MZ, Islam MR (2003) Field efficiency of biofertilizers on the growth of okra (*Abelmoschus esculentus* [(L.) Moench]. J Plant Nutr Soil Sci 166: 764-770.

21. Lalitha R, Fathima K, Ismail SA (2000) The impact of biopesticeide and microbial fertilizers on productivity and growth of *Abelmoschus esculentus*. Vasundara: The Earth 1: 4-9.

Assessment of Calotropis Procera Aiton and Datura alba Nees Leaves Extracts as Bio-Insecticides Against Tribolium castaneum Herbst in Stored Wheat Triticum Aestivum L.

Attia Batool Abbasi[1]*, Azhar Abbas Khan[2], Rehana Bibi[2], Muhammad Shahid Iqbal[2], Javairia Sherani[2], and Arif Muhammad Khan[3]

[1]Department of Entomology, University of Sargodha, Sargodha, Pakistan
[2]College of Agriculture D.G. Khan, sub-campus University of Agriculture, Faisalabad, Pakistan
[3] National Institute for Biotechnology and Genetic Engineering, Faisalabad, Pakistan

Abstract

Plant materials have been used for pest control for centuries but recently, preservation of cereals products in storage has relied upon chemical insecticides to control stored grain pests but having drawbacks of toxicity to non-target organisms, human health hazards, development of pest resistance and environmental pollution. The present study was conducted to control the serious cereal's (wheat) stored grains pest *Tribolium castaneum* Herbst. *Calotropi sprocera* Aiton (Ak) and *Datura alba* (Dhatura) have been reported by many researchers as natural insecticides against stored grain pests. Five concentrations of leaf extracts of *C. sprocera* and *D. alba* (20%, 40%, 60%, 80% and 100%) were compared to see their efficiency to control *Tribolium castaneum*. Different parameters like repellency, growth inhibition, mortality rate, infestation/spoilage of grains, insect population and losses in grain weight were tested after application of treatments. Three months storage of wheat grains showed maximum repellency, mortality at higher concentrations of *C. sprocera* and *D. alba*, but there was less infestation/spoilage, loss in grain weight and insect population at higher concentrations (80%,100%).

Keywords: *Tribolium castaneum*; *Triticum Aestivum*; Stored wheat; *Bio-pesticide; Calotropis Procera*; *Datura alba*

Introduction

Losses due to insect infestation are the most serious threat in grain storage, particularly in developing countries, where poor sanitation and inappropriate storage facilities encourage insect attack [1,2]. It was estimated that more than 20,000 species of field and storage pests destroy approximately 1/3rd of the world's food production, valued more than US$ 100 billion annually, among which the highest losses (43% of potential production) occur in developing Asian and African countries [3]. In Pakistan it has been estimated that 5 to 7% loss of food grains occurs due to poor storage conditions [4]. *Tribolium castaneum* is found in stored grains of different cereals. Control of this insect relies mainly on the use of synthetic insecticides and fumigants that lead to problems such as disturbances of environment, increasing cost of application, pest resurgence, resistance to pesticides and lethal effect on non-target organisms in addition to direct toxicity to users [5]. Tissues of plants from several families contain chemical compositions that are considered as defensive substances against their enemies. They include oils, alkaloids, organic acids and other compounds [6]. The insecticidal and acaricidal properties of number of plants have been discovered long ago, and some of the plants can compete with synthetic means of control [7]. To minimize the use of pesticides and to avoid environmental pollution, natural antifeedant, deterrent and repellent substances have been searched for pest control during recent times [8,9] reported the successful results of crude extracts of *C. procera* and *D. alba* against the termites in the sugarcane crop. *Calotropis Procera* and *Datura alba* have been reported by many researchers as natural insecticide against stored grain pests [10,11]. There is a possibility to develop a method of stored product protection without or with reduced use of synthetic chemicals [12]. Present study was undertaken to check the efficacy of *Calotropis Procera* and *Datura alba* extracts against the *Tribolium castaneum* in stored wheat. The leaves of these medicinal plants, growing under regional environmental conditions were used to investigate their potential in controlling insect infestation. Alam et al., [13] reported significant insecticidal activity of root bark of Calotropis gigantea and its chloroform and petroleum ether solutions against larvae and adults of *Tribolium castaneum*.

Materials and Methods

The study was conducted in Entomological Laboratory of College of Agriculture, Dera Ghazi Khan, Pakistan during the year 2010. Adult *Tribolium castaneum* were collected from food storage Godowns of Punjab Food Department and from local grain markets of D. G. Khan District of province Punjab, Pakistan.

Rearing of Insects

Insects were reared in already damaged grains of local wheat variety (*Triticum Aestivum* L.) under controlled conditions of 27 ± 1°C temperature and 55 ± 5% RH in the growth chamber. Each three liter jar was filled with 800 gm of wheat flour and damaged wheat grains and about 150 beetles were added to each jar. The jars were covered with muslin cloth, tied with rubber bands to avoid the escape of beetles and to prevent any unwanted entry. The insects were placed in the chamber for two months for rearing. Adults of first generation were shifted to next jars for collection of eggs and these eggs were further transferred to damaged grains in new jars. Adult insects were shifted to the experimental and control jars for data recording.

***Corresponding author:** Department of Entomology, University of Sargodha, Sargodha, Pakistan, E-mail: azhar512@gmail.com

Preparation of Botanical Extracts

Mature plant leaves of *C. procera* and *D. alba* were collected from surrounding villages of D.G. Khan District, dried for 3-4 days under shade and then was chopped into 2 cm pieces with electric fodder cutter (local made). Chopped leaves were soaked in distilled water in ratio of 1:10 for 24 hours [14]. The extract was filtered with the help of fine sieve of 10 and 60 meshes gradually and boiled at 100°C to evaporate water to decrease volume upto 20 times. The grains were treated with concentrations (20%, 40%, 60%, 80% and 100%) of botanical extract solution. Insects were released in jars @ 16 insects/100 g of wheat grains [15]. Jars were covered with muslin cloth and rubber band.

Preparation of different concentrations

From a stock solution with 100% concentration, different concentrations (20%, 40%, 60%, 80% and 100%) of *D. alba* and *C. sprocera* were prepared. Total volume of water required for pasting the grains was 100 ml.

1. 20 ml botanical extract + 80 ml distilled water
2. 40 ml botanical extract + 60 ml distilled water
3. 60 ml botanical extract + 40 ml distilled water
4. 80 ml botanical extract + 20 ml distilled water

Each concentration was replicated thrice in experimental jars. For repellency studies, these concentrations were applied to a filter paper in the petri dishes. Data regarding bio-insecticides applied to grains was obtained after three months storage in experimental jars while for repellency studies, data was recorded with the interval of 9 hrs and 16 hrs consecutively for 5 days.

Infestation/spoilage (%)

Infestation of stored wheat grains by *T. castaneum* was measured by calculating the percentage grains weight loss. At the end of trials completion after three months, *C. procera* and *D. alba* treated jars with different concentrations 20%, 40%, 60%, 80% and 100% of leaf extracts containing wheat samples, were weighed and compared with control samples.

Loss in grain weight (gm)

Loss in grain weight caused by *T. castaneum* was measured by calculating the difference between grain weight before releasing the insects and after trial completion in *C. sprocera* and *D. alba* treated jars containing wheat samples, at different concentrations 20%, 40%, 60%, 80% and 100% of leaf extracts was done.

Mortality Percentage (%)

The mortality of *T. castaneum* was measured by using following formula.

$$\text{\% age mortality} = \frac{\text{No of insects alive intest} \times 100}{\text{No of insects alive in control}}$$

Mortality of *Tribolium castaneum* was measured after releasing the specific no. of insects in the treated jars of wheat grains of *C. procera* and *D. alba* at different concentrations 20%, 40%, 60%, 80% and 100% of leaf extracts, and also no. of insects in the control wheat grain jars.

Insect population count

Insect were counted by releasing a specific number of insects in treated wheat grains along with control. The effect of *C. procera* and *D. alba* was observed at their different concentration level of 20%, 40%, 60%, 80% and 100% of leaf extracts.

Repellency studies

Repellency studies were carried out by using paper strip method. Filter paper (Whatmann No. 1) was cut into strips of 8×10 cm. Half filter paper strips was treated with 1 ml of extract and was allowed to air dry for 10 minutes. Each treated half strip was then attached lengthwise, edge-to-edge, to a control half strip with adhesive tape and was placed in a petri dish. Twenty adult insects were released in the middle of each filter paper circle in a petri dish. Insects settled on each half of the filter paper strip were counted continuously for 5 hours with the interval of 1 hour. Repellency against red flour beetle on the basis of time interval of one hour at different concentrations (20%, 40%, 60%, 80% and 100%) of *C. procera* and *D. alba* were examined.

Result and Discussion

The Data presented in table 1 shows that number of alive adult *T. castaneum* in control treatments were maximum (40) and at 20% concentration, no. of alive individuals were (36) and (37) of *C. procera* and *D. alba* leaf extracts respectively and at 40% concentration, no. of alive insects were (30.33) and (34) of *C. procera* and *D. alba* leaf extracts respectively. At 60% concentration, no. of alive insects were (23.33) and (27.67) of *C. procera* and *D. alba* leaf extracts respectively and at 80% concentration, no. of alive insects were (20) and (19.33) of *C. procera* and *D. alba* leaf extracts respectively. However at 100% concentration of *C. sprocera* and *D. alba*, numbers of alive insects were (26.94) and (29.67) respectively. It is clear from the given data that the no. of insects alive in *C. procera* treated jars was less than *D. alba* as compared to control.

Loss in grains weight

The data presented in table 2 shows that maximum loss in grain weight was observed at control (51 and 47.33) caused by *T. castaneum*. Minimum grain weight loss was observed at 60% concentration of *C. procera* and *D. alba* (5) and (7.67) respectively and at 40% concentration, loss in grain weight caused by *T. castaneum* was (23.67)

Concentrations	*C. sprocera*	*D. alba*	*Total*
Control	40a	40a	40a
20 %	36b	37ab	36.5b
40 %	30.33cd	34bc	32.17c
60 %	23.33e	27.67d	23.5d
80 %	20ef	19.33f	29.67e
100%	12g	20ef	16f
Total	26.94b	29.67a	

Table 1: Means comparisons for the insects count of *Tribolium castaneum* at different concentration of *C. sprocera* and *D. alba* leaf extract

Concentrations	*C. sprocera*	*D. alba*	*Total*
Control	51a	47.33a	49.17a
20 %	30.67b	33.33b	32b
40 %	23.67bc	12cdef	21c
60 %	5f	7.67ef	17.83c
80 %	18.33bcd	23.67bc	16.67c
100 %	11.33def	22bcd	6.33d
Total	23.33a	24.33a	

Table 2: Means comparisons for Loss in grains weight at different concentration of *C. sprocera* and *D. alba* leaf extract caused by *Tribolium castaneum*

and (12) of *C. procera* and *D. alba* leaf extracts respectively. At 60% concentration, loss in grain weight caused by *T. castaneum* were (5) and (7.67) of *C. procera* and *D. alba* leaf extracts respectively and at 80% concentration, loss in grain weight caused by *T. castaneum* were (18.33) and (23.67) of *C. sprocera* and *D. alba* leaf extracts respectively. At 100% concentration, loss in grain weight was (11.33) and (22) in *C. procera* and *D. alba* treated jars respectively. Comparatively loss in grain weight caused by *T. castaneum* was less in *C. procera* treated jars than *D. alba* treated jars.

Infestation/ Spoilage of grains

The data presented in table 3 shows that maximum infestation/ spoilage of grains (20.49, 18.93) was observed in control treatments. Minimum infestation/spoilage (2, 2.8) was observed at 60% concentration of *C. procera* and *D. alba* respectively and at 40% concentration, infestation/spoilage of grains caused by *T. castaneum* were (9.5) and (3.5) of *C. procera* and *D. alba* extracts respectively. At 60% concentration, infestation/spoilage of grains caused by *T. castaneum* were (2) and (2.8) of *C. procera* and *D. alba* leaf extracts respectively and at 80% concentration, infestation/spoilage of grains caused by *T. castaneum* were (7.33) and (10.8) of *C. procera* and *D. alba* leaf extracts respectively. At 100% concentration infestation/spoilage was (4.53) and (8.8) in *C. procera* and *D. alba* treated jars respectively. Comparatively infestation/spoilage of grains was less in *C. procera* treated jars than *D. alba* treated jars.

Mortality Percentage

The data presented in table 4 shows that maximum mortality of *T. castaneum* (70, 50) was observed at 1000% concentration of *C. procera* and *D. alba* respectively. Minimum mortality of *T. castaneum* (0.0) was observed at control treatments.

Mortality of *T. castaneum* at 20% concentration was 10 and 7.5 of *C. procera* and *D. alba* leaf extract respectively, and at 40% concentration, mortality of *T. castaneum* were (24.17) and (15) of *C. procera* and *D. alba* leaf extracts respectively. At 60% concentration, mortality of *T. castaneum* were (41.67) and (30.83) of *C. procera* and *D. alba* leaf extracts respectively and at 80% concentration, mortality of *T. castaneum* were (50) and (51.67) of *C. procera* and *D. alba* leaf extracts respectively. At 100% concentration, mortality of *T. castaneum* was (70) and (50) in *C. procera* and *D. alba* treated jars respectively. Comparatively mortality of *T. castaneum* was less in *C. procera* treated jars than *D. alba* treated jars. At 100% concentration of *C. procera* treated jars *T. castaneum* adults were found alive underside of the muslin cloth but not in the treated grains, similar results were observed for *D. alba* treated jars.

Concentrations	*C. sprocera*	*D. alba*	*Total*
Control	20.4a	18.93a	19.67a
20 %	12.3b	13.33b	12.8b
40 %	9.5bc	3.5ef	9.1c
60 %	2f	2.8ef	6.67c
80 %	7.33cde	10.8bc	6.47c
100%	4.53def	8.8bcd	2.4d
Total	9.33a	9.69a	

Table 3: Means comparisons for Infestation/Spoilage of grains at different concentration of *C. sprocera* and *D. alba* leaf extracts caused by *Tribolium castaneum*

Concentrations	*C. sprocera*	*D. alba*	*Total*
Control	0	0	0
20 %	10f	7.5fg	8.75e
40 %	24.17de	15ef	19.58d
60 %	41.67c	30.83d	36.25c
80 %	50bc	51.67b	50.83d
100 %	70a	50bc	60a
Total	32.63a	25.83b	

Table 4: Means comparisons of Mortality Percentage of *Tribolium castaneum* at different concentration of *C. sprocera* and *D. alba* leaf extracts

Repellency Studies

The data presented in tables 5 and 6 show that maximum repellency of *T. castaneum* (16.97) was observed at 40% concentration of *C. sprocera* and *D. alba*. Minimum repellency of *T. castaneum* (6.97) was observed at 80% concentration while at 20% concentration of *C. procera* and *D. alba* leaf extracts, repellency of *T. castaneum* was (10.07) respectively. At 60% concentration, repellency of *T. castaneum* were (41.67) and (30.83) of *C. procera* and *D. alba* leaf extracts respectively and at 80% concentration, repellency of *T. castaneum* were (50) and (51.67) of *C. procera* and *D. alba* leaf extracts respectively. At 100% concentration, repellency of *T. castaneum* was (14.67). Comparatively repellency of *T. castaneum* was less in *C. procera* (9.92) than *D. alba* treated (13.87) treatments.

The data packed in table 7 shows that maximum repellency of *T. castaneum* (15.93) was observed after 5 hour interval. Minimum repellency of *T. castaneum* (9.37) was observed after 1st hour interval. Whereas after 2nd hour interval, repellency showed by the *T. castaneum* was (10.13), while repellency in 3rd hour interval was (11.73) and 4th hour interval (12.3) was observed. The results also supported to clarify that with the passage of time after application of extracts, repellency increased and maximum repellency of *T. castaneum* (15.93) was observed after 5th hour.

Concentrations	Means
20 %	10.07d
40 %	16.97a
60 %	10.8c
80 %	6.97e
100%	14.67b

Variance of Concentrations = 1.53

Table 5: Means comparisons for the Repellency of *Tribolium castaneum* at different concentration of *C. sprocera* and *D. alba* leaf extract Table of Means

Extracts	Means
D. alba	13.87a
C. sprocera	9.92b

Table 6: Repellency of *Tribolium castaneumat* different concentration of *C. sprocera* and *D. alba* leaf extract

Time Interval	Means
1 hour	9.37d
2 hour	10.13c
3 hour	11.73b
4 hour	12.3b
5 hour	15.93a

Variance of Time interval = 1.53

Table 7: Means comparisons for the Repellency of *Tribolium castaneum* at different concentration of *C. sprocera* and *D. alba* leaf extract with respect to time interval

References

1. Talukder FA, Islam MS, Hussain MS, Rahman MA, Alam MN (2004) Toxicity effects of botanicals and synthetic insecticides on *Tribolium castaneum* (Herbst) and Rhizopertha dominica (F). Bengladesh J Environ Sci 10: 365-371.

2. Talukhdar FA (2005) Insects and insecticide resistance problem in post harvest agriculture. Proceedings of International Conference, Post harvest Technology and Quality Management in Arid Tropics, Sultan Qaboos University.

3. Ahmed S, Grainge M (1986) Potential of neem tree (Azadirachta indica) for pest control and rural development. Eco Botany 40: 201-209.

4. Jilani G, Ahmad H (1982) Safe storage of wheat at farm level. Progressive Farming 2: 11-15.

5. OKenkwo EV, Okoye WI (1996) The efficacy of four seed powders and essential oils as protectants of cow pea and maize grains against infestation by Collosobruchus maculantus and Sitophillus zeamais in Nigeria. Intl J Pest Manag 42: 143-146.

6. Beck SD, Schoonhoven LM (1980) Insect behavior and plant resistance: In Maxwell, FG & PR Jennings (Edition) Breeding plants resistance to insects. Willey and Sons, Inc USA.

7. Hedin PA, Hollingworth RM (1997) New Application of phytochemical pest controls Agents, In: Medin PA, Hollingworth RM, Maseler EP (Edition). Phytochemicals for pest control, American Chemical Society, Washington.

8. Govindachari TR, Suresh G, Gopalakrishnan G, Wesley SD (2000) Insect antifeedant and growth regulating activities of neem seed oil - the role of major tetranortriterpenoids. J Appl Ent 124: 287-291.

9. Ahmad S, Fiaz S, Riaz MA, Hussain A (2005) Comparative efficacy of crude extracts of *Calotropi sprocera*, *Datura alba* and imidacloprid on termites in sugarcane at Faisalabad. Pak Entomol 27: 11-14.

10. Jacob S, Sheila MK (1993) A note on the production of stored rice from the lesser grain borer by indigenous plant products. Indian J Entomol 55: 337-779.

11. Rahman MM, Islam W (2007) Effect of acetonic extracts of Calatropi sprocera R Br-in (Ait) on reproductive potential of Flat grain beetle Cryptolestespusillus. Bangladesh J Sci Ind Res 42: 157-162.

12. Talukhdar FA (2006) Plant products as potential stored product insects management agents: A mini review. Emir J Agric Sci 18: 17-32.

13. Alam MA, Habib MR, Nikkon F, Khalequzzaman M, Karim MR (2009) Insecticidal activity of root bark of Calotropis gigantea L. against *Tribolium castaneum* (Herbst) World J Zool 4: 90-95.

14. Iqbal J, Cheema ZA (2009) Response of purple nuts edge (Cyperus rotundus L.) to crop extracts prepared in various solvents. Allelopathy J 23: 445-452.

15. Collins PJ (1998) Resistance to grain protectants and fumigants in insects pests of stored products in Australia. Proceedings of Australia post harvest technical conference.

Inhibitory Effect of Buprofezin on the Progeny of Rice Weevil, *Sitophilus oryzae* L. (Coleoptera: Curculionidae)

Gopal Das*

Department of Entomology, BangladeshAgricultural University, Mymensingh-2202, Bangladesh

Abstract

The rice weevil *Sitophilus oryzae* (L.) (Coleoptera: Curculionidae), is an important pest of stored grains throughout the world. Different biopesticides and synthetic insecticides have been used for a long time to minimize the rice weevil infestation but their efficacy is not satisfactory yet. The climate and storage conditions are highly favorable for quick progeny which may be responsible for lower efficiency of insecticides and higher level of infestation. Insect growth regulators (IGRs) like buprofezin are the semi-synthetic insecticides, mimic the insect-produced hormones and don't kill the insects directly but reduces pest populations by affecting mating behavior, reproduction, egg viability, pre and post embryonic development etc. Laboratory experiments were conducted from August to November 2012, to evaluate the efficacy of buprofezin on the mortality and the suppression of progeny production of rice weevil. Ten adult rice weevils were exposed on three types of rice grains (long, medium and short) treated with buprofezin at 100, 200 and 300 ppm. Mortality was counted at 15, 21 and 28 days after treatment while adult progeny was counted at 6, 7 and 8 weeks after buprofezin treatment to get a new generation. The data showed that buprofezin had no direct effect on the mortality of rice weevils regardless the concentrations. Buprofezin at 300 ppm in rice grains significantly inhibited progeny production while lower doses (200 and 100 ppm) had no significant effect but virtually reduces progeny number. Types of grains were not factor for increasing or decreasing the rice weevils populations. In conclusion, buprofezin caused decreasing progeny productivity by *S. oryzae* (L.) with increasing concentrations regardless the types of rice grains.

Keywords: *Sitophilus oryzae*; Buprofezin; Mortality; Progeny inhibition

Introduction

Stored grain pests are a great problem in Bangladesh and about 5-8% of the food grains, seeds and different stored products are lost annually due to stored pests infestation [1]. About 13 species of insects have been recorded from Bangladesh which loss about 15% of stored rice [2]. Among them, rice weevil, *Sitophilus Oryzae* (L.) (Coleoptera: Curculionidae) that is widely distributed, is one of the most destructive insect pests which cause severe economic loss. Rice weevils can cause losses to grain in storage, either directly through consumption of the grain or indirectly by producing 'hot spots' causing loss of moisture and thereby making grain more suitable for their pests [3]. Therefore, the effective control of this insect has long been the goals of entomologists [4]. Uncertain climatic condition, storage with debris, poor sanitation and inappropriate storage facilities encourage insect attack. In tropical countries like Bangladesh, the climate and storage conditions are highly favorable for insect growth and development [5].

Various synthetic insecticides with different mode of action are currently used to control rice weevil infestation in stored condition but their efficacy as well as safety is really questionable. Recently, different biopesticides have been used to control the rice weevil infestations [6-8]. Concern about the impact of pesticides on both health and environment has resulted in the search for alternative control measures for stored-product insect pests. Among such alternatives, insect growth regulators (IGRs), a class of biorational compounds that mimic insect-produced hormones and potentially reduces pest populations through endocrine disruption. Because of their selectivity these compounds appear to fit the requirements for third generation insecticides, environmentally benign and safer grain protectants. There are three groups of insect growth regulators e.g. juvenoids or anti-juvenoids, ecdysone inhibitor and chitin synthesis inhibitor (CSI). IGRs are potentially used to control insect pests either in the field or in stored condition. Like other conventional insecticides, IGRs don't kill the insect directly but reduces pest populations by affecting mating behavior, reproduction, egg sterility, pre and post embryonic development etc.Methoprene and hydroprene are commonly used IGRs against stored-product pests [9-11]. In recent years, buprofezin became a dominant IGR in different Asian countries like China, Japan, and India to control the field pests effectively [12-14]. The exact mode of action of buprofezin in the insect body is not known yet but adult females that are sprayed directly or come into contact with wet residues, lay sterile eggs and down reproduction capability, ultimately reduces the number of individuals in the next generation [15]. In addition, buprofezin has been shown to have vapor activity, which means that insects not directly treated may still be killed [15]. There is limited informationabout the efficacy of buprofezin on the mortality as well as inhibition of progeny production by *S. oryzae* in stored condition.

Therefore the present study was conducted to elucidate the efficacy of buprofezin (a chitin synthesis inhibitor) on the adult mortality as well as inhibition of progeny production by *S. oryzae* (L.).

Materials and methods

Culturing of *S. oryzae*

The initial adult populations of *S. oryzae* were obtained from the

***Corresponding author:** Gopal Das, Department of Entomology, Bangladesh Agricultural University, Mymensingh-2202, Bangladesh
E-mail:gopal_entom@yahoo.com

stock culture of the laboratory of entomology, BAU, Mymensingh. For the continuous supply of experimental insects, adult rice weevils were cultured in the same laboratory at 60-70% relative humidity and 28-32°C temperature. Before culturing, all equipments and glassware's were disinfected with 70% ethyl alcohol to protect stock culture from microbial infection and safeties of the insects. Adults of *S. oryzae* were released into 200g of disinfected rice grains in plastic containers @25 numbers/container of mixed age and sex. Mouth of the container was capped with muslin cloth fastened with rubber band to ensure proper ventilation as well as prevent the escape and was incubated for 15 days. On the 15th day, the released adults were removed by sieving and the rice grains were kept undisturbed for 2 months for emergence of fresh adults. Emerged adults were collected daily to maintain the same age and released in separate containers for continuous mass culturing. Adults of 5-7 days old age were used for the experimentation.

IGR and doses

Three doses of buprofezin 100, 200 and 300 ppm will be tested and accordingly concentrations were made using distilled water.

Bioassay

Three types of rice grains were used in this study as follows; long, medium and short grain.From each rice type, three 40 g subsamples were weighed and each 40 g was sequentially treated with three concentrations 100, 200 and 300 ppm using micro-sprayer (1 ml/40 g rice) and air dried for 10 minutes.Therefore, a total of 9 treatments will be made for 3 rice types using 3 concentrations. Three replications were maintained for each concentration and arranged following Completely Randomized Design (CRD). Each rice types were also treated with distilled water those were considered as negative control. Then 40 g of buprofezin-treated or water-treated grains were confined with 10 freshly emerged adults in 150 cm diameter petridishes for 10 days for oviposition and kept them undisturbed.

Data collection

The adult mortality was assessed from each treatment combination at 14, 21 and 28 days after buprofezin treatment. After 28-day mortality observation, dead and live adults were removed by sieving from each petridishes and commodity was left as the previous conditions. 40 days after their confinement in petridishes (or 40 days after treatment), number of F1 adults present in each treatment was counted to record the progeny build up and counting was continued for an additional period of 20 days. No data were collected on egg, larva or pupae either from treated or control rice grain.

Data analysis

All observations were corrected by using the Abbott's formula [16]. Data obtained were subjected to analysis of variance (ANOVA) and means were separated by Fishers protected least significant difference (LSD) test. Values were represented as mean ± SEM.

Results

Effect of buprofezin on the mortality of adult *S. oryzae*

Buprofezin-treated as well as water-treated (control) mortality of *S. oryzae* has shown in Table 1. Data clearly showed that buprofezin had no significant effect on the mortality of adult rice weevil in comparison with that in the control. Compared to the control (long grain, 4.53 ± 1.0; medium grain, 3.43 ± 1.0; short grain, 3.97 ± 0.6) the mean mortality level was gradually increased with the increase of concentration while the highest mortality was recorded from 300 ppm for all grain types (long grain, 6.06 ± 1.3; medium grain, 6.20 ± 0.3; short grain, 6.37 ± 0.6) but the differences were not significant between control and buprofezin-treated rice grains. On the other hand, mortality level was not increased with the increase of time for all types of rice grains. It was noted that the larval and pupal mortality was not recorded in the present study.

Effect of grain types on the buprofezin-induced mortality of *S. oryzae*

It was assumed that progeny production by *S. oryzae* in different types of rice grains may be influenced by the action of buprofezin. To confirm this hypothesis, three types of rice grains (e.g. long, medium and short) have been selected in this study to elucidate which grain type is suitable for buprofezin action and hence buprofezin-treated mortality is also increased. There has no significant difference among the rice grain types based on the mortality (Table 1). In all types of rice grains, the mortality level was little higher at the highest concentration (300 ppm) which was followed by 200 and 100 ppm, respectively in comparison with that in the water-treated grains. But the differences

Table 1: Mortality (%) of adult *S. oryzae* when exposed on three rice types treated with 100, 200 and 300 ppm of buprofezin.

Rice grain types	Concentrations (ppm)	Mortality (%) at different days after buprofezin treatment			
		14 days	21 days	28 days	Mean
Long grain	0 (Control)	4.0 ± 0.3	4.4 ± 0.8	5.2 ± 1.9	4.53 ± 1.0
	100	3.8 ± 1.5	3.0 ± 0.4	4.5 ± 0.2	3.77 ± 0.7
	200	4.1 ± 0.9	3.5 ± 0.5	4.6 ± 0.3	4.06 ± 0.6
	300	6.1 ± 1.1	5.8 ± 1.6	6.3 ± 1.2	6.06 ± 1.3
Significant level		NS	NS	NS	NS
Medium grain	0 (Control)	3.9 ± 1.2	4.1 ± 1.5	2.3 ± 0.3	3.43 ± 1.0
	100	5.1 ± 0.9	4.5 ± 1.1	4.8 ± 0.6	4.80 ± 0.9
	200	5.2 ± 0.8	5.9 ± 0.5	5.3 ± 0.9	5.47 ± 0.7
	300	6.5 ± 0.2	6.2 ± 0.3	5.9 ± 0.5	6.20 ± 0.3
Significant level		NS	NS	NS	NS
Short grain	0 (Control)	4.9 ± 0.5	3.0 ± 0.2	4.0 ± 1.1	3.97 ± 0.6
	100	5.0 ± 0.5	4.9 ± 0.7	4.6 ± 0.5	4.83 ± 0.6
	200	6.2 ± 0.5	4.8 ± 1.5	4.9 ± 1.5	5.30 ± 1.2
	300	5.9 ± 0.5	6.8 ± 0.2	6.4 ± 1.0	6.37 ± 0.6
Significant level		NS	NS	NS	NS

Values are represented here as Mean ± SEM. Buprofezin (any of the concentration) had no significant effect on the mortality of adult *S. oryzae* versus control or water-treated rice grain. Types of rice grain also had no effect on the mortality of adult *S. oryzae*. NS=Not significant.

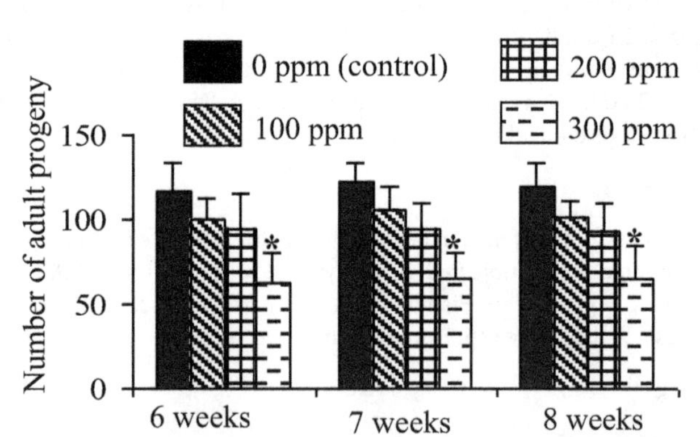

Figure 1: Progeny suppression of S. oryzae when long rice grains were treated with different concentrations of buprofezin. Data were counted from 6 weeks to 8 weeks after treatment. *P<0.05 vs. control, data were expressed as mean ± SEM."

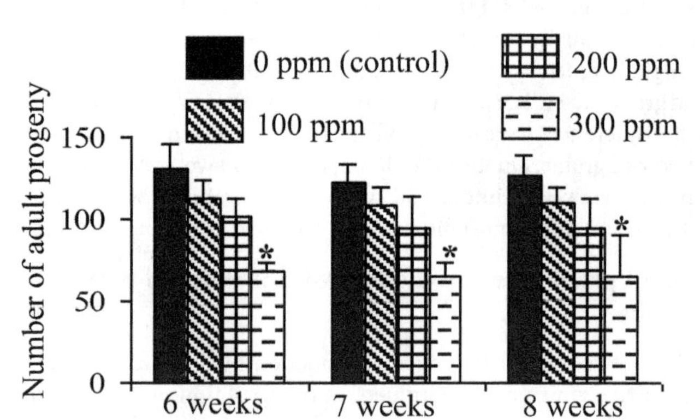

Figure 2: Progeny suppression of S. oryzae when medium rice grains were treated with different concentrations of buprofezin. Data were counted from 6 weeks to 8 weeks after treatment. Progeny was significantly suppressed only at 300 ppm and the lower doses had no significant effect. *P<0.05 vs. control, data were expressed as mean ± SEM."

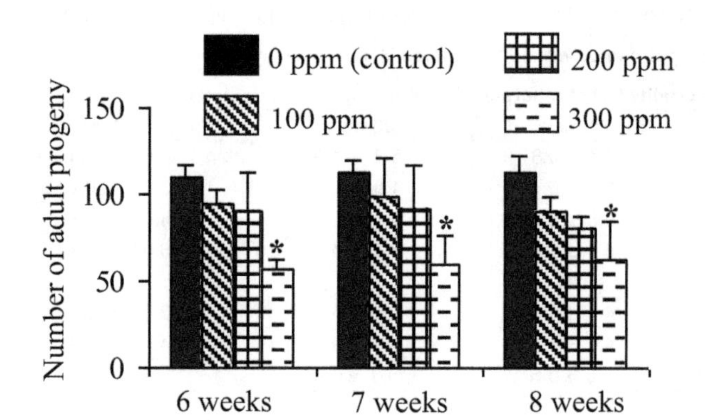

Figure 3: Inhibitory effect of buprofezin on progeny production of S. oryzae when short rice grains were treated with different concentrations. Data were counted from 6 weeks to 8 weeks after treatment. *P<0.05 vs. control, data were expressed as mean ± SEM."

were quite insignificant between controls (water-treated) as well as buprofezin-treated rice grains with different concentrations from 100 to 300 ppm. It is possible that more than 300 ppm concentration will be able to suppress rice weevils progeny significantly although this dose has not been tested in the current study.

The same number of adult rice weevils was carefully tested for all the buprofezin doses to get an accurate mortality level for all the treatments.

Adult progeny production by *S. oryzae* in buprofezin-treated long rice grain

As mentioned in the methodology, 10 pairs of adults weevils were released in buprofezin-treated (100, 200 and 300 ppm) long rice grain to observe whether progeny production is inhibited by different concentrations of buprofezin. Progeny production was also observed in water-treated rice grains to confirm the buprofezin effect on the progeny production by *S. oryzae*. It is noted that the larval and pupal mortality was not recorded in the present study. Adult progeny were counted at 6, 7 and 8 weeks after adults released in buprofezin-treated rice grains. Progeny production of *S. oryzae* was decreased with increasing buprofezin concentrations (Figure 1). Progeny production was reduced by both 100 and 200 ppm of buprofezin but the difference was insignificant when compared with water-treated control. On the other hand, mean number of adult progeny was significantly ($p<0.05$) suppressed when rice grains were treated with 300 ppm of buprofezin in comparison with that in the water-treated control regardless the time after treatment. Moreover, the level of progeny suppression differed insignificantly when compared among the concentrations.

Adult progeny production by *S. oryzae* in buprofezin-treated medium rice grain

Adult progeny were counted at 7, 8 and 9 weeks after adults released in buprofezin-treated rice grains. Like as long grain, progeny production was significantly inhibited ($p<0.05$) by buprofezin when rice grains were treated with highest concentration (300 ppm) (Figure 2) regardless the time. On the other hand, lower doses (100 and 200 ppm) had no significant effect on the suppression of progeny production by *S. oryzae* (Figure 2) although number was reduced virtually compared to the control.

Adult progeny production by *S. oryzae*in buprofezin-treated short rice grain

The *S. oryzae*progeny was significantly inhibited ($p<0.05$) by buprofezin only when rice grains were treated with 300 ppm and the lower doses (100 and 200 ppm) had no significant effect on the suppression of progeny production by *S. oryzae* (Figure 3) regardless the length of time.

Progeny production versus types of rice grains

Three types of rice grains were used in this study to know the better action of buprofezin as well as tendency of progeny production among the rice grain types. Number of progeny production by *S. oryzae* was not affected by morphological differences of rice grain. The results also showed that the tendency of production of progeny by adult rice weevil was similar to the length of time. A control group (water-treated) was included in the experiment to make a comparison with buprofezin-treated group.

Discussion

The post-harvest losses of rice grains by rice weevil, *Sitophilus oryzae* are significant. Treatment with chemical insecticide has great health hazard as well as environmental contaminations. Insect growth regulators are reported to environmentally safe, biodegradable and non-toxic to human and they are considered as third generation insecticides. Various insect growth regulators are potentially using to control the field pests while their uses are mostly limited against stored-pests [12-14]. There are three groups of IGRs; juvenile hormone analogue, ecdysteroid receptors agonists or antagonists and chitin synthesis inhibitors. IGRs are directly not toxic to insects but potentially reduce pest populations by disrupting normal growth and development through multiple targets like reduction of egg viability and hatchability, affecting pre or post embryonic development, abnormal larvae or pupae formation as well as disruption of chitin synthesis [17,18]. Buprofezin has multiple targets to reduce pest populations but the main target is disruption of chitin synthesis [17-20]. Buprofezin is potentially used in the field condition to control lepidoptera, diptera or coleopteran pests [12,21-23] while against stored-pests are not well reported.

The current study was conducted to elucidate the efficacy of buprofezin on the mortality and inhibition of progeny production on different types of rice grains. The major finding of this study is that buprofezin has no direct effect on the mortality of adult *S. oryzae* but it significantly inhibited progeny production. It was interesting that progeny production was inhibited significantly ($P<0.05$) only at 300 ppm of buprofezin while lower doses (100 and 200 ppm) had no significant effect although number of progenies were virtually reduced either by 100 or 200 ppm. The mechanism of progeny inhibition by buprofezin is not clear yet but it has been reported that adult females that are sprayed directly or come into contact with wet residues of buprofezin, lay sterile eggs or down reproduction capability, ultimately reduces the number of individuals in the next generation [15]. It was not also clear yet why progeny was potently suppressed only at 300 ppm but not at 200 or 100 ppm, the possible cause may be penetration level. Maximum penetration of buprofezin molecules might be occurred at 300 ppm by vaporization because buprofezin has been shown to have vapor activity [15]. Either mortality or progeny suppression was not affected by morphological features of rice grains which also raises the possibility that rice hull for each type was similar thickened or with similar features. Arthur [11] also stated that progeny production of *Rhizopertha dominica* was not affected by rice grain types when adult *R. dominica* was exposed on methoprene-treated rice grains.

In Bangladesh, rice weevils take about 35 to 40 days to reach in adult after oviposition although it depends on the prevailing temperature. Therefore, the first progeny production was counted at 6 weeks after treatment and counting was continued for an additional period of 2 weeks to assess whether progeny production really changed or not over time. Data clearly showed that mean number of progeny production was almost stable for 6, 7 or 8 weeks after treatment which raises the possibility that adults weevils were laid eggs almost in the same time after exposure on the treated rice grains. It is noted that culturing of *S. oryzae* were done carefully so that almost same aged adults can be released onto treated rice grains. IGRs like methoprene, hydroprene, pyriproxyfen are widely using against rice weevils and other stored pests but the use of buprofezin against stored-pests especially rice weevils is not well reported. Despite the demonstrated effectiveness of some IGRs, their practical application presents certain problem against rice weevils because this insect spend a large part of their life cycle inside the grains kernel and the insects are surrounded by food, are not easily accessible to control insects. Current study clearly showed that buprofezin was not effective to kill adult weevils even with higher concentration and this might be caused due to little contact of insects with buprofezin. In contrary, buprofezin significantly inhibited progeny build up at 300 ppm but not at 100 or 200 ppm. As reported previously, buprofezin has volatile properties, and therefore, the highest concentration of buprofezin was able to contact with insects and inhibited progeny production possibly through reduction of mating, egg viability, hatchability as well as affecting the pre and post embryonic development.

IGRs have several advantages over neurotoxic insecticides. They affect development at an early stage and, therefore, can slow down population build up, particularly where there is no constant pressure of re-infestation. They can be used in minute amounts and pose no residues problems. They are reported to be non-toxic to mammals and produce no teratogenic or mutagenic effects in warm-blooded animals even with high concentrations [24]. Therefore, uses of IGRs especially buprofezin may be the potential alternatives of various toxic insecticides against rice weevils infestation in Bangladesh. However, it needs further investigations to explore the molecular mechanisms of suppression of progeny build up by *S.oryzae* using buprofezin.

References

1. Alam MZ (1971) Pests of Stored Grains and Other Stored Products and Their Control. Agril Inf Serv.
2. Khan RA (1991) Crop loss and waste assessment USAID, BRAC Checchiand Company Consulting Incorporate, Dhaka.
3. Longstaff BC (1981) Biology of the grain pest species of the genus *Sitophilus*(Coleoptera: Curculionidae): a critical review. ProtecEcol 3: 83-130.
4. Mondal KAMSH, Parween S (2000) Insect growth regulators and their potential in the management of stored-product insect pests. Integr Pest Manage Rev 5: 255-295.
5. Jacobson M (1989) Botanical insecticides: Past, present and future: Insecticides of Plant Origin. (Arnason, JT, Philogene, BJR and Morand, P. edn), Symposium Series No. 387, American Chemical Society, Washington DC, 1-10.
6. Asawalam EF, Ebere UE, EmeasorKC(2012)Effects of some plants products on the control of rice weevil *Sitophilus oryzae*(L.) Coleoptera: Curculionidae. J Med Plant Res 6: 4811-4814.
7. Saljoqi AUR, Afridi MK, Khan SH, Rehman S (2006) Effects of six plant extracts on rice weevil *Sitophilus oryzae*L. in the stored wheat grain. J AgricBiolSci 1: 1-5.
8. Yankanchi SR, Gadache AH (2010) Grain Protectants efficacy of certain plant extracts against rice weevil, Sitophilus oryzae L. (Coleoptera: Curculionidae). J Biopesticides 3: 511-512.
9. Loschiavo SR (1976) Effects of the Synthetic Insect Growth Regulators Methoprene and Hydroprene on Survival, Development or Reproduction of Six Species of Stored-products Insects. J Econo Entom 69: 395-399.
10. Arthur FH (1996) Grain protectants: current status and prospects for the future. J Stored prod Res 32: 293-302.
11. Arthur FH (2004) Evaluation of methoprene alone and in combination with diatomaceous earth to control *Rhyzoperthadominica*(Coleoptera: Bostrichidae) on stored wheat. J Stored Prod Res 40: 485-498.
12. Uchida M, Asai T, Sugimoto T (1985) Inhibition of cuticle deposition and chitin biosynthesis by a new insect growth regulator buprofezin in *NilaparvatalugensStal*. AgricBiolChem 49: 1233-1234.
13. Konno T(1990) Buprofezin: A reliable IGR for the control of rice pests. Society of Chemical Industry 23: 212 - 214.
14. Izawa Y, Uchida M, Sugimoto T, Asai T (1985) Inhibition of Chitin Biosynthesis by buprofezin analogs in relation to their activity controlling *Nilaparvatalugens*. Pesticide BiochemPhysiol24: 343-347.
15. Cloyd R (2006) Insect growth regulators. OFA bulletin, No. 898, 4p.
16. Abbott WS (1925) A method of computing the effectiveness of an insecticide. J Econo Entom 18: 266-267.
17. Merzendorfer H, Zimoch L (2003) Chitin metabolism in insects: structure, function and regulation of chitin synthases and chitinases. J Expt Biol 206: 4393-4412.
18. Palli SR, Retnakaran A. 1990. Molecular and bio-chemical aspects of chitin synthesis inhibition: Chitin and Chitinases.(Jolles, P., Muzzarelli, RAA edn.), BirkhauserVerlag 85-98.

19. Khater HF (2012)Ecosmart Biorational Insecticides: Alternatives Insect Control Strategies, Insecticides-Advances in Integrated Pest Management, (Dr. FarzanaPerveenedn), ISBN: 978-953-307-780-2, In Tech.

20. Zoebelein G, Hammann I, Sirrenberg W (1980)BAY-SIR-8514, a new chitin synthesis inhibitor. J ApplEntomol 89: 289-297.

21. Nasr HM, Badawy EI, Rabea EI (2010) Toxicity and biochemical study of two insect growth regulators, buprofezin and pyriproxifen, on cotton leafworm*Spodopteralittoralis*. Pesticide BiochemPhysiol 98: 198-205.

22. Ragaei M, Sabry KH (2011) Impact of spinosad and buprofezin alone and in combination against the cotton leafworm, *Spodopteralittoralis*under laboratory condition. J Biopest 4: 156-160.

23. Eisa AA, El-Fatah MA, El-Nabawi A, El-Dash AA (1991) Inhibitory effects of some insect growth regulators on the developmental stages, fecundity and fertility of the Florida wax scale, *Ceroplastes floridensis*. Phytoparasitica 19: 49-55.

24. Antognini J(1972) Insect growth regulators and sex attractants in pest control. Invitational paper presented at the 56th annual meeting of the Pacific Branch Entomological Society of America, Victoria, B.C.

Effects of Compost and Inorganic NP Rates on Growth, Yield and Yield Components of Teff (*Eragrotis teff* (Zucc.) Trotter) in Girar Jarso District, Central Highland of Ethiopia

Alemu Assefa[1]*, Tamado Tana[2] and Jemal Abdulahi[2]

[1]*Fitche Soil Research Center, Oromia Agricultural Research Institute, Ethiopia*
[2]*Department of Plant Science, Haramaya University, Ethiopia*

Abstract

Teff is the major crop produced in study area whose productivity is being affected by low soil fertility and organic matter depletion. Hence, an experiment, having factorial combination of 0, 2.5, 5 and 7.5 t ha^{-1} of compost and 0/0, 16/11.5, 32/23 and 64/46 kg ha^{-1} of N/P_2O_5, was conducted in 2014/15 main cropping season to assess the effect of rates of compost and NP fertilizers on growth, yield and yield components of teff and to determine their economical rates. Treatments were laid out in RCBD in three replications. The results indicated that the main effect of compost rate significantly increased dry biomass and straw yield and decreased harvest index. Except on number of productive tillers, the main effect of NP fertilizer was significant on all the parameters measured. The shortest days to heading, highest; lodging percentage, plant height, panicle length and dry biomass were recorded at 64/46 kg ha^{-1} of N/P_2O_5, all showing increasing trend with increasing NP The highest grain and straw yield and net farm benefit were recorded at interaction of 64/46 N/P_2O_5 kg ha^{-1} and 7.5 tons ha^{-1} of compost. Generally, the study suggested that, the use full dose of nationally blanket recommended NP rate (64/48 kg of N/ P_2O_5) with 7.5 tons ha^{-1} of compost is likely combination to attain the optimum grain yield and profit and can be alternative approach for integrated soil fertility management measure instead of the sole application of inorganic fertilizers.

Keywords: Grain yield; Interaction effect; Partial budget analysis; Straw yield

Introduction

Cereals are an important dietary protein and energy source throughout the world [1]. Teff is grown as important cereal in Ethiopia [2]. It is national obsession and is grown by an estimated 6.3 million farmers [3]. It has also recently been receiving global attention particularly as a 'health food' due to the absence of gluten and gluten-like proteins in its grains [4].

Teff has significantly highest share in Ethiopian in area of production. It was reported that teff covered 22.23% of the total area under cereal production followed by maize (16.39%) in 2013 [5]. According to the same report, teff is also the major crop grown in North shoa (study area) covering more than 28% of the total area under grain production.

Teff performs well at an altitude of 1800-2100 m a s l, annual rainfall of 750-850 mm, growing seasons rainfall of 450-550 mm, and a temperature of 10°C-27°C although it can adapt wide range of agro-climatic conditions [6]. Moderately fertile clay and clay loam soils are ideal for teff. It can also withstand moderate water logged conditions [7].

Regardless of its wider adaptation, productivity of teff is low in the country with the national average grain yield of 1.379 tons ha^{-1} [5]. This is mainly because of low soil fertility [8] and severe organic matter depletion [9] aggravated by low rate of chemical fertilizer application. The rate of chemical fertilizer application is low in the country due to unaffordable price for resource-poor smallholder farmers [10]. The continued use of chemical fertilizers is also not recommendable as it causes for health and environmental hazards such as ground and surface water pollution by nitrate leaching [11].

One of the possible options to make use of low rate of chemical fertilizer application without nutrient deficiency of the soil could be recycling of organic wastes. But it is also difficult to attain sustainable productivity neither by inorganic fertilizers nor organic sources alone [12,13]. The best remedy for soil fertility management is, therefore, a combination of both inorganic and organic fertilizers, where the inorganic fertilizer provides nutrients and the organic fertilizer mainly increases soil organic matter and improves soil structure and buffering capacity of the soil [13]. The combined application of inorganic and organic fertilizers is also widely recognized as a way of increasing yield and/or improving productivity of the soil sustainably [14]. Several researchers [15] have demonstrated the beneficial effect of integrated nutrient management in mitigating the deficiency of many secondary and micronutrients. There are also some research reports in Ethiopia that revealed the combined effect of organic (compost and manure) and chemical (NP) fertilizer enhanced the yield of teff and reduced the amount of recommended chemical fertilizer by half [16,17] Though there is a huge variation in crop response to different NP fertilizer rates, 64/46 N/P_2O_5 kg ha^{-1} was given by Ministry of Agriculture and Rural Development as national blanket recommendation [18].

Farmers in Ethiopia have also awareness about compost and have been preparing and using huge amount especially in central highlands of Ethiopia. According to North Shoa zone agriculture department annual report (2014), about 5,056,260 m^3 of compost has been prepared and

***Corresponding author:** Alemu Assefa, Fitche Soil Research Center, Oromia Agricultural Research Institute, Ethiopia,
E-mail: think2greeneconomy@yahoo.com

Color	pH	OC%	Particle size distribution			Textural class	Available P in ppm	TN (%)
Black	6.4	1.3	Clay=45	Silt= 24	Sand=31	Clay	8	0.12

Table 1: Physico-chemical properties of experimental soil.

				Exchangeable bases (cmol/kg)				Available Micronutrients and Phosphorus (mg/kg)						
EC	pH	%OC	TN	Ca	Mg	Na	K	Ca	Mn	Fe	Cu	Zn	B	S	P
2.32	7.22	17.81	0.96	62.5	20.5	10	0.96	20.14	31.05	61.2	1.42	16.75	4.56	113.06	382

Table 2: Constituents of experimental compost determined during incorporation to the soil.

used in 2013/14 cropping season. However, there is little information about the rate of application of compost and chemical fertilizer in the study area either to apply in sole or in combination. Therefore this study was undertaken with the objectives: to assess the effect of rates of compost and inorganic NP fertilizers on growth, yield and yield components of teff; and to determine economically appropriate rate of compost and inorganic NP fertilizers for teff production.

Materials and Methods

Experimental site description

Field experiment was conducted on farmer's field in 2014/15 main cropping season in Girar Jarso District of North Shoa Zone of Oromia Regional State. The site is geographically located at 09°45.121'N latitude and 038°46.728'E longitude and at an altitude of 2677 meters above sea level. The area receives mean maximum and minimum temperature of 22.13°C and 10.26°C, respectively and average long term annual rainfall of 1000 mm. Some of soil physical and chemical properties of the trial field determined during planting are given in Table 1. Accordingly the initial status of the experimental soil is classified as slightly acidic in soil reaction [19] low in total N [20], medium in P content [21] and medium in organic matter content [22].

Experimental materials

Compost: The experimental compost was collected from five farmers pit and was thoroughly mixed to be representative of the compost currently prepared and being used by the farmers around the study area. One composite sample from the mixture was taken and submitted to JEJE analytical Testing Laboratory for chemical analysis. The weight of the compost was measured for each level at air dried (at 10.5% moisture) bases right before application (Table 2).

Crop/variety: The teff variety named Quncho (DZ-Cr-387), which was developed and released by Debrezeit Agricultural Research Centre in 2006 was used for the experiment. It is a high yielding white-seeded cultivar adapted to a wide range of altitudes. The certified seed was used for the trial collected from Ethiopian seed enterprise.

NP sources: Urea (46% N) and DAP (18% N and 46% P_2O_5) were used as a source of nitrogen and phosphorus.

Treatment and layout

A total of sixteen treatment combinations (0, 2.5, 5, and 7.5 tons ha^{-1} of compost combined with 64/46, 32/23, 16/11.5 and 0/0 kg ha^{-1} of NP) were used.

The experiment was laid out in randomized complete block design (RCBD) in factorial arrangement and treatments were replicated three times. The gross plot size was 2 m × 3 m (6 m^2) and the net plot size was 2.5 × 1.5 m (3.75 m^2). Compost was incorporated to the soil on prepared seedbeds twenty one days before planting and chemical fertilizers (NP) were applied during planting and seeds were broadcast at the rate of 25 kg ha^{-1} (15 g/plot). All other cultural practices were uniformly applied as per the recommendations.

Data collection, measurements and analysis

Phenological, Growth, Yield and Yield Component data were collected according to the descriptor for teff.

Days to 50% panicle emergence: This parameter of the plant was determined by counting the number of days from sowing to the time when 50% of the plants started to emerge the tip of panicles through visual observation.

Days to 90% physiological maturity: Days to physiological maturity was determined as the number of days from sowing to the time when 90% of the plants in a plot reached maturity based on visual observation. It was indicated by senescence of the leaves as well as free threshing of grain from the glumes when pressed between the forefinger and thumb.

Plant height: Plant height was measured at physiological maturity from the ground level to the tip of panicle from ten randomly pre-tagged mother plants in each plot.

Panicle length: It is the length of the panicle from the node where the first panicle branches emerge to the tip of the panicle which was determined from an average of ten randomly pre-tagged mother plants per plot.

Number of productive tillers: The numbers of effective tillers was determined by counting the tillers from an area of 0.25 m × 0.25 m by throwing a quadrat into the middle portion of each plot.

Biomass yield: At maturity, the whole plant parts, including leaves, stems and kernels from the net plot area was harvested and after sun drying for five days the biomass was measured.

Thousand kernel weight: Thousand kernels from the bulk of threshed yield were counted from each net plot and weight was recorded using a sensitive balance.

Grain yield: Grain yield was measured by harvesting the crop from the net plot area of 2.5 × 1.5 m excluding border effects.

Straw yield: After threshing and measuring the grain yield, the straw yield was measured by subtracting the grain yield from the total above ground biomass yield.

Harvest index: Harvest index was calculated by dividing grain yield by the total above ground air dry biomass yield.

$$\text{Harvest Index (\%)} = \frac{\text{Seed Yield per plot}}{\text{Above Ground Biomass per plot}} \times 100$$

Lodging index: Lodging percentage was taken as the sum of the product of each scale of lodging (0-5 scale) and its respective percentage divided by five where 0 stands for upright stand, 1 for slightly slant, 2 for medium slant, 3 for very slant and 4 for extremely slant and 5 stands for 100% plants lodged.

Data analysis

The collected data was analyzed by general linear model (GLM) procedures using GenStat Release 15 software [23]. Means of significant treatment effects were separated using the Fishers' protected Least Significant Difference (LSD) test at 5% level of significance. Finally, economic analysis is made following CIMMYT methodology [24].

Results and Discussion

Days to Heading

Days to heading was highly significantly ($P<0.01$) affected by the main effects of chemical fertilizer rates. But this parameter was not affected either by compost rates or interaction effects of compost with chemical fertilizer.

Generally, as the rate of NP increased, the number of days elapsed to heading was shortened. Hence, the longest days (78.4) to heading was recorded at a control plots while the shortest days to heading (67) was recorded at the highest (64/46 N/P_2O_5 kg ha^{-1}) rate of NP fertilizer (Table 3). The hastened heading as a result highest rate of NP could be due to the fact that plots receiving the highest rates of nutrients encouraged for early establishment, rapid growth and development promoted by nitrogen as explained by Ref. [25]. In contrary, delayed heading at lower rates of the nutrients (NP) could result due to longer time required to establish, grow and complete the vegetative growth. In line with this result, Ref. [26] found that N and P_2O_5 at the rates of 64/46 kg ha^{-1} significantly shortened days to heading of teff than the control. Likewise Ref. [27] and Ref. [28] reported that as the rate of N fertilizer increased to 90 kg ha^{-1}, tasseling of maize was significantly hastened than the control and lower rates of N. In contrast to this result, Ref. [29] found increased days to heading of wheat at the rate of 80/60 of NP than the control and other lower rates. Similarly, Ref. [30] reported that the application of N at the rate of 46 kg ha^{-1} delayed heading of teff than the control.

Treatment	Days to heading
Compost (t ha^{-1})	
0.0	74.75
2.5	73.58
5	74.50
7.5	74.25
LSD (0.05)	NS
N/P_2O_5 (kg ha^{-1}) 0/0	78.42a
16/11.5	76.17b
32/23	73.58c
64/46	68.92d
LSD(0.05)	1.81
CV (%)	2.9

Table 3: Days to 50% heading of teff as affected by the main effect of NP and compost.

Compost	Chemical Fertilizer (N/P_2O_5) rates (kg ha^{-1})			
(t ha^{-1})	0/0	16/11.5	32/23	64/46
0.0	135.3e	144.3abc	143.7abc	138.3cde
2.5	139cde	138.7cde	139cde	143.7abc
5.0	145.7ab	147.3a	141.7a-d	136.3de
7.5	141.7a-d	147a	141.7a-d	140.0b-e
LSD (0.05)	6.260			
CV (%)	2.7			

Table 4: Days to maturity of teff as affected by the interaction of NP fertilizer and compost rates.

Treatments	Plant height (cm)	Panicle length (cm)	Lodging (%)
Compost (t ha^{-1})			
0.0	70.37	28.18	25.00
2.5	70.82	28.45	27.08
5.0	68.58	27.29	33.33
7.5	69.12	27.48	33.33
LSD (0.05)	NS	NS	NS
N/P_2O_5 (kg ha^{-1})			
0/0	55.59d	23.45c	8.33d
16/11.5	62.99c	24.96c	22.92c
32/23	72.18b	28.64b	37.5b
64/46	88.13a	34.34a	50.0a
LSD(0.05)	4.999	1.786	7.9
CV (%)	6.8	7.7	32.2

Table 5: Plant height, panicle length, and lodging index of teff as affected by the main effects of NP fertilizer rates.

Days to Maturity of teff was significantly ($P=0.023$) affected by the main effect of NP fertilizer rates while compost rate was not significant to affect this parameter. Interaction of compost and NP rates was also significantly ($P=0.014$) affected maturity days of teff. Regardless of significant effect, there was no consistent trend in increasing or decreasing in days to maturity with the interaction effects of compost and NP rates. However, as the level of compost increased while NP is constant, the number of days to maturity increased in most cases (Table 4). The shortest days to maturity (135 days) was recorded at the control plots while the longest (147.3 and 147 days) were recorded at (16/11.5 N/P_2O_5 kg ha^{-1} and 5 tons ha^{-1}) and (16/11.5 N/P_2O_5 kg ha^{-1}, 7.5 tons ha^{-1}) of NP and compost combination.

The non-consistency could have resulted due to opposite action of N and P on maturity; N may cause delayance while P hastened crop maturity as also indicated by [31] stating that application of N and P significantly influenced days to 75% maturity; N fertilizer prolonged days to 75% maturity of teff, whereas the reverse trend was seen with applied P. The delay in maturity with increase in the rate of compost keeping NP constant could have resulted because of more vegetative growth. This result was in agreement with the findings of [32]. It could also been resulted because of effect of compost in retaining soil moisture. Similar finding was also reported by [33] as residue treated plots delayed in maturity of maize with the same justification given above and the significant increased soil water holding capacity (WHC) from compost and cow dung treated plots than the control [34].

Plant Height was highly significantly ($P<0.01$) affected by the main effects of NP rates. However, it was not significantly affected either by the compost rates or by the interaction of compost and NP fertilizer. Generally, as the rate of NP fertilizer increased to the highest rate, from 0/0 to 64/46 N/P_2O_5, markedly a linear increase in plant height was observed and thus, the highest plant height of 88.13 cm was recorded at the highest N/P_2O_5 rate of 64/46 kg ha^{-1} while the shortest height (55.59 cm) was noted from the control plot (Table 5). The increase in plant height with increasing NP could have resulted due to sufficient supply of nutrient which encourages plant growth: nitrogen plays critical role in the structure of chlorophyll, while P is main element involved in energy transfer for cellular metabolism in addition to its structural role [35]. The result was in agreement with the findings of Ref. [36,37].

The non-responsiveness of the compost might be due to the slow release of nutrients at early crop vegetative growth stages. In line with this result, Yihenew reported that the plant height in plot that received 6.5 months old compost prepared from cereal straw performed even less than the control plots reasoned as the type of composting material

Treatments	NPT	TKW(g)	BY (Kg ha^{-1})	Harvest index
Compost (t ha^{-1})				
0.0	5.03	0.300	3259c	0.2908a
2.5	5.23	0.297	3732b	0.2775a
5.0	4.72	0.304	4276a	0.2426b
7.5	4.63	0.301	4458a	0.2428b
LSD (0.05)	NS	NS	192.189	0.0232
N/P_2O_5 (kg ha^{-1})				
0/0	4.7	0.2925b	2246d	0.286a
16/11.5	5.07	0.3062a	3380c	0.271ab
32/23	4.79	0.306a	4827b	0.235c
64/46	5.04	0.296ab	5274a	0.2616b
LSD	NS	0.0107	192.189	0.0232
CV (%)	32.2	4.3	5.9	10.6

Where, NPT=number of productive tillers, TKW=thousand kernel weight, BY=total biomass yield, NS= non-significant. Means sharing the same letter under the same column are not significant at P=0.05 according to the LSD test.

Table 6: Number of productive tillers per plant, thousand kernel weights, biomass yield and harvest index of teff as affected by the main effects of compost and NP fertilizer rates.

and duration of composting has affected it [38]. Similarly, Abdalla et al. reported that application of different organic fertilizers: compost, vermicompost and farm yard manure, at different rates was not able to affect plant height in the first year of experiment [39]. In contrast to this, Medhn et al. reported significant and superior plant height of teff in plots received organic fertilizers as compared to conventional once [40].

Panicle length

Panicle length was highly significantly (P<0.01) affected by the main effects of NP fertilizer rates. However, neither compost rates nor its interaction with NP fertilizer significantly affected this parameter.

Similar to plant height, panicle length also increased with increasing NP fertilizer rates. The highest panicle length (34.34 cm) was recorded at the highest N/P_2O_5 rate of 64/46 (Table 5). This could be due to similar reason as that of plant height. The positive correlation between N fertilizer and teff panicle length has also been reported by Giday et al. [36]. The higher the panicle length may have also positive contribution to the grain and straw yield since it has a positive correlation to grain yield. In line with this result, Fayera et al. reported that the application of balanced fertilizer and efficient utilization of nutrients leads to high photosynthetic productivity and accumulation of high dry matter, which ultimately increases panicle length and grain yield [41].

Lodging index

Lodging index was highly significantly (P<0.001) affected by the main effects of NP fertilizer, however, the main effect of compost rates did not show significant effect on lodging. NP fertilizer and compost also did not interact to affect lodging.

The lodging index was increased with increasing NP rates. The highest lodging (50%) was recorded from the highest N/P_2O_5 rate (64/46) while the lowest lodging index (8.3%) was recorded from the control (Table 5). The increasing lodging index with increasing NP fertilizer could be because of the increase in plant height which in turn resulted from abundant supply of nutrients. This result was in line with the findings of Shiferaw who reported highest lodging of teff (74%) at N/P_2O_5 rate of 64/46 kg ha^{-1} [26]. Likewise, Fayera et al. reported the highest lodging percentage (79.74%) of teff was recorded in the highest rate of NPK application though the rate the authors reported is much higher (138 kg N/ha+55 kg P/ha) than the present result [41]. Kebebew and Tams et al. also confirmed that abundant supply of nutrients in the soil can contribute to the process of lodging [42,43].

It has been reported that lodging is the serious problem of teff production that causes high yield reduction because of the use of high amount and unbalanced different rates of NPK fertilizers practiced in the country. On average, lodging accounted about 11-22% total grain yield losses [6]. Though susceptibility to lodging is characteristics of all varieties of teff [44] it could be induced by both external and internal factors like wind, rain, and morphological traits of the crops or by their interactions. Where, NS=non-significant, t=tons. Means sharing the same letter under the same column are not significantly different at P=0.05 according to the LSD test.

Number of productive tillers

Number of productive tillers was not affected neither by the main effects nor the interactions of compost and NP fertilizer.

Generally, there was no difference in number of productive tiller between the levels of compost and NP fertilizer and it did not show any increasing or decreasing trend (Table 6). This result was in agreement with the findings of Shiferaw who reported a non-significant difference in productive tillers between the higher N/P_2O_5 (64/46) rates and the control [26]. In contrast with the result of this study, Fayera et al. found that number of the productive tillers of teff was significantly increased with the increase in the rates of NPK [41]. Giday et al. also reported positive and significant increase in number of productive tillers with increasing rates of N fertilizer on teff [36].

Thousand kernel weight

Thousand kernel weight was significantly (P=0.025) affected by the main effects of NP fertilizer rate but not significantly affected by the compost rates and interaction effects of compost and NP rates. Plots received 16/11.5 and 32/23 N/P_2O_5 showed significantly higher thousand kernel weight (0.31 g) than control plots (0.29 g), However, plots treated with higher dose of N/P_2O_5 (64/46) and control were similar in performance (Table 6) indicating increase in thousand kernel weight to certain rates of applied NP and decreasing for further application dosage (beyond 32/23 N/P_2O_5).

The increase in thousand kernel weight with increasing rate of NP from 0/0 to 32/23 N/P_2O_5 could be related to plant growth, the higher the plant growth the higher the photosynthetic area and so photosynthesis, the higher assimilate translocation to the sink. In another way, the reduction in thousand kernel weight with increasing applied rates of both NP beyond 32/23 N/P_2O_5 might probably be the result of insufficient supply of carbohydrates to individual spikelets due to competition effect resulted by vigorous plant growth and the increased number of its spikelets. This result agreed to the findings of Heluf and Mulugeta who stated as only application of 13.2 kg ha^{-1} of P fertilization significantly increased thousand kernel weight of rice but N had no effect on this parameter [45]. Similarly, Hasegawa et al. reported that increased number of spikelets and vigorous growth of rice due to high rates of N fertilizer application induced competition for carbohydrate available for grain filling and spikelet formation [46].

Total above ground dry biomass

Total above ground dry biomass was highly significantly (P<0.001) affected by the main effects of NP and compost, but not by their interaction. Generally as the compost rates increased to 7.5 tons ha^{-1}, the total above ground dry biomass yield was also proportionally increased to 4.458 tons ha^{-1} from 3.259 tons ha^{-1} in control plots (Table

6). This indicates the about 37% higher biomass yields advantage at 7.5 tons ha^{-1} than the control. The significant increase of above ground dry biomass yield without increasing the growth parameters (plant height, panicle length and tillers) could be attributed from the secondary branches and leaf number and size which were grown even during grain filling period. The result was in agreement to the findings of Medhn et al. who indicated that a significantly higher mean biomass yield (5.12 tons ha^{-1}) of organic farming compared to the conventional farming (4.01 tons ha^{-1}). In contrary Dejene and Lemlem found no significant variation among the application of 7 tons ha^{-1} of compost and FYM each separately as compared to the control [16]. The finding from this experiment and the similar results indicated here may indicate that organic fertilizer like compost had a positive contribution in increasing the total above ground biomass yield of teff.

Similarly, increasing the rate of NP from 0/0 to 64/46 N/P_2O_5, the dry above ground biomass yield was also increased to 5.274 tons ha^{-1} from 2.246 tons ha^{-1} indicating 134.82% higher than the control (Table 6). The higher total above ground dry biomass yield may be attributed due to the proportional vegetative growth (especially the plant height) as a result of NP. The result was also in line with the findings of Fissehaye et al. who reported as the application of N and P increased the above ground biomass yield to the level of 69 N and 46 P_2O_5, kg ha^{-1} separately though not significantly different from 46 and 23 kg ha^{-1} of N and P_2O_5, respectively [39]. However, unlike the findings of Getachew et al. who reported that the application of 50% recommended NP rate and 50% manure and compost as inorganic N equivalence on teff crop resulted in total biomass increments of 113% compared to the control treatment (23/10 kg N/P ha^{-1}), the two component treatments (NP and compost) did not interact to affect the biomass yield in the current study [47].

Grain yield

Grain yield was highly significantly (P<0.01) affected by the main effect of NP fertilizer and the interaction of NP fertilizer and compost, but not significantly affected by the main effect of compost rates. Generally at the rate of 0/0 and 64/46 N/P_2O_5, the increase in compost rate from 0 to 7.5 tons ha^{-1}, brought about increasing trend in mean grain yield though not all increases are statistically significant. However, at the rate of 16/11.5 and 32/23 of N/P_2O_5, the increase in compost rate from 0 to 7.5 tons ha^{-1} showed no difference in changing the mean grain yield (Table 7). Numerically highest grain yield (1.395 tons ha^{-1}) of teff was obtained at the combination of the highest rate of NP (64/46 kg ha^{-1} N/P_2O_5) fertilizer and the highest rate of compost (7.5 tons ha^{-1}) while the lowest grain yield of 0.447 tons ha^{-1} was recorded from non-treated (zero rate of NP and compost) plots. However, the highest grain yield (1.395 tons ha^{-1}) recorded at 7.5 tons ha^{-1} of compost and 64/46 kg ha^{-1} N/P_2O_5, was not significantly different from the yield

	Fertilizer (N/P_2O_5) rates (kg ha^{-1})			
Compost (t ha^{-1})	0/0	16/11.5	32/23	64/46
0.0	0.447h	0.899ef	1.172bc	1.293ab
2.5	0.563h	0.932def	1.154bc	1.382a
5.0	0.713g	0.915ef	1.038cde	1.391a
7.5	0.829fg	0.886f	1.078cd	1.395a
LSD (0.05)	0.149			
CV (%)	8.9			

Table 7: Grain yield (tons ha^{-1}) of teff as affected by the interaction of compost and NP fertilizer rates.

	Fertilizer (N/P_2O_5) rates (kg/ha)			
Compost (t/ha)	0	16/11.5	32/23	64/46
0.0	1.178h	2.132g	2.854de	3.063d
2.5	1.296h	2.223fg	3.524c	3.855bc
5.0	1.965g	2.580ef	4.254a	4.248a
7.5	1.992g	2.954de	4.233ab	4.467a

LSD (0.05)=0.384; CV(%)=5.9; Means in rows and columns sharing the same letter (s) do not differ significantly at P=0.05 according to the LSD test.

Table 8: Straw yield (tons ha^{-1}) of teff as affected by the interaction of compost and NP fertilizer rates.

Means in columns sharing the same letter do not differ significantly at P=0.05 according to the LSD test.

obtained from 0, 2.5 and 5 tons ha^{-1} of compost combined with 64/46 kg ha^{-1} N/P_2O_5 (Table 5) indicating the more importance of inorganic NP than the compost in affecting the grain yield.

The increase in grain yield due to interaction effects could be attributed to the positive effects of compost in increasing the efficiency of chemical fertilizer by preventing losses of the nutrients through denitrification, volatilization and leaching and releasing with the passage of time. Compost could also improve the soil structure which leads to better root development which may result in more nutrient up take from the soil in addition to its gradual/slow release of macro and micro nutrients by itself [39]. Unlike the findings of Dejene and Lemlem (2012) who reported that the application of half recommended dose of NP and FYM (11.5 kg ha^{-1} N 15 kg ha^{-1} P and 3.5 tons ha^{-1} of FYM) gave the highest grain yield of teff than sole application of either FYM or NP, the increase in grain yield as a result of interaction effect was not much higher than the main effect of NP at recommended rate. Getachew et al. also reported that the application of 50% recommended NP (30/10 N/P) rate and 50% compost (3.25 tons ha^{-1} of compost) resulted in grain yield increments of 122% compared to the control on teff which is comparable to the full NP dose [47]. Likewise, Quansah also indicated that the use of organic (Household waste and Poultry manure Compost) and/or inorganic fertilizers (NP) increased maize grain yield separately, but, the yields obtained by the combined treatments were significantly higher than their sole treatments [48].

Straw yield

Straw yield was highly significantly (P<0.01) affected by the main effects of fertilizer (NP) and compost rates and significantly (P<0.05) by the interaction effects of the two treatments.

The straw yield increased with the increased rate of the compost rate from 0 to 5 tons ha^{-1} and NP rates from 0/0 to 32/23 N/P_2O_5 kg ha^{-1}. The highest straw yield (4.467 tons ha^{-1}) was recorded at 64/46 N/P_2O_5 kg ha^{-1} and 7.5 tons ha^{-1} compost though statistically in par with (5 tons, 32/23 kg), (7.5 tons, 32/23 kg) and (5 tons, 64/46 kg) compost and N/P_2O_5 while the lowest straw yield (1.178 tons ha^{-1}) was recorded from the control (Table 8). The result also indicated that the application of compost beyond 5 tons ha^{-1} and NP beyond 32/23 kg ha^{-1} had no significant effect on the straw yield. The increase in straw yield as result of the interaction of compost and NP fertilizer could have resulted due to the positive effects of compost that might have enhance the efficiency of chemical fertilizer.

The result was in line with the findings of Dejene and Lemlem who reported that the highest straw yield was obtained at a combination of half recommended dose for both organic (compost and farm yard manure) and inorganic (NP) fertilizers than the full dose application of each at 7 tons ha^{-1} of compost and farm yard manure and 23N and 30P though sole application of both compost and farmyard manure

Treatment (C,N/P_2O_5)	Total variable Cost (USD ha^{-1})	Grain yield (Kg/ha)	Straw yield (kg/ha)	Gross benefit (USD ha^{-1})	Net benefit (USD ha^{-1})
0, 0/0	0	446.82	1177.80	371.0711	371.0713
2.5, 0/0	22.374	563.04	1295.63	447.609	425.2349
0, 16/11.5	32.97221	898.70	2132.41	721.2991	688.3269
2.5, 16/11.5	55.34621	932.17	2223.39	749.365	694.0188
0, 32/23	65.94442	1172.02	2853.58	948.3368	882.3924
2.5, 32/23	78.89779	1153.84	3524.03	777.6109	921.0542
2.5, 64/46	154.2628	1382.36	3854.97	117.0179	1016.123
5, 64/46	174.2817	1391.11	4248.00	1216.862	1042.581
7.5, 64/64	194.3005	1395.00	4467.48	1242.267	1047.967

Where, C=compost t ha^{-1}, N/P_2O_5=kg ha^{-1}

Table 9: Partial budget analysis for NP and Compost rates.

at different rates did not show any improvement over inorganically treated plots for straw yields [16]. However, in contrast to this Edwards et al. indicated that the application of 5-15 tons ha^{-1} was able to increase the straw yield of teff by 53.8 and 8.14% than the control and 120 kg ha^{-1} mixture of urea and DAP, fertilizer respectively. Similarly, Medhn et al. reported as the organic fertilizer increased straw yield of teff than the conventional and reasoned that could be due to higher plant height of organic teff than the conventional [40]. Many studies [49,50] also reported proportional increase of straw yield of teff with N in sole application.

Straw yield has to be considered while evaluation any agronomic practice as its importance has become as equal as its grain yield as it is preferred as animal feed during dry period and also sold at reasonable price.

Harvest index

Harvest index was highly significantly (P<0.001) affected by the main effects of both compost and chemical fertilizer (NP) rates, but the two factors did not show significant interaction. The harvest index was decreased from 29.08% to 24.26% with increasing compost rates from 0 to 5 tons ha^{-1} but remained constant at compost rate of 7.5 tons ha^{-1} (Table 6). Similarly, though the decrease was not consistent, the harvest index showed decreasing trend with increasing NP rates from 0 to 32/23 kg ha^{-1}, hence, the lowest harvest index (23.5%) was recorded at N/P_2O_5 rates of 32/23 while the highest harvest index (28.6%) was recorded at control (0/0 NP).

The decrease in harvest index with increasing rates of both compost and NP fertilizer could be due to the fact that compost and NP fertilizer encouraged more vegetative growth than the grain since the harvest index is the ration of grain yield to dry biomass yield. In line with this result, Heluf and Mulugeta reported that the harvest index consistently declined with increasing levels of applied N up to the highest level (150 kg) of N kg ha^{-1} though they found that harvest index increased with the application of P fertilizer at the rate of 26.4 kg ha^{-1} while further increase beyond resulted in highly significantly reduction in rice crop [45]. Likewise, Medhn et al. also indicated low harvest index of teff with organic fertilization than the convention [40]. In contrary to this result Gafar et al. reported that the increased harvest index of rice to 12.89% with application of 15 tons ha^{-1} compost, 83 kg ha^{-1} of N and 50 kg ha^{-1} superphosphate from 5.38% in the control [17].

Partial budget analysis

Partial budget analysis is a method of organizing experimental data and information about the costs and benefits of various alternative treatments. As it is indicated in Table 9, the net farm benefit was calculated taking possible field variable costs and all benefits (grain and straw yield). The maximum farm benefit (22,248 Birr/ha) was recorded at the maximum N/P_2O_5 rates of64/46 kg ha^{-1} combined with 7.5 tons ha^{-1} of compost, though not much higher than 22133.99 Birr/ha and 21572.29 Birr/ha benefit obtained at the next two lower rate (5 tons ha^{-1} of compost and 64/46 kg ha^{-1} N/P_2O_5) and (2.5 tons ha^{-1}, 64/46 kg ha^{-1} N/P_2O_5), respectively. It was also indicated that the relative benefit was declining after the combined level of 2.5 compost tons ha^{-1} and 64/46 N/P_2O_5 to the increasing rate of compost.

Variable costs are:

Urea (N-source)=56.52 USD/100 kg, DAP (both source of N and P)=70.65 USD/100 kg, Compost estimated as 0.71 USD/100 kg, Compost application cost (labor cost) estimated as 4 man/day/7.5 tons, labor cost for fertilizer application=2 man days/(64/46 kg/ha), labor cost/man/day=2.36 USD and gross output (grain and straw): price of teff grain=0.61 USD/kg, price of straw=0.12 USED/kg [51-54].

Conclusions and Recommendations

Teff is an important cereal crop in Ethiopia particularly in North Shoa whose productivity is being affected by low soil fertility and organic matter depletion. Hence factorial experiment consisting of different levels of compost and inorganic NP fertilizer was conducted in Girar Jarso district in 2014/5 cropping season.

The main effects of compost rates positively affected only total above ground dry biomass, straw yield while negatively the harvest index. Except number of productive tillers, the main effect of NP fertilizer had significantly affected all parameters measured and recorded. Heading was hastened and all other parameters increased proportionally with increasing NP rates with the exception to harvest index which inversely decreased and highest thousand kernel weight attained at two middle rates of NP. This indicates inorganic NP is most important to affect the growth, yield and yield attributes of teff than the compost. Interaction of compost with NP fertilizer also significantly affected: days to maturity, grain yield and straw yield of teff. Numerically the maximum grain and straw yields of 1.395 kg ha^{-1} and 4.467 tons ha^{-1} were recorded at the highest rates of both NP and compost. Generally this experimental result fail to support previous reports that the use of compost reduced the recommended NP rates by half, rather the use of full dose of nationally blanket recommended NP rate (64/48 kg of N/P_2O_5) with 7.5 tons ha^{-1} of compost is most likely combination to attain the optimum grain yield and optimum profit and can be alternative approach for integrated soil fertility management measure instead of the sole application of inorganic fertilizers.

Acknowledgements

I would like to express my heartfelt gratitude to Fitche Soil Research Center staff and Mr. Mulualem Diba, Henok Girma and Dereje Getahun for their cooperation during field work and soil laboratory analysis. I also express my gratitude to Oromia Agricultural Research Institute for offering me the opportunity of this study with partial fund.

References

1. Bos C, Juillet B, Fouillet H, Turlan L, Dare S, et al. (2005) Postprandial metabolic utilization of wheat protein in humans. American Journal of Clinical Nutrition 81: 87-94.
2. Abeba T (2009) Material transfer agreements on Teff and Vernonia, Ethiopian Plant Genetic Resources. Journal of Politics and Law 2: 77-89.
3. Claire P, Elissa J (2014) Move over quinoa, Ethiopia's teff poised to be next big super grain. The Guardian News and Media Limited.
4. Spaenij DL, Kooy WY, Koning F (2005) The Ethiopian cereal tef in celiac disease. New England Journal of Medicine 353: 1748-1750.
5. CSA (Central Statistical Agency) (2013) Report on: Area and Production of Major Crops (Private Peasant Holdings, Meher Season) Statistical Bulletin, 532, Addis Ababa, Ethiopia.
6. Ketema S (1993) Tef (Eragrosits tef), Breeding, agronomy, genetic resources, utilization and role in Ethiopian agriculture. Institute of Agricultural Research.
7. National Soil Service (1994) Training materials on Soils for Use by development personnel. Watershed Development and Land Use Department MONRDEP, Addis Ababa, Ethiopia.
8. Negassa W, Abera Y (2013) Soil Fertility Management Studies on Teff. In: Assefa K, Chanyalew S, Tadele Z (eds.) Achievements and Prospects of Tef Improvement, University of Bern, Switzerland.
9. IFPRI (International Food Policy Research Institute) (2010) Report on Fertilizer and Soil Fertility Potential in Ethiopia: Constraints and opportunities for enhancing the system.
10. Endale K (2011) Fertilizer Consumption and Agricultural Productivity in Ethiopia. Ethiopian Development Research Institute, Addis Ababa, Ethiopia.
11. Pimentel D (1996) Green Revolution and chemical hazards. Science and Total Environment 188: 86-98.
12. Satyanarayana V, Prasad PV, Murthy VRK, Boote KJ (2002) Influence of integrated use of farmyard manure and inorganic fertilizers on yield and yield components of irrigated lowland rice. Journal of Plant Nutrition 25: 2081-2090.
13. Godara AS, Gupta US, Singh R (2012) Effect of integrated nutrient management on herbage, dry fodder yield and quality of oat (Avena sativa L.). Forage Research 38: 59-61.
14. Mahajan A, Bhagat RM, Gupta RD (2008) Integrated nutrient management in sustainable rice-wheat cropping system for food security in India. Journal of Agriculture 6: 29-32.
15. Singh R, Agarwal SK (2001) Growth and yield of wheat (Triticum aestivum L.) as influenced by levels of farmyard manure and nitrogen. Indian Journal of Agronomy 46: 462-467.
16. Kassahun D, Mekonnen LS (2012) Integrated agronomic crop managements to improve teff productivity under terminal drought, water stress. Ismail Md. Mofizur R (eds.), InTech, improve-teff-productivity-under-terminal-drought.
17. Agegnehu G, vanBeek C, Bird MI (2014) Influence of integrated soil fertility management in wheat and teff productivity and soil chemical properties in the highland tropical environment. Journal of Soil Science and Plant Nutrition 14: 532-545.
18. Yadeta K, Ayele G, Negatu W (2001) Farming research on teff: Small holders production Practices. In: Teffera H, Belay G, Sorrels M (eds.) Narrowing the Rift: Teff Research and Development. Proceeding of the International Work shop on teff genetics and improvement, Addis Ababa, Ethiopia, pp: 9-23.
19. Abdalla EM, Sarra AM, Saad S, Ibrahiem IS (2012) Utility and nutritive values of organic and inorganic fertilization on teff grass (Eragrostis tef Zucc. Trotter) growth and some soil chemical properties. International Journal of Sudan Research 2: 23-39.
20. Berhanu D (1980) The physical criteria and their rating proposed for land evaluation in the highland region of Ethiopia. Land Use Planning and Regulatory Department, Ministry of Agriculture, Addis Ababa, Ethiopia.
21. Olsen SR, Cole CV, Watanabe FS, Dean LA (1954) Estimation of available phosphorus in soils by extraction with sodium bicarbonate.
22. Tadese T (1991) Soil, plant, water, fertilizer, animal manure and compost analysis. International Livestock Research Center for Africa, Addis Ababa, Ethiopia.
23. GenStat (2012) GenStat Procedure Library. VSN International Ltd.
24. Derek B (1988) From Agronomic Data to Farmer Recommendations: An economics Training Manual. Completely Revised Edition, Mexico.
25. Tucker MR (1999) Essential Plant Nutrients: their presence in North Carolina soils and role in plant nutrition. NCDA and SC, USA.
26. Tolosa S (2012) Effects of inorganic fertilizer types and sowing methods of variable.
27. Cassman K, Roberts B, Bryant D (2003) Dry matter production and productivity of maize as influenced by residual fertilizer nitrogen and legume green manuring. Journal of Soil Science Society of America 56: 823-830.
28. Orkaido O (2004) Effects of nitrogen and phosphorus fertilizers on yield and yield components of maize (Zea mays L.) on black soils of Regede, Konso. School of Graduate Studies, Alemaya University, Alemaya, Ethiopia.
29. Rehman S, Khalil KS, Muhammad F, Rehman A, Amir ZK, et al. (2010) Phenology, leaf area index and grain yield of rainfed wheat influenced by organic and inorganic fertilizer. Pakistan Journal of Botany 42: 3671-3685.
30. Abraha A (2013) Effects of rates and time of Nitrogen fertilizer application on Yield and Yield Components of Teff [(Eragrostis Tef (Zucc.) Trotter]. In: Habro District, eastern Ethiopia.
31. Mirutse F, Haile M, Kebede F, Tsegay A, Yamoah C (2009) Response of teff [Eragrostis tef) Trotter] to phosphorus and nitrogen on a vertisol at north Ethiopia. Journal of the Drylands 2: 8-14.
32. Khan S, Khalil SK, Amanullah, Shah Z (2013) Crop stand and phenology of wheat as affected by integrated use of organic and inorganic fertilizers. Asian Journal of Agricultural Biology 1: 141-148.
33. Dahal S, Tika BK, Lal PA, Birendra KB (2014) Tillage, residue, fertilizer and weed management on phenology and yield of spring Maize in Terai, Nepal. International Journal of Applied Science and Biotechnology 2: 328-335.
34. Vengadaramana A, Jashothan PTJ (2012) Effect of organic fertilizers on the water holding capacity of soil in different terrains of Jaffna peninsula in Sri Lanka. Journal of Natural Product and Plant Resource 2: 500-503.
35. Wiedenhoeft AC (2006) The Green world: Plant Nutrition. Chelsea house publisher, Yew York, USA.
36. Giday O, Gibrekidan H, Berhe T (2014) Response of teff (Eragrostis tef) to different rates of slow release and conventional urea fertilizers in vertisols of southern tigray, ethiopia. Advances in Plants & Agriculture Research.
37. Lemlem A, Adenew T, Gugsa L (2002) Toward farmers' participatory research: attempts and achievements in the central high land of Ethiopia. Proceeding of client oriented research evaluation workshop. Holetta Agricultural research Center, Holetta, Ethiopia.
38. Gebresilassie Y (2009) Decomposition dynamics and inorganic fertilizer equivalency of compost prepared from different plant residues. Ethiopian Journal of Natural Resource 11: 1-16.
39. Arshad M, Khalid A, Mahmood MH, Zahir ZA (2004) Potential of nitrogen and L-tryptophan enriched compost for improving growth and yield of hybrid maize. Pakistan Journal of Agricultural Science 41: 16-24.
40. Berhane M, Kebede F, Fitiwy I, Abreha Z (2013) Comparative productivity and profitability of organic and conventional teff [Eragrostis tef (Zucc.) Trotter] production under rainfed condition: Tigray, Northern Ethiopia. World Journal of Agricultural Sciences 1: 303-311.
41. Asefa F, Debela A, Mohammed M (2014) Evaluation of teff [Eragrostis tef (Zuccagni) Trotter] responses to different rates of NPK along with Zn and B in Didessa district, southwestern Ehiopia. World Applied Sciences Journal 32: 2245-2249.
42. Asefa K (1991) Effects of some synthetic plant growth regulators on lodging and other agronomic and morphological characters of teff [Eragrostis tef, (Zucc.) Trotter]. Alemaya University of Agriculture, Alemaya, Ethiopia.
43. Tams AR, Mooney SJ, Berry PM (2004) The effect of lodging in cereals on

morphological properties of the root-soil complex. 3rd Australian New Zealand Soils Conference, University of Sydney, Australia.

44. Bekabil F, Befekadu B, Simons R, Tareke B (2011) Strengthening the Teff value chain in Ethiopia. Ethiopian Agricultural transformation Agency, Addis Ababa, Ethiopia.

45. Gebrekidan H, Seyoum M (2006) Effects of mineral N and P fertilizers on yield and yield components of flooded lowland rice on vertisols of Fogera plain, Ethiopia. Journal of Agriculture and Rural Development in the Tropics and Subtropics 107: 161-176.

46. Hasegawa T, Koroda Y, Seligma NG, Horie T (1994) Response of spikelet number to plant nitrogen concentration and dry weight in paddy rice. Agronomy Journal 86: 673-676.

47. Farah GA, Yassin MID, Samia OY (2014) Effect of different fertilizers (bio, organic and inorganic fertilizers) on some yield componants of rice (Oryza Sativa L.). Universal Journal of Agricultural Research 2: 67-70.

48. Quansah GW (2010) Effect of Organic and Inorganic Fertilizers and their Combinations on the growth and yield of maize in the semi-deciduous forest Zone of Ghana. Kwame Nkrumah University of Science and Technology, Kumasi, Ghana.

49. Geleto T, Tanner DG, Mamo T, Gebeyehu G (1996) Response of rainfed bread and durum wheat to source, level and timing of nitrogen fertilizer at two vertisol sites. In: the ninth regional wheat workshop for eastern, Central and Southern Africa. Addis Ababa, Ethiopia.

50. Selamyihun K, Tanner DG, Mamo T (1999) Effect of nitrogen fertilizer applied to tef on the yield and N response of succeeding tef and durum wheat on a highland Vertisol. African Journal of Crop Science 7: 35-46.

51. Edwards S, Arefayne A, Araya H, Egziabher TBG (2007) Impact of compost use on crop yields in Tigray, Ethiopia. Food and Agriculture Organization of the United Nations (FAO), Rome, Italy.

52. USDA Circular 939: 1-19.

53. Journal of Animal and Plant Sciences 23: 258-260.

54. Abebe B, Abebe A (2016) Seed rates on yield and yield components of tef [Eragrostis tef (Zucc.)Trotter] in Ada'a Woreda, Central Ethiopia. Haramaya University, Haramaya, Ethiopia.

Biosorption Potential of the Microchlorophyte *Chlorella vulgaris* for Some Pesticides

Mervat H Hussein[1], Ali M Abdullah[2*], Noha I Badr El Din[1] and El Sayed I Mishaqa[2]

[1]*Botany Department, Faculty of Science, Mansoura University, Mansoura, Egypt*
[2]*Reference Laboratory for Drinking Water, Holding Company for Water and Wastewater, Cairo, Egypt*

Abstract

Nowadays, pollution of either surface or ground water with pesticides is considered as one of the greatest challenges facing Humanity and being a national consideration in Egypt. Agricultural activities are the point source of pesticides that polluting water bodies. The present study investigated the potentiality of *Chlorella vulgaris* for bioremoval of pesticides mixture of 0.1 mg/mL for each component (Atrazine, Molinate, Simazine, Isoproturon, Propanil, Carbofuran, Dimethoate, Pendimethalin, Metoalcholar, Pyriproxin) either as free cells or immobilized in alginate. Two main experiments were conducted including short- term study having 60 min contact time using fresh free and lyophilized cells and other long-term study having five days incubation period using free and immobilized cells. In the short-term study, the presence of living cells led to bioremoval percentage ranged from 86 to 89 and the lyophilized algal biomass achieved bioremoval ranged from 96% to 99%. In long-term study, the presence of growing algae resulted in pesticides bioremoval ranged from 87% to 96.5%. The main mechanism behind the removal of pesticides in water phase is proposed to be biosorption onto the algal cells. This conclusion is based on the short duration required for removal to occur. The obtained results encourage using microalgae in bioremediation of pesticides polluted water.

Keywords: Biosorption; *Chlorella vulgaris*; Algal immobilization; Pesticides; Bioremoval efficiency; LC-MS/MS

Introduction

Recently, application of pesticides is known in everywhere all over the world resulting in exposing the general population to low concentrations of pesticides used in agriculture as herbicide, insecticides and fungicides for controlling plant pests as well as contamination of air, water and foods [1,2]. Pesticides contamination of water has been well documented worldwide to be considered as a potential risk for the ecosystem. Pesticide residues are frequently present in the aquatic environments according to surface runoff, leaching from surface pesticides applications and via industrial activates and/or domestic sewage as founded by Miliadis, Tikoo and Priyadarshani [3-5]. This is why we are in an urgent need for developing some efferent bioprocesses for remediation of pesticides pollutants. Biosorption process is one of the bioremediation mechanisms which is favorable, using living microorganisms as fungi, microalgae as well as bacteria for recovery process that have low costs as suggested by Naturvårsverket [6,7].

In the near future, water reusing will become very important in densely populated arid areas where there is an increasing demand to supply water from limited supplies. Human well-being in a future world will depend mostly upon this sustainable resource and the characterization of emerging contaminants will become important for ecological and human health risk assessments and commodity valuation of water resources [8,9]. Egypt characterized with developing agricultural activities accounted 28% of the total national income, and nearly half of the country's work force is dependent on the agricultural subsector for its livelihood.

There are anaggravating chemical environmental contamination by attributed to using organo-chlorinated pesticides, herbicides, fungicides as well as insecticides that are anticipated along the Nile Delta, which is referred to as "Green Lungs of Egypt" [10]. Moreover, chemical industries in Egypt is one of the main sources of hazardous wastewater. Barakat suggested that, water pollution is exacerbated by agricultural pesticides, raw sewage, and urban and industrial effluents [11]. Consequently, remediation of pesticides from water bodies as well as ground water is very urgent, especially bioremediation by microalgae.

Chemical properties of the pesticide such as molecular weight, functional groups and toxicity affect the metabolic degradation of it [5]. Algae appear to be more able to metabolize organic compounds with low molecular weights than larger molecules [12-14].

The main objective of this study is to investigate the capability of the microalga *C. vulgaris* either free or immobilized cells for bioremoval of ten pesticides mixture.

Materials and Methods

Microalgae

Fresh water *Chlorella vulgaris* was isolated from water sample from river Nile. Culture purification was according to Andersen [15] and the alga was identified according to Philipose [16]. *Chlorella vulgaris* was grown in axenic cultures at 27 ± 2°C under continuous illumination 3600 lux in 500 ml Erlenmeyer flasks, containing 200 ml BG11 medium for 5 days incubation period in an IlluminatedMemmert incubator (Memmert GmbH+Co. KG, Germany) [17]. The starting inoculum size of *C. vulgaris* was 10% (v/v) taken from 5 day-old culture, supplemented as biomass pellet

***Corresponding author:** Ali M Abdullah, Botany Department, Faculty of Science, Mansoura University, Mansoura, Egypt
E-mail: dr2252000@dr.com (or) tsm.hcww@gmail.com

after centrifugation (3000 g, 15 min, Bench-top - TD5B, Germany). Lyophilized algal biomass was prepared from 5 day-old culture pellets that washed once with distilled water and lyophilized in a freeze dryer for 24 h. The lyophilized biomass was stored under dark conditions at room temperature, while living biomass was produced under the same growth conditions.

Selected pesticides

Custom standard mixture (Atrazine, Molinate, Simazine, Isoproturon, Propanil, Carbofuran, Dimethoate, Pendimethalin, Metoalcholar, Pyriproxin) 0. 1 mg/mL for each in methanol was purchased from Accustandard Inc., USA. The standard was obtained from The Reference Laboratory for Drinking Water, Cairo, Egypt. Standard solution containing the 10 microcontaminants in methanolic solution was added to each flask (final water or medium volume of 0. 1 L) to obtain a final concentration of 2 μg L^{-1} and 10 μg L^{-1}. The concentration 10 μg L^{-1} waskeptin high concentration level for further detection of pesticides in agricultural surface water following a runoff or spray drift events [18,19].

Experimental set-up

The short-term study: An initial concentration of 2.0 μg L^{-1} and 10 μg L^{-1} was obtained by adding the pesticide mix to sterile Milli Q water. The experiments included Lyophilizedalgal biomass, living algalbiomass and a control without any biomass, with three replicates per experiment. The amount of biomass (living biomass or lyophilized biomass) added to each replicate corresponded to 10% (v/v) taken from 5 day-old culture. There were three replicates per treatment and the total volume of each culture was 100 ml. The treatments were stirred on anorbital shaker at a speed of 380 rpm for 1 h at room temperature (Thermo Scientific™ MaxQ™ 4450 Benchtop Orbital Shakers, USA). After one hour, the biomass was removed from the aqueous phase by centrifugation and the samples were stored in the freezer at -20°C until analysis.

The long-term study: Final pesticides concentrations of 2.0 μg L^{-1} and 10 μg L^{-1} was obtained by supplementing the pesticides mixture to sterile BG11 with inoculum volume of 10% (v/v) of a five-day old culture with total volume of 100 ml. The experiments were kept under the growing condition described above for 5 days. After the experiment the biomass was removed by centrifugation and samples of the aqueous phase were taken and stored in the freezer until analysis.

Preparation of immobilized algae-alginate beads: After 7 days culturing, algal cells at their log phase were harvested by centrifugation at 5000 rpm for 10 min at 48°C, the pellet was washed and re-suspended in sterilized deionized water (20 ml). This concentrated algal suspension was then mixed with 3% (w/v) sodium alginate solution in 1:3 volume ratios to yield a mixture of algal-alginate suspension which was dropped into calcium chloride solution (2.5%) using magnetic stirrer (CORNING BC620D, USA) to form uniform algal beads. The algal beads were left in $CaCl_2$ solution for 12 h forhardening. Blank alginate beads were prepared in the same way as the algal beads without adding algal cell suspension.

Chromatographic analyses: Samples (50 ml) from the aqueous solution were sent to The Reference Laboratory for Drinking Water, Cairo, Egypt for chromatographic analyses. Reference method EPA 536, were used to conduct the pesticides analysis, which is based on a combination of liquid chromatography (LC) and mass-spectroscopy (MS) specifically called LC-MS/MS (tandem-MS) [20]. Tandem-MS (Xevo-TQ-S, Waters Corporation, Milford, MA, USA) provides low detection limits and very high security, which means that more substances can be tracked at lower level [21].

Results

Short-term study

As illustrated from Figure 1, the highest pesticides bioremoval activity of *Chlorella vulgaris* living cells was recorded to the herbicide atrazine (0.213 μg/l) with initial concertation 2 μg/l, while the herbicide isoproturon recoded the minimum biosorption (0.291 μg/l). Starting with 10 μg/l pesticide mixture, the maximum absorption was documented to the herbicide molinate and the minimum to pendimethalin (1.112 μg/l and 1.687 μg/l respectively).

Concerning 2 μg/l initial concentration, the highestlyophilized *C. vulgaris* biosorption activity was documented for the herbicide isoproturon, while the minimum activity was confirmed to herbicide molinate. Initial concentration 10 μg/l induced maximum bioremoval of the herbicidecarbofuran (0.2178 μg/l), whereas the minimum activity was confirmed to atrazine (0.3712 μg/l) as illustrated in Figure 2.

Starting with 2 μg/l pesticide mixture, living *C. vulgaris* exhibited good biosorption efficiency ranged from (85. 60% to 88. 15%) which was documented for Atrazineand Isoproturn respectively. Concerninginitial conc 10 μg/l *C. vulgaris* lyophilized cells showed high biosorption efficiency ranged from (83.13% to 88.88%) for Molinateand Pendimethalin respectively (Figure 3).

Starting with pesticide mixture concentration (2 μg/l) lyophilized *Chlorella vulgaris* exhibited good biosorption efficiency ranged from (98.6% to 99.36%) which was documented for Isoproturn and molinate respectively. Concerning initial concentration 10 μg/l *C. vulgaris* lyophilized cells showed high biosorption efficiency ranged from (96.29% to 97.822%) for carbofuran and atrazine respectively as indicated from Figure 4.

Long-term study

Long-term study by living *C. vulgaris* biomass: Long term experiment with living *C. vulgaris* biomass (Figure 5) reviled that, bioremoval activity reached the maximum level (0.065 μg/l) with the herbicide simazine, while the minimum bioremoval activity was indicated to pendimethalin (0.243 μg/l) with the initial pesticides concentration 2 μg/l. Although beginning with 10 μg/l, the maximum

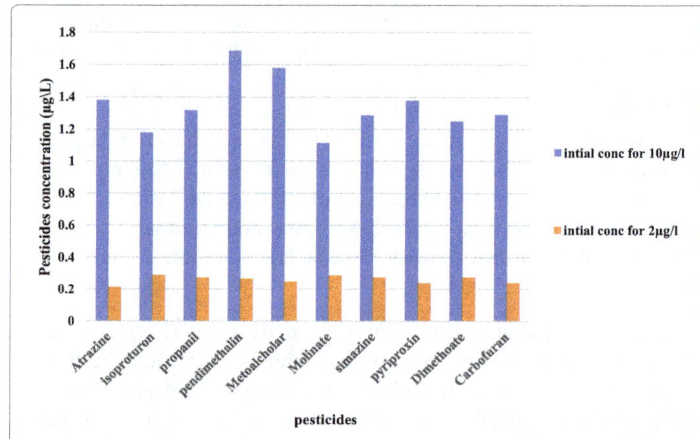

Figure 1: Graph showing amount of trehalose (mg/μl) treated with aqueous extract in adult *C. chinensis*. All values are mean ± SE of six replicates and significant at p ≤ 0.05 level of significance.

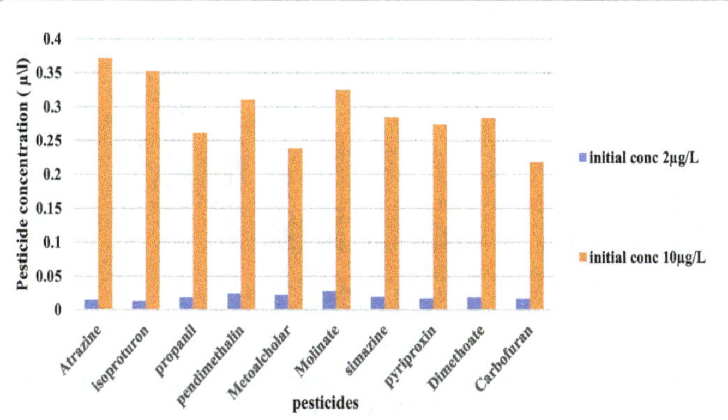

Figure 2: Residual pesticides concentration after biosorption by lyophilized cells of *Chlorella vulgaris* starting with two initial concentrations (2 µg/l and 10 µg/l) after one hour contact time.

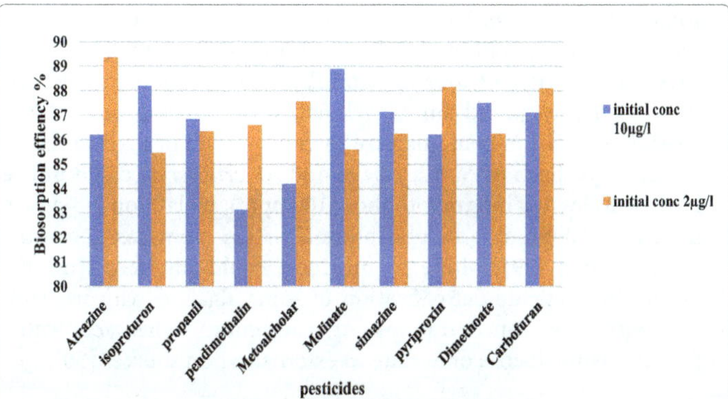

Figure 3: Biosorption efficiency of living cells of *Chlorella vulgaris* for pesticide mixture starting with two initial concentrations (2 µg/l and 10 µg/l) after one hour contact time.

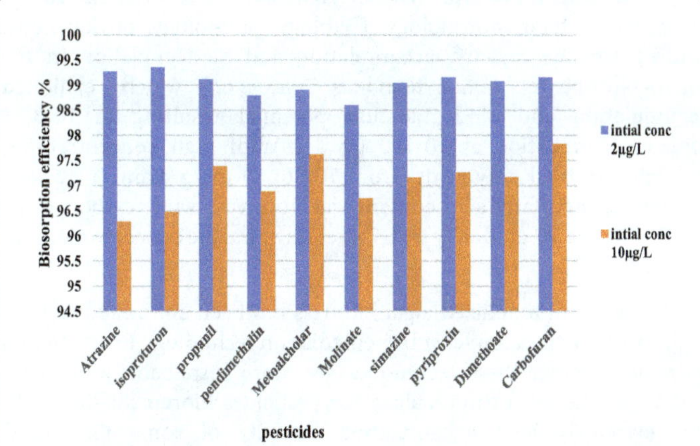

Figure 4: Biosorption efficiency of lyophilized cells of *Chlorella vulgaris* for pesticide mixture starting with two initial concentrations (2 µg/l and 10 µg/l) after one hour contact time.

biosorption activity (0.43 µg/l) was restricted to simazine and the minimum activity (1.186 µg/l) was illustrated to propanil.

Long term study by *C. vulgaris*-alginate beads: Immobilization of *C. vulgaris* cells alginate showed pronounced pesticide bioremoval activity with the two initial concentrations 10 µg/l and 20 µg/l through 5 days incubation period. Figure 6 illustrated maximum (0.157 µg/l) and minimum (1.026 µg/l) bioremoval activity of the two pesticides pendimethalin and propanil respectively for 10 µg/l as initial concentration whereas (0. 003 µg/l and 1.25 µg/l) represented the maximum and minimum bioremoval activity recorded for carbofuran and isoproturon, respectively starting with 20 µg/l.

Long term experiment with living *C. vulgaris* biomass (Figure 7) demonstrated that, biodegradation efficiency reached its maximum level (96.75%) with the herbicide simazine, while the minimum absorption activity was restricted to pendimethalin (87.85%) with the initial pesticides concentration 2 µg/l. Although beginning with 10 µg/l, the maximum biosorption efficiency (95.7%) was recorded to simazine and the minimum activity (88.14%) was illustrated to propanil.

In respect to biosorption potentiality and biodegradation, immobilization of *C. vulgaris* cells in alginate maintained distinct pesticide biosorption potentiality with the two initial concentrations 10 µg/l and 20 µg/l for 5 days incubation period. Figure 8 illustrated maximum (98.43%) and minimum (89.74%) biosorption activity of the two pesticides pendimethalin and propanil respectively for 10 µg/l as initial concentration whereas (87.5% to 99.97%) represented the maximum and minimum biosorption efficiency recorded for carbofuran and isoproturon, respectively starting with 20 µg/l.

Discussion

Recently, the spreading use of pesticides in the exhaustive agricultural activities and the modern daily life as well resulted in urgent environmental complications needing non-conventional treatment strategies as using of microalge for bioremediation processes. Present study indicated that, short-term study using both living and lyophilized biomass of *C. vulgaris* achieved high removal percentages as illustrated in Figures 1-4. The bioremoval of these pesticides over a period of 60 minutes suggested that the probable mechanism as biosorption according to the finds of Komárek, whereas, in Long-term study, there was sufficient time for some mentalizations processes occurred and the algae may either have biosorbed, metabolized or facilitated the degradation of pesticides, or it can be attributed to a combination of all these mechanisms [22]. Exhaustive agricultural activates which depends on using agrochemicals as fertilizers and pesticides, also they increase the global food production but also the at the same time contaminate the environment extensively as suggested by Cáceres, Singh and Walker [23,24].

Algal cell size, density, morphology and physiological activates can be attributed to biosorption and removal of pesticide. Microalgae is characterized by processing high surface area to biovolume ratio which provide high potential for sorption and the following inter action with pesticides. microalgae can utilize pesticide at the nontoxic levels. This was explained by Butler who reported that some macroalgal species (*Chlorella, Monoraphid, Actinastrum, Scenedemu, Nitzschia*) had the ability to degrade herbicides as following:1 ppm of carbaryl and diazinon, and 0.01 ppm of methoxychlor and 2,4-D [25]. Microgreen algae *C. vulgaris* and *Scenedesmus bijugatus* could metabolize organophosphorus insecticides (mon ocrotophos and quinalphos) while some cyanobacteria (*Synechococcus elongatus, Phormidium tenue* and *Nostoc linckia*) could metabolized these pesticides in the range of 5 to 50 ppm via 30 days as demonstrated by Megharaj [26]. *Anabaena* sp. and *Aulosira fertilissima* stored DDT, fenitrothion and chlorpyrifos in their cells as indicated by Lal, who reported that *Anabaena* sp. Absorbed 1568 ppm DDT, 3467 ppm fenitrothion and 6779 ppm chlorpyrifos, while *A. fertilissima* stored 1429 ppm DDT, 6651 ppm fenitrothion and 3971 ppm chlorpyrifos; where as these cyanobacteria species could metabolize DDT to DDD and DDE [27].

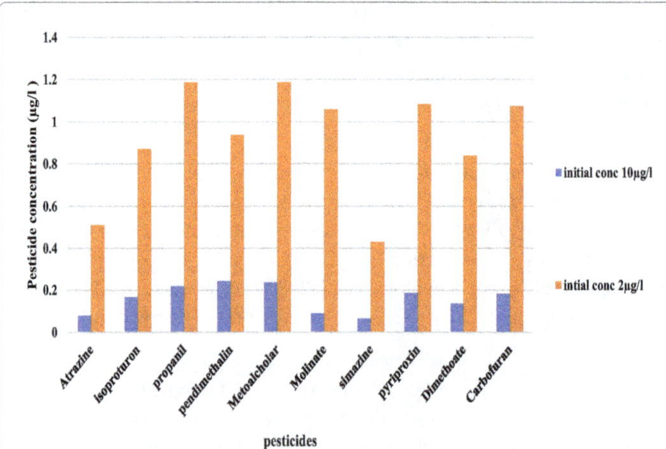

Figure 5: Residual pesticides concentration after biosorption by living cells of *C. vulgaris* starting with two initial concentrations (2 µg/l and 10 µg/l) after incubation period of 5 days.

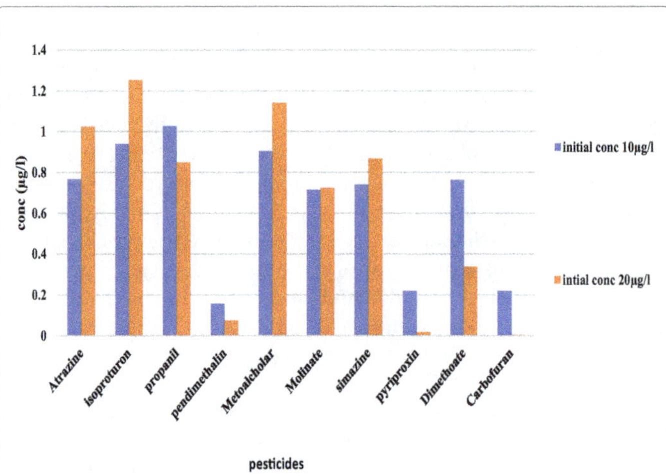

Figure 6: Residual pesticides concentration after biosorption by *C. vulgaris*-alginate beads starting with two initial concentrations (10 µg/l and 20 µg/l) after incubation period of 5 days.

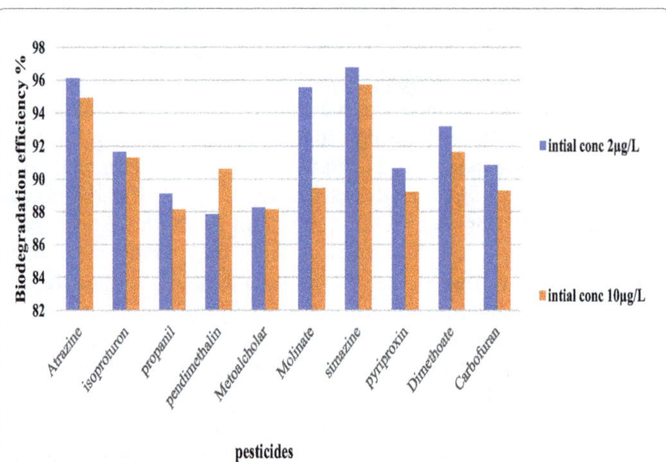

Figure 7: Biodegradation efficiency of pesticides for *Chlorella vulgaris* living cells starting with different initial concentrations (2 µg/L and 10 µg/L) after incubation period of 5 days.

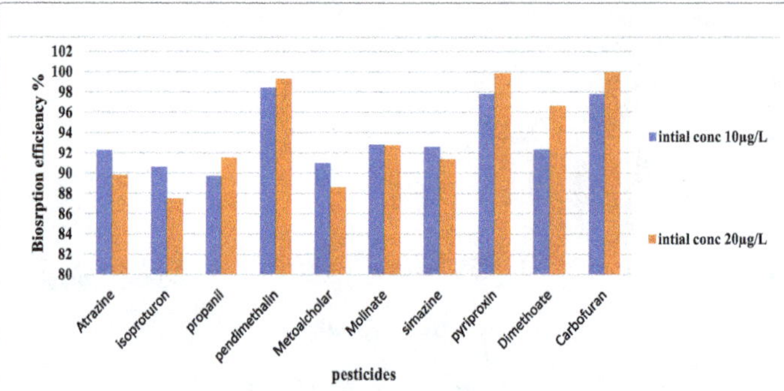

Figure 8: Biodegradation potential of pesticides for *C. vulgaris* - alginate beads starting with two initial concentrations (10 µg/L and 20 µg/L) after incubation period of 5 day.

Megharaj, indicated that greenalgae (*C. vulgaris and S. bijugatus*) and cyanobacteria (*N. linckia, Nostoc muscorum, Oscillatoria animalis* and *Phormidium foveolarum*) could metabolized the organophosphorus insecticide methyl parathion, and as a phosphorus source [28]. After 30 days of incubation, *O. animalis* and *P. foveolarum* totally biosorbed 20 µg mL^{-1} methyl parathion as well as its hydrolysis product PNP, whereas *N. muscorum* could oxidize the nitro group of PNP to nitrite via fifteen days. Each of *N. muscorum* and *A. fertilissima* could utilize each of the following monocrotophos (100 ppm), malathion (75 ppm), dichlorovos (50 ppm) and phosphomidon (25 ppm) as suggested by Subramanian [29]. Mode of nutrition could interfere with the ability of the pesticide detoxification of some algae. *C. vulgaris* with stand poisonous concentrations of carbofuran when cultivated ixotrophicallyon glucose or acetate as external cabon sources [30].

Concerning atrazine, Tang found that green algae as *Chlamydomonas sp., Chlorella sp., Pediastrum sp.,* and *S. quadricauda* could biosorbe more atrazine compared with diatoms (*Cyclotella gamma, Cyclotella meneghiniana, Synedra acus* and *Synedra radians*) [31]. The differential selectivity of algae species to atrazine can be attributed to their morphology, Cytology, physiology, phylogenetic, studied atrazine toxicity, accumulation and biodegradation in the microchlorophyte *Chlamydomonas mexicana* which exhibited accumulation and biodegradation potential resulting in 14-36% atrazine degradation at 10-100 µg L^{-1}. With high concentrations, reduction in total fatty acids (from 102 to 75 mg/g^{-1}) and increasing the unsaturated fatty acid content was observed, while carbohydrate content increased gradually with increasing atrazine concentrations up to 15% [32,33].

Living or immobilized algal biomass has been confirmed to have high capability as a low-cost bioremediation technology for pesticides bioremoval, since these techniques are more sustainable and might encourage the use of microalgae for pesticides bioremediation [34]. The extremely high accumulation capacity of some microalgae for potentially dangerous substances has been also exploited for bioremediation techniques for water [35,36].

Conclusion

It is concluded that the lyophilized biomass achieved removal percentages reached up to 99% of pesticides which was higher than living *C. vulgaris* biomass at the short-term experiments. On the other hand, long-term experiments proved the ability of growing *Chlorella vulgaris* for the removal of pesticides. The present results indicate the possibility of water bioremediation using microalgae for removal of organic pollutants (pesticides) in water.

References

1. The 2012 European Union Report on Pesticide Residues in Food (2014) European Food Safety Authority.
2. Trunnelle KJ, Bennett B, Deborah H, Ahn P, Chang K, et al. (2014) Concentrations of the urinary pyrethroid metabolite 3-phenoxybenzoic acid in farm worker families in the MICASA study. Environ Res 131: 53-159.
3. Miliadis GE (1994) Determination of pesticide residues in natural waters of Greece by solid phase extraction and gas chromatography. Bull Environ Contam Toxicol 52: 25-30.
4. Tikoo V, Shales SW, Scragg AH (1996) Effects on Pentachlorophenol on the Growth of Microalgae. Environmental Technology 17: 39-44.
5. Priyadarshani I, Sahu D, Rath B (2011) Microalgal bioremediation: current practices and perspectives. Journal of Biochemical Technology 3: 299-304.
6. Avloppsreningsverkens förmåga att ta hand om läkemedelsrester och andra farliga ämnen (2008) Redovisning av regeringsuppdrag.
7. Zaini MA, Amano Y, Machida M (2010) Adsorption of heavy metals onto activated carbons derived from polyacrylonitrile fiber. Journal of Hazardous Materials 180: 552-560.
8. Blasco C, Picó Y (2009) Prospects for combining chemical and biological methods for integrated environmental assessment. TrAC Trends in Analytical Chemistry 28: 745-757.
9. Young JG, Eskenazi B, Gladstone EA, Bradman A, Pedersen L, et al. (2005) Association between in utero organophosphate pesticide exposure and abnormal reflexes in neonates. Neurotoxicology 2: 199-209.
10. Mansour SA (2004) Pesticide exposure-Egyptian scene. Toxicology 198: 91-115.
11. Barakat AO (2004) Assessment of persistent toxic substances in the environment of Egypt. Environment International 30: 309-322.
12. Semple KT, Cain RB, Schmidt S (1999) Biodegradation of aromatic compounds by microalgae. FEMS Microbiology letters 170: 291-300.
13. Juhasz AL, Ravendra N (2000) Bioremediation of high molecular weight polycyclic aromatic hydrocarbons: a review of the microbial degradation of banzo[a]pyrene. Int Biodeterior Biodegrad 45: 57-88.
14. Ghasemi Y, Rasoul-Amini S, Fotooh-Abadi E (2011) Review: The Biotransformation, Biodegradation, and Bioremediation of Organic Compounds by Microalgae. Journal of Phycology 47: 969-980.
15. Andersen RA (2005) Algal Culturing Techniques. Academic Press, USA.
16. Philipose MT (1967) Chlorococcales. Volume 8, Indian Council of Agricultural Research, New Delhi, India, pp: 31-41.
17. Stanier RY, Kunisawa R, Mandel M, Cohen-Bazire G (1971) Purification and properties of unicellular blue-green algae (order Chroococcales). Bacteriol Rev 35: 171-205.
18. Fielding G (1995) Leaching of phenoxy alkanoic acid herbicides from farmland. Science of the total environment 168: 11-18.
19. Davis AM, Thorburn PJ, Lewis SE, Bainbridge ZT, Attard SJ, et al. (2013) Environmental impacts of irrigated sugarcane production: Herbicide run-off dynamics from farms and associated drainage systems. Agriculture, ecosystems and environment 180: 123-135.
20. Smith GA, Pepich BV, Munch DJ (2007) EPA Method 536: Determination of Triazine Pesticides and Their Degradates in Drinking Water by Liquid Chromatography Electrospray Ionization Tandem Mass Spectrometry (LC/ESI-MS/MS).
21. Jansson C, Kreuger J (2010) Multiresidue analysis of 95 pesticides at low nanogram/liter levels in surface waters using online preconcentration and high performance liquid chromatography/tandem mass spectrometry. Journal of AOAC International 93: 1732-1747.
22. Komárek M, Čadková E, Chrastný V, Bordas F, Bollinger JC (2010) Contamination of vineyard soils with fungicides: a review of environmental and toxicological aspects. Environment international 36: 138-151.
23. Cáceres L, Escudey M, Fuentes E, Báez ME (2010) Modeling the sorption kinetic of metsulfuron-methyl on Andisols and Ultisols volcanic ash-derived soils: Kinetics parameters and solute transport mechanisms. Journal of hazardous materials 179: 795-803.
24. Singh BK, Walker A (2006) Microbial degradation of organophosphorus compounds. FEMS Microbiology Reviews 30: 428-471.
25. Butler GL, Deason TR, O'Kelley JC (1975) Loss of five pesticides from cultures of twenty-one planktonic algae. Bull Environ Contam Toxicol 13:149-52.
26. Megharaj M, Venkateswarlu K, Rao AS (1987) Metabolism of monocrotophos and quinalphos by algae isolated from soil. Bull Environ Contam Toxicol 39:251-6.
27. Lal S, Lal R, Saxena DM (1987) Bioconcentration and metabolism of DDT, fenitrothion and chlorpyrifos by the blue-green algae Anabaena sp. and Aulosira fertilissima. Environ Pollut 46:187-96.
28. Megharaj M, Madhavi DR, Sreenivasulu C, Umamaheswari A, Venkateswarlu K (1994) Biodegradation of methyl parathion by soil isolates of microalgae and cyanobacteria. Bull Environ Contam Toxicol 53: 292-297.
29. Subramanian G, Sekar S, Sampoornam S (1994) Biodegradation and utilization of organophosphorus pesticides by cyanobacteria. Int Biodeterior Biodegrad 33:129-43.
30. Megharaj M, Pearson HW, Venkateswarlu K (1993) Toxicity of carbofuran to soil isolates of Chlorella vulgaris, Nostoc linckia and N. muscorum. Appl Microbiol Biotechnol 39: 644-648.
31. Tang J, Hoagland KD, Siegfried BD (1998) Uptake and bioconcentration of atrazine by selected freshwater algae. Environ Toxicol Chem 17: 1085-1090.
32. Debelius B, Forja JM, Del Valls A, Lubián LM (2008) Effect of linear alkylbenzene sulfonate (LAS) and atrazine on marine microalgae. Marine Pollution Bulletin 57: 559-568.
33. Kabra AN, Ji MK, Choi J, Kim JR, Govindwar SP, et al. (2014) Toxicity of atrazine and its bioaccumulation and biodegradation in a green microalga, Chlamydomonas mexicana. Environmental Science and Pollution Research 21: 12270-12278.
34. Hultberg M, Jönsson HL, Bergstrand KJ, Carlsson AS (2014) Impact of light quality on biomass production and fatty acid content in the microalga Chlorella vulgaris. Bioresource Technology 159: 465-467.
35. Maeda S, Mizoguchi M, Ohki A, Takeshita T (1990) Bioaccumulation of zinc and cadmium in freshwater alga, Chlorella vulgaris. Part I. Toxicity and accumulation. Chemosphere 21: 953-963.
36. Greene B, Bedell GW (1990) Algal gels or immobilized algae for metal recovery. Introduction to Applied Phycology. SPB Academic Publishing, The Netherlands, pp: 137-149.

Efficacy Testing of *Acetobacter* and *Azospirillum* Isolates on Maize cv. GM-3

Jhala YK*, Shelat HN and Panpatte DG

Department ofAgricultural Microbiology, Anand Agricultural University, Anand-388 110, Gujarat, India

Abstract

Endophytic bacteria residing within living plant tissues without substantially harming plants have found a large number of applications in today's agriculture such as nutrient cycling, tolerance to biotic and abiotic stress as well as promotion of plant growth. Endophytic bacteria namely, *Acetobacter* and *Azospirillum* promised to be practically used in agriculture are nowadays thought of as most active components in association with cereals. So, in present study, endophytic bacteria mainly belonging to genera *Acetobacter* and *Azospirillum* were isolated from surface sterilized plant parts of species *Cynodon dactylon* (Durva), *Pothos scandens* (Money plant), *Ipomea batata* (Sweet potato), *Saccharum officinarum* (Sugarcane) *cv.* CO.LK-8001 and CO.-84135, *Musa paradica* (Banana) and *Zea mays* (maize) *cv.* GM-6 by using LGIP and Nitrogen free bromothymol blue media selective for growth of *Acetobacter* and *Azospirillum,* respectively. Inoculation of maize *cv.* GM-3 with endophytic bacterial isolates in combination with half and full Recommended dose (RD) of urea recorded significant increase in all the growth parameters wherein A-9 was found the best for growth stimulation as compared to other treatments.

Keywords: Endophytic; *Acetobacter*; *Azospirillum*; Maize

Introduction

It is well known that "corn" crop is considered among the most important cereal crops either all over the world that consumes huge quantities of chemical nitrogenous fertilizers (16 million tones/year). Many attempts have been tried to replace a part of those harmful chemical fertilizers by bio fertilizers to get yield of a good quality without loss in its quantity. Bio fertilizers are the microbial inoculants which help in increasing crop productivity by way of biological nitrogen fixation, increased availability or uptake of nutrients through solubilization or increased absorption, stimulation of plant growth through hormonal action or antibiosis, earlier days the term biological nitrogen fixation (BNF) was restricted to the endophytic bacteria of genus *Rhizobia* which can form mutually beneficial symbiotic nitrogen fixing association with leguminous plants. But now recent researches have intensely brought down the difficulties and increased the scope for nitrogen fixation in cereals by means of endophytic nitrogen fixation [1]. Kado defined endophytic bacteria that reside within plant tissue without doing substantial harm or gaining benefit other than securing residency. Among endophytic bacteria *Acetobacter* is interesting because it carries out nitrogen fixation under aerobic conditions, because it requires oxygen for the production of large quantities of ATP required for nitrogen fixation. *A. diazotrophicus* is of special interest because it can excrete almost half of the fixed nitrogen in the form potentially available to plants. *Azospirillum,* an associative microaerophilic organism can live in association with diverse group of plants. Associative nitrogen fixation, capability to produce plant growth promoting antifungal / antibacterial substances and their effect on root morphology are the principal mechanisms responsible for the observed promotion of crop yield. Inoculation with *Azospirillum* results in enhanced assimilation of mineral nutrients (N, P, K, Fe^{2+}), and water and offers resistance to pathogens. The utilization of inoculants containing biofertilizer is becoming more popular due to increasing reports of expressive gains in grain yields. Many workers have proved the role of biofertilizers in reducing nitrogen fertilizer requirement of the corn crop. Fukami et al. [2] reported in field experiments that inoculation with *A. brasilense* allowed for a 25% reduction in the need for N fertilizers. So, present investigation was carried out with an objective to find out effective native biofertilizer strains to reduce chemical nitrogen fertilizer input in maize crop.

Materials and Methodology

Source of isolates

In present study, isolation of endophytic bacteria was attempted from various plant parts (root, stem and leaves) of species *viz. Cynodon dactylon* (Durva), *Pothos scandens* (Money plant), *Ipomea batata* (Sweet potato), *Saccharum officinarum* (Sugarcane) *cv.* CO.LK-8001 and CO.-84135, *Musa paradica* (Banana) and *Zea mays* (maize) *cv.* GM-6.

Isolation procedure for endophytic bacteria

For isolation, enrichment culture technique was employed. After collection, the plant was thoroughly washed with tap water to remove any soil particles adhered to it and was cut in to small pieces approximately of 0.5 cm size and surface sterilized by treatment of 1% $HgCl_2$ (Mercuric chloride) followed by treatment of 70% ethyl alcohol and finally washing with sterile distilled water. These surface sterilized pieces were then slightly burned in the low flame of burner to alleviate any surface adhering microorganism and then inoculated in to sterilized LGIP and NFB medium [3] (Cavalcante and Dobereiner), selective for growth of *Acetobacter* and *Azospirillum,* and incubated at 28 ± 2°C for one week on shaker to ensure good growth of organisms. After giving two successive transfers in same selective medium, 100 μl aliquot was inoculated in semisolid LGIP and NFB media to check pellicle formation. Pellicles forming bacterial cultures were streaked over the surface of solid LGIP and NFB medium to obtain colonial growth of organism. Single colony with typical morphology of *Acetobacter* and *Azospirillum* as described by Dobereiner [4] were picked from these plates to subculture by re-streaking onto LGIP and NFB plates and used for further study (Table 1).

***Corresponding author:** Jhala YK, Department of Agricultural Microbiology, Anand Agricultural University, Anand-388 110, Gujarat, India,
E-mail: yogeshvari.jhala@gmail.com

In vitro efficacy testing of endophytic bacterial isolates on maize *cv.* GM-3

In vitro efficacy of isolates on maize seeds was tested on solid water agar in petri plates. Maize seeds of variety GM-3 were surface sterilized by treatment of 0.1% $HgCl_2$ solution 3 times for time interval of 15 mins followed by treatment with ethyl alcohol and finally washing with sterile distilled water. These surface sterilized seed were then inoculated with 0.01 ml. of previously grown starter cultures of *Acetobacter and Azospirillum* isolates in LGIP and NFB broth and allowed to stand for 30 mins. Control seeds without treatment were also used as check. Five treated seeds were now allowed to grow on petri plates containing sterilized 1% water agar medium under dark conditions. After one week of incubation the plantlets were removed carefully from water agar and root length, shoot length, fresh weight and dry weight were measured.

In vivo efficacy of endophytic bacterial isolates on maize *cv.* GM-3

Characteristics of soil used for pot trial: The soil of the experimental pot was sandy loam, locally known as "Goradu". The soil was well drained and retentive of moisture. It responded well to irrigation and manuring and was reasonably suitable for maize cultivation. Ultimate Physico-chemical condition of the experimental soil was analyzed at Department of Soil Science, BACA, AAU, Anand is given in Table 2.

Seed inoculation efficacy of endophytic bacterial isolates on maize *cv.* GM-3: This trial was undertaken with graded doses of urea i.e., recommended dose (120 kg/ha) and half of the recommended dose (60 kg/ha) for demonstration of nitrogen savings in maize which is highly exhaustive crop. GM-3 seeds were inoculated with previously grown starter cultures of *Acetobacter* and *Azospirillum* isolates in LGIP and NFB broth having 10^8-10^9 bacterial counts at the rate of 5 ml/kg seeds and allowed to stand for 30 mins. Five seeds were sown at the depth of 5 cm in pots having 10 kg soil collected from agronomy farm, AAU, Anand. Treatments were set as per point mentioned in treatment details. Agronomic practices were common for all the treatments. Factorial completely Randomized Design (F-CRD) was applied with 22 treatments and 3 replications (Table 3).

S No	Name of isolate	Source of organism		
		Scientific name	Common name	Plant part
1.	A-1	*Pothos scandens*	Money plant	Leaf
2.	A-2	*Cynodon dactylon*	Durva	Leaf
3.	A-3	*Ipomea batata*	Sweet potato	Root
4.	A-4	*Saccharum officinarum cv.* CO.-84135	Sugarcane	Stem
5	ACG-1	*Acetobacter diazotrophicus* from *Saccharum officinarum*	Sugarcane	Stem
6.	A-6	*Pothos scandens*	Money plant	Leaf
7.	A-7	*Cynodon dactylon*	Durva	Leaf
8.	A-8	*Musa paradica*	Banana	Root
9.	A-9	*Saccharum officinarum cv.* CO.LK-8001	Sugarcane	Root
10	ASA-1	*Azospirillum lipoferum* from *Pennisetum glucam*	Bajara	Root

Table 1: Endophytic bacterial isolates of different plant parts and species.

S No	Element	Properties of soil
1.	Total nitrogen (%) from soil	0.021
2.	Organic Carbon (%)	0.24
3.	Available P_2O_5 (kg ha^{-1})	22.78
4.	Available K_2O (kg ha^{-1})	176.88
5.	Electrical Conductivity (dSm^{-1} at 25°C)	0.14
6.	pH	8.41

Table 2: Chemical characteristic of experimental soil.

Tr. No.	Nitrogen (N)			Bacterial isolate (B)
T_1	N_1	60 kg/ha (½ RD)	B_0	No biofertilizer (control)
T_2			B_1	A-1
T_3			B_2	A-2
T_4			B_3	A-3
T_5			B_4	A-4
T_6			B_5	ACG-1
T_7			B_6	A-6
T_8			B_7	A-7
T_9			B_8	A-8
T_{10}			B_9	A-9
T_{11}			B_{10}	ASA-1
T_{12}	N_2	120 kg/ha (RD)	B_0	No biofertilizer (control)
T_{13}			B_1	A-1
T_{14}			B_2	A-2
T_{15}			B_3	A-3
T_{16}			B_4	A-4
T_{17}			B_5	ACG-1
T_{18}			B_6	A-6
T_{19}			B_7	A-7
T_{20}			B_8	A-8
T_{21}			B_9	A-9
T_{22}			B_{10}	ASA-1

Note: RD=Recommended dose

Table 3: Treatment details.

Observations Recorded

Plant height at 30 DAS, 60 DAS and 90 DAS

At the time of 30, 60 and 90 days after sowing, plant height of five plants was measured and average was reported as plant height.

Stem girth and internode length at 60 DAS

At the time of 60 days after sowing, internode length and stem girth of five plants were measured and average was reported as internode length and stem girth.

Fresh and dry shoot and root weight

After harvesting fresh shoot and root weight of 5 individual excised plants was recorded and their average was reported as fresh shoot and root weight. Shoots and roots were dried under natural conditions and weight of 5 individual plants was recorded and their average was reported as dry shoot and root weight.

Chemical analysis of soil

Organic carbon (%) by Walkley and Black's titration method: 1.0 gm of the soil sample (finely ground to pass through 0.5 mm. sieve) was taken into a 500 ml Erlenmeyer flask. To this, 10 ml of normal potassium dichromate solution was added, followed by 20 ml. of conc. H_2SO_4. This mixture was shaken for half an hour. At the end of the period, 200 ml of water was added to the flask with the addition of 10 ml 85% H_3PO_4 and 10 ml of the indicator diphenylamine. When the contents of the flask attain a dark blue color, titration was done against 0.5 N ferrous sulphate solution till the contents attain a brilliant green color.

Calculation:

$$\%\,\text{of total nitrogen} = \frac{10(B-T)}{B} \times \frac{0.003 \times 100B}{\text{Wt.of soil(g)}}$$

% of total nitrogen=% OC × 0.0862

***Acetobacter* and *Azospirillum* soil counts:** Soil samples were collected before sowing and at the time of harvest pot wise separately and stored in polythene bags and kept in refrigerator till processed. *Acetobacter* and *Azospirillum* counts were done by taking 1 gm soil sample in sterile 100 ml D/W and shaken it for 1 hour and 0.1 ml sample was taken aseptically from it and transferred in 4.5 ml D/W containing dilution tube to make up to 10^{-8} dilutions by serial dilution method and spreaded on LGIP and N free BTB agar plates and incubated for 48 hours. After 48 hours, counts were taken by calculating cfu/gm.

$$\text{Final Count(cfu/ gm)} = \frac{\text{Number of well} - \text{isolated colonies}}{\text{Dilution factor} \times \text{Aliquot taken}}$$

Results and Discussion

Source of isolates

In all, total 24 strains were isolated from different plant parts and species. Out of these, total 10 isolates were selected on the basis of their appearance and vigor to grow on NFB (5-isolates) and LGIP (5-isolates) medium. Standard strains of *Azospirillum lipoferum* (ASA-1) and *Acetobacter diazotrophicus* (ACG-1) were collected from Department of Agriculture Microbiology, BA College of Agriculture, Anand Agricultural University, Anand and used as positive check during entire investigation. After getting pure culture of organism series of morphological, physiological and biochemical tests were carried out and which have well established similarities of isolates A-1 to A-5 with genus *Acetobacter* and isolates A-6 to A-10 with genus *Azospirillum* So, we can classify them according to [5] Bergey's manual of systematic bacteriology as *Acetobacter* and *Azospirillum* species.

In vitro efficacy testing of endophytic bacterial isolates on maize *cv.* GM-3

Inoculation with all the bacterial strains had significant effect on the development of maize *cv.* GM-3 after 14 days of inoculation. Isolate A-9 showed significantly higher root (15.8 cm) and shoot length (13.2 cm), fresh (0.9 g) and dry biomass (0.5 g) which was at par with isolate A-7 and ASA-1 except shoot length (Table 4). All the bacterial inoculants showed maximum root hairs (Figure 1) as compared to non-inoculated control.

Treatment	Root length (cm)	Shoot length (cm)	Fresh biomass weight (g)	Dry biomass weight (g)
Control	6.7^{f}	6.5^{fg}	0.5^{d}	0.1^{d}
A-1	13.2^{bc}	10.9^{b}	0.9^{a}	0.5^{a}
A-2	8.4^{e}	10.9^{b}	0.7^{c}	0.4^{abc}
A-3	13.3^{bc}	5.6^{g}	0.7^{c}	0.3^{c}
A-4	12.4^{cd}	8.7^{cd}	0.7^{bc}	0.3^{c}
ACG-1	11.1^{d}	7.5^{ef}	0.6^{c}	0.3^{c}
A-6	13.3^{bc}	9.7^{bc}	0.7^{c}	0.3^{c}
A-7	14.8^{ab}	10.5^{b}	0.9^{a}	0.4^{bc}
A-8	13.7^{bc}	7.9^{de}	0.8^{ab}	0.5^{ab}
A-9	15.8^{a}	13.2^{a}	0.9^{a}	0.5^{a}
ASA-1	14.8^{ab}	10.8^{b}	0.8^{ab}	0.5^{ab}
SEM	0.5	0.4	0.03	0.03
CD at 5 %	1.4	1.1	0.1	0.1
CV %	6.6	6.9	7.8	16.0

Note: Treatment means with the letter/letters in common are not significant by Duncan's New Multiple Range Test at 5% level of significance.

Table 4: *In vitro* efficacy testing of isolates on maize *cv.* GM-3.

Seed inoculation efficacy of *Acetobacter* and *Azospirillum* isolates on maize *cv.* GM-3

A pot trial was carried out with recommended dose (120 kg N/ha) and half of the recommended dose (60 kg N/ha) of urea to study the effect of various endophytic bacterial isolates on maize *cv.* GM-3 and also to demonstrate nitrogen savings in maize due to bio-inoculants.

Effect on plant height, internode length and stem girth

Data regarding changes in the plant height at 30, 60 and 90 DAS and internode length and stem girth at 60 DAS as well as root and shoot fresh and dry weight due to inoculation of endophytic bacteria along with recommended and half of the recommended doses of nitrogen are presented in Table 5. The results revealed that seed inoculation of all the isolates significantly influenced plant height at 30, 60 and 90 DAS (Figures 2 and 3). Among the different strains of endophytic bacteria isolate A-9 recorded maximum plant height (65.0, 87.8 and 116.9 cm) at 30, 60 and 90 DAS respectively which was significantly superior over uninoculated control (51.9, 64.0 and 66.8 cm). Application of urea significantly influenced plant height. Treatment with R.D. gave 59.2, 81.2 and 106.1 cm. average plant height at 30, 60 and 90 DAS which was significantly superior over half R.D. (56.7, 76.1 and 93.8 cm). However, interaction between endophytic bacteria and doses of nitrogen proved to be non-significant at 30, 60 and 90 DAS suggesting that at any level of nitrogenous fertilizer bacterial inoculation improves growth of plant. Swędrzyńska et al. [6] reported that maize (*Zea mays sp. Saccharata* L.) inoculated with *Azospirillum brasilense* showed 27% increase in yield and higher cob mass than uninoculated control under different cultivation

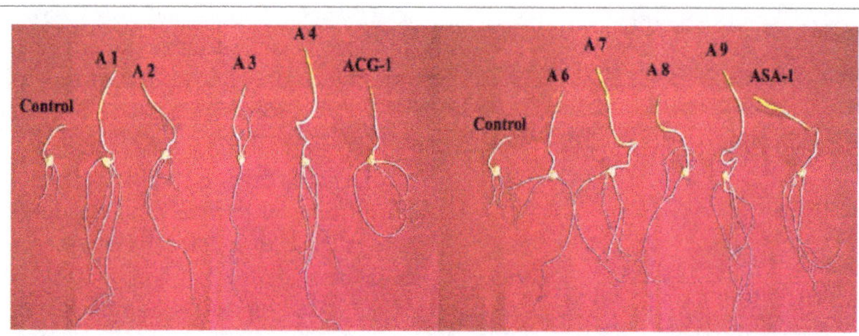

Figure 1: *In vitro* efficacy testing of isolates on maize *cv.* GM-3.

Treatment	Plant height (cm)			Internode length (cm)	Stem girth (cm)	Root fresh weight (g)	Root dry weight (g)	Shoot fresh weight (g)	Shoot dry weight (g)
	30 DAS	60 DAS	90 DAS	60 DAS		90 DAS			
N levels									
N_1 (60 kg/ha)	56.7	76.1	93.8	6.9	4.5	94.7	60.3	81.5	32.1
N_2 (120 kg/ha)	59.2	81.2	106.1	9.5	4.9	108.3	70.7	89.2	34.9
SEM	0.55	0.9	2.1	0.2	0.1	3.0	1.2	0.9	1.1
C.D. at 5%	1.57	2.6	6.1	0.6	0.1	8.6	3.3	2.6	3.1
B levels									
B_0: No biofertilizer	51.9	64.0	66.8	6.5	4.3	61.0	32.4	74.1	26.1
B_1 : A-1	57.1	78.6	96.8	7.7	4.6	79.6	45.8	82.4	29.7
B_2 : A-2	58.9	82.2	98.6	7.1	4.7	77.2	43.3	88.0	29.9
B_3: A-3	61.0	84.7	104.8	8.4	4.9	126.0	85.7	89.6	39.0
B_4: A-4	56.0	80.0	98.5	7.1	5.1	93.7	62.0	84.1	26.5
B_5: ACG-1	55.3	69.3	94.9	6.0	4.5	98.1	64.3	78.2	31.5
B_6: A-6	53.0	75.7	101.2	7.7	4.4	119.0	82.9	85.9	36.2
B_7: A-7	55.0	84.7	102.2	8.0	4.9	90.8	53.8	90.3	37.4
B_8: A-8	62.5	84.1	114.7	8.5	4.6	127.3	87.5	87.0	37.1
B_9: A-9	65.0	87.8	116.9	8.1	5.2	131.5	88.3	98.5	43.2
B_{10}: ASA-1	61.3	74.5	104.0	8.0	4.9	112.0	74.5	83.4	31.9
SEM	1.3	2.3	5.0	0.2	0.12	7.1	2.7	2.1	2.6
C.D. at 5%	3.7	4.5	14.2	0.6	0.34	20.1	7.7	6.0	7.3
Interaction effect									
N × B	NS	NS	NS	NS	0.48	NS	NS	NS	NS
CV %	5.5	7.1	12.2	16.1	6.2	17.0	10.1	6.0	18.7

Table 5: Effect on plant height, internode length and stem girth.

Figure 2: Effect of bacterial isolates along with half RD (60 kg N/ha) on maize *cv.* GM-3.

Figure 3: Effect of bacterial isolates along with full RD (120 kg N/ha) on maize *cv.* GM-3.

conditions. Riggs et al. [7] conducted greenhouse experiment in maize without N fertilizer. Inoculation of *G. diazotrophicus* Pal-5 significantly increased dry weight of maize genotypes Mo17, B14 and B84 by 42.6, 25.2 and 15.6 percentage, respectively. In field trail where N @ 224 kg/ha was applied, maize genotypes B73xMo17, 36H36 and 3905 showed increase in yield by 25.3, 23.4 and 14.4 percentage, respectively. Moreover, the results also revealed that seed inoculation of all the isolates significantly influenced inter node length at 60 DAS. Among the different strains of endophytic bacteria isolate A-8 recorded highest inter node length (8.5 cm) 60 DAS which was significantly superior over uninoculated control (6.5 cm), closely followed by isolate A-3 (8.4 cm) and A-9 (8.1 cm). Similarly application of endophytic bacterial cultures with half R.D. and full R.D. significantly influenced stem girth at 60 DAS. Isolate A-9 reported significantly higher stem girth (5.2 cm) as compared to control (4.2 cm). The data also indicated that stem girth per plant also significantly influenced due to application of urea. N_2: 120 kg/ha recorded higher stem girth (4.9 cm). Here, interaction between endophytic bacterial inoculants with and without urea also significantly affected stem girth of maize at 60 DAS. These findings are confirmed by Osmar et al. [8] who had reported maize Cargil - 909 seeds inoculated with *Azospirillum* sp. RAM-7 and RAM-5 strains can reduce 40% of the recommended N fertilizer under field conditions. Mehnaz et al. [9] studied effect of *G. azotocaptan* strain DS1 isolated from maize rhizosphere and *G. diazotrophicus* strain Pal-5 and *nif* D mutant strain on different maize cultivars and reported that *G. diazotrophicus* Pal-5 and *G. azotocaptan* DS-1 were resulted in significant increase in shoot weight of corn variety 39D82 and 39M27, respectively in sand experiment. *G. azotocaptan* DS-1 significantly increased root weight of corn variety 39H84 and *G. diazotrophicus nif* D significantly increased shoot weight of corn variety 39M27 soil experiment.

Effect on root and shoot fresh and dry weight at harvesting

Data regarding changes in the root and shoot fresh and dry weight at 90 DAS are presented in Table 5. Among different strains of endophytic bacteria isolate A-9 recorded highest root fresh weight (98.5 g) and root dry weight (43.2 g), shoot fresh weight (131.5 g) and shoot dry weight (88.3 g), as compared to uninoculated control, closely followed by A-8, A-3 and A-6 (Figure 4). Interaction between endophytic bacteria and doses of urea proved to be non-significant for root and shoot fresh and dry weight at 90 DAS. Overall, from the above results it is ascertained that bacterial inoculation had significant impact on growth and growth attributes of maize *cv.* GM-3 and can substitute about 50% nitrogen fertilizer in maize crop (Figure 5). These results are in conformity with Cohen et al. [10] who had reported that maize (*Zea mays*) plants inoculated with *A. brasilense* Sp-7 and CD-1 significantly increased dry weight of plants (i.e., 26 and 20% respectively) after 4 weeks of inoculation as compared to uninoculated control.

Figure 4: Effect of bacterial isolates on root and shoot length of maize.

Figure 5: Saving of 50 % nitrogenous fertilizer in maize *cv.* GM-3 due to biofertilizer.

Treatment	% O.C.	% N	*Acetobacter* counts (cfu/g)	*Azospirillum* counts (cfu/g)
T_1	0.292	0.025	2.6×10^3	4.8×10^3
T_2	0.250	0.022	5.5×10^4	-
T_3	0.216	0.019	6.8×10^4	--
T_4	0.291	0.025	5.8×10^4	-
T_5	0.195	0.017	3.9×10^4	-
T_6	0.294	0.025	4.8×10^4	-
T_7	0.217	0.019		5.5×10^4
T_8	0.302	0.026		4.0×10^4
T_9	0.299	0.026		5.0×10^4
T_{10}	0.217	0.019		6.8×10^4
T_{11}	0.214	0.018		4.1×10^4
T_{12}	0.238	0.020	1.8×10^4	4.0×10^4
T_{13}	0.315	0.027	6.7×10^4	-
T_{14}	0.291	0.025	4.0×10^4	-
T_{15}	0.317	0.027	5.0×10^4	-
T_{16}	0.215	0.019	7.7×10^4	-
T_{17}	0.220	0.019	5.2×10^4	-
T_{18}	0.210	0.018	-	7.5×10^4
T_{19}	0.313	0.027	-	5.0×10^4
T_{20}	0.215	0.018	-	6.8×10^4
T_{21}	0.165	0.014	-	5.2×10^4
T_{22}	0.317	0.027	-	6.3×10^4
Before	0.240	0.021	2.1×10^3	4.3×10^3

Table 6: Effect on soil organic carbon, nitrogen and bacterial counts.

Effect on soil nutrient content and bacterial counts

Data regarding changes in soil organic carbon, total nitrogen and soil count at 90 DAS are presented in Table 6. Data revealed that T_{22} recorded maximum O.C. (0.317%) and total nitrogen content (0.027%) as compared to T_1 (0.292% O.C. and 0.025% N) and T_{12} (0.238% O.C. and 0.020% N). T_{16} recorded highest *Acetobacter* counts (7.7×10^4 cfu/g) and T_{20} recorded highest *Azospirillum* counts (7.5×10^4 cfu/g) and making soil fertile for further cultivation. Above results, indicates that application of endophytic bacteria by seed treatment not only improves the growth of treated crop but also benefits the subsequent crop to be taken by improving soil nutrient content and bacterial counts. These findings are in agreement with those of Das and Saha [11] who also reported that *Azotobacter* and *Azospirillum* alone and in combination with recommended dose of N increased total Nitrogen content in soil. Rao and Charyulu [12] reported that *A. brasilense* inoculation +40 kg N ha^{-1} gave Total N 3.94 mg N/g as compared to control (2.98 mg N/g) in foxtail millet. Smith et al. reported that *A. brasilense* inoculation showed Total N 1.34 mg N/g as compared to control (1.2 mg N/g) in sorghum. Overall results indicated that all the isolates noticeably increased root and shoot length, fresh and dry biomass of maize *cv.* GM-3 in laboratory. In a pot efficacy testing, all the bacterial treatments have eye catching impact on growth and growth attributes of maize *cv.* GM-3. Isolate A-9 recorded significantly higher plant growth parameters under pot trial condition.

References

1. Dobereiner J (1995) Isolation and identification of aerobic nitrogen-fixing bacteria from soil and plants. In: Methods in Applied Soil Microbiology and Biochemistry. Landon: Academic Press. pp: 134-141.
2. Cohen E, Okon Y, Kigel J, Nur I, Henis Y (1980) Increase in dry weight and total nitrogen content in *Zea mays* and *Seratia italic* associated with nitrogen fixing *Azospirillum* species. Plant Physiol 66: 746-749.
3. Rao KV, Charyulu P (2005) Evaluation of effect of inoculation of *Azospirillum* on the yield of *Setaria italica* (L.). African J of Biotec 4: 989-995.
4. Riggs PK, Chelius MK, Iniguez AL, Kaeppler SM, Triplett EW (2001) Enhanced maize productivity by inoculation of diazotrophic bacteria. Aus J Plant Phyisol 28: 829-836.
5. Das AC, Saha D (2003) Influence of diazototrophic inoculations on nitrogen nutrition of rice. Aus J of Soil Res 41: 1543-1554.
6. Cavalcante VA, Dobereiner J (1988) A new acid tolerant nitrogen fixing bacterium associated with sugarcane. Plant Soil 108: 23-31.
7. Bergey's Manual of Systematic Bacteriology (1983) Williams ST, Sharpe ME, Holt JG (eds.). Williams and Wilkins, Baltimore, USA.
8. Osmar RD, Carlos RS, Pedro RJ, Hernández RF, Alvarez GLM, et al. (2004) Effects of inoculation of *Azospirillum* sp. in maize seeds under field conditions. Food Agri & Environ 2: 238-242.
9. Mehnaz S, Lazarovits G (2006) Inoculation effect of *Pseudomonas putida, Gluconcacetobacter azotocaptans* and *Azospirillum lipoferum* on corn plant growth under greenhouse conditions. Microb Eco 51: 326-335.
10. Swędrzyńska D, Sawicka A (2001) Effect of inoculation on population numbers of *Azospirillum* bacteria under Winter wheat, oat and maize. Polish J of Environ Studies 10: 21-25.
11. Kado CI, Balows A, Truper HG, Dworkin M, Harder W (1992) Plant pathogenic bacteria. In The Prokaryotes. Springer. New York, USA. pp: 660-662.
12. Fukami J, Nogueira MA, Araujo RS, Hungria M (2016) Accessing inoculation methods of maize and wheat with *Azospirillum brasilense.* AMB Expr 6: 1-14.

Characterization and Utilization of Bioslury from Anaerobic Digester for Fertilizer in Crop Production

Kefale Wagaw*

Faculty of Chemical and Food Engineering, Bahir Dar University, Bahir Dar, Ethiopia

Abstract

In biogas plant there is a release of mainly two products. Methane gas which use for different purpose like cooking, lighting and slurry which can be used as organic fertilizer. The main target of this study was characterizing the slurry by measuring the relative amount of macro and micro nutrients like Nitrogen, Phosphorus, Potassium Calcium, Magnesium, and Manganese by using standard methods. And comparative study of the organic fertilizer at different treatments with commercial fertilizer on selected crops. Potassium and Calcium (part of plant cell wall) show positive deviation when it was composted.

Keywords: Biogas slurry; Compare; Compost; Macro; Micro-contents

Introduction

Agriculture contributes for more than 46% of the GDP and 90% of the export earnings, and supports 85% of the labour force of Ethiopia. Most farmers in the country use traditional way of agriculture using chemical fertilizer [1]. This chemical fertilizer is costly and decrease fertility of soil from time to time. To overcome this problem Organic manures such as cow dung, poultry manure, crop residues and biogas slurry in liquid and composted form can be used as organic fertilizer [2]. Bioslury obtained from the biogas plant may be considered as a good source of organic fertilizer as it contains considerable amounts of both macro and micronutrients, so using organic fertilizer (bioslury) can have economical value chain in crop production. Bioslury is an anaerobic digested organic material released as a digestant from the biogas plant after production of combustible biogas for cooking and lighting. Biogas slurry is a good source of plant nutrients and can improve soil fertility and properties. These nutrients are mainly nitrogen, phosphorus and potassium (macro) and magnesium, calcium and manganese (micro) elements and characterizing these nutrients using ICP, Spectrophotometer and Kjeldahl method [3] will be the major task of the project.

The value of bio-slurry as a fertilizer depends on the nutrient contents, e.g., the amount of nitrogen, phosphorus, potassium, calcium and magnesium [4], the ratio between nutrients in the bio-slurry, e.g., the N/P-ratio and/or the N/K-ratio [5]; and the availability of the nutrients, which is determined by the compounds that contain the nutrients. Farmers can use bioslury as liquid form ancompost form. Compost is an aerobically decomposed organic material [6] derived from plants and animal source.

Composting is a natural process of decomposition of organic matter by microorganisms under controlled conditions. During composting there will be loss of important nutrients like ammonia the basic source of Nitrogen and it is better to characterize before and after composting. Therefore characterizing the slurry and trying to analyze its content will motivate societies of our country to implement biogas plant in small scale which have triple value: energy interms of heating or lighting, organic fertilizer, and clean environment. After characterizing both soil and compost the demand gap of nutritional content was investigating. Finally this experiment was validated by growing crops based on characterization results [7].

When plant materials or animal manure is added to the soil, it does not stay in its original form for long. It is immediately attacked by a host of different soil organisms and undergoes a complex series of biochemical steps leading ultimately to its complete breakdown. The bulk of the material undergoes an oxidation or burning' process in which the carbon and hydrogen which make up about half of the dry weight of organic matter combine with oxygen to produce carbon dioxide and water. Energy is released in the process, and this is what is used by bacteria and other soil microorganisms for their survival and growth.

As the basic structure of the plant material is broken down (the breakdown may start from a series of intermediary steps like the digestive system of living creatures and anaerobic fermentation process or it may start from the soil itself if these materials are returned to the soil as such), nutrients such as nitrogen, phosphorous, potash, sulphur, etc., are released from their original organic form. Part of These may become soluble, and therefore be immediately available to growing plants. Most are, however, taken up by microorganisms and stored in their tissues as they grow and multiply [8].

These are only released when original plant matter has been used up and the organisms themselves start to die off and decompose. A variety of complex organic products accumulate in the soil as the process of decomposition continues. These include lignins and other materials that are resistant to decomposition as well as polymers derived from microbial products [9]. This more or less stable fraction is called humus. It is usually dark in color and persists in soil for many years, degrading very slowly but being replenished each year by the new additions of organic materials. The breakdown of organic matter depends on a variety of soil and site conditions. Nutrient and pH status, moisture content and temperature, and the availability of oxygen for soil microorganisms affect the rate of breakdown of organic materials [10].

***Corresponding author:** Kefale Wagaw, Faculty of Chemical and Food Engineering, Bahir Dar University, Post Box 26, Bahir Dar, Ethiopia
E-mail: wkefale@yahoo.com

Scope

This research was confined to the effects of digested slurry and compost on crop production in the existing agro-ecological conditions and the prevailing socio-cultural milieus in piccolo and adjacent areas with similar agro-ecological and socio-cultural environment [11].

Materials and Methods

Raw material

Main raw materials to do this project are animal manure, biogas slurry liquid type and solid type (compost).

Chemicals

Metallic salts such as calcium chloride (Figure 1) hydrated with two molecules of water, manganese sulphate hydrated with 1 molecule of water, potassium nitrate and magnesium nitrate with 6 molecule of water were used to prepare standard solution to Ca, Mn, K and Mg. During this experiment we used concentrated nitric acid, peroxide ammonium, Molybdenum antimony potassium tartrate [12], ascorbic acid and KH_2P0_4 to determine nitrogen we used Kjeldahl method.

Equipment

During the experiment in this project we used inductively coupled plasma atomic emission spectroscopy ICP-AES also known as ICP-OES as the main equipment to analyze amount of metals such as K, Mg, Ca and Mn of the manure [13], the biogas slurry and compost. The other equipment we used to measure the nutrient was spectrophotometer to measure the amount of non-metallic nutrients like phosphorus. To determine nitrogen it was used Kjeldahl method. During this method equipments like flask, stove and condenser were used, and the third equipment used was colony counter to measure the bacterial load in the sample. In this experiment we used homogenizer, sterilizer, and test tube. In this project we used volumetric flasks with volume 50 ml, 100 ml, 500 ml and 1000 ml and electronic balance.

Methods

Sample collection

Three samples from three different treatment stages were collected from piccolo, west Gojjam, Ethiopia. Each sample was collected in separate plastic bags and labelled.

Macro and micro nutrients determination standard preparation

To measure metallic nutrients standards was prepared and Nitric acid was used to digest the solid sample [14]. To prepare standard solution to determine calcium, manganese potassium and magnesium concentration in each samples their salts $CaCl_2.2H_2O$, $MnSO_4.H_20$, KNO_3, Mg $(NO_3)_2$ respectively are used.

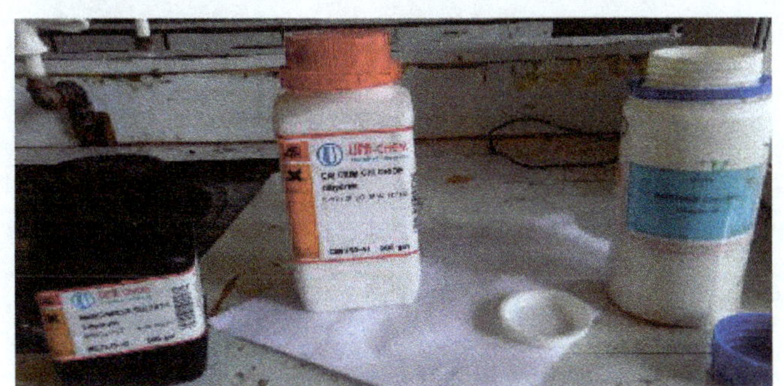

Figure 1: Salts used for standard solution preparation.

Digestion of samples

Since the ICP analyzes the solution in pure liquid form the solid and turbid liquid samples were digested using concentrated nitric acid [15]. 1 gm of each sample (manure, slurry and compost) was measured and 8 ml of HNO_3 was added to each samples. This mixture of sample was digested with microwave digester at 200°C for 30 minutes.

Working solution preparation

There were three samples for each of four metallic elements. These were at 5 ppm, 10 ppm and 20 ppm by 100 ml volumetric flask. These three samples of four elements (total of 12 samples) were prepared by 100 ml of volumetric flask with standard solution [16]. These working solutions were given to the ICP and the result was read as mg/l.

Phosphorus concentration determination Stoke solution preparations

Stock solution A: 12 gm of ammonium molybdate was dissolve in 250 ml of distilled water and 0.2908 gm antimony potassium tartrate was dissolved in 100 ml of distilled water. Finally both of these solutions were added to 1000 ml of 5N H_2SO_4 This solution was made 2000 ml with distilled water.

Solution B: 1.056 gm of ascorbic acid was dissolved in 200 ml of solution A and mixed thoroughly (Figure 2).

Digestion: 1 gm of each sample was weighed and 20 ml of HNO_3 was added to the samples. This was heated at 130°C until white dense fume appear then 10 ml of H_2O_2 was added to each sample and farther heated at 200°C.

Calorimetric determination of phosphorous

Digested sample was filled to 50 ml flask. 10 ml of solution B was added and the flask was filled to 50 ml with distilled water finally read the solution at 880 maxima using spectrophotometer (Ultima-2).

Nitrogen concentration determination

The method consists of heating a substance with sulphuric acid, which decomposes the organic substance by oxidation to liberate the reduced nitrogen as ammonium sulphate. In this step potassium sulphate was added to increase the boiling point of the medium (from 337°C to 373°C) Chemical decomposition of the sample is complete when the initially very dark-colored medium has become clear and colorless. The solution was then distilled with a small quantity of sodium hydroxide, which converts the ammonium salt to ammonia. The amount of ammonia present, and thus the amount of nitrogen present in the sample, is determined by back titration. This done by Kjeldahl method.

Digestion: The sample to be analyzed was weighed on an analytical balance into the digestion flask. Then the sample is digested by concentrated H_2SO_4.

Distillation: After digestion was completed, the content in the flask was diluted by water and a concentrated NaOH is added to neutralize the acid and to make the solution slightly alkaline.

Titration: If boric acid is used, the titration is called an indirect titration, because the ammonia that is chemically equivalent in converting the boric acid (ammonia bound to an equivalent of borate ion) is directly titrated with standard acid (0.1 N HCI).

$NH_3+H_3BO_3$-----$NH_4^++H_2BO_3^-$

Liberation of ammonia $(NH_4)2SO_4+2NaOH$---$Na_2SO_4+2H_2O+2NH_3$

Capture of ammonia $B(OH)^3+H_2O+NH_3^+$---$NH_4+B(OH)_4^-$

% $N_2=(V_{Hcl} \times N_{Hcl} \times m_{wn}/wt) \times 100$

Results and Discussion

Elemental composition of bioslury before and after treatment

As it was investigated in results of metallic nutrient determination K and Ca (part of plant cell wall) show positive deviation when it was composted. This is because of that decomposed vegetables are This is because of that decomposed vegetables are sources of both Ca and K during composting While Magnesium and Manganese show negative deviation from slurry to compost (Table 1).

Nitrogen decrease (33%) during composting (Table 2) this is due to volatilization of N_2 in the form of NH_3 and NH_4. This phenomenon will increase at low PH value. Orthophosphate which is found in the soil is the main source of phosphorous. Due to this P value increases during composting. According to the results of this project it is possible to conclude that biogas slurry has considerable amount of both macro and micro nutrients which have significant role to minimize the consumption of chemical fertilizer. During composting the phosphorus concentration increase from 0.2 (mg/l) to 0.3 (mg/l) (Table 3).

Table 4 shows that the biogas slurry was not able to achieve higher incremental onion yield (Figure 3) as compared to compost and commercial fertilizer. But both compost and commercial fertilizer have equal productivity for onion cultivation. Compost, in fact gave the highest incremental yield of chills as shown in Figure 4 (even higher than that of chemical fertilizer). The researcher comments: "From these simple field trials my observations do not confirm with the results other scientists found in other parts of the world. Even if biogas slurry has higher nitrogen content its productivity on selected crops is low. There

Figure 2: Standard solution preparation.

Sample	Potassium concentration [mg/l]	Calcium Concentration [mg/l]	Magnesium Concentration [mg/l]	Manganese concentration [mg/l]
Manure	66.92	0.84	5.11	56.50
Slurry	107.56	0.69	2.3	46.16
Compost	146.11	2.22	4.91	55.35

Table 1: Metallic composition of bioslury sources of both Ca and K during composting while Mg and Mn show negative deviation from slurry to compost.

S. No.	Sample	V_{HCl} (ml)	ᴺHCl	M_{wn} (g)	Wt (g)	N_2%
1	Compost	3.6	0.1	14.1	0.3	1.68
2	Manure	4.5	0.1	14.1	0.3	2.10
3	Slurry	5.4	0.1	14.1	0.3	2.52

Table 2: Nitrogen composition of bioslury.

S. No.	Standard	Absorbance		Sample	Absorbance		Phosphorus Concentration (mg/l)
		690 nm	880 nm		690 nm	880 nm	
1	0.0 ppm	0.00	0.00	Manure	0.095	0.099	0.22
2	0.5 ppm	0.02	0.02	Slurry	0.089	0.093	0.2
3	1.0 ppm	0.03	0.04	Compost	0.092	0.097	0.33

Table 3: Phosphorous composition of bioslury.

A. Compost B. Fertilizer C. Slurry

Figure 3: Productivity of organic fertilizer on onion.

A. 1. Observing productivity with fertilizer

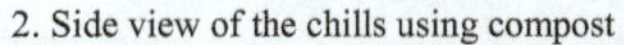

2. Side view of the chills using compost

B. 1. Sectional view of chills prod by compost 2. Side view on chills prod by slurry

Figure 4: Productivity of organic fertilizer on Chills.

	Fertilizer Type	Land Area [m²]	Produced Crop [kg]	Yield [kg/m²]	Earned money
Onion	Slurry	2 × 4	5	0.83	232 ETB
	Compost	2 × 4	12	1.5	
	Commercial fertlizer	2 × 4	12	1.5	
Chills	Commercial fertlizer	2 × 4	28	3.5	1560 ETB
	Compost	2 × 4	30	3.75	
	Slurry	2 × 4	20	2.5	

Table 4: The yield of chills and onion.

may be several factors which were not looked upon in detail during these trials. This particular experiment did not furnish information on the form of slurry used, the possible toxic effect of biogas slurry, presumably in its fresh liquid from. The claim of toxicity is not verified by further research in Ethiopia.

Acknowledgements

The research fund was granted from the research and post graduate office of Bahir Dar Institute of Technology, Bahir Dar University, Ethiopia.

References

1. Chawla OP (1984) Manurial Aspects Advances in Biogas Technology. New Delhi: ICAR, India.
2. Acharya CN (1952) Preparation of Fuel Gas and Manure by Anaerobic Fermentation of Organic Materials. New Delhi: Indian Agriculture Research institute.
3. Acharya CN (1953) Cow Dung Gas Plants. Indian Fmg 3: 16-18.
4. Dhussa AK (1985) Biogas Plant Effluent Handling and Utilisation. Changing Villages 7: I.
5. Kanthaswamy V (1993) Effect of Bio-Digested Slurry in Rice. Biogas Slurry Utilsation.
6. Riggle D (1997) Anaerobic Digestion Gets New Life on Farms. pp: 74-78.
7. Rosenberg G (1951) Methane Production from Farm Wastes as Source of Tractor Fuel. Journal of Min Agric 58: 487-494.
8. Dohne E (1990) Optimal Bioslurry Fertilisation in West Germany. Biogas Forum.
9. Sharma UP (1981) Complete Recycling of Cattle-Shed Wastes Through Biogas Plant. Janata Biogas Technology and Fodder Production.
10. MoARD (2010) Ethiopia's Agricultural Sector.
11. Policy and Investment Framework (PIF) Draft Office. Addis Ababa, Ethiopia.
12. Martens DA (1998) Management and crop residue influence soil aggregate stability. Journal of Environmental Quality 29: 723-727.
13. Assefa AG (2015) Evaluation of Alternative Soil Amendments to Improve Soil Fertility and response to bread Wheat (Tritium aestivum) Productivity in Ada'a District, Central Ethiopia. Addis Ababa University.
14. Hailu A (2010) The effect of compost on soil fertility enhancement and yield increment Under smallholder farming: A case of Tahatai Maci chew district of Tigray region, Ethiopia. University of hohenheim.
15. Sherchan DP, Gurung GB (1996) Effect of five years continuous application of organic and inorganic fertilizers on crop yields and physico-chemical properties of soil under rainfed maize/millet cropping pattern. AGRIS 168: 1-13.
16. Robert C, Edward JC, Michael L, Hans S (1981) Food Policy Issues in Low-Income Countries.

17

Isolation of Phosphate Solubilizing Bacteria from White Lupin (*Lupinus albus* L.) Rhizosphere Soils Collected from Gojam, Ethiopia

Dereje Haile[1*], Firew Mekbib[2] and Fassil Assefa[3]

[1]*Biology Department, Hawassa University, Hawassa, Ethiopia*
[2]*Department of Plant Science, Haramaya University, Dire Dawa, Ethiopia*
[3]*Department of Microbiology, Addis Abeba University, Addis Abeba, Ethiopia*

Abstract

Phosphorus is an essential macronutrient. A great portion of phosphorus from chemical fertilizers becomes insoluble and unavailable to plants because of its conversion into salts. Phosphate solubilizing microorganisms as a phosphorus biofertilizer improve soil fertility by solubilizing insoluble phosphate salts and increase crop production. This research aimed in isolation and characterization of phosphate solubilizing bacteria. Sixty-three soil samples collected from seven different parts of white lupin growing areas of Gojam, Ethiopia. Halo zone formation and plate screening method used for isolation. Based on halo zone formation; 152 phosphate solubilizing bacterial isolates obtained. Based on their solubilization index four isolates selected for subsequent experiments and characterization. The four selected isolates tested for their solubilization efficiency on solid media. Isolates HUPSB-35 and HUPSB-45 appeared maximum solubilization index (4.5 mm). Isolate HUPSB-57 was unique with its colony formation; salt tolerance (able to grow up to 10%) and utilization of wide range of carbon sources (utilize all the tested sugars). Isolate HUPSB-27 showed a wide range of pH and temperature preference (100% growth at 10, 15, 20, 25 and 30°C in all tested pH).

Keywords: Biofertilizer; Isolates; Phosphorous; PSB

Introduction

Phosphorus is one of the least available and the least mobile mineral nutrients in soil for plant growth. However, it is vital for plants and they absorb only inorganic form of phosphorous but the level of inorganic phosphorus is very low in soil because most of the phosphorous is present as insoluble forms [1,2]. Unlike nitrogen, there is no large atmospheric source that can be made biologically available to plants [1]. P chemical fertilizer is widely applied in agricultural production [2]. However, a large proportion of these fertilizers are also converted to insoluble form leads to low-fertilizer efficiency where the volume of phosphatic fertilizers as low as 15-20% are only utilized by plants due to fixation of P in acidic and alkaline soils [3].

Phosphorus deficiency is the most important problem of Ethiopian soils and more than 70-75% of highland soils are characterized by phosphorus deficiency [4]. Around 70% of Ethiopian vertisols have available phosphorus below 5 ppm, which is very low for supporting good plant growth; and, fixation in vertisols is related more to calcium, than Al^{3+} and Fe^{3+} [5]. Yield is usually low under traditional farming system because of poor cultural practices that is compounded with poor soil fertility. Though chemical fertilizers play a role in agricultural sector, the ever-increasing prices of chemical fertilizers have made them unfavorable for most farmers. Consequently, farmers are applying fertilizers below the recommended rates with yields lower than the potential of their cultivated crops. Thus, there is an urgent need to search for supplementary and cheaper source of nutrients, particularly those of nitrogen and phosphorus because the two nutrients often limit crop production.

One of the cheap sources for improving phosphorus nutrition is manipulation of phosphate solubilizing microbes (PSM) alone or together with cheap rock phosphate [6]. Different studies showed that selected microorganisms are found to be promising as important components of biofertilizer production towards the efforts of integrated soil fertility management [7]. Phosphate solubilizing microorganisms (PSMs) or phosphate solubilizing bacteria (PSB) can convert insoluble phosphates into soluble forms [8]. Population in the soil of PSMs varies from soil to soil and ranges from less than 10^2 colony forming units (cfu) g^{-1} of soil to 3×10^6 cfu g^{-1} of soil [9] reviewed that the most commonly found phosphate solubilizing bacteria (PSB) were belonging to genus *Pseudomonas, Bacillus* and some of other bacteria such as: *Mycobacterium, Micrococcus, Flavobacterium, Achromobacter* sp. of these genera, *Pseudomonas* sp. is the most efficient phosphate solubilizer.

In Ethiopia, the hitherto researches on the manipulation of microorganisms for soil fertility management have concentrated on understanding and improving nitrogen fixing microorganisms. However, previous researches also indicated that phosphate solubilizing microbes (especially bacteria) are very important in response to phosphorus deficiency in soil [4,10] evaluated the effect of co-inoculation of *Bradyrhizobium japonicum* and phosphate solubilizing *Pseudomonas spp* on soya bean, and got a promising result in Assossa area. Some of the root nodule bacteria were also effective phosphate solubilizers, there is no extensive research undertaken on the status of PSB in lupin growing areas. Hence, the objective was to isolate and characterize efficient phosphate solubilizing bacteria from white lupin rhizosphere by evaluating their efficiency on solid media.

Materials and Methods

Study area

The sampling areas covered 63 sampling sites from seven different

***Corresponding author:** Dereje Haile, Biology Department, Hawassa University, Hawassa, Ethiopia, PO Box 05, Hawassa, Ethiopia,
E-mail: gaginager93@gmail.com

parts of Gojam, Amhara Region, Ethiopia. Areas include: Dangila, Debre-Marekose, Rob-Gebeya, Dembecha, Kossober, Merawi and Tilili that are located between 1110'59.880"N, 3953'60.000" E and known for lupin production. White lupin was selected for rhizosphere soil sample collection due to its ability to grow on different soil types, area and high soil fertility maintenance capacity. Farmers in the study area cultivate lupin with intercropping basically for fertility maintenance other than crop production. The research conducted at Holeta Agricultural Research Center.

Study design

A cross sectional (for soil sampling) and laboratory based (for experiment) design were followed. Soil samples were randomly excavated in triplicates from a depth of 0-30 cm under the rhizosphere of lupin plants, sub-sampled and separately collected in sterile plastic bags from Jan 2011 - Feb 2012. Soils were taken to Chemistry Laboratory for analysis and to Holeta Agricultural Research Center for isolation of phosphate solubilizing bacteria.

Isolation of phosphate solubilizing bacteria from soil

Each soil sample was grind in to smaller parts with mortar and pestel and sieved with 2 mm mesh size. 10 gm of fine soil mixed thoroughly with 90 ml of sterilized distilled water for about 15 min. From the suspension, one ml transferred to 9 ml of sterilized 0.85% of normal saline solution for serial dilution up to 10^{-3}(v/v). From the last dilution (10^{-3}) [11], 100 µl was spread on plates of Pikovskaya's agar medium (PVK-medium). Pikovskaya's agar medium composition; Glucose, 10 gm/l; Tricalcium phosphate, 5 gm/l; Ammonium sulphate, 0.5 gm/l; Potassium chloride, 0.2 gm/l; Magnesium sulphate, 0.1 gm/l; Manganous sulphate, Traces (0.0002); Ferrous sulphate, Traces (0.0002); Yeast extract, 0.5 gm/l; Agar, 18 gm/l; pH, 7.0 and autoclave 121°C for 15 min. Plates were incubated at 30°C for 5-8 days. Bacterial colonies showing clear zone were then sub-cultured, purified, and preserved at 4°C in the Microbial Biotechnology Lab. of Holeta Agricultural Research Center. Isolates designated as; HUPSB followed with serial numbers.

Screening for phosphate-solubilization in petri plates

Estimation of phosphate solublization ability of isolates was undertaken using plate screening methods. Hundred micro-liters of pure isolates from 48 hrs broth culture spread evenly under aseptic conditions on Pikovskaya's agar [12], and incubated at 30°C. The halo and colony diameter measured at 2, 4, 6 and 8 days of incubation. To calculate their solubilzation index [13].

SI=(HZD+CD)/CD

Where; SI=Solubilization index, HZD=Halo zone diameter, CD=Colony diameter.

Characterization of isolates

Based on their solubilization index 4 bacterial isolates, 3 rhizosphere bacteria with high solubilization index (HUPSB- 27, HUPSB-35, HUPSB-45) and one isolate with low index (HUPSB-57) were selected for biochemical and physiological characterization. Gram staining, motility, cultural characteristics (colony color, texture and shape), biochemical (growth on different carbon sources), pH and temperature tests of PSB isolates carried out according to the procedure of Ref. [14].

Mean generation time: The standard curve of isolates was plotted by measuring Optical Density at wavelength of 600 nanometer and relating with colony forming units (CFUs) obtained from plates. Mathematically mean generation time (g) calculated according to Ref. [15],

$g=t/n$,

$$n = \frac{\log N_t - \log N_0}{0.301}$$

Where, n=The number of generations in time t

N_0=The initial population number

N_t=The population at time t

Utilization of other substrates by isolates: Molasses, urea, and Egyptian rock phosphate (ERP) were evaluated as alternative sources for carbon, nitrogen, and tricalcium phosphate (TCP) respectively incorporated in PVK agar. Loopful of 48 hrs grown culture of each isolate was streaked and incubated at 30°C for 5 days.

pH vs. temperature test: Isolates were tested for their ability to grow at different pH in PVK liquid medium adjusted to pH 4, 5, 6, 7, 8, 9, and 10 in relation with different temperature of 5, 10, 15, 20, 25, 30, 35, 40 and 45°C [16].

Salt tolerance test: Isolates were grown on PVK medium with different concentrations of salt (NaCl). The tested concentrations were 0.5, 1, 2, 3, 4, 5, 6, 7, 8, 9 and 10%. The result was recorded after 5 days incubation at 30°C at neutral pH (7.0).

Data analysis

Upon eight days incubation isolate's colony diameter and halozone formation were measured. These used to calculate SI for each isolate and set their average values. Data from OD measurements and CFU counts for selected isolates were analyzed using SAS 9.1.3 version with Fisher's LSD (significance set at $P \leq 0.05$) and presented in table.

Results

Isolation of PSB

A total of 152 PSB isolates were obtained; some of them indicated in Table 1 including colony diameter, halo zone diameter and solubilization index (SI) upon 2- 8 days of incubation. The largest halo zone (15 mm) recorded from isolates HUPSB-57Mer$_1$ followed by HUPSB-17D/M$_1$ (10 mm); whereas the smallest halo zone (3 mm) was measured from isolate HUPSB-34T$_1$. Likewise, from tested isolates, the highest solubilization index (SI) was shown by isolates HUPSB-35 and HUPSB-45 (4.5 mm) and the smallest SI from isolate HUPSB-57 (2.2 mm). In general, the average SI of isolates found to be 2.8 (Table 1). More over different isolates could be isolated from different parts of the world with different solubilization efficiency. Variation in their SI could possibly the deference in; isolates type, natural environment, physico-chemical properties of soil, soil management and agricultural practices. In addition, the importance of isolation from local environment is to avoid competition with indigenous microbes.

Selection and characterization of PSB

Based on SI four isolates were selected for further tests and experimental activities. From the selected four isolates, all except HUPSB-35 formed halo zone starting from day two, and increased in halo zone size from 1-2 mm to 5-15 mm upon 8 days of incubation. They also displayed Solubilization Index (SI) of 2-2.5 up to 2.7-4.5. The most effective isolate, HUPSB-45 showed SI of 2.5 at day two and 4.5 upon 8 days incubation. Isolate HUPSB-35, which failed to show any

Isolates	Day 2			Day 4			Day 6			Day 8		
	Diameter (mm)			Diameter (mm)			Diameter (mm)			Diameter (mm)		
	C	H	SI	C	H	SI	C	H	SI	C	H	SI
HUPSB-3-RG$_1$				2	3	2.5	3	4	2.3	5	7	2.4
HUPSB-5-RG$_2$				1	2	3	2	5	3.5	3	6	3
HUPSB-5-RG$_8$				1	2	3	2	4	3	2	4	3
HUPSB-5-RG$_{10}$				1	2	3	3	5	2.7	4	7	2.8
HUPSB-5-RG$_{11}$				1	2	3	3	5	2.7	4	6	2.5
HUPSB-6-RG$_3$				2	2	2	2	4	3	3	5	2.7
HUPSB-7-RG$_2$				1	2	3	2	3	2.5	2	4	3
HUPSB-8-RG$_I$				2	3	2.5	3	6	3	4	7	2.8
HUPSB-9-RG$_1$	1	1	2	1	2	3	3	4	2.3	4	6	2.5
HUPSB-13-D/M$_2$				4	6	2.5	4	6	2.5	4	6	2.5
HUPSB-13-D/M$_I$				3	5	2.7	3	7	3.3	3	7	3.3
HUPSB-14-DEM$_1$				1	2	3	3	6	3	4	7	2.8
HUPSB-15-D/M$_I$				2	3	2.5	3	5	2.7	4	6	2.5
HUPSB-16-D/M$_{II}$				1	2	3	2	4	3	3	5	2.7
HUPSB-17-D/M$_1$				3	6	3	4	8	3	6	10	2.7
HUPSB-18-D/M$_1$				1	1	2	2	3	2.5	3	6	3
HUPSB-19-DEM$_3$				1	2	3	2	4	3	2	4	3
HUPSB-24-DEM$_I$				2	3	2.5	2	3	2.5	3	5	2.7
HUPSB-26-DEM$_I$				2	4	3	3	6	3	3	7	3.3
HUPSB-27-DEM$_1$	1	1	2	1.5	3	3	2	5	3.5	2	5	3.5
HUPSB-27-DEM$_2$	1	1	2	1	3	4	2	5	3.5	3	6	3
HUPSB-30-T$_1$	1	1	2	2	4	3	3	5	2.7	5	8	2.6
HUPSB-30-T$_2$				1	2	3	2	5	3.5	3	6	3
HUPSB-31-T2$_1$				2	2	2	2	3	2.5	3	7	3.3
HUPSB-32-T$_1$				1	2	3	1	2	3	2	4	3
HUPSB-33-T$_1$				1	1	2	2	5	3.5	3	5	2.7
HUPSB-34-T$_1$				1	1	2	1	1	2	2	3	2.5
HUPSB-34-T$_2$				2	2	2	2	3	2.5	3	5	2.7
HUPSB-35-T$_1$				**1**	**2**	**3**	**2**	**5**	**3.5**	**2**	**7**	**4.5**
HUPSB-36-T$_2$	1	1	2	2	4	3	2	4	3	2	4	3
HUPSB-39-K$_1$	1	1	2	2	2	2	3	4	2.3	4	6	2.5
HUPSB-40-K$_1$	1	1	2	2	3	2.5	3	5	2.7	4	6	2.5
HUPSB-40-K$_2$	1	1	2	2	4	3	2	5	3.5	3	5	2.7
HUPSB-44-K$_1$				3	5	2.7	4	7	2.8	6	9	2.5
HUPSB-44-K$_2$				2	4	3	3	5	2.7	3	7	3.3
HUPSB-44-K$_3$				1	1	2	2	3	2.5	4	6	2.5
HUPSB-45-K$_2$	**1**	**1.5**	**2.5**	**1**	**2**	**3**	**2**	**5**	**3.5**	**2**	**7**	**4.5**
HUPSB-48-DAN$_1$	1	1	2	2	4	3	3	5	2.7	4	6	2.5
HUPSB-48-DAN$_3$	1	1	2	2	3	3	2	5	3.5	3	5	2.7
HUPSB-50-DAN$_1$				1	3	4	2	3	2.5	3	4	2.3
HUPSB-50-DAN$_2$				1	2	3	2	3	2.5	2	4	3
HUPSB-52-DAN$_1$				1	1	2	2	4	3	2	4	3
HUPSB-55-MER$_4$				1	3	4	3	5	2.7	4	7	2.8
HUPSB-57-MER$_2$				3	3	2	4	5	2.3	5	6	2.2
HUPSB-57-MER$_4$				2	4	3	3	5	2.7	3	6	3
HUPSB-57-MER$_5$				4	4	2	4	5	2.3	4	5	2.3
HUPSB-57-MER$_1$	2	2	2	7	9	2.3	8	10	2.5	9	15	2.7
HUPSB-57-MER$_{IV}$	1	1	2	2	4	3	3	6	3	4	7	2.8
HUPSB-60-MER$_1$				2	4	3	3	5	2.7	2	4	3
HUPSB-61-MER$_1$				2	2	2	2	4	3	4	6	2.5
Average				1.7	2.9	2.7	2.8	4.9	2.7	3.7	6.2	**2.8**

NB: DAN: Dangila; D/M: Debre Markose; DEM: Dembecha; K: Kosober; MER: Merawi; RG: Robgebeya; T: Tilili; C: Colony; H: Halo zone; SI: Solubilization index; HUPSB: Haramaya University Phosphate Solubilizing bacteria.

Table 1: Measurement of Halo and Colony Diameters of Isolates upon 8 Days Incubation.

Isolates	Colony color	Colony shape	Gram stain and cell shape	Motility	Mean Generation time (hrs)
HUPSB -27	White	Round, RM	- R	+	5.8
HUPSB -35	Yellow	Round, RM	- R	+	4.1
HUPSB -45	Yellow	Round, RM	- R	+	4.7
HUPSB -57	White	Round, IM	- R	+	5.1

NB: RM: Regular margin; IM: Irregular margin; R: Rod shape; '-' gram negative; '+' motile.

Table 2: Colony color, shape, gram stain, generation time and motility of the PSB isolates.

Isolates	CFU (10^{-7})	OD	Incubation Time (hrs)
HUPSB-27-DEM$_1$	0.3333^d	0.008000^a	0
HUPSB-35-T$_1$	1.0000^c	0.009333^a	
HUPSB-45-K$_2$	2.6667^b	0.013000^a	
HUPSB-57-MER$_1$	4.3000^a	0.014333^a	
LSD	0.2728	0.0071	
CV	6.982410	33.70624	
Mean	2.075000	0.011167	
HUPSB-27-DEM$_1$	7.6667^b	0.55633^a	6
HUPSB-35-T$_1$	2.7000^c	0.07367^c	
HUPSB-45-K$_2$	4.3000^c	0.05467^c	
HUPSB-57-MER$_1$	20.3333^a	0.40333^b	
LSD	1.9896	0.0637	
CV	12.07685	12.44084	
Mean	8.750000	0.272000	
HUPSB-27-DEM$_1$	53.6667^a	0.85333^a	12
HUPSB-35-T$_1$	7.0000^d	0.43533^c	
HUPSB-45-K$_2$	25.3333^c	0.51533^b	
HUPSB-57-MER$_1$	40.3333^b	0.82100^a	
LSD	1.0775	0.0647	
CV	1.811956	5.237784	
Mean	31.58333	0.656250	
HUPSB-27-DEM$_1$	73.333^b	1.48667^a	18
HUPSB-35-T$_1$	53.333^d	1.35533^a	
HUPSB-45-K$_2$	60.667^c	1.11000^b	
HUPSB-57-MER$_1$	103.000^a	1.40000^a	
LSD	4.7883	0.1897	
CV	3.503735	7.531417	
Mean	72.58333	1.338000	
HUPSB-27-DEM$_1$	106.000^b	1.57133^a	24
HUPSB-35-T$_1$	88.333^c	1.66667^a	
HUPSB-45-K$_2$	87.667^c	1.25333^b	
HUPSB-57-MER$_1$	151.000^a	1.57333^a	
LSD	3.957	0.1223	
CV	1.941420	4.284632	
Mean	108.2500	1.516167	
HUPSB-27-DEM$_1$	132.333^d	2.18533^{ba}	36
HUPSB-35-T$_1$	185.000^b	2.28767^a	
HUPSB-45-K$_2$	151.000^c	1.90700^c	
HUPSB-57-MER$_1$	203.333^a	2.14833^b	
LSD	7.4525	0.1138	
CV	2.357189	2.834303	
Mean	167.9167	2.132083	
HUPSB-27-DEM$_1$	147.000^d	2.41000^a	48
HUPSB-35-T$_1$	197.000^b	2.47767^a	
HUPSB-45-K$_2$	176.000^c	2.16333^b	
HUPSB-57-MER$_1$	219.333^a	2.33467^a	
LSD	5.1849	0.1464	
CV	1.489875	3.313546	
Mean	184.8333	2.346417	

Table 3: OD and CFU value of the four selected isolates.

solubilization zone on 2nd day of incubation showed a SI of 4.5 similar to isolate HUPSB-45. On the contrary, isolate HUPSB-57, which showed fastest solubilization at day two with larger SI of two did not show considerable change in SI (2.7) upon 8 days of incubation (Table 1). The data showed that there is steady increase in solubilization as incubation time increases and also HUPSB-27, HUPSB-35, and HUPSB-45 showed similar pattern of solubilization within 4-6 days.

Cultural and growth characteristics of isolates

Four selected isolates were taxonomically characterized based on their cultural, morphological, and growth characteristics (Table 2). Accordingly, all isolates were gram negative, rod shaped motile bacteria with round colony margin on solid media, except isolate HUPSB-57 (round irregular margin). With regard to colony colour, isolates HUPSB-27 and HUPSB-57 appeared to be white, whereas isolates HUPSB-35 and HUPSB-45 showed a yellow colour. Although isolates HUPSB-27 and HUPSB-57 were white, they differ in their transparency; in that; the latter was opaque and mucoid whereas the former was transparent and spreading. The isolates mean generation time range from 4.1 to 5.8 hrs and their OD values in relation to CFUs presented in Table 3. Relatively HUPSB-35 showed shorter doubling time (4.1 hrs) while HUPSB-27 had longer doubling time (5.8 hrs).

Sugar utilization of psb isolates

BTB (bromothyml blue) indicator used to identify sugar utilization. Isolate HUPSB-57 was the most efficient that utilized all of the given carbon sources whereas isolates HUPSB-45 and HUPSB-27 utilized fewer sugars (Table 4). Glucose, maltose and sucrose utilized by all of the isolates whereas lactose was utilized only by HUPSB-57.

Growth of isolates on different substituted media

All the experimental isolates were able to grow on the substituted media of PVK: molasses agar, urea agar and PVK with ERP.

Growth of isolates at different pH and temperature

Isolates showed different response of growth in combinations of different medium pH (pH 4-10) and incubation temperature (5°C-45°C) (Table 5). No isolate was able to grow in any combination of pH and temperature of 45°C (Table 4). All isolates were found to grow at temperature of 10°C-25°C, at all tested pH values, except isolate HUPSB-57 that failed to grow under pH 10 and at 25°C. The most resilient isolate was HUPSB-27 that was able to grow at 30°C with all pH adjustments of the media (pH 4-10), and at 35°C and 40°C at pH 6 and pH 7 (Table 4). Isolates HUPSB-35 and HUPSB-45 showed similar pattern of growth at 30°C with pH 4-8, and 40°C at pH 7. In general, isolate HUPSB-27 performed better in different pH and temperature stress factors, whereas the two isolates (HUPSB-35 and HUPSB-45) showed similar pattern of tolerance. Isolate HUPSB-57 was relatively the most sensitive isolate to grow at different combinations of pH and incubation temperature.

Growth of isolates at different salt concentration

Salt tolerance of isolates was tested at different salt concentrations.

Isolates	Glu	Gala	Mal	Mani	Ara	Suc	Xyl	Fru	Lac	Dex
HUPSB-27	+	-	+	+	+	+	-	+	-	-
HUPSB-35	+	+	+	+	-	+	+	-	-	+
HUPSB-45	+	+	+	-	-	+	+	-	-	+
HUPSB-57	+	+	+	+	+	+	+	+	+	+

NB: Glu: Glucose; Gala: Galactose; Mal: Maltose; Mani: Manitole; Ara: Arabinose; Suc: Sucrose; Xyl: Xylose; Fru: Fructose; Lac: Lactose; Dex: Dextrose; '+' presence and '-' absence of growth.

Table 4: Sugar test of PSB isolates.

Isolate	pH							Temperature (°C)
	4	5	6	7	8	9	10	
HUPSB-27	-	+	+	+	+	+	+	5
	+	+	+	+	+	+	+	10
	+	+	+	+	+	+	+	15
	+	+	+	+	+	+	+	20
	+	+	+	+	+	+	+	25
	+	+	+	+	+	+	+	30
	-	-	+	+	-	-	-	35
	-	-	+	+	-	-	-	40
	-	-	-	-	-	-	-	45
HUPSB-35	-	+	+	+	-	-	-	5
	+	+	+	+	+	+	+	10
	+	+	+	+	+	+	+	15
	+	+	+	+	+	+	+	20
	+	+	+	+	+	+	+	25
	+	+	+	+	+	-	-	30
	-	-	-	+	-	-	-	35
	-	-	-	+	-	-	-	40
	-	-	-	-	-	-	-	45
HUPSB- 45	-	+	+	+	-	-	-	5
	+	+	+	+	+	+	+	10
	+	+	+	+	+	+	+	15
	+	+	+	+	+	+	+	20
	+	+	+	+	+	+	+	25
	+	+	+	+	+	-	-	30
	-	-	-	+	-	-	-	35
	-	-	-	+	-	-	-	40
	-	-	-	-	-	-	-	45
HUPSB-57	-	+	+	+	-	-	-	5
	+	+	+	+	+	+	+	10
	+	+	+	+	+	+	+	15
	+	+	+	+	+	+	+	20
	+	+	+	+	+	+	-	25
	+	+	+	+	+	-	-	30
	-	-	-	+	-	-	-	35
	-	-	-	+	-	-	-	40
	-	-	-	-	-	-	-	45

NB: '+' presence and '-' absence of growth.

Table 5: Growth of isolate HUPSB-27, HUPSB-35, HUPSB-45 and HUPSB-57 at different pH and temperature.

All isolates were found to grow up to 2% salt concentration (0.5%, 1% and 2%) (Table 6). The most sensitive isolates were HUPSB-27 and HUPSB-35, that able to grow only up to 2% salt concentration. However, isolate HUPSB-57 showed a broad growth range and was able to grow in all tested salt concentrations (0.5-10%), followed by isolates HUPSB-45 grown up to 5% salt concentration.

In general, isolates showed different cultural, morphological, growth, and eco-physiological characteristics. Isolates HUBSP-35 and HUBSP-45 were similar as they imparted yellow color on growth medium, the same pattern of SI of 4.5 after 8 days of incubation; temperature and pH tolerance, they differed in mannitol utilization, and salt tolerance. It can be argued that these four isolates can be tentatively grouped into genus *(Pseudomonas)* under different species. Isolates efficiency and symbiotic effectiveness were evaluated on soil under greenhouse condition. They showed significance difference over control in shoot height, shoot fresh and dry weight of white lupin as well as variation in total P content of white lupin and rhizosphere soil (data not shown).

Discussion

Formation of clear zone is an indication of inorganic P-solubilization by isolates and measurement of SI on a solid medium is a very important and reliable tool for a preliminary screening of phosphate solubilizing microorganisms [17-19]. In Ethiopia, Keneni et al. reported that the largest clear zone diameter of 4.5 mm from three different PSB isolates (isolated from Mehalmeda) and Esubalew, reported the maximum SI of 1.8 by some isolates from Gojam soil.

All the tested isolates belong to fast growing bacteria according to Somasegaren and Hoben, who stated that fast growers would develop pronounced turbidity in liquid media within 2-3 days. Similarly, Esubalew, discussed white lupin nodulating and phosphate-solubilizing isolates displayed different doubling time ranging from 2-5. 3 hrs. The difference in all above cultural and growth characteristics may be due to genetic variation among isolates. Baon and Prescott, reported that bacteria growing on solid surfaces such as agar could form quite complex and complex colony shapes (or variation in colony shapes) [15,18].

On testing utilization of carbohydrates by white lupin nodulating bacteria including phosphate solubilizing and nodulating bacteria, Esubalew, stated that 95% of isolates grew in the presence of glucose. Poonam and Ghosh, also reported all of their isolates produced acid from glucose but only two out of five isolates were able produce acid from lactose. Nautiyal argued that the most efficient strain is the one that is capable of utilizing a wide range of carbon and nitrogen sources.

Experimental isolates were unable to grow after 80°C (data not shown). Similarly, Poonam and Ghosh, reported none of the isolates able to grow at high temperature (52°C). Prescott, discussed that cardinal temperatures for a particular species are not rigidly fixed but often depend to some extent on other environmental factors such as pH and available nutrients and also it varies between microorganisms.

The most sensitive isolates for salt tolerance were HUPSB-27 and HUPSB-35, which were able to grow at only 2% salt concentration. Similarly, Esubalew, reported that phosphate solubilising root nodule bacteria from lupin were able to grow at 2%. However, isolate HUPSB-57 was able to grow 0.5- 10%, which was similar to the work of; Poonam and Ghosh, who found that isolates could tolerate up to 10%; [20], where strain solubilize P at 10% and Zhu who found an isolate that could tolerate high salt concentration (20%).

Isolates	[salt]%										
	0.5	1	2	3	4	5	6	7	8	9	10
HUPSB-27	+	+	+	-	-	-	-	-	-	-	-
HUPSB-35	+	+	+	-	-	-	-	-	-	-	-
HUPSB-45	+	+	+	+	+	+	-	-	-	-	-
HUPSB-57	+	+	+	+	+	+	+	+	+	+	+

Table 6: Growth at different salt concentration.

Conclusion

In order to apply PSM/PSB as a biofertilizers, there need to be isolated and characterized. In this study, different phosphate solubilizing bacteria were isolated from different places with different solubilization efficiency and with different characters. The selected isolates were effective in solubilization of insoluble phosphorus on solid media. Phosphorous solubilization increased along the incubation days increased. This is true for plate screening (halo zone increased with increasing of colony diameter and resulted in SI increased).

Acknowledgements

First, we would like to thank Hawassa University, Haramaya University and the MoE for granting study opportunity, financial support and sponsorship for the study. Our great thanks go to Holeta Agricultural Research Center especially DSWRM, the technical assistants of Microbial Biotech. and Chemistry laboratories.

References

1. Kannaiyan S, Kumar K, Govindarajan K (2004) Biofertilizer technology for rice based cropping system. Scientific Pub, Jodhpur, India.
2. Richardson AE (2001) Prospects for Using Soil Microorganisms to Improve the Acquisition of Phosphorus by Plants. Australian J Plant Phy 28: 897-906.
3. Poonam AS, Ghosh AK (2011) Characterization, Identification and Cataloguing of Agriculturally Important Microorganisms Isolated from Selected Wetland and Rain-Fed Ecosystem of Bihar. Asian J Expr Bio Sci 2: 577-582.
4. Keneni A, Fassil A, Prabu PC (2010) Isolation of Phosphate Solubilizing Bacteria from the Rhizosphere of Faba Bean of Ethiopia and Their Abilities on Solubilizing Insoluble Phosphates. J Agri Sci Tech 12: 79-89.
5. Mamo T, Haque I, Kamara CS (1988) Phosphorus Status of Some Ethiopian High Land Vertisols. In: Management of Vertisols in Subsaharan Africa, proceedings of a conference held at ILCA, 31 August-4 September 1987. Addis Ababa, Ethiopia.
6. Haile W, Fassil A, Asfaw H (1999) Studies on Phosphate Solubilizing Ability of Bacteria Isolated from Some Ethiopia Soils. Proc of the 9th Annual Conference of the Bio Soci of Ethiopia, Awassa, Ethiopia.
7. Elkoca E, Kantar F, Ahin F (2008) Influence of nitrogen and phosphorus solubilizing bacteria on the nodulation, plant growth and yield of chickpea. J Plant Nutr 31: 157-171.
8. Chen YP, Rekha PD, Arun AB, Shen FT, Lai WA, et al. (2006) Phosphate solubilizing bacteria from subtropical soil and phosphate solubilizing abilities. App Soil Eco 34: 33-41.
9. Son HJ, Kim YG, Lee SJ (2003) Isolation, identification and physiological characteristics of biofertilizer resources, insoluble phosphate-solubilizing bacteria. Korean J Micr 39: 51-55.
10. Anteneh A (2012) Evaluation of Co-inoculation of Bradyrhizobium japonicum and Phosphate Solubilizing Pseudomonas spp Effect on Soybean (Glycine max L. (Merr.) in Assossa Area. J Agri Sci Tech 14: 213-224.
11. Somasegaran P, Hoben HJ (1994) Hand Book for Rhizobia – Methods in Legume Rhizobium Technology. Springer-Verlag, Heidelberg, Germany.
12. Pikovskaya RI (1948) Mobilization of Phosphate in Soil in Connection with their Vital Activities of Some Microbial Species. Microbiologiya 17: 362-370.
13. Alikhani H, Salehrastin N, Antoun A (2006) Phosphate solubilizing activity of rhizobia native to Iranian soils. Plant soil 287: 35-41.
14. Benson HJ (2001) Microbiological Applications: Laboratory Manual in General Microbiology. McGraw Hill, Boston, 8th edn, p: 455.
15. Prescott HK (2002) Microbiology. 5th edn. The McGraw-Hill, p: 1147.
16. Zhu F, Qu L, Hong X, Sun X (2011) Isolation and Characterization of Phosphate-Solubilizing Halophilic Bacterium Kushneria sp. YCWA18 from Daqiao Saltern on the Coast of Yellow Sea of China. Research Article Evidence-Based Complementary and Alternative Medicine.
17. Rodriguez H, Fraga R (1999) Phosphate solubilizing bacteria and their role in plant growth promotion. Biotech Advance 17: 319-339.
18. Baon JB, Wedhastri S, Kurniawan A (2012) The Ability of Phosphate Solubilizing Bacteria Isolated from Coffee Plant Rhizosphere and Their Effects on Robusta Coffee Seedlings. J Agri Sci Tech 2: 1064-1070.
19. Ghosh U, Subhashini P, Dilipan E, Raja S, Thangaradjou T, et al. (2012) Isolation and Characterization of Phosphate-Solubilizing Bacteria from Seagrass Rhizosphere Soil. J Ocean Uni China 11: 86-92.
20. Nautiyal CS, Bhadauria S, Kumar P, Lal H, Mondal R, et al. (2000) Stress induced phosphate solubilization in bacteria isolated from alkaline soils. Fed Eur Micr Societies 182: 291-296.

18

Stems Extract of Kemuning cina (*Catharanthus roseus*) as Biofungicides against White Root Fungal (*Rigidoporus microporus*) of Rubber Trees (*Hevea brasiliensis*)

Hazwani Mohd Zaini and Normala Halimoon*

Department of Environmental Sciences, Faculty of Environmental Studies, Universiti Putra Malaysia, 43400 UPM Serdang, Selangor, Malaysia

Abstract

White root disease which cause by *Rigidoporus Microporus* fungus are the most destructive and serious disease among the three major root disease in rubber plantation. Biological control using stems extract of kemunting cina (*Catharanthus roseus*) have been used as a healing agent of infected rubber trees. The objectives of the study are to determine the ability of stems extract from kemunting cina to control white root disease of rubber trees and to observe the plants performance towards the fungus. Stems of the plant were soaked in dichloromethane (DCM) solution because the solvent was found as the best extraction for the treatments. Three batches of rubber trees were prepared in the research, which consists of five trees. Group A was a control, while the trees in Group B were planted with fungus and the trees in Group C was applied with the stems extract together with the fungus. Five hundred ml of extracts was used as biofungicides against white root disease. The chlorophyll, diameter, height and number of leaves were recorded. Among the three groups of rubber trees, Group C shows the ability of extract to inhabit against *Rigidoporus Microporus* growth refers to the improvement of growth performance of the plants. The stem extracts of kemunting cina were antagonistic inhibited against *Rigidoporus microporus* fungus.

Keywords: White root disease; *Rigidoporus Microporus*; Kemunting cina (Catharanthus roseus); Dichloromethane; Biological control

Introduction

The agriculture sector was indeed the most important sector for developing countries like Malaysia. In fact, it was one of the most important features that differentiate status between developing countries and developed countries. These sectors contributed to the foundation of Malaysian economy in the post-independence era where the majority of the population-based activities focused on agriculture and mining. The Prime Minister of Malaysia, Datuk Seri Abdullah bin Hj. Badawi announced that the agricultural sector will developed in the Ninth Malaysia Plan in 2005 (9MP) [1]. Since the announcement, many government agency and non-government organization get involved in this sector. However, the rapid development and high progressive in the sector, excessive agricultural activities will cause disturbance to the ecosystem stability. Therefore, some arguments regarding to the issues could threaten the health of ecosystem in Malaysia.

Agriculture sector are closely related to the environment and the most important sectors to generate food security for human and animals. Plant infected with fungus can reduce the productivity of agriculture. Fungicides are able to treat the plant disease and improve crop quality and supplies. Fungicide is an agent that been used to kill fungi, bacteria and viruses that generally harm human and environment as well. Most of the fungicide that commonly used was chemical fungicides. However, prolonged usage of the chemical fungicide not only kills the fungus but also other wildlife. Apart from that, it can trigger human health. Chemical fungicide also can cause cancer and lead to birth defects on baby. Therefore, the researches on substances with antimicrobial activity are frequently studied, and medicinal plants have been considered used to remedies for many infectious diseases [2].

Hevea brasiliensis or well known as rubber trees was the most common plant that have been planted since 1877 in Kuala Kangsar, Perak. Rubber trees are well known due to its important in producing latex that is useful for industrial use in Malaysia. Although, the production of rubber was highly at one time, but due to the attack of pest and plant diseases, the number of rubber trees present nowadays is decreasing. Large number of pest and disease problems occurs mainly in man-made forest [2]. Monoculture plantations of rubber trees also show clear evidence to promote higher outbreaks of pests and disease compared to mixed culture plantations [3]. According to Wingfield et al. [4], most of the failures occurs in the early phase of forest plantation development has been linked to the serious problems with root disease. White root disease which caused by *Rigidoporus microporus* fungi are known to be the most serious and dangerous to plants as it can spread the disease very quickly and are often fatal.

Kemunting cina (*Catharanthus roseus*) is a tropical plant was important as medicinal plant from *Apocynaceae* family. It was cultivated mainly for its alkaloids due to the presence of indispensable anti-cancer, drugs, vincristine and vinblastine. The content of alkaloid in root of the plant is the main source of anti-hypertension [5]. It also known as ornamental plants among the community as there are commonly have two varieties of plant based on the flower color which is pink and white in color [6]. The plants have a lot of advantages especially in the medical field due to the ability to treat several types of disease. Some parts of the plants like the leaves and stems are the sources of dimeric alkaloids, vinacristine and vinblastine that are indispensable cancer drugs. And

***Corresponding author:** Normala Halimoon, Department of Environmental Sciences, Faculty of Environmental Studies, Universiti Putra Malaysia, 43400 UPM Serdang, Selangor, Malaysia,E-mail: mala_upm@upm.edu.my

its roots have antihypertensive, ajmalicine and serpentine [7]. It also was used as a folk remedy for diabetes in Europe for centuries. In India, juice from the leaves was used to treat wasp stings. In Hawaii, the plant was boiled to make a poultice to stop bleeding. In China, it was used as an astringent, diuretic and coughs remedies [8]. Furthermore, its leaves can be used to produce the Jamaican tea that can treat diabetes.

Catharanthus roseus also has been categorized as poisonous plants due to the present of catharanthine, leurosine, norharman, lochnerine, tetrahydroalstonine, vindoline, vindolinine, akuammine, vincamine, vinleurosin and danvinrosidin. It can only be taken in dosages that have been defined and allowed only. However, there are a few studies that revealed the ability of the plant to treat fungus diseases for infected plant. Junaid et al. [9] have studied about *Catharanthus roseus*, an important drug: its application and production. Antibiogram of *Catharanthus roseus* extracts also have been analyzed by Sathiya et al. [10]. According to all of the researches above, *Catharanthus roseus* are able to apply as effective biological tool in treating the plant disease [11].

The objectives of the research are to identify the ability of *Catharanthus roseus* stems extract to control white root rot disease of rubber trees and to observe the performance of rubber plant towards the *Rigidoporus microporus*. The Catharanthus roseus stems extract was selected as a healing and control agent for *Rigidoporus microporus*fungus, which causes white root diseases to all rubber trees. The research was conducted in a small scale study using the same height of rubber trees. The chlorophyll, diameter, height and number of leaves were recorded to reveal the effectiveness of the extracts.

Material and Methods

Samples collection

The fungus of *Rigidoporus microporus* and rubber trees species (Hevea brasiliensis) has been provided from Forest Research Institute of Malaysia (FRIM) for the research purposes. The healthy and fresh stems of *Catharanthus roseus* were collected from Terengganu coastline and from the residential area located near Universiti Putra Malaysia (UPM).

Preparation of *Catharanthus roseus* extracts

The stem samples of *Catharanthus roseus* were air dried for almost 4 week until the stems were fully dry. The stems of *Catharanthus roseus* were used in this study to determine the potential of stems part as biopesticide. The samples were grinded to powder using mortar. Then, the powder samples were soaked with dichloromethane (DCM) almost for four days to make sure all of bioactive compounds dissolved in the solvent. The solutions were filtered with whatman No. 1 filter paper to separate between the powdered and liquid. Then, rotary evaporator was used to remove DCM solvent from the extracts. The crude extracts were dissolved using sterile distilled water to a final concentration of 20 g/L. The 20 g/L of crude plant extract of *Catharanthus roseus* species was used in the study because the concentration are capable of inhibiting the growth of *Rigidoporus microporus.*

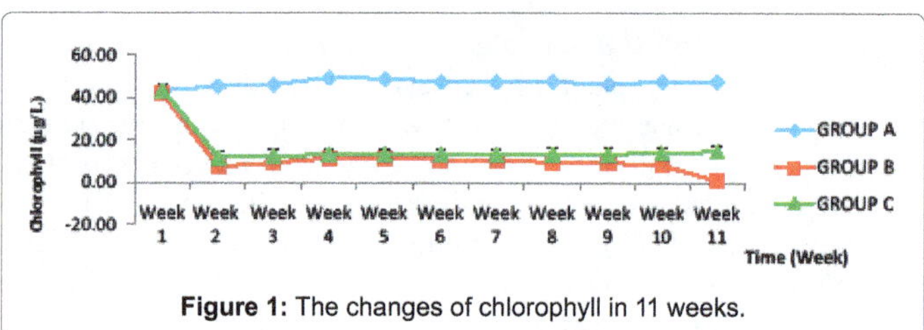

Figure 1: The changes of chlorophyll in 11 weeks.

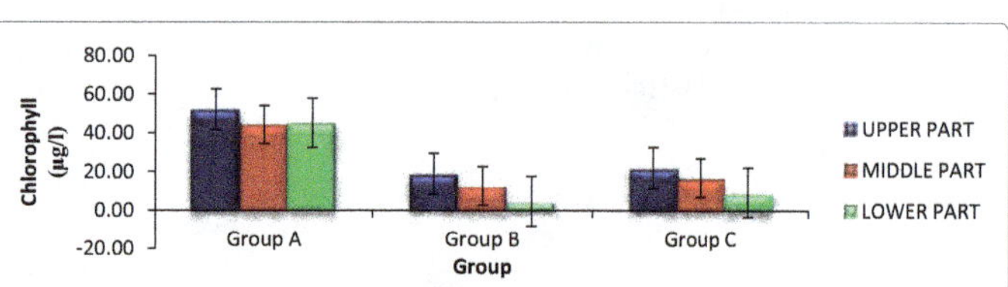

Figure 2: The distribution of chlorophyll based on the different part of the leaves.

Preparation of inoculum blocks

According to Lee and Sikin [12], young rubber branches were cut into wood blocks measuring approximately 2.0-3.0 cm diameter by 8 cm length. Then, it were placed in vertical condition in autoclavable plastics bags for the colonization process to take place. One to four fungal discs were taken from the fungus specimen and inserted into the autoclavable plastics bags using sterilized needle. The plastics bags were then stoppered back with cotton wool and sealed with parafilm to avoid from outside contamination. The blocks were incubated in the dark at room temperature for one month to allow for the colonization process with the fungi to occur.

Planting of rubber trees with colonize wood blocks

Five trees per group were prepared for the study. The rubber trees in group B and C were replanted together with the colonized blocks. Eight inoculum blocks were tied together and placed in contact with the taproot.

Application of *Catharanthus roseus* stems extract to rubber trees

Group A of rubber tree as a control without any application of fungus and stems extract, while Group B, the trees were planted with fungus as mention above without any extracts application. The trees in Group C were applied with the stem extracts together with the fungus. About 500 ml of the extract were applied to each rubber tree in Group C three times before replanting procedure. The extracts were applied before the fungus was fully reproduced on the root. The *in-situ* parameter of measurement for plants such as chlorophyll, height, diameter and number of leaves were measured using chlorophyll meter SPAD-502, measuring tape and vernier caliper through observation. The measurements of each parameter were carried out for 11 weeks and the data were collected once a week at the same time. All of 15 rubber trees were placed in the nursery of Forestry Faculty, Universiti Putra Malaysia.

Statistical analysis

Bar chart was used to provide clearer view and to show general trends of the data collected. Based on the statistical analysis using ANOVA single factor, it generates p-values, one for each parameter independently.

Chlorophyll

The chlorophyll reading in Group C was increased at the end of the experiment as shown in Figure 1. The stems extract was ability to inhibit the growth of fungus rather than Group B, which shows negative results due to the fungus attack. While plants in Group A, shows the constant reading and not much different throughout the 11 weeks of observation time because there is no disturbance to the plants. The chlorophyll reading was based on the concentration of the green pigment, which determines the color of leaves. There are significant differences ($p \leq 0.05$) in chlorophyll reading between all of the plant.

Figure 2 shows the upper part of leaves shows the highest reading followed by the lower one and the middle leaves. The upper part of plant consist of the youngest leaves, the middle and lower part consist of the older leaves. Based on the previous study [13], the concentration of the green pigment were different based on the different in color of the leaves.

Height

Figure 3 shows the changes of height were increased in Group A throughout the observation weeks. The result shows that the plants have enough nutrient and resources for their growth without any disturb from fungus infection. Meanwhile, height of the plant in Group C showed the increased at the end of the experiment. The presence of stems extract was capable to inhabit the growth of fungus and improved the growth of the rubber trees. The changes of height were constant in Group B where the fungus is become dominant and affect the growth of rubber trees. Thus, the growth of plant was retarded and finally, the rubber trees become death. The presences of the fungus will harm the rubber trees because fungus relied on the host plant for their supply of food [14]. The result shows that there was significant difference ($p \leq 0.05$) in height of plant between the entire plant groups.

Diameter

According to Figure 4, the diameter changes in Group A and C were increased throughout the 11 week of study. However, the diameter is almost the same every week. The stems extract was much influenced in diameter changes of the plants. Meanwhile, the result in Group B shows the diameter changes are remains constant starting from week 6 onwards. The plants begin to dieand the growth was stunted due to the fungus attacked. There was a significant difference ($p \leq 0.05$) in diameter between all the tree groups.

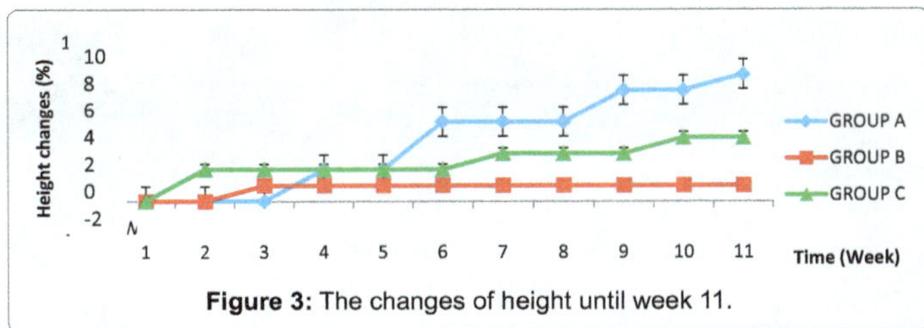

Figure 3: The changes of height until week 11.

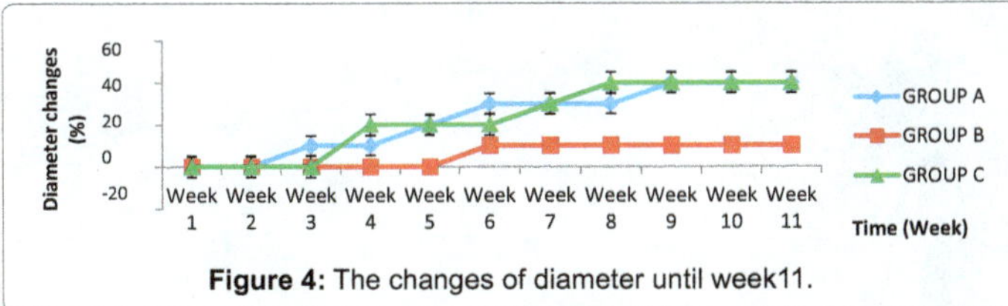

Figure 4: The changes of diameter until week11.

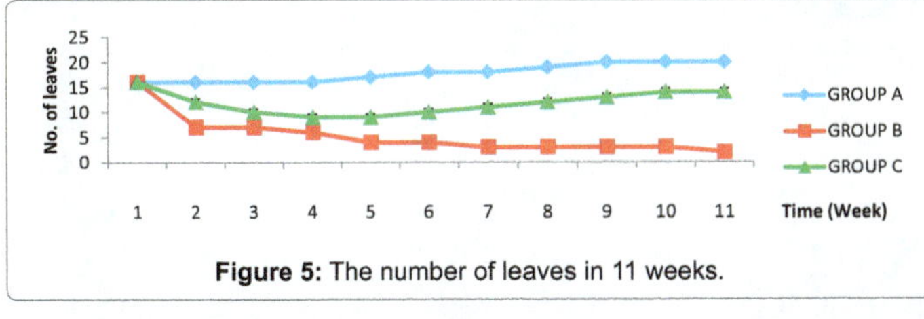

Figure 5: The number of leaves in 11 weeks.

Number of leaves

According to Figure 5, the result shows that the number of leaves in Group A and C was increased until the end of the experiments. Application of the stems extract seems possible to enhance the shoot growth of rubber trees after the infection even though at the beginning showed there was reduction in the number of leaves. In Group B, there was reduction in number of leaves throughout the week. The reduction occurs due to fungus infection and also water stress. There was significant difference in number of leaves between all the groups. The symptoms appeared were wilting and yellowing of the leaves, defoliation and white mycelium on the root system after infection.

Discussion

According to Lee and Sikin [12], usually the fungus will start to attack after 2 week the inoculum been planted, but it depends on the plants as every plant have its own defense to the diseases. Moreover, since the *Rigidoporus microporus* do not contain chlorophyll, so it cannot produce their own food and they need to rely on their host (rubber plants) to provide them with food. The color of leaves also influence the chlorophyll reading for the whole plant. When the leaves was old and there was damaged to the leaves, it may turn yellow or orange in color. The chlorophyll was translocated out of the leaves and appeared yellow before death. Meanwhile for Group C, the chlorophyll reading shows slightly increased until the end of the observation. This proved that the extracts were able to act as an agent to reduce the effect of white root diseases. The early application of the extract makes the plants to build its own resistance towards the diseases.

The plant height in Group B shows that no changes at the first3weeks until the end of observation. *Rigidoporus microporus* attacks at its root system influence the performance and growth of the tree. The root system got damaged as the fungus become dominant. In order for the fungus to grow and live, they need to rely on the other plants to get the nutrients thus inhibit the plant growth. But, the rubber trees were capable to keep growing due to the presence of the *Catharanthus roseus* extracts (Group C). Thus, the result proved that the *Catharanthus roseus* stems extract was capable to improve the life span of the rubber trees.

The diameter of tree stems changes with time are the first been used to detect plant growth [15]. The changes are influence by a lot of factors and one of the major factors is water content and water tension inside the rubber plants. These changes have been detected since the late 19th century and their connection with environmental factors have been reported. The funguses that have been attacked to the rubber tree in Group B were influenced to the diameter changes. Plants in Group B were trying to adapt to the new environment due to the presence of *Rigidoporus microporus* fungi which absolutely give changes to the life span of the tree as well as the tree diameter. The fungus gets their resources to survive and the rubber plants were become as their host for food and water. Moreover, fungus usually loves to lives on moist or damp environment. Finally, the fungus was become dominant and able to conquer that particular area. The plants in Group C were trying to fight against the *Rigidoporus microporus* fungus in the presence of stems extract. After week 3, the diameter of plants was increased where the plants tried to manage the stems extract as their weapon to avoid and slow down the fungus activity.

The numbers of leaves were increased form the plants in Group C have been recorded throughout the 11 weeks of observation time. Several new shoots have been observed after week 4 shows the rubber plant tried to recover. The presence of stems extract inhabits the growth of fungus at the plant root. However, the plants in Group B shows almost lost their leaves week by week. Starting from week 2 the number of leaves decreased dramatically. The fungal attacked to the plant root causes loss their function to carry water and nutrients from soil. The fungus was harmed the life span of the plants.

Chlorophyll contents in the leaves also influence the leaves performance. If the level of chlorophyll decreases, the leaves will become yellow. The possibility of the leaves to fall is higher. Weather and climate also play a huge role in causing the leaves to fall. Sudden changes in temperature can lead the leaves to turn yellow or brown and thus cause the leaves to drop.

Conclusion

White root disease caused by *Rigidoporus microporus* fungus can be recognized by the presence of white rhizomorphs on the tree root surface. The stems extract of *Catharanthus roseus* medicinal plant are able to reduce the effects of white root diseases. The results confirmed that stems extract were antagonistic inhabited against the diseases. The presence of the stem extracts showed that the plants have the ability to improve the rubber trees performance after infection refer to positive result of chlorophyll density, height, diameter and number of leaves as shown in this study. The plants in Group C tried to survive after exposed with fungus and *Catharanthus roseus* stems extract compared to Group B without extract. Thus, the stems extract of *Catharanthus roseus* was suitable to become as one of the biological control of plants diseases especially for root rot diseases.

Acknowledgement

The authors would like to thank to Universiti Putra Malaysia throughout the place and equipment used for the research.

References

1. (2006) Speech by the Prime Minister in the Dewan Rakyat. Ninth Malaysia Plan 2006-2010.
2. Bakshi BK, Reddy MAR, Puri YN, Sujan Singh (1972) Forest disease survey (final technical report 1967-1972).
3. Speight MR, Wylie FR (2001) Insects pests in Tropical forestry. CABI Publishing, Wallingford, Oxon, UK. 249.
4. Wingfield MJ (1999) Pathogens in Exotic plantation forestry. International Forestry Review 1: 163-168.
5. Jaleel CA, Panneerselvam R (2006) Variations in the antioxidative and indole alkaloid status in different parts of two varieties of *Catharanthus roseus*. Chinese journal of Pharmacology and Toxicology 21: 487-494.
6. Jaleel CA, R.Gopi, Panneerselvam R, Manivannan P (2008) Soil salinity alters the morphology in *Cathanranthus roseus* and its effects on endogenous mineral constituents. EurAsia J BioSci 2: 18-25.
7. Kulkarni RN, BaskaranK, Chandrashekara RS, Kumar S (1999) Inheritance of morphological traits of periwinkle mutants with modified contents and yields of leaf and root alkaloids. Plant breed 118: 71-74.
8. Farnsworth NR (1961) The pharmacognosy of the periwinkles: *Vinca* and *Catharanthus*. Lloydia. 24: 105-138.
9. Junaid A, Sheba HK, Zahid HS, Zohra F, Mehpara M, et al. (2010) Catharanthus roseus (L.) G. Don. An Important Drug: It's Applications and Production. Pharmacie Globale (IJCP) 4
10. Sathiya S, Karthikeyan B, Jaleel CA, Azooz MM, IqbalM (2008) Antibiogram of *Catharanthusroseus* extracts. Global journal molecular science 3: 01-07.
11. Mumtaz (2010). Potential of Water Lettuce and Water Hyacinth as an Effective Nutrient Removal in Wastewater. Thesis, Universiti Putra Malaysia.
12. Lee SS, Noraini SY (1999) Fungi associated with heart rot of Acacia mangium trees in Peninsular Malaysia and East Kalimantan. Journal of Tropical Forest Science 11: 240-254.
13. Nielsen UB, Hansen JK, Kromann HK (2011) Impact of site and provenance on economic return in Nordmann fir Christmas tree production. Scand J For Res 26: 7489.
14. Salter PJ and KA (1980) Plant diseases, an imagricultural" Know& How to grow vegetables" Oxford University Press.
15. Leikola M (1969) Influence of environmental factors on the diameter growth of forest trees: Auxanometric study. Acta Forestalia Fennica 92: 24-30.

Nitrogen Fertilizer Effects on Nitrous Oxide Emission from Southwest Brazilian Amazon Pastures

André Mancebo Mazzetto[1]*, Arlete Simões Barneze[2], Brigitte Josefine Feigl[2], Carlos Eduardo Pellegrino Cerri[1] and Carlos Clemente Cerri[2]

[1]*Escola Superior de Agricultura Luiz de Queiroz, Universidade de São Paulo, Avenida Pádua Dias, 11, 13400-000, Piracicaba, São Paulo, Brazil*
[2]*Centro de Energia Nuclear na Agricultura, Universidade de São Paulo, Avenida Centenário, 303, 13400-970, Piracicaba, São Paulo, Brazil*

Abstract

Beef production is one of the most important agricultural activities in Brazil. In order to increase production without increasing deforestation, farmers are intensifying breeding and pasture improvements. The main technique for increasing pasture improvement is the application nitrogen fertilizer, but this action can result in emission of nitrous oxide (N_2O). We assessed the impact of nitrogen fertilizer application on GHG emissions and pasture yield in a pasture located at Southwest Brazilian Amazon. Agronomic recommended rates of nitrogen fertilizer (NF) and higher rates, as two times (2NF) and four times (4NF) the recommended rate were applied. A control treatment with no fertilizer was also analysed. The experiment had 30 days duration, where we observed the baseline emissions from all treatments, including control. Nitrogen fertilizer application resulted in high N_2O emissions. We found no differences between NF and 2NF treatments, but all treatments were different from control. The higher forage yield leads to low N_2O emission per kg of forage in 4NF treatment. According to our study, the best (agro-environmental benefits) practice is the application of 100 kg N ha^{-1} (2NF treatment) in the region studied.

Keywords: Beef production; Tropical climate; N_2O; Pasture yield

Introduction

Brazil is the second largest beef exporter, responsible for 15% of worldwide production [1]. The typical system of beef production in Brazil is pasture-based, predominantly occurring on unimproved pastures. Pastures occupy three-quarters of the national agricultural area, about 180 million hectares [2]. There are projections of increased demand for beef in the order of 2.5% per year by 2017-2018. In order to meet this demand, Brazilian farmers must develop a more intensive system in order to produce more beef without increase deforestation [3]. This intensive system must have higher beef production per unit area, with low emissions of greenhouse gases (GHGs) per kg of beef produced. If farmers do not adopt sustainable options for pasture intensification, deforestation could be increased, increasing GHG emissions from the sector. The improvement of the whole system of beef production is a key component to reduce emissions from all relevant sources, including land use, land use change and livestock [4].

In order to improve the beef system, intensification methods must be applied, such as the use of nitrogen fertilizer. The main limiting-nutrients for grass growth in Brazilian conditions are phosphorus (P) and nitrogen (N). The application of nitrogen fertilizer enhances the availability of N to the plant and microorganisms, but an excess of N can result in nitrous oxide (N_2O) emissions through nitrification and denitrification processes [5]. The effect of N fertilizer on N_2O emission is well reported in the literature [6,7]. However, there are very few studies in tropical climates examining N fertilizer application in pastures [8,9] with respect to GHG emissions and they do not cover the range of edaphoclimatic conditions.

We measured the effect of N fertilizer application on soil N_2O emission. To simulate the intensification practices, we tested the effect of the application of the currently recommended levels of fertilizer on GHG emissions. There is anecdotal evidence that farmers usually apply more than the recommended rate. Therefore, we also tested higher rates of N fertilizer to verify the impacts on GHG intensities.

Materials and Methods

The experiment was carried out on a permanent pasture, covered by *Brachiaria* and was not grazed by livestock before or during the experiment. The studied site was split in 4 paddocks (0.1 ha). Plots consisted of an area fertilizer application (0.05 ha). The experiment was carried out from 09 November to 10 December of 2012 (summer) at Agropecuária Nova Vida, Rondônia state, Brazil (10°10'05"S and 62°49'27"W) under tropical climatic conditions (Aw-Köppen climatic classification). Soil is an Oxisol (Ustox), and its texture is sandy loam. Soil properties (upper 10 cm) at the start of the experiment are showed in Table 1. Meteorological data were recorded at the nearest meteorological station (rainfall and air temperature), which was within 1 km of the field site.

We studied the application of ammonium nitrate (NH_4NO_3) at rates of 0, 50, 100 and 200 kg N ha^{-1} (treatments C, NF, 2NF and 4NF, respectively), with five replicates to each treatment, in a complete randomized block design. The fertilizer was applied in the first day of the experiment, right before the first sampling.

Sand	Clay	Silt	pH	Bulk density	Total C	Total N
----------------	%	----------------	$CaCl_2$	g m^{-3}	---------g	kg^{-1}---------
65.0	26.7	8.3	5.1	1.5	27.3	3.0

Table 1: Soil properties (0-10 cm) at the beginning of the field experiment.

***Corresponding author:** André Mancebo Mazzetto, Escola Superior de Agricultura Luiz de Queiroz, Universidade de São Paulo, Avenida Pádua Dias, 11, 13400-000, Piracicaba, São Paulo, Brazil,
E-mail: andremmazzetto@hotmail.com

The closed static chamber technique [10] was used to collect gas samples. At the field, unvented chambers (28 cm diameter, 13 cm height) were placed two days before the first gas sampling. The chambers were inserted to a depth of up to 3 cm to ensure an airtight seal. The volume enclosed by the chamber was approximately 11 L. At the time of sampling, lids were placed on top of the chambers and a seal was achieved via a water-filled groove on the chamber that the lid fitted in to. Gas sampling was normally carried out between 9:00 and 12:00. Samples were collected at 0, 10, 20 and 30 minutes after the chambers were closed. A 20 ml syringe was used to collect the gas samples from the chambers, which were then transferred to pre-evacuated 13 ml headspace vials using a hypodermic needle. The glass vials had a chlorobutyl rubber septum (Chromacol). The pre-evacuation was carried out using a vacuum pump.

Gas sampling was carried out daily during the first week, then twice a week for 3 weeks, and then once a week until the end of the experiment. The samples were analysed within 2 weeks of collection using gas chromatography (GC - Shimadzu 2014). The N_2O was detected with an ECD (electron capture detector).

The flux of N_2O was calculated using the linear change in the concentration as a function of the incubation time within the chamber. Gas fluxes were calculated from the time vs. concentration data using linear regression. These data were used to calculate the cumulative emissions over the experimental period by the linear interpolation of data points between two successive days and numerical integration of the area under the curve using the trapezoid rule [11]. The emission factors were calculated using the recommendations of the IPCC [12].

Designated soil sampling plots were installed adjacent to each gas chamber, which also received the same fertilizer rate. Soil (0-10 cm) was sampled on days 1, 7, 14, 21 and 28. Soil mineral N concentrations were determined by extraction with 2 M KCl with a 1:2 ratio of soil and extractant [13]. Soil extracts were filtered and stored at 4°C. Concentrations of NH_4^+ and NO_3^- in the extracts were determined by automated flow injection analysis (FIA) [14].

Gravimetric moisture contents were determined after drying at 105°C for 48 h. Grasses from each chamber were cut at 4-5 cm height at the end of the experiment. The green matter was transferred to a pre-weighed paper bag and dried at 70°C for 1 week. After that, dry matter weight was recorded.

Statistical analyses

Total GHG emissions were estimated by calculating cumulative fluxes over an experimental period of 30 days. Data were verified for normal distribution and treatment means for daily N_2O fluxes and cumulative fluxes over the period of the experiment were compared using one-way analysis of variance. To determine the statistical significance of the mean differences, Turkey tests were carried out at 0.05 probability level.

Results

The average air temperature and total precipitation were 29°C (varying from 25 to 33°C) and 250 mm (varying from 3 to 107 mm) (Figure 1). Those conditions are representative of the summer season of the southwestern part of the Brazilian Amazon. Due to the periodical rainfall, the water-filled-pore-space (WFPS) was high during all the experiment, ranging from 69 to 86% (Table 2).

Nitrous oxide emissions were variable over the study period (Figure 2). Our results showed that the application of N fertilizer increased N_2O emissions, changing the pasture status from a net sink to a net source of N_2O (Table 3). The increase in N fertilization had a high correlation with N_2O emission (r^2=0.99). The emission factors calculated were 0.45, 0.12 and 0.17 for NF, 2NF and 4NF treatments, respectively.

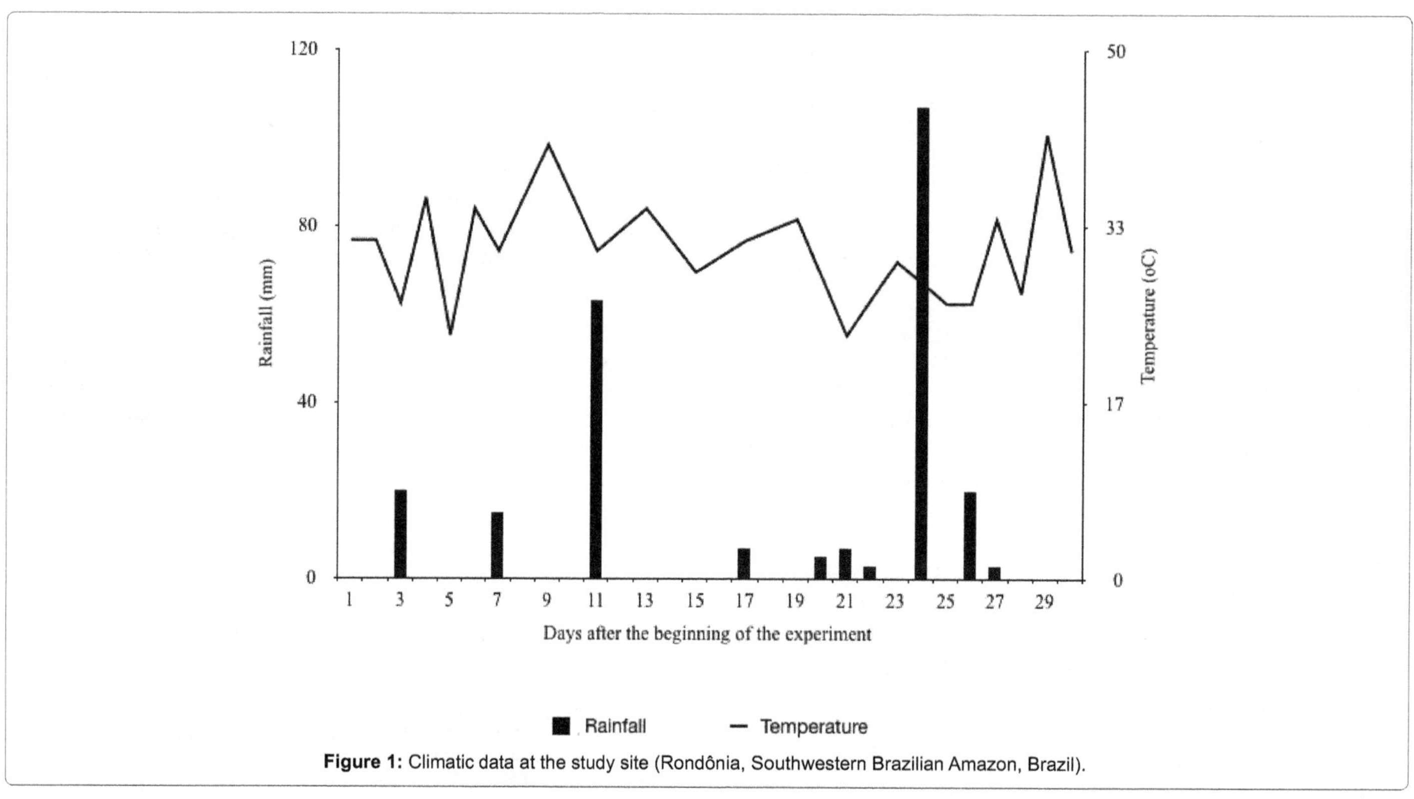

Figure 1: Climatic data at the study site (Rondônia, Southwestern Brazilian Amazon, Brazil).

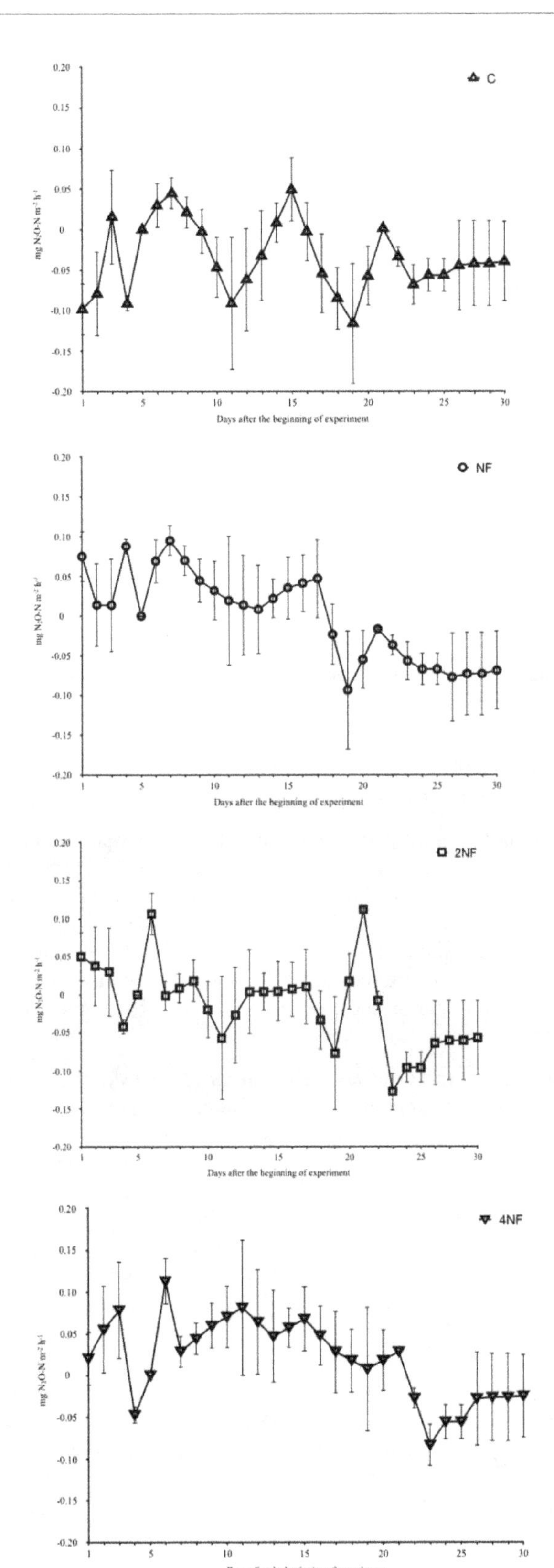

Figure 2: N_2O emissions from the studied site. C: Control; NF: application of 50 kg N ha^{-1}; 2NF: application of 100 kg N ha^{-1}; 4NF: application of 200 kg N ha^{-1}. The error bars denote the standard deviation.

Soil NH_4^+ levels in the NF treatment increased rapidly after the application of N fertilizer, with levels different from the control ($p<0.05$) throughout the experiment until day 21 (Figure 3). Soil NH_4^+ levels in the 2NF and 4NF treatments increased after day 7, and remained high until day 21 (Figure 3). Soil NO_3^- concentrations in all treatments were significantly higher than the control during throughout the experiment.

The forage yield increased with N fertilizer application (Table 3). While all N-treated plots had significantly higher N_2O emissions than the control, especially 4NF (Table 3), the increase in forage yield with N fertilizer lead to lower N_2O emission per kg of forage (166, 59 and 105 mg N_2O kg dry matter^{-1} for NF, 2NF and 4NF, respectively).

Discussion

The increase in N_2O emission was expected, since the application of ammonium nitrate increases the availability of nitrate in soil (Figure 3). The available N can be quickly taken up by plants or lost as N_2O in a few days [15]. The assimilation of NH_4^+ is energetically more efficient than NO_3^- [16]. *Brachiaria* grasses are well adapted to the N poor soils from Brazil. When both NH_4^+ and NO_3^- are available in soil, the plant absorbs NH_4^+ preferably, leading NO_3^- that can be denitrified in soil [17].

The relationship between the N_2O emission and the ammount of N fertilizer applied was not linear. Other studies showed that the expoential curve fits better [18]. In our study the N_2O emission from NF and 2NF treatments did not show statistical difference. In this case was not possible to test wich curve would fit better, since there was only 2 possible points (NF × 4NF or 2NF × 4NF).

Soil moisture is a key factor for N_2O emission [19]. During this study the soil showed high WFPS (Table 2) due to the periodical rainfall, typical from the summer in the Amazon region. The temperature is also an important factor in the studied situation, since the high temperature of the tropics can stimulate N turnover, consuming O_2 and creating an anaerobic environment, ideal for the denitrification process. The interaction betweeen soil moisture and temperature can provide a possible explanation for the negative fluxes observed, mainly in the control treatment. The increase in temperature has a positive impact in the N mineralization, combined with the ideal anaerobic environment

	%		SD	CV
Day 1	69.6	b	7.8	11.2
Day 7	72.4	b	8.3	11.5
Day 14	84.4	a	8.8	10.4
Day 21	82.3	a	10,4	12.6
Day 28	86.3	a	9.1	10.5

Table 2: Soil water filled pore space (WFPS-%) in the soil (0-10 cm) at the field experiment. SD: Standard deviation; CV: Coefficient of variation; Means followed by the same letters in columns are not statistically different (Tukey, pB 0.05).

	N-N_2O				Yield			
	mg m^{-2}				kg DM m^{-2}			
Treatment	CE		SD	CV	Average		SD	CV
⁰N (control)	-18.7	c	4.9	26.4	0.08	d	0.008	10.6
NF	3.9	b	1.2	26.9	0.14	c	0.009	6.9
2NF	-7.0	b	0.4	6.1	0.19	b	0.023	11.7
4NF	15.4	a	6.2	40.2	0.32	a	0.025	8.0

Table 3: Cumulative N_2O emission from field study and the effect of nitrogen fertilizer application on pasture yield. C: Control; NF: application of 50 kg N ha^{-1}; 2NF: application of 100 kg N ha^{-1}; 4NF: application of 200 kg N ha^{-1}; CE: Cumulative emission; SD: Standard Deviation; CV: Coefficient of Variation; Means followed by the same letters in columns are not statistically different (Turkey, p ≤ 0.05).

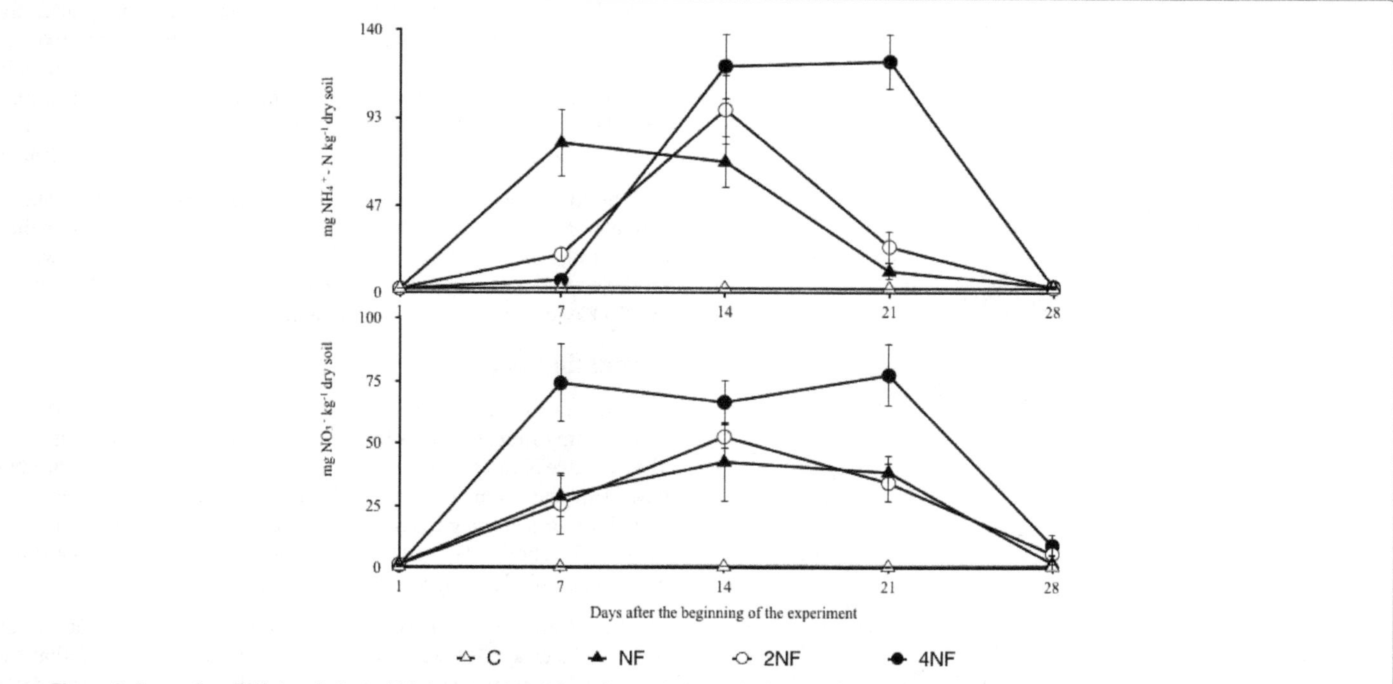

Figure 3: Ammonium (NH_4^+) and nitrate (NO_3^-) content in soil from the studied site. C: Control; NF: application of 50 kg N ha^{-1}; 2NF: application of 100 kg N ha^{-1}; 4NF: application of 200 kg N ha^{-1}. The error bars denote the standard deviation.

created by the high WFPS. Is such situation, most of the N of the soil is consumed by microorganisms or completed denitrificated to N_2. Wetlands can act as N_2O sinks [20]. The high WFPS can contribute to complete denitification, reducing even further the N_2O emission. One possible explanation for the negative fluxes is the use of N_2O as an final electron aceptor in the absence of other source [21]. Recently, more studies have been reported negative fluxes of N_2O [22-24] but there are no consensus of wich mechanism is reponsible for the N_2O consume. More studies must be done to address this gap, specially those in controlled conditions with different types of soil, moisture content and temperature, associated with a profile of the microbial communities.

The Biological Nitrification Inhibition (BNI) is other factor that can have a high impact on N_2O emission from Brazilian pastures. In pastures covered by *Brachiaria* grasses the flow of N from NH_4^+ to NO_3^- is restricted by a natural root exudate (brachialactone), and NH_4^+ accumulates in soil [17]. In such situation, the BNI keeps NH_4^+ in the soil and the plant gradually absorbs this nutrient, while the excess is nitrified to NO_3^-. Such process must be investigated, since *Brachiaria* is the main type of grass in Brazilian pastures.

The emission factors obtained in this study are significantly lower than the recommended by the IPCC (1%), regardless the amount of N fertilizer applied. There is a lack of studies on N_2O emissions from fertilizers in tropical pastures. The study of Morais et al. [9] was conducted in Rio de Janeiro, with a different source of N (urea) and a different type of grass (elephant grass), resulting in a higher emission factor (0.51%) than the obtained in our study. Other studies also reported that N_2O fluxes are larger when ammonium nitrate is used as an N source compared to other mineral or organic fertilizers [14,25]. Cardenas et al. [26] showed higher N_2O emissions in wetter regions of the UK. Soil temperature influences N_2O emissions, increasing the nitrification and denitrification processes [27]. These differences in N source, rainfall and temperature can significantly change the N dynamics in soil. Therefore, our recommendation is that the emission factors for Brazilian conditions must be specific for the different sources of N, soil type and regions or biomes.

It is only recently that molecular-based analyses of microbial diversity have been combined with measurements of N_2O production and process rates [28]. There are few studies that offer a rigorous assessment of the microbial community and N_2O emissions, most of them with conflicting results [29-31]. We think that more studies concerning the microbial diversity and N_2O emission must be encouraged and performed in order to obtain a better picture of this relationship.

Although our study was a short-term (30 days), we noticed that baseline emissions were achieved after this period of time. Furthermore, usually in Brazil the N fertilizer is applied in fractions during the rainy season. This one-month result completely fits in the interval of application. Our study showed that these fractions must not be higher than 100 kg N ha^{-1} (2NF treatment), since this dose increased the forage yield (agronomic benefits) and decreased N_2O emission (environmental benefits) (Table 3), resulting in the lowest N_2O emission per kg of Dry matter (agro-environmental benefits) in the studied area.

Conclusion

The application of N fertilizer resulted in an increase in N_2O emission under the studied conditions. Our study showed that the application of N fertilizer at doses above the recommended rate increases the N_2O emission even further. In order to improve grassland management, we advise the application of a maximum fertiliser rate of 100 kg N ha^{-1}.

Acknowledgements

This work was supported by Fundação de Amparo à Pesquisa do Estado de São Paulo (FAPESP 2010/00554-0). We would like to thank the owners and employees of Agropecuária Nova Vida and the team from Biogeochemical Laboratory (CENA-USP).

References

1. Food and Agriculture Organization (2012) The State of Food and Agriculture. Sales and marketing group, Viale delle Terme di Caracalla, 00153 Rome, Italy, pp: 1-182.
2. IBGE (2006) National Agricultural Census. Farmers and Agricultural Enterprises, United nations organization, Brazil.
3. Martha GB, Alves E, Contini E (2012) Land-saving approaches and beef production growth in Brazil. Agricultural Systems 110: 173-177.
4. Bowman MS, Filho BS, Merry FD, Nepstad DC, Rodrigues H, et al. (2012) Persistence of cattle ranching in the Brazilian Amazon: A spatial analysis of the rationale for beef production. Land Use Policy 29: 558-568.
5. Wrage N, Velthof GL, Beusichem ML, Oenema O (2001) Role of nitrifier denitrification in the production of nitrous oxide. Soil Biology and Biochemistry 33: 1723-1732.
6. Beek CL, Pleijter M, Jacobs CMJ, Velthof GL, Groenigen JW, et al. (2009) Emissions of N_2O from fertilized and grazed grassland on organic soil in relation to groundwater level. Nutrient Cycling in Agroecosystems 86: 331-340.
7. Jassal RS, Black TA, Roy R, Ethier G (2011) Effect of nitrogen fertilization on soil CH_4 and N_2O fluxes, and soil and bole respiration. Geoderma 162: 182-186.
8. Sanhueza E, Cárdenas L, Donoso L, Santana M (1994) Effect of plowing on CO_2, CO, CH_4, N_2O and NO fluxes from tropical savannah soil. Journal of Geophysical Research 99: 16429-16434.
9. Morais RF, Boddey RM, Urquiaga S, Jantalia CP, Alves BJR (2013) Ammonia volatilization and nitrous oxide emissions during soil preparation and N fertilization of elephant grass (Pennisetum purpureum Schum.). Soil Biology and Biochemistry 64: 80-88.
10. Jones SK, Rees RM, Skiba UM, Ball BC (2005) Greenhouse gas emissions from a managed grassland. Global Planetary Change 47: 201-211.
11. Whittaker ET, Robinson G (1967) Trapezoidal and parabolic rules. 4th edn. The Calculus Observation: A Treatise of Numerical Mathematics.
12. IPCC (2007) Climate change: the physical science basis. In: Solomon S, Qin D, Manning M, Chen Z, Marquis M, et al. (Eds.), Contribution of Working Group I to the Fourth Assessment Report of the Intergovernmental Panel on Climate Change. Cambridge University Press, Cambridge, UK, p: 996.
13. Bremmer JM, Keeney DR (1966) Determination and isotope-radio analysis od different forms of nitrogen in soils. Exchageable ammonium, nitrate and nitrite by extraction-distillation methods. Soil Science Society of American 30: 577-582.
14. Ruzicka J, Hansen EH (1981) Flow injection analysis. Interscience, Wiley Online, New York, USA.
15. Jones SK, Rees RM, Skiba UM, Ball BC (2007) Influence of organic and mineral N fertiliser on N_2O fluxes from a temperate grassland. Agriculture, Ecosystems and Environment 121: 74-83.
16. Salsac L, Chaliou S, Morot-Gaudry J, Lesaint C (1987) Nitrate and ammonium nutrition in plants. Plant Physiology and Biochemistry 25: 805-812.
17. Subbarao GV, Sahrawat KL, Nakahara K, Rao IM, Ishitani M, et al. (2013) A paradigm shift towards low-nitrifying production systems: the role of biological nitrification inhibition (BNI). Annals of Botany 112: 297-316.
18. Shcherbak I, Millar N, Robertson GP (2014) Global metal analysis of the nonlinear response of soil nitrous oxide (N_2O) emissions to fertilizer nitrogen. PNAS USA 111: 9199-9204.
19. Butterbach BK, Bags EM, Dannenmann M, Kiese R (2013) Nitrous oxide emissions from soils: how well do we understand the porcesses and their controls. Philosophical Transactions of Royal Society B 368: 1-13.
20. Audet J, Hoffmann CC, Andersen PM, Baattrup PA, Johansen JA, et al. (2014) Nitrous oxide plants in undisturbed riparian wetlands located in agricultural catchments: emission, uptake and controlling factors. Soil Biology and Biochemistry 68: 291-299.
21. Lardy L, Wrage N, Metay A, Chotte JL, Bernoux M, et al. (2007) Soils, a sink for N_2O. A review. Global Change Biology Bioenergy 13: 1-17.
22. Syakila A, Kroeze C, Slomp CP (2010) Neglecting sinks for N_2O at the earth's surface: does it matter. Journal of Integrative Environmental Sciences 7: 79-87.
23. Wu D, Dong W, Oenema O, Wang Y, Trebs I, et al. (2013) N_2O consumption by low-nitrogen soil and its regulation by water and oxygen. Soil Biology and Biochemistry 60: 165-172.
24. Schlesinger WH (2013) An estimate of the global sink for nitrous oxide in soils. Global Change Biology Bioenergy 19: 2929-2931.
25. Dobbie KE, Smith KA (2003) Impact of different forms of N fertilizer on N_2O emissions from intensive grassland. Nutrient Cycling in Agroecosystems 67: 37-46.
26. Cardenas LM, Thorman R, Ashlee N, Butler M, Chadwick D, et al. (2010) Quantifying annual N_2O emission fluxes from grazed grassland under a range of inorganic fertiliser nitrogen inputs. Agriculture, Ecosystems and Environment 136: 218-226.
27. Skiba UM, Sheppard LJ, Macdonald J, Fowler D (1998) Some key environmental variables controlling nitrous oxide emissions from agricultural and semi-natural soils in Scotland. Atmospheric Environment 32: 3311-3320.
28. Butterbach BK, Baggs EM, Dannenmann M, Kiese R, Zechmeister-Boltenstren S (2013) Nitrous oxide emissions from soil: how well do we understand the processes and their controls? Philosophical Transactions of the Royal Society B 368: 1621.
29. Philippot L, Cuhel J, Saby NBA, Cheneby D, Chronakova A, et al. (2009) Mapping field-scale spatial patterns of size and activity of the denitrifier community. Environmental Microbiology 11: 1518-1526.
30. Henry S, Texier S, Hallet S, Bru D, Dambreville C, et al. (2008) Disentangling the rhizosphere effect on nitrate reducers and denitrifiers: insight into the role of root exudates. Environmental Microbiology 10: 3082-3092.
31. Cuhel J, Simek M, Laughlin RJ, Bru D, Cheneby D, et al. (2010) Insights into the effect of soil pH on N_2O and N_2 emissions and denitrifier community size and activity. Applied Environmental Microbiology 76: 1870-1878.

Phycoremediation of Some Pesticides by Microchlorophyte Alga, *Chlorella* Sp.

Mervet H Hussein[1], Ali M Abdullah[2*], Eladl G Eladal[1] and Noha I Badr El-Din[2]

[1]Botany Department, Faculty of Science, Mansoura University, Mansoura, Egypt
[2]Holding Water Company for Water and Wastewater, Cairo, Egypt

Abstract

Every year, pesticides are found in surface and ground waters in Egypt. Pesticides are uncommon usage and applied in high amounts in agricultural activities. The present study investigated the possible removal of some herbicides from water using the microalgae *Chlorella vulgaris* Microorganisms are capable of decomposing a range of organic pollutants and the main focus in previously published studies has been on bacteria and fungi. Microalgae are microorganisms that have different morphological, physiological, and genetic traits that confer the ability to produce different biologically active metabolites. Because of the high capacity of microalgae in biosorbing heavy metals, most of their studies concentrated on this advantage, but fewer studies reported the removal of organic pollutants such as pesticides. The experiments were conducted as the following; the first was long-term experiment (5 days) using growing cells, and the second was short-term experiment (60 min) using dead and living cells. In the long-term experiment, the presence of growing algae resulted in removal percentages of pesticides ranged from 87% to 96.5%, while in the short-term study, the presence of live algae cells led to removal percentages ranged from 86 to 89% and dead algae biomass achieved removal ranged from 96% to 99%. The main mechanism behind the removal of pesticides in the water phase is proposed to be biosorption onto the algal cells. This conclusion is based on the short duration required for removal to occur.

Keywords: Phycoremediation; *Chlorella vulgaris*; Pesticides; Bioremoval efficiency; Algae biomass; LC-MS/MS

Introduction

In the near future, water reuse will become especially important in densely populated arid areas where there is an increasing demand to supply water from limited supplies. Human well-being in a future world will depend more heavily upon this sustainable resource and the characterization of emerging contaminants will become important for ecological and human health risk assessments and commodity valuation of water resources [1,2]. Egypt is an agricultural country. Agricultural activities account for 28% of total national income, and nearly half of the country's work force is dependent on the agricultural subsector for its livelihood. An increase in environmental contamination by various chemicals such as OCPs and herbicides are anticipated along the Nile Delta, which is referred to as "Green Lungs of Egypt" [3]. Furthermore, the chemical industry in Egypt is, by far, the main source of hazardous waste release in developed regions. These industries have encountered frequent problems in disposing of the hazardous waste they generate. In addition to the foregoing pollution, water pollution is exacerbated by agricultural pesticides, raw sewage, and urban and industrial effluents [4]. Consequently, pesticides residues in water, plants and grasses may be ingested by herbivorous animals and eventually find their way into tissues [5]. Thus, the remediation of pesticides is very urgent, especially bioremediation by microalgae.

Chemical properties of the pesticide such as molecular weight, functional groups and toxicity affect the metabolic degradation of it [6]. Algae appear to be more able to metabolize organic compounds with low molecular weights than larger molecules [7-9].

Atrazine have effects on health that classified in three groups developmental reproductive and cancerous US Department of Health and Human Services (USDHHS). In developmental causes post implantation losses, decrease in fetal body weight in complete bone formation, neuro development effects, delayed puberty and impaired development of reproductive system. The effects harmful on reproductive system include pre-term delivery, miss carriage and various birth defects. The cancerous effects include Non-Hodgkin's lymphoma, prostrate, brain, testes, breast and ovarian cancer [10] Atrazine used widespread and toxicity necessitates search for remediation technology. Several methods are available for remove atrazine from contaminated soil, water and wastewater such as chemical treatment, incineration, adsorption, phytoremediation and biodegradation. Biodegradation of atrazine is a complex process depends on nature and amount of atrazine in soil or water. The biodegradation of atrazine in environment is limited by microorganism's available [11]. The major steps of atrazine degradation pathway are Hydrolysis, dealkylation, deamination and ring cleavage. Process dealkylation of amino groups to give 2–chloro 4–hydroxyl -6- amino– 1, 3, 5 triazine is unknown.

In hydrolysis, atrazine degradation occurs by hydrolytic pathway is consist of three enzymatic steps catalyzed AtzA, AtzB and AtzC that hydrolysis the bound between c-cl and then Ethyl and isopropyl groups catalyzed and in the end of this process producing of cyanuric acid that convert to ammonia and carbon dioxide by AtzD, AtzE and AtzF enzymes [12].

The main objective of the present study is to examine the possibilities for utilization of the microalgae *Chlorella vulgaris* to simultaneously remove a number of herbicides, pesticides and insecticides in concentrations representative of their residue values in monitoring reports in water.

***Corresponding author:** Ali M Abdullah, PhD, Holding Water Company for Water and Wastewater, Cairo, Egypt, E-Mail: dr2252000@dr.com

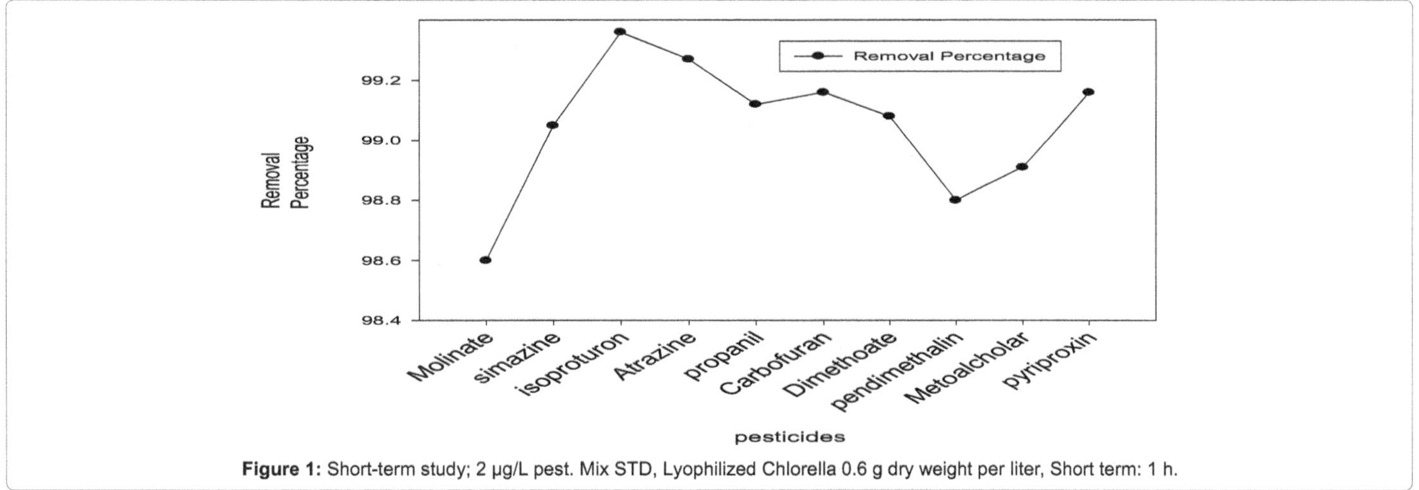

Figure 1: Short-term study; 2 µg/L pest. Mix STD, Lyophilized Chlorella 0.6 g dry weight per liter, Short term: 1 h.

Materials and Methods

Algal strain isolation, identification and culture conditions

Fresh water *Chlorella vulgaris* was isolated from water sample from river Nile. Culture purification was according to [13] and the alga was identified according to Ref. [14]. *Chlorella vulgaris* was grown in axenic cultures at 27 ± 2°C under continuous illumination 3600 lux in 500 Ml Erlenmeyer flasks, containing 200 mL BG11 medium [15] for 5 days incubation period in an Illuminated Memmert incubator.

Pesticides

Custom standard mixture (Atrazine, Molinate, Simazine, Isoproturon, Propanil, Carbofuran, Dimethoate, Pendimethalin, Metoalcholar, Pyriproxin) 0.1 mg/mL for each in methanol was purchased from Accustandard Inc., USA. The standard was obtained from The Reference Laboratory for Drinking Water, Cairo, Egypt. Standard solution containing the 10 micro contaminants in methanolic solution was added to each flask (final water or medium volume of 0.1 L) to obtain a final concentration of 2 µg L^{-1} and 10 µg L^{-1}. The concentration 10 µg L^{-1} was kept in high concentration level for further detection of pesticides in agricultural surface water following a runoff or spray drift events [16,17].

Short-term study

Lyophilized biomass was prepared by cultivating *Chlorella vulgaris* under certain conditions described in the previous section for Five days. Collecting the biomass using centrifugation (3000 g, 15 min, Bench-top - TD5B, Germany), after washing once with distilled water, the pellet was lyophilized in a freeze dryer for 24 h. Storing the lyophilized biomass in dark conditions at room temperature.

The lyophilized biomass was stored under dark conditions at room temperature whereas under similar growth conditions were used to produce the living biomass. After five days, an equal amount of live biomass was centrifuged. The dry biomass was crushed by a small mortar to a powder, ahead of the experiment. An initial concentration of 2.0 µ gL^{-1} and 10 μgL^{-1} (Figures 1 and 2) was obtained by adding the pesticide mix to sterile Milli Q water. The experiments included Lyophilized algal biomass, living algal biomass and a control without any biomass, with three replicates per experiment. The amount of biomass (lyophilized or live) added to each replicate corresponded to 0.6 g dry weight per liter (6×10^7 cells ml^{-1}). There were three replicates per treatment and the total volume of each replicate was 100 ml. The treatments were stirred on a shaker orbital at a speed of 380 rpm for 1 h at room temperature. After one hour, the biomass was removed from the aqueous phase by centrifugation (4000 g, 20 min, Bench-top - TD5B, Germany) and the samples were stored in the freezer at -20°C until analysis (Thermo Scientific™ MaxQ™ 4450 Benchtop Orbital Shakers, USA), the experiments were conducted. After one hour, by centrifugation (4000 g, 20 min, Bench-top - TD5B, Germany) the biomass was removed from the aqueous phase and the samples were kept in the freezer at -20°C until analysis. The pH values at the end of the experiment were measured and found to be 6.2 ± 0.06 in treatment with dead algae, 6.4 ± 0.2 in live algae and 6.4 ± 0.4 in the control, The pH was measured using a Hach® HQ40d pH meter.

Long-term study

A final concentration of 2.0 μgL^{-1} and 10 µg L^{-1} was obtained by adding the pesticide mix to sterile BG11. The experiments consisted of one experiment with growing *Chlorella* and a control without any biomass. There were three replicates per experiment and the total volume of each replicate was 100 ml. In the treatment with *Chlorella.* and inoculum of 10% (v/v) of a five-day old culture was added which resulted in a starting density of 3×10^6 cells ml^{-1}.

The control treatment received an inoculum of 10% (v/v) of sterile BG11. The experiments were kept under the growing condition described above for 5 days. After the experiment the biomass was removed by centrifugation (4000 *g*, 20 min, Bench-top - TD5B, Germany) and samples of the aqueous phase were taken and stored in the freezer until analysis. After the experiment, the cell density was 3×10^7 cells ml^{-1} and the pH was measured to 7.6 ± 0.03 in the treatment with algae and 7.43 ± 0.03 in the control.

Chromatographic analyses

Samples (50 ml) from the aqueous solution were sent to The Reference Laboratory for Drinking Water, Cairo, Egypt for chromatographic analyses. Reference method EPA 536 (EPA 536, 2007), were used to conduct the pesticides analysis, which is based on a combination of liquid chromatography (LC) and mass spectroscopy (MS) specifically called LC-MS/MS (tandem-MS). Tandem-MS (Xevo-TQ-S, Waters Corporation, Milford, MA, USA) provides low detection limits and very high security, which means that more substances can be tracked at lower level.

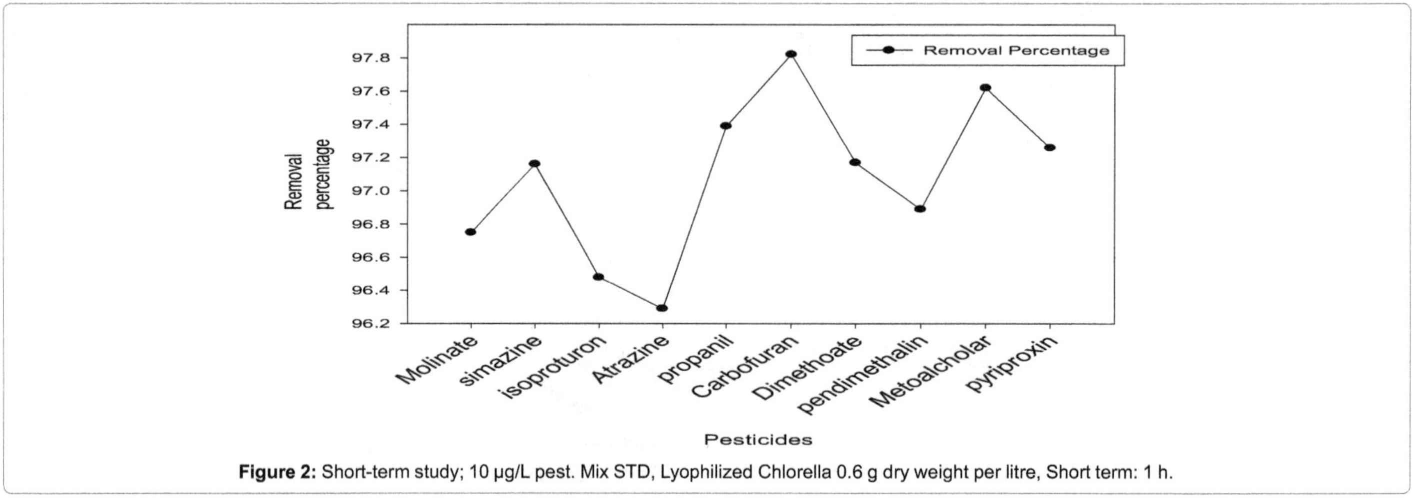

Figure 2: Short-term study; 10 µg/L pest. Mix STD, Lyophilized Chlorella 0.6 g dry weight per litre, Short term: 1 h.

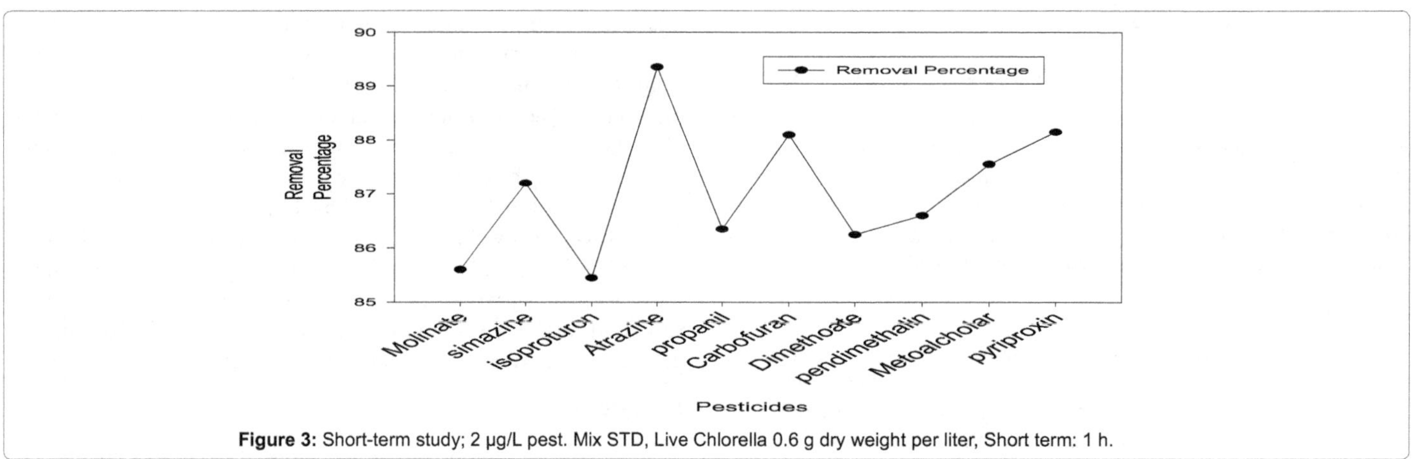

Figure 3: Short-term study; 2 µg/L pest. Mix STD, Live Chlorella 0.6 g dry weight per liter, Short term: 1 h.

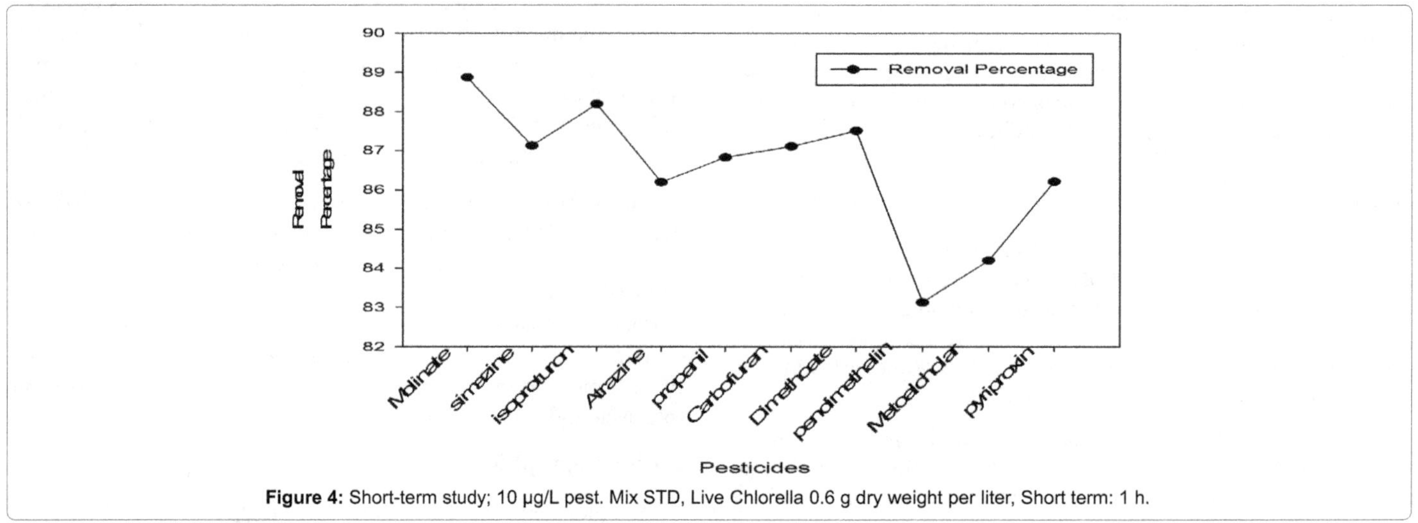

Figure 4: Short-term study; 10 µg/L pest. Mix STD, Live Chlorella 0.6 g dry weight per liter, Short term: 1 h.

Results

Biosorption/Short-term study

Both lyophilized and living biomasses of *C. vulgaris* achieved a good removal percentages for the two concentrations of pesticides 2 µg/L and 10 µg/L. In the short-term study, Figure 1 shows removal percentages ranged from 98.6 to 99% by lyophilized algae.

Figure 2 shows that: in the short – term study, the presence of lyophilized algae at the concentration of 10 mg/L led to removal percentages range 96% to 97.8 5%.

Figure 3 shows that: In the short-term study, the presence of live algae cells at the concentration of 2 µg/L led to removal percentages ranged from 86 to 89.6%.

Figure 4 show that: In the short-term study, the presence of live

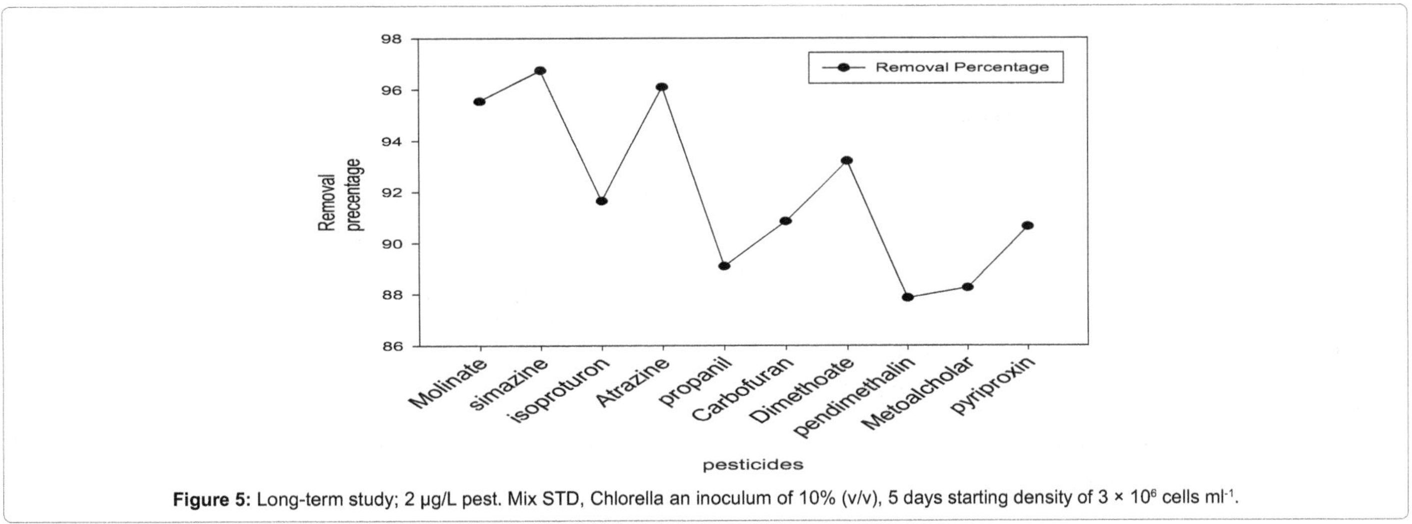

Figure 5: Long-term study; 2 μg/L pest. Mix STD, Chlorella an inoculum of 10% (v/v), 5 days starting density of 3×10^6 cells ml^{-1}.

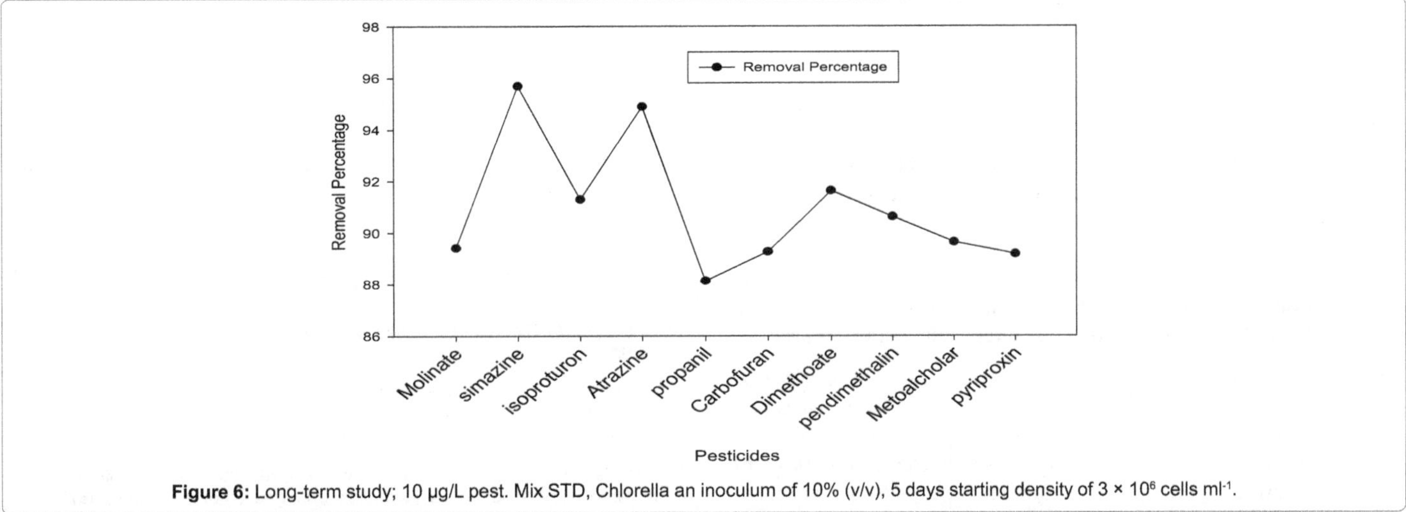

Figure 6: Long-term study; 10 μg/L pest. Mix STD, Chlorella an inoculum of 10% (v/v), 5 days starting density of 3×10^6 cells ml^{-1}.

algae cells at the concentration of 2 μg/L led to removal percentages ranged from 86 to 89%.

Long-term study

For the long-term study, it was observed that the removal of pesticides had been achieved with high percentage, by the growing microalgae as illustrated in Figures 5 and 6.

Discussion

Pesticides are used worldwide. The general population can be exposed to low concentrations of agricultural pesticides through contamination of air, water, food supplies [18] and also through household use [19]. Due to the application of pesticides in agriculture or for the purposes of protection of public health such as malaria prevention, high exposures are linked to these compounds. Pesticides 'contamination of water has been well documented worldwide to be considered as a potential risk for the ecosystem. Pesticide residues are commonly existing in the aquatic environment as a result of surface runoff, leaching from surface pesticides' applications, careless disposal of empty containers, and through industry and domestic sewage [20,21].

Consequently, developing efficient treatment systems is necessary for remedying these pesticides in polluted water bodies or catching them in wastewater treatment before they pollute the environment. Ranged from 87% to 96.5%.

Short-term study

Both lyophilized and living biomasses of *C. vulgaris* achieved a good removal percentages for the two concentrations of pesticides 2 μg/L and 10 μg/L, In the short-term study, the presence of live algae cells led to removal percentages ranged from 86 to 89% and dead algae biomass achieved removal ranged from 96 to 99%.

Because of the short period (60 min) needed for the removal of pesticides, it implies that biosorption as the proposed mechanism. Since, there is insufficient time for active uptake or mentalization processes to occur.

Long-term study

At the end of the five-day experiment, the removal percentages ranged from 87 to 96.5% for all the pesticides were used. The removal percentages were near that was achieved by the living Chlorella in short-term experiment, which suggest that the pesticides removal in the long-term experiment is the same as in the short-term experiment, namely biosorption. However, in the long-term treatment there was sufficient time for some mentalizations processes to happen by the algae. The algae may either have biosorbed, metabolized or facilitated the degradation of pesticides, or it can be due to a combination of those.

The recovery percentages of the control in the long-term experiment is the same as in the short-term experiment equal to 99%. These results demonstrate that the pesticides under study are very stable to hydro- and photolysis in aqueous media through all the experiments conducted.

This data indicates that it is a complex system where many effects take place simultaneously. Even though the herbicide concentration is not a significant factor by itself, it interacts with the other variables. The result of most practical interest is the high removal when algae is used, which indicates that the combination of an adsorption mechanism by the biological activity of algae to degrade herbicides, achieves more than 90% removal after 5 days of treatment.

Application of microalgae for treatment

In the present work it was proved that it was possible to remove not only the heavy metals as the previous studies achieved, but also pesticides from water by short time remediation with algal cells. The easily produced species *Chlorella vulgaris* is a promising organism to work with for the removal of heavy metals and pesticides from polluted water bodies.

This offers an idea to make a filter of dead algae biomass to be used for removal of pesticides from polluted water. As stated previously, using dead biomass instead of live has the advantages that the product will be stable and no risk for damaging the cells is expected. Dead biomass has also the ability to be recycled [22]. The biosorbed pollutants could be washed away from the algal biomass [23] and processed in a safe way [24] and thereafter the biomass filter itself could be reused.

Another idea is to use live *Chlorella vulgaris* in remediation systems, providing metabolization in addition to biosorption. *C. vulgaris* can grow in autotrophic, mixotrophic and heterotrophic modes [25] which gives the algae competitive advantages over bacteria and fungi in treatment of organic pollutants in certain environments, which acts as great advantage of using living cells [26].

Conclusion

In this study, pesticides removal by microalgae *Chlorella vulgaris* was investigated. Two main experiments were conducted: short-term for dead and living *Chlorella vulgaris*, to examine their ability for removal, and long-term treatment by growing *Chlorella vulgaris*. The results showed that the lypholized biomass achieved removal percentages reached up to 99% of pesticides and higher than living *Chlorella vulgaris* at the short-term experiments. On the other hand, long-term experiments proved the ability of growing *Chlorella vulgaris* for the removal of pesticides, which ranged from 87 to 96.5%. These results confirm that it is possible to remove more than 90% of these herbicides.

References

1. Blasco C, Picó Y (2009). Prospects for combining chemical and biological methods for integrated environmental assessment. TrAC Trends in Analytical Chemistry 28: 745-757.
2. Young JG, Brenda E, Eleanor AG, Asa B, Lesley P, et al. (2005) Association between in utero organophosphate pesticide exposure and abnormal reflexes in neonates. Neurotoxicology 26: 199-209.
3. Mansour SA (2004) Pesticide exposure - Egyptian scene. Toxicology 198: 91-115.
4. Barakat AO (2004) Assessment of persistent toxic substances in the environment of Egypt. Environment International 30: 309-322.
5. World Health Organization (1990) Public health impact of pesticides used in agriculture. Geneva: World Health Organization.
6. Priyadarshani I, Sahu D, Rath B (2011) Microalgal bioremediation: current practices and perspectives. Journal of Biochemical Technology 3: 299-304.
7. Semple KT, Cain RB, Schmidt S (1999) Biodegradation of aromatic compounds by microalgae. FEMS Microbiology letters 170: 291-300.
8. Juhasz AL, Ravendra N (2000) Bioremediation of high molecular weight polycyclic aromatic hydrocarbons: a review of the microbial degradation of banzo[a]pyrene. Int Biodeterior Biodegrad 45: 57-88.
9. Heredia-Arroyo T, Wei W, Ruan R, Hu B (2011) Mixotrophic cultivation of Chlorella vulgaris and its potential application for the oil accumulation from non-sugar materials. Biomass and Bioenergy 35: 2245-2253.
10. Abigail MEA, Nilanjana D (2012) Microbial degradation of atrazine, commonly used herbicide. International Journal of Advanced Biological Research 2: 16-23.
11. Singh P, Suri CR, Cameotra SS (2004) Isolation of a member of Acinetobacter species involved in atrazine degradation. Mol Cell Biol Res Commun 317: 697-702.
12. Crawford JJ, Sims GK, Mulvaney RL, Radosevich M (1998) Biodegradation of atrazine under denitrifying conditions. Appl Microbiol Biotechnol 49: 618-623.
13. Andersen RA (2005) Algal Culturing Techniques. Academic Press, USA.
14. Philipose MT (1967) Chlorococcales. Indian Council of Agricultural Research, New Delhi, India 8: 31-41.
15. Rippka R, Deruelles J, Waterbury JB, Herdman M, Stanier RY (1979) Generic assignments, strain histories and properties of pure cultures of cyanobacteria. Microbiology 111: 1-61.
16. Felding G (1995) Leaching of phenoxyalkanoic acid herbicides from farmland. Science of the total environment 168: 11-18.
17. Davis AM, Thorburn PJ, Lewis SE, Bainbridge ZT, Attard SJ, et al. (2013) Environmental impacts of irrigated sugarcane production: Herbicide run-off dynamics from farms and associated drainage systems. Agriculture, ecosystems & environment 180: 123-135.
18. EFSA (2014) The 2012 European Union Report on pesticide residues in food. EFSA J, p: 12.
19. Trunnelle KJ, Deborah HB, Ahn KC, Marc BS, Tancredi DJ, et al. (2014) Concentrations of the urinary pyrethroid metabolite 3-phenoxybenzoic acid in farm worker families in the MICASA study. Environ Res 131: 153-159.
20. Miliadis GE (1994) Determination of pesticide residues in natural waters of Greece by solid phase extraction and gas chromatography. Bulletin of environmental contamination and toxicology 52: 25-30.
21. Tikoo V, Shales SW, Scragg AH (1996) Effects on Pentachlorphhenol on the Growth of Microalgae. Environmental Technology 17: 1139-1144.
22. Aksu Z, Dönmez G (2006). Binary biosorption of cadmium (II) and nickel (II) onto dried Chlorella vulgaris: co-ion effect on mono-component isotherm parameters. Process Biochemistry 41: 860-868.
23. Smith GA, Pepich BV (2007) EPA Method 536: Determination Of Triazine Pesticides And Their Degradates In Drinking Water By Liquid Chromatography Electrospray Ionization Tandem Mass Spectrometry (LC/ESI-MS/MS).
24. Friesen-Pankratz BB, Doebel CC, Farenhorst AA, Gordon Goldsborough L (2003) Interactions between algae (Selenastrumcapricornutum) and pesticides: implications for managing constructed wetlands for pesticide removal. Journal of Environmental Science and Health, Part B 38: 147-155.
25. Jansson C, Kreuger J (2010) Multiresidue analysis of 95 pesticides at low nanogram/liter levels in surface waters using online preconcentration and high performance liquid chromatography/tandem mass spectrometry. Journal of AOAC International 93: 1732-1747.
26. Subashchandrabose SR, Ramakrishnan B, Megharaj M, Venkateswarlu K, Naidu R (2013) Mixotrophic cyanobacteria and microalgae as distinctive biological agents for organic pollutant degradation. Environment International 51: 59-72.

Response of Grain Sorghum to Split Application of Nitrogen at Tanqua Abergelle Wereda, North Ethiopia

Sofonyas Dargie[2]* Efriem Tariku[2], Meresa Wslassie[2] and Gebrekiros Gebremedhin[1]*

[1]*Mekelle Soil Research Center, Mekelle, Ethiopia*
[2]*Abergelle Agricultural Research Center, Abi Adi, Ethiopia*

Abstract

In *Tanqua Abergelle*, where the study has conducted, application of fertilizer particularly nitrogen is done once in either of the crop growth stages. To this fact, it is known that nitrogen is an easy going chemical fertilizer that can simply undertake nitrification and be washed away by heavy rainfall without meeting the required objectives. Not only this but also, the sensitive fertilizer requirement of the crop growth stage is not known which this varies even from crop to crop. Hence, the objective of this study was to improve the nitrogen utilization efficiency of sorghum by split application of nitrogen there by obtaining higher yields while simultaneously identifying the best time of fertilizer application. The research was undertaken in *Abergelle* agricultural research center testing site (*Mearey*). There were seven treatments replicated three times and for doing so, randomized complete block design was used. Split application of nitrogen at different growth stages of sorghum didn't brought a statistically significant difference in plant height, panicle length, biomass yield and thousand seed weight. Moreover, application of nitrogen at initial, development, mid, Initial Dev, Initial Mid and Development Mid were statistically the same in grain yield. However, there was statistically significant difference ($p<0.05$) in yield between application of nitrogen at development growth stage (application at once) and application of nitrogen at the three growth stages (Initial, Development and Mid) in splitting form. Application of nitrogen during development growth stage provides the highest grain yield (3.2 t/ha) followed by initial (3.1 t/ha) and initial mid (2.98 t/ha). The lowest grain yield was obtained in Initial, Development and Mid growth stages (2.4 t/ha) i.e., with application of the same amount of nitrogen during initial, development and mid growth stages in split form. Even though, sorghum (Meko1) grain yield poorly responds to spilt application of nitrogen, further multiple years' research data is needed to reach at strong conclusions.

Keywords: Initial; Development; Mid; Split application; Growth stage

Introduction

The large need of plants for nitrogen and the limited ability of soils to supply available nitrogen cause nitrogen to be the most limiting nutrient for crop production on the globe [1]. Worldwide interest associated with increasing cereal grain protein has added an attention on improving the utilization of nitrogen in cereals [2]. Moreover, the concern of ground water contamination, cost of manufacturing and distribution has pressurized farmers to use nitrogen more efficiently [3].

Proper nitrogen application time and rates are critical to meet crop needs and indicate considerable opportunities for improving nitrogen use efficiency [4]. The growth stage of crops at which fertilizer is applied determines the nitrogen use efficiency; however, the response can vary by genotype. Luxuriant application of nitrogen fertilizer at sowing increases the emergence of broad leaf weeds, thereby the labor requirements for hand weeding, hence, split application of nitrogen is considered as more economical both in terms of weed management and nitrogen use efficiency for optimizing grain yield [5].

The most commonly used practice in improving nitrogen use efficiency of crops is split application of fertilizers, selection of crop growth environment (soil type and climate) and management practices (sowing date and rate of nitrogen application). The efficiency of the applied nitrogen in satisfying the demand of the crop depends on the type of fertilizer, timing of application, crop sequence, the supply of residual and mineralized nitrogen and seasonal trends [4,6].

Worldwide, nitrogen use efficiency (NUE) of cereal crops (such as wheat, maize, rice, barley, sorghum, millet and oat) is approximately 33% [7]. Loss of fertilizer nitrogen results from gaseous plant emission, soil denitrification, surface runoff, volatilization, and leaching. Increasing cereal NUE is unlikely, unless a system approach is implemented that uses numerous strategies such as use of nitrogen sources, timing of application, slow release fertilizer, placement techniques and nitrification inhibitors have been devised to reduce nitrogen losses and improve fertilizer use efficiency [8,9]. Therefore, this study was conducted for evaluating the response of grain sorghum [*Sorghum bicolor* (L) Moench] to nitrogen time of application of each growth stages.

Limitations of the study

This study was conducted at one agricultural research testing site and doesn't have replication across different locations. Soil and leaf sample were not collected for nitrogen use efficiency analysis, but was indirectly evaluated using the yield response of the crop to the applied nitrogen per each growth stages. Moreover, the blanket recommendation of nitrogen per hectare (46 kg of N) was equally split in to the crop growth stages, while actually each growth stage has different fertilizer requirements.

***Corresponding author:** Gebrekiros Gebremedhin, Abergelle Agricultural Research Center, Abi Adi, Ethiopia, E-mail: kiroskey@gmail.com

Materials and Methods

Area description

The experiment was undertaken at *Tanqua Abergelle Wereda* at *Abergelle* agricultural research center testing site. *Tanqua Abergelle* is found in central zone of Tigray regional state. It is located at about 120 kilometers away from the capital city of Tigray regional state, Mekelle, to the west direction. The research was undertaken at a specific site called *Mearey* (Figure 1).

It is located at 13° 14' 06" N latitude and 38° 58' 50" E longitudes and agro ecologically characterized as hot warm sub- moist low land (SM1- 4b) below 1500 masl. Textural class of Mearey site soil in Tanqua Abergelle district was sandy loam and it was sandy dominated soil. The pH of the experimental site was 7.5. According to FAO the preferable and productive soils pH ranges for most crops ranges from 4 to 8. Thus, the pH of the experimental site soil was within the range of productive soils. Both the organic matter and total nitrogen content [10] of the experimental site Mearey were low (Table 1).

The low organic matter and total nitrogen contents of soils of the experimental site could be due to the properties of the soil which is sandy dominated that accelerates rate of oxidation of organic matter. Soil texture influences the rate of soil organic matter (SOM) decomposition. Soils with high clay content may have higher soil organic matter content, due to slower decomposition of organic matter.

The rainfall pattern of the district is monomodal with a wet season of about two months occurring in July to August. The mean annual rain fall and temperature are 350-700 mm and 24-41°C respectively.

There is high interannual variability of rainfall in the area. As the experiment was held in 2011, the recorded annual rainfall during that year was very low (Figure 2) and even erratic. Loss of fertilizer particularly nitrogen in such areas which are manifested by erratic rainfall could most likely be due to volatilization and even can be washed away when it rains heavy rainfall.

Research design

For conducting this research the total growing period of sorghum were used as a milestone to commence the experiment. The sorghum variety that was used in this experiment was Mekol. According to food and agricultural organization, the estimated total growing days of sorghum is 130 days in warm semi-arid climates. These total growing days were then, categorized in to different growth stages called initial, development, mid and late. Therefore, the initial, development, mid and late growing days of the vegetables was stratified in the following ways.

Having this information, split fertilizer application was done in the three crop growth stages (Table 2), but, application of fertilizer at the late growth stage is not recommended and hence, was excluded from our study. Treatments were replicated three times and randomized complete block design with plot size of 4.5 × 5 m (22.5 m^2) was used. The spacing between plots, rows, blocks and plants were 50, 75, 100 and 20 cm respectively. The recommended amount of nitrogen applied in the area is 46 kg per hectare, accordingly, per 22.5 m^2 sizes of land 103.5 gram, and hence, in splitting form the application was 51.75 gram for two times that is at two growth stages and 34.5 gram for three times (three growth stages). The treatments were arranged in the following ways.

Treatments

Application of nitrogen at:

T1: Initial growth stage (103.5 gram)

T2: Development growth stage (103.5 gram)

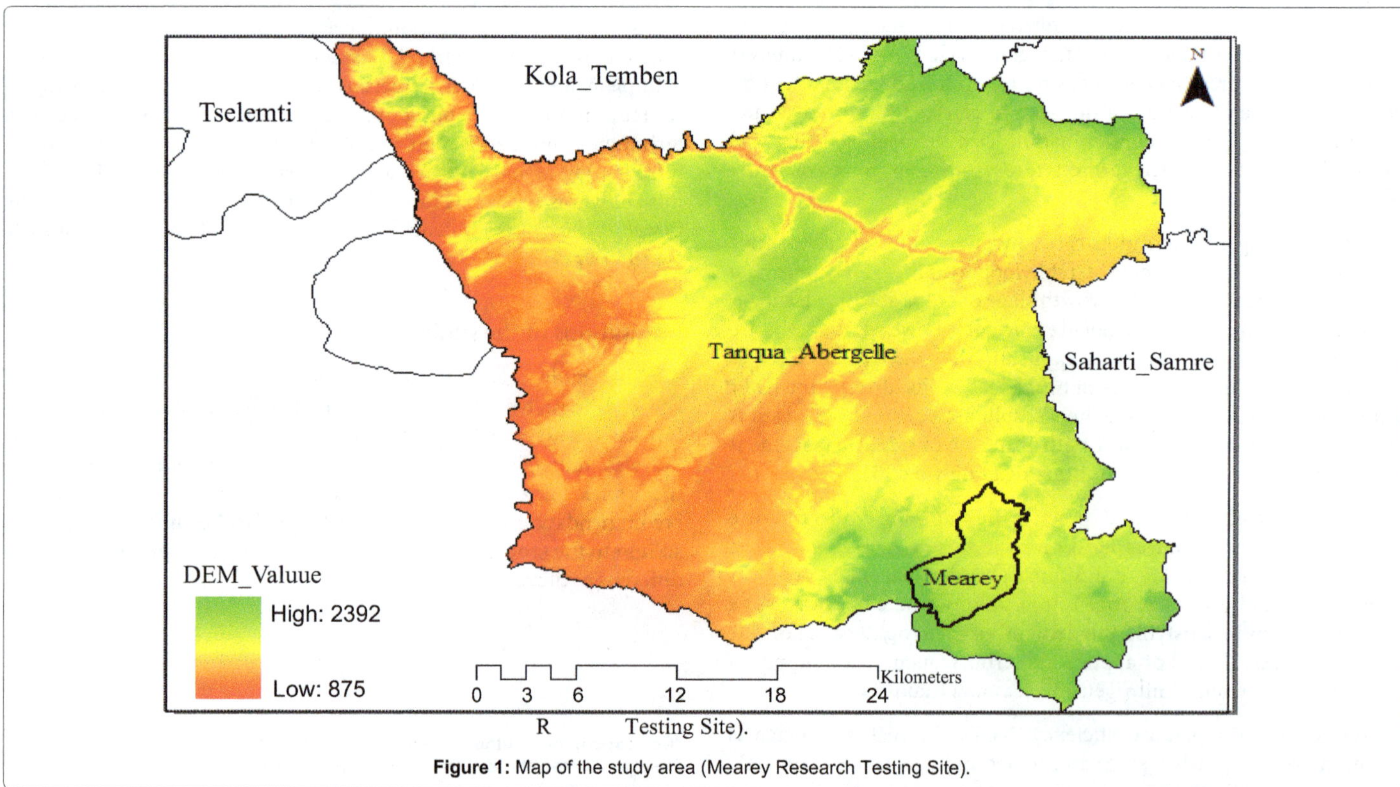

Figure 1: Map of the study area (Mearey Research Testing Site).

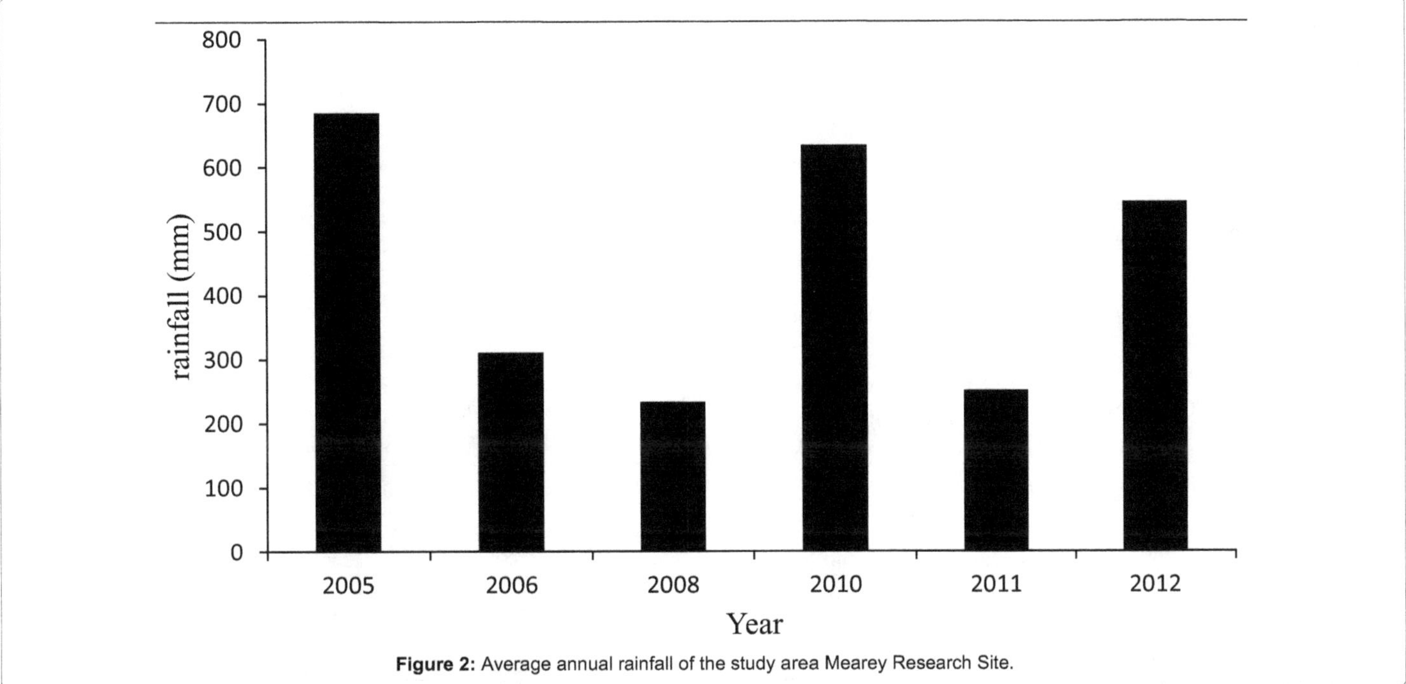

Figure 2: Average annual rainfall of the study area Mearey Research Site.

Soil Parameters	Mearey Experimental Site
Sand (%)	71
Silt (%)	13
Clay (%)	16
Textural Class	Sandy loam
pH (1:2.5 H_2O)	7.5
Organic Matter (%)	0.89
Total Nitrogen (%)	0.012
Available Phosphorous (mg kg^{-1})	7
Available K (ppm)	**78**

Table 1: Soil physicochemical properties of Mearey research testing site.

Crop type	Initial	Development	Mid	Late	Total Growing days
Sorghum	20	85	45	90	130

Table 2: Growing days of sorghum bicolor by each growth stages.

T3: Mid growth stage (103.5 gram)

T4: Initial (51.75 gram) and development (51.75 gram) growth stages

T5: Initial (51.75 gram) and mid (51.75 gram) growth stages

T6: Development (51.75 gram) and mid (51.75 gram) growth stages

T7: Initial (34.5 gram), development (34.5 gram) and mid (34.5 gram) growth stages (Table 3)

Fertilizer can be applied either in banding or broadcasting method, but, for this experiment banding method of application was used to the soil so as to minimize loss of fertilizer which can be due to competition of weeds or other factors. The recommended amount of DAP (92 t/ha) was applied during planting.

Data collection and analysis

The yield and yield components of sorghum which are plant height, panicle length, biomass, grain yield and TSW (thousand seed weight), sowing date, fertilizer application date and weeding frequency were collected. The collected data were statistically analyzed using analysis of variance (ANOVA) procedures with LSD mean separation at 5% level of significance.

Results and Discussion

Phenology and grain yield of sorghum

Split application of nitrogen at different growth stages of sorghum didn't brought a statistically significant difference in plant height, panicle length, biomass yield and thousand seed weight (Table 4). Moreover, application of nitrogen at initial, development, mid, indev, inmid and devmid were statistically the same in grain yield. However, there was statistically significant difference ($p<0.05$) in yield between application of nitrogen at development growth stage (application at once) and application of nitrogen at the three growth stages (initial development and mid) in splitting form.

Application of the recommended nitrogen during development growth stage provides the highest grain yield (3.2 t/ha) followed by initial (3.1 t/ha) and initial mid (2.98 t/ha). The lowest grain yield was obtained at initial development and mid growth stages (2.4 t/ha) i.e., with application of the same amount of nitrogen during initial, development and mid growth stages in split form. Similar to the grain yield, though not statistically significant, the highest biomass yield was recorded in development growth stage (10.321 t/ha) while the lowest (7. 657 t/ha) was in initial development and mid.

It was initially hypothesized that, instead of applying the recommended amount of nitrogen to an area at once, split application of nitrogen would increase the grain yield of sorghum. However, what we found here was in contrary to our hypothesis. The reason why this happens could be due to the allocation of equal amount of nitrogen fertilizer to each growth stages, while, practically their nitrogen requirement is different. The authors in Ref. [11,12] similarly stated that split application of nitrogen had a little effect on yield. In contrary, Ref. [13] found that time of nitrogen application had significant effect on

Treatments		Amount of nitrogen applied and application frequency		
S No		05/08/2011	14/07/2011	24/08/2011
1	Initial	103.5 g	-	-
2	Development	-	103.5 g	-
3	Mid	-	-	103.5 g
4	Initial development	51.75 g	51.75 g	-
5	Initial mid	51.75 g	-	51.75 g
6	Development mid	-	51.75 g	51.75 g
7	Initial development mid	34.5 g	34.5 g	34.5 g

Table 3: Split application of nitrogen at different growth stages of sorghum.

		Crop parameters				
S/N	**Treatments**	**Plant height (cm)**	**Panicle length (cm)**	**Biomass (t/ha)**	**Yield (t/ha)**	**TSW (gm)**
1	Initial	176	24.83	8.59	3.10 ab	36.03
2	Development	179	24.83	10.32	3.18 a	34.93
3	Mid	173	24.00	8.45	2.59 ab	35.67
4	Inidev	174	24.27	8.69	2.73 ab	34.97
5	Inimid	175	23.00	9.48	2.98 ab	39.83
6	devmid	171	22.99	9.14	2.86 ab	34.83
7	Inidevmid	172	24.25	7.66	2.48 b	34.83
	LSD	**0.11**	**3.3**	**2.95**	**0.7**	**7.96**
	CV (%)	**0.19**	**7.7**	**18.66**	**13.83**	

Table 4: Response of Sorghum Yield and yield components to Split Application of N.

sorghum yield attributes. Application of nitrogen in split form worth's a lot than applying the whole doze at once; however, there are many challenges that hinder effectiveness of this practice such as availability of soil moisture during the time of nitrogen application [14].

Conclusion

In the area where this research has undertaken, application of fertilizer particularly nitrogen is done once in either of the crop growth stages. To this fact, it is known that nitrogen is an easy going chemical fertilizer that can simply undertake nitrification, denitrification, surface runoff, volatilization and leaching without meeting the required objectives. Hence, it was initially hypothesized that, instead of applying the recommended full doze of nitrogen at once, split application of nitrogen would increase the grain yield of sorghum through improving the nitrogen use efficiency of the crop. However, split application of nitrogen at different growth stages of sorghum didn't brought a statistically significant difference in plant height, panicle length, biomass yield and thousand seed weight of sorghum. Moreover, application of nitrogen at initial, development, mid, indev, inmid and devmid were statistically the same in grain yield. Application of nitrogen during development growth stage provided the highest grain yield followed by initial and initial mid with the least in the split application (Inidevmid).

Recommendations

Even though this research finding is very crucial, further research should be done using different rates of nitrogen to the different crop growth stages, which helps to know the nitrogen sensitive growth stage of the crop and hence, recommend the estimated rate of applications for each growth stages. On top of this, further research should be held using different crop types and varieties for multiple years to reach at strong conclusions.

References

1. Foth HD, Ellis BG (1988) Soil fertility. 1st edn. John Wiley and Sons Inc., New York, USA.
2. Desai RM, Bhatia CR (1978) Nitrogen uptake and nitrogen harvest index in durum Wheat cultivars varying in their grain protein concentration. Euphytica 27: 561-566.
3. Jeremy L (2007) Nitrogen management for wheat protein and yield in the Esperance Port Zone. State of Western Australia, pp: 14-20.
4. Blankenau K, Olfs HW, Kuhlmann H (2002) Strategies to improve the use efficiency of mineral fertilizer nitrogen applied to winter wheat. Journal Agronomy and Crop. Sci 188: 146-154.
5. Gelato T, Tanner DG, Mamo T, Gebeyehu G (1996) Response of rainfed bread and durum wheat to source, level and timing of nitrogen fertilizer on two Ethiopian Vertisols: II. N uptake, recovery and efficiency. Proceeding the Ninth Regional Wheat Workshop for Eastern, Central and Southern Africa. Addis Ababa, Ethiopia 44: 195-204.
6. Borghi B (2000) Nitrogen as determinant of wheat growth and yield. In: Satorre EH, Slafer GA (eds). Wheat Ecology and Physiology of Yield Determination. Food Products Press, New York, USA, pp: 67-48.
7. Raun WR, Johnson GV (1999) Improving Nitrogen Use Efficiency for Cereal Production. Agron Journal 91: 357-363.
8. Slanger JHP, Kerkhoff P (1984) Nitrification Inhibitors in agriculture and horticulture. Fertilizer Res 5: 77-98.
9. Freney JR, Smith CJ, Mosier AR (1992) Effect of a new nitrification inhibitor (Wax Coated Carbide) on irrigated wheat. Fertilizer Res 32: 1-11.
10. Debele B (1980) The physical criteria and their rating proposed for land evaluation in the highland region of Ethiopia. Land Use Planning and Regulatory Department, Ministry of Agriculture, Addis Ababa, Ethiopia.

11. Ayoub M, Lussier S, Smith DL, Guertin S (1994) Timing and levels of nitrogen fertility effects on wheat yield in Eastern Canada. Crop Sci 34: 748-756.

12. Aleminew A (2015) Yield Response of Local Long Maturing Sorghum Varieties to Timing of Nitrogen Fertilizer Application in Eastern Amhara Region, Ethiopia. Sirinka Agricultural Research Center. Journal of Biology Agriculture and Healthcare 6: 14.

13. Tadesse T, Assefa A, Liben M, Tadesse Z (2013) Effects of nitrogen split-application on productivity, nitrogen use efficiency and economic benefits of maize production in Ethiopia. International Journal of Agricultural Policy and Research 1: 109-115.

14. Mesfin T, Tesfahunegn GB, Wortmann CS, Nikus O, Mamo M (2009) Tied ridging and fertilizer use for sorghum production in semi-arid Ethiopia. Nutr Cycl Agroecosys 85: 87-94.

22

Isolation of *Trichoderma* Spp. from Desert Soil, Biocontrol Potential Evaluation and Liquid Culture Production of Conidia Using Agricultural Fertilizers

Montoya-Gonzalez AH[1,2], Quijano-Vicente G[1], Morales-Maza A[3], Ortiz-Uribe N[2] and Hernandez-Martinez R[4]*

[1]*Laboratorio de Biotecnología, SPR de RI. Bustamante-Parra y Asociados, Km 8 Carretera a Riito. San Luis Rio Colorado, México 83430*
[2]*Departamento de Posgrado, Universidad Estatal de Sonora, Unidad Académica San Luis Rio Colorado, Carretera Sonoyta Km 6.5, San Luis Río Colorado, México 83450*
[3]*Departamento de Cultivos Protegidos, Instituto Nacional de Investigaciones Forestales, Agrícolas y Pecuarias. Campo Experimental Valle de Mexicali, Baja California. México, Carretera a San Felipe Km. 7.5, Colorado Dos, Mexicali, México 21700*
[4]*Departamento de Microbiología, Centro de Investigación Científica y de Educación Superior de Ensenada Baja California (CICESE), Carretera Ensenada-Tijuana 3918, Ensenada, México 22860*

Abstract

Three *Trichoderma* isolates were obtained from sandy soils collected at the "Gran Desierto de Altar" in the northwest of Mexico and characterized by morphologic and molecular analyses as *Trichoderma harzianum* 8.4, *Trichoderma asperellum* 12-2 and *Trichoderma asperellum* BP60. Isolate *T. asperellum* BP60 inhibited *Setophoma terrestris*, grew above 50°C, and produced chitinases and siderophores, therefore it was chosen to obtain enough biomass and conidia for field applications. Conidia production was intended in liquid culture fermentation using food grade ingredients and agricultural fertilizers. Assays were done using baffled Erlenmeyer flasks containing 75 mL of culture media, kept under constant agitation at 150 RPM, with initial pH adjusted to 6.5 (NaOH 1N) at 28 ± 2°C and evaluated at 3, 6, 9 and 12 days after inoculation (DAI). Among the carbon sources, sucrose and vinaze; the former induced higher yields of biomass and conidia. Regarding nitrogen sources, the fertilizer $(NH_4)NO_3$ induced higher conidia yield. V8 juice (V8) induced the highest effect on production of both biomass and conidia. Therefore, maximum yield was 1.06 × 10^9 conidia.mL^{-1}, with the formulation with 5 g of KH_2PO_4 (MKP, Greenhow®), 1.3 g of $MgSO_4 \cdot 7H_2O$ (Sul-Mag, Peñoles®), 20 mg of $FeCl_3 \cdot 6H_2O$ (Fermont®), 150 ml of V8, 10 ml of vinaze and 2.5 g.L^{-1} of $(NH_4)NO_3$. Results presented here prove the potential for using an alternative, low cost, liquid media to produce conidia of *T. asperellum*.

Keywords: Biocontrol; Liquid culture media; Conidia production

Introduction

Horticultural crops in the northwest of Mexico have increased in area and diversity. An example is the valley of San Luis Rio Colorado; where agriculture has diversified from traditional crops such as cotton and wheat, to vegetables including green onions, celery, broccoli, radish, asparagus, watermelon, and Brussels sprouts. The eastern section of the valley of San Luis Rio Colorado, is located within the limits of the "Gran Desierto de Altar" [1]. In this area, green onions have a high economic impact, since most of the production is exported to United States of America and the United Kingdom [2]. On average this crop is grown on 5,000 ha yearly, generating an approximated income of $12,800.00 USD.$ha^{-1}$ [3]. Pink root rot in green onion is caused by *Setophoma terrestris* (H.N. Hansen) Gruyter, Aveskamp and Verkley, comb. nov. MycoBank MB514659. The disease is enhanced by high temperatures and it is especially important in soils with low organic matter, high plant densities and minimal crop rotations [4,5]. According to the growers, the disease in the valley of San Luis Rio Colorado has increased from 2008 to 2014 growing seasons, which accounts for losses of up to 15% of the crop yield, and therefore has great economic impact. Traditional chemical control of pink root rot is not an option for growers in this area, because to export vegetables to international markets there are standards on food quality and food safety to fulfill; and thus it encourages the use of alternative techniques with less environmental impact. The practices of integrated agricultural management, where chemicals are replaced by bioproducts, are the most suitable option [6]. Biological control agents (BCA) are bioproducts based on microorganisms that cause harmful alterations to plant pathogens by chemical or physical processes [7,8]. BCA differ from chemical agents in that to be effective, they need to grow and successfully colonize and therefore they need to be applied in high and frequent quantities [9]. Fungi from the genus *Trichoderma* spp. have a long history of successful control of plant diseases [7,10,11]. Several mechanisms have been described as responsible for their biocontrol activity, including competition for space and nutrients, biofertilization and stimulation of the plant defense systems, rhizosphere modification, secretion of chitinolotic enzymes, mycoparasitism and production of inhibitory compounds [7,8,12]. Since all these mechanisms produce an effective control after the colonization of plant roots; the ability to suppress a disease is directly proportional to their population density. *Trichoderma* spp. produce three types of propagules: hyphae, chlamydospores and conidia [13,14]. All can be used in formulated bioproducts, however since hyphae cannot withstand some scale-up processes, chlamydospores and conidia are used as the active ingredients; normally the production focuses on conidia, because of their higher production [15,16]. For an agroindustry producing their own *Trichoderma*-based products, liquid fermentation has several advantages, among others, conidia

***Corresponding author:** Hernandez-Martinez R, Departamento de Microbiología, Centro de Investigación Científica y de Educación Superior de Ensenada Baja California (CICESE), Carretera Ensenada-Tijuana 3918, Ensenada, México, 22860, E-mail: ruhernan@cicese.mx

are produced in a short time, their production does not require much space and labor, and any contamination can be controlled easily [15]. Studies on growth media for *Trichoderma* spp. showed that nitrogen sources are essential for the production and germination of conidia, but the effect of carbon sources on conidial and biomass production, differ among isolates easily [16]. The use of agricultural fertilizers as ingredients in culture medium for the growth of *Trichoderma spp.* is uncommon, but they have been used in microalgae scaling processes [17,18]. In *Trichoderma*, nitrogen fertilizers stimulate growth and conidia production and, in large amounts may have a synergistic effect on biocontrol effectiveness [19]. In general, the use of food grade ingredients for BCA is uncommon because of the potential risk of promoting growth of undesirable microorganisms [20], however since these ingredients are low-cost alternatives to media formulations, here we tested agricultural fertilizers and food grade ingredients to develop a media capable of inducing a high yield of conidia from a selected *Trichoderma* sp.

Materials and Methods

Area of study

Soil samples were collected on the agricultural farm "Los Pivotes del Desierto" located at the "Gran Desierto de Altar" (32°19'58.4" N, 114°52'08.4" W), near the city of San Luis Rio Colorado, Sonora, Mexico. Average temperatures range from a maximum of 32.0°C to a minimum of 13.5°C, with an annual average of 22.7°C. Although, in summer temperatures can reach up to 50°C [21]. Soil composition is sandy, belonging to the Entisol group in the Soil Taxonomy classification of USDA [22].

Isolation and identification of fungal strains

An isolate of *S. terrestris* was obtained from green onion showing pink root rot disease, and identified morphologically after cultivation on PDA media. For *Trichoderma* spp. isolation, soil samples were taken using a soil auger of 2.5 cm diameter from the 0-15 cm top layer, fifteen samples were mixed to make a composite sample of around 0.5 Kg. Fourteen composite samples were obtained, placed in plastic bags, kept cool and brought to the laboratory of phytopathology of the Center for Scientific Research and Higher Education of Ensenada (CICESE). From each composed sample, 10 g of soil was weighed and mixed 1:1 w/v with sterile distilled water and a two drops of Tween20, vortexed (3500 RPM), to suspend soil particles and let stand for five min. From the supernatant, serial dilutions were done up to 10^{-3}. From each dilution, 100 μL were plated onto PDA amended with rose Bengal (25 mg.L^{-1}) and chloramphenicol (10 mg.L^{-1}). Plates were maintained at 28 ± 1°C, for 10 days in darkness and checked daily. Well-defined colonies that had the characteristic features of the genus *Trichoderma* spp. were recovered on fresh PDA plates. To confirm *Trichoderma* identity, morphology was observed using a Zeiss inverted microscope (100X). Mycelium with conidia of putative *Trichoderma* strains were harvested by scraping, suspended in sterile distilled water and stored at 4°C for further use. For long- term preservation, strains were placed in 5 mL.L^{-1} of glycerol (20%) and kept at -80°C. To molecularly identify *Trichoderma*-like isolates, DNA was extracted using a commercial kit (Qiagen DNeasy kit®), DNA integrity was verified by gel electrophoresis and PCR the elongation factor-1α (*EF1a*) and their Internal transcribed spacer (ITS) were amplified. Primers used were EF1 (ATGGGTAAGGA(A/G) GACAAGAC) and EF2 (GGA(G/A)GTACCAGT(G/C)ATCATGTT) for *EF1a* [23], and ITS1 (TCCGTAGGTGAACCTGCGG) and ITS4 (TCCTCCGCTTATTGATATGC) for ITS [24]. DNA amplification was done using 50 ng of DNA, 5 μl of 10 × Taq polymerase buffer, 0.5 U of Taq polymerase, 10 mM of dNTPs mix, 1 mM of reverse and forward primers and deionized water to complete a total volume of 50 μl. For the amplification of EF1a the BioRad Thermocycler was programmed with an initial denaturing at 94°C for 4 min, followed by 30 cycles of denaturation at 94°C for 30 sec, annealing at 57°C for 30 sec and extension at 72°C for 1 min and the final extension at 72°C for 7 min. For ITS, PCR amplification was done as for EF1a, except that the annealing temperature was adjusted to 59°C. PCR products were analyzed by gel electrophoresis and amplified fragments purified using QIAquick PCR Purification Kit (Qiagen Inc., Valencia, Calif, USA), according to the manufacturer's indications. Cleaned PCR products were sent to Clemson University Genome Institute (CUGI) for sequencing. Resulting sequences were compared using the NCBI GenBank database.

Biocontrol potential evaluation

Biological control capabilities of the isolated strains against *S. terrestris* were evaluated by performing volatile organic compounds, non-volatile organic compounds, and competition assays on agar plates according to the methods reported by Dennis and Webster, Dennis and Webster and Royse and Ries, respectively [25-27]. Also, Siderophore Siderophore [28] and chitinase [29] production were evaluated. Finally, to test for thermic tolerance, isolates were placed on PDA, and grown at 30 ± 1°C, 40 ± 1°C, 50 ± 1°C and 60 ± 1°C, for 5 days in full darkness. Obtained results were analyzed to select an isolate for further experiments.

Inoculum recovery for conidia production assay

From the selected isolate, conidia were harvested by scraping the surface of a 7 day-old culture grown on a PDA plate. The concentration was adjusted to 1×10^6 conidia mL^{-1} in 10 mL centrifuge tubes using sterile deionized water. In all following assays, 1% (v/v) of this initial concentration was used as starting inoculum.

Selection of culture media

All tested media contained a basic formulation per liter consisting in: KH_2PO_4 (MKP, Greenhow®), 5 g; $MgSO_4 \cdot 7H_2O$ (Sul-Mag, Peñoles®), 1.3 g; and $FeCl_3 \cdot 6H_2O$ (Fermont®), 20 mg. An amendment of 150 mL.L^{-1} V8 juice (8 Verduras, Herdez®) was used. Assayed carbon sources were 8 g.L^{-1} of sucrose (Sulka®), and 10 mL.L^{-1} of vinaze, a residual product from *Agave tequilana* blue Weber from the tequila distillation process (Tecno Ferti-V, Vida Verde®). Nitrogen sources used were, $KNO_3 \cdot K_2SO_4$ (Nitro K Sul, Greenhow®), $(NH_4)NO_3$ (Sulfonit, ISAOSA®) and $(NH_4) \cdot 2SO_4$ (granulated ammonium sulfate, ISAOSA®) at 10 g.L^{-1}.

Assay 1

To select from the suitable ingredients, an assay was stablished using all the possible combinations, thus treatments were as follows: T1 (V8/sucrose/$KNO_3 \cdot K_2SO_4$), T2 (V8/sucrose/$(NH_4)NO_3$), T3 (V8/sucrose/$(NH_4) \cdot 2SO_4$), T4 (V8/vinaze/$KNO_3 \cdot K_2SO_4$), T5 (V8/vinaze/$(NH_4)NO_3$), T6 (V8/vinaze/$(NH_4) \cdot 2SO_4$), T7 (NA/sucrose/$KNO_3 \cdot K_2SO_4$), T8 (NA/sucrose/$(NH_4)NO_3$), T9 (NA/sucrose/$(NH_4) \cdot 2SO_4$), T10 (NA/vinaze/$KNO_3 \cdot K_2SO_4$), T11 (NA/vinaze/$(NH_4)NO_3$) and T12 (NA/vinaze/$(NH_4) \cdot 2SO_4$). After adding all ingredients the pH was adjusted to 6.5 using NaOH 1N. Autoclaved media was distributed into Erlenmeyer flasks with three baffles (AvitroLab®, 250 mL) placing 75 mL in each. Flasks were maintained under continuous agitation at 150 RPM and 28 ± 2°C and placed in a complete randomized array. Sampling to estimate biomass and conidia production was done at 3, 6, 9 and 12 days after inoculation (DAI) as described below, and pH was measured directly for each experimental unit. All experiments were done in triplicate.

Assay 2

Once results were evaluated, and in order to increase conidia production, a second assay was done by selecting the culture media that showed the higher yield in the first assay. All growing conditions, basic culture media, evaluation of conidia and biomass production were performed exactly as in the first experiment.

Evaluation of conidia and biomass production

To evaluate the produced conidia in each experimental unit, a 1 mL aliquot was taken into a microcentrifuge tube and counting was performed directly using an improved Neubauer Chamber (Housser Scientific®) as described before [15,25-31]. Concentration values are expressed in conidia.mL^{-1}. To quantify the produced biomass, the rest of the culture was passed through a Whatman® grade 4 qualitative filter paper, using a vacuum pump at 24 ± 1 in Hg vac (Gast™ V4BG608X, Fisher Scientific Inc.). Filters were dried at 95°C overnight to obtain a constant weight (± 0.005 g) and cooled at room temperature in a desiccator. Dry weight values were obtained (Explorer® OHAUS®), and reported as g.L^{-1} [13,15,32].

Statistical analysis

Data was analyzed by a factorial analysis of variance (ANOVA) and Tukey HSD test was applied when ANOVA revealed significant differences ($P \leq 0.05$). Software used was Statistica 12©, StatSoft, Inc. 1984-2013.

Results and Discussion

Isolation, identification and evaluation of biocontrol potential

Three strains were isolated from desert sandy soils and identified by morphological and molecular analysis as *T. harzianum* 8.4, *T. asperellum* 12-2, and *T. asperellum* BP60 (Figure 1A-1C), respectively. While *Trichoderma* has been reported as the most abundant fungi in soil [8,11,25,26], the dry environment and lack of organic matter in the site might be reducing the occurrence and diversity of this species. Similarly, low *Trichoderma* recovery was obtained before in sandy soils of Egypt [33] and in beaches of Brazil [34]. Isolates grew well at 30°C, 40°C and 50°C without noticeable changes, *T. asperellum* BP60 also grew at 60 ± 1°C although it showed reduced growth. Inhibition percent against *S. terrestris* were on average 33, 35 and 30%, respectively. In the evaluation of the production of volatile compounds, all *Trichoderma* isolates reduced the size of *S. terrestris* when compared with the control, also the color changed from purple to white, especially in the presence of *T. asperellum* T12-2. When evaluating the production of volatile compounds, only *T. asperellum* BP60 and *T. harzianum* T8.4 reduced the size of the colony of *S. terrestris*, but it was not significantly different from the control. All three isolates were capable of produce chitinases, but only *T. asperellum* BP60 produced siderophores. Among the isolates, *T. asperellum* BP60 showed the most promising characteristics as a biological control agent. It presented adverse effects on *S. terrestris* (Figure 1D), produces quitinases and siderophores (Figure 1E and 1F), grows at higher temperature, as well as producing volatile compounds and to inhibit *S. terrestris* growth. Therefore isolate BP60 was chosen to perform the next experiments. Previous reports indicate *that T. asperellum* had biological control activity over *Phytophthora capsici, P. megakarya* and *Rhizoctonia solani* [35-37]. *T. asperellum* BP60 was also capable of growing in a broad range of temperatures, since temperatures in the zone can reach up to 50°C, the isolate is well adapted to the predominant conditions and could therefore tolerate and prevail in the region.

Selection of culture media for conidia production of *T. asperellum* BP60

Assay 1: Higher biomass and conidia production of *T. asperellum* BP60 were obtained in all treatments containing V8 juice. Regarding biomass, T4 was the more efficient media containing V8, where *T. asperellum* BP60 reached 4.18 g.L^{-1} at 6 DAI (Figure 2A); this represents an increase of 79.6% when compared to T8, which showed the lowest value among the treatments without V8 (0.85 g.L^{-1} at 9 DAI) (Figure 2B). No statistical differences were detected between T1 and T2; and between T3 and T5 (Figure 2A). Meanwhile, maximum conidia production was 2.88 × 10^{8} conidia.mL^{-1} in T5 at 12 DAI (Figure 2C). Among treatments without V8, maximum conidia concentration obtained was 4.38 × 10^{6} conidia.mL^{-1} in T11 at 9 DAI (Figure 2D).

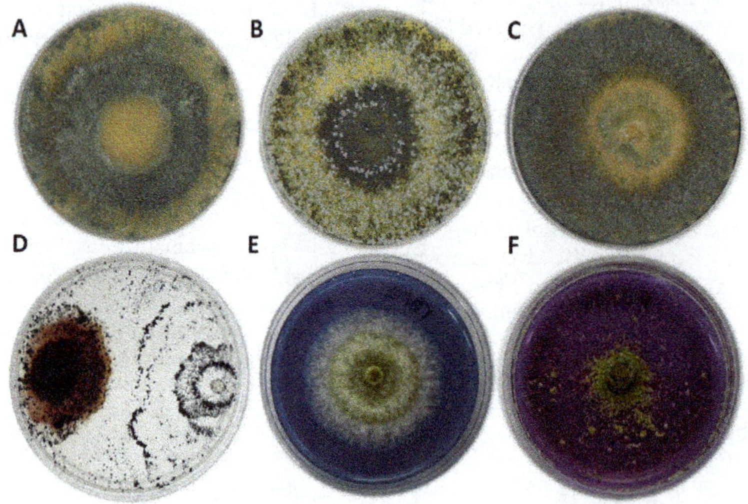

Figure 1: Native *Trichoderma* isolated from agricultural sandy soils of "Gran Desierto de Altar" in Sonora, Mexico (32°19'58.4"N 114°52'08.4"W) growth after 7 days on PDA at 28°C: (A)*Trichoderma harzianum* 8.4, (B) *T. asperellum* 12-2, and (C) *T. asperellum* BP60. (D) Competition assay of *T. asperellum* BP60 with *Setophoma terrestris* after 10 days of growth, (E) production of siderophores and (F) chitinase production.

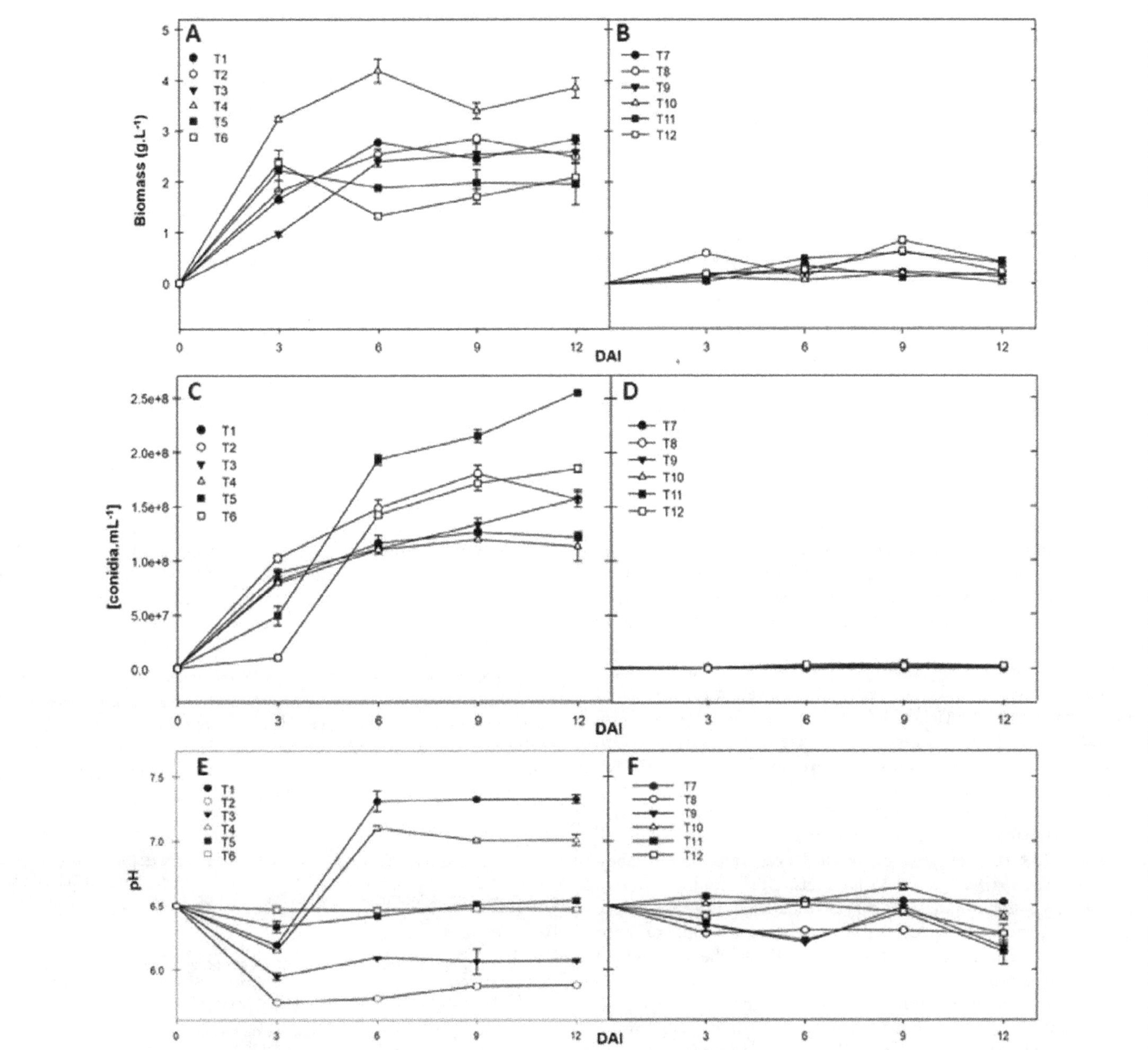

Figure 2: Biomass and conidia production and pH changes of *T. asperellum* in a fertilizer-based liquid media culture containing V8 juice (A, C and E) and without V8 juice (B, D and F). As carbon source treatments T1(●), T2(O), T3(▼), T7(●), T8(O) and T9(▼) contained sucrose; T4(△), T5(■), T6(□), T10(△), T11(■) and T12(□) contained vinaze. Used nitrogen sources were $KNO_3 \cdot K_2SO_4$ (T1, T4, T7 and T10), $(NH_4)NO_3$ (T2, T5, T8 and T11) and $(NH_4) \cdot 2SO_4$ (T3, T6, T9 and T12). All treatments were inoculated with 1 × 10^6 conidia.mL^{-1} at 1% (v/v), growth conditions were maintained during 12 days after inoculation (DAI) at 28 ± 2°C, 150 RPM and initial pH adjusted to 6.5 (NaOH 1N). Each data point indicates the mean (± S.E) of three replicates.

Comparing treatments with the lowest concentrations of conidia (T6 and T11) with or without V8, conidial concentration was increased by 98.2%. Treatments that contained $(NH_4)NO_3$ as the nitrogen source have higher conidia concentrations (T5 and T2). pH was nearly constant amongst the group without V8 (Figure 2E) with a minimal value of 6.12 for T11 at 12 DAI and a maximum of 6.64 for T10 at 9 DAI. In contrast, the treatments with V8 showed a pH ranging from 5.74 for T6 at 3 DAI, up to 7.32 for T3 at 12 DAI (Figure 2F). Among all the ingredients used to formulate the culture media, V8 juice had a major effect over biomass and conidia production, becoming the most valuable ingredient in the formulation. All treatments without V8, showed a minimal growth in conidial concentration.

Assay 2: To improve the treatment T5, which yielded the highest conidia concentrations in assay 1, and in an attempt to decrease the use its ingredients; V8 juice and vinaze were first assayed at 50, 100 or 150 mL.L^{-1} and at 10, 15 or 20 mL.L^{-1}, respectively. As before, all treatments were inoculated with 1 × 10^6 conidia.mL^{-1} at 1% (v/v), maintained during 12 days at 28 ± 2°C, 150 RPM and initial pH adjusted to 6.5 (NaOH 1N). In all treatments, pH varied from 6.54 to 6.63. The highest biomass of 1.62 g.L^{-1} was obtained when using 100 ml of V8 and 15

Treatment	V8 ($mL.L^{-1}$)	Vinaze ($mL.L^{-1}$)	Biomass ($g.L^{-1}$)	Conidia.mL^{-1}	pH
T5.1	50	10	0.54a	4.65×10^7a	6.42bc
T5.2	50	15	0.78b	4.84×10^7a	6.34a
T5.3	50	20	0.35a	5.36×10^7a	6.34a
T5.4	100	10	1.50de	1.05×10^8b	6.63f
T5.5	100	15	1.07c	9.63×10^7b	6.52e
T5.6	100	20	1.62e	1.01×10^8b	6.39b
T5.7	150	10	0.83b	1.63×10^8d	6.46cd
T5.8	150	15	0.74b	1.46×10^8c	6.46cd
T5.9	150	20	1.32d	9.20×10^7b	6.42bcd

All treatments were inoculated with 1×10^6 conidia.mL^{-1} at 1% (v/v), growth conditions were maintained during 12 days after inoculation at 28 ± 2°C, 150 RPM and initial pH adjusted to 6.5 (NaOH 1N). Values indicate the means of four samples with three replicates each. Letters indicate significant differences by Tukey HSD ($P \leq 0.05$).

Table 1: *T. asperellum* BP60 production of biomass, conidia concentration and pH response in liquid culture media containing different concentrations of V8 juice and vinaze.

ml of vinaze (T5.5), while maximum conidia yields of 1.63×10^8 conidia.mL^{-1} where obtained using 150 ml of V8 and 10 ml of vinaze (Table 1), since conidia are preferred over biomass, treatment T5.7 was considered the best. Consequently, to evaluate for the suitable amount of $(NH_4)NO_3$ in T5.7, concentrations of 10, 7.5, 5.0 or 2.5 $g.L^{-1}$ were used. With 10 $g.L^{-1}$, conidia concentrations were statistically lower (1.6×10^8) than when using 7.5 $g.L^{-1}$ (1.06×10^9), 5.0 $g.L^{-1}$ (8.57×10^8) or 2.5 $g.L^{-1}$ (1.06×10^9). No statistical differences were found among the last three treatments, This indicates that only 2.5 $g.L^{-1}$ of $(NH_4)NO_3$ are needed in the media. Final improved liquid medium is prepared with 5 g of KH_2PO_4 (MKP, Greenhow®), 1.3 g of $MgSO_4 \cdot 7H_2O$ (Sul-Mag, Peñoles®), 20 mg of $FeCl_3 \cdot 6H_2O$ (Fermont®), 150 ml of V8, 10 ml of vinaze and 2.5 $g.L^{-1}$ of $(NH_4)NO_3$. Studies showed that *T. asperellum* grows using different solid substrates [37-40]. In this work, media containing fertilizers and food grade ingredients provided good yields of conidia. In the lack of information on *T. asperellum* grown in liquid media, *T. harzianum* reports were used for comparison purposes. Biomass values obtained here were lower than the reported before for *T. harzianum* [13,15], but conidia concentrations for *T. asperellum* BP60 in liquid medium reached 1.06×10^9 conidia.mL^{-1}; which are comparable to concentrations of 1×10^8 UFC.ml^{-1}, 2×10^7 conidia.ml^{-1}, 1×10^9 UFC.g^{-1} and 1×10^7 UFC.g^{-1} reported in labels of leading *Trichoderma* spp. liquid products. Other studies reported yields of 2.28×10^7 conidia.mL^{-1} for *T. viride* using vegetable waste media [41]. It should be noted that direct counts, accounts only for supernatant conidia, and therefore might subestimate the total conidia concentration obtained in our assays. Between the carbon sources evaluated, vinaze induced more biomass than sucrose and a higher concentration of conidia. Evaluation of different concentrations of vinaze in the culture media indicates that adding 10 $ml.L^{-1}$ increased conidial concentration. Formulations containing carbon sources with non-defined composition has shown good results on *T. harzianum*, reaching concentrations up to 1×10^9 spore.g^{-1} with molasses adjusted to 37% of total sugars on media [42]; also, media using molasses from sugarcane are used to induce *T. asperellum* to produce extracellular lignocellulosic enzymes [43]. Regarding the nitrogen source, the highest biomass was induced with $KNO_3 \cdot K_2SO_4$, whereas the lowest values were obtained with $(NH_4)NO_3$, and $(NO_4) \cdot 2SO_4$. Monga et al. (2001) used NH_4Cl and KNO_3 as N-sources in minimal concentrations of 2.0 $g.L^{-1}$, and reported poor sporulation for *T. viride,* in contrast, *T. harzianum* had excellent response [44]. $(NH_4)NO_3$ induced the highest production of conidia, intermediate and lower concentrations were obtained using $(NO_4) \cdot 2SO_4$ and $KNO_3 \cdot K_2SO_4$, respectively (Figure 2B). Treatments with $KNO_3 \cdot K_2SO_4$ showed a tendency to have a neutral pH; $(NO_4) \cdot 2SO_4$ pH remained constant throughout the growing time, whereas $(NH_4)NO_3$ pH showed a minimal increase over time. Among all the experiments presented in this study, pH was little affected over time and probably did not influence conidia or biomass production, as stated from Lewis and Papavizas for *T. harzianum*, where conidial production was not influenced by the initial pH of the media or their continuous maintenance at pH 4 or 7 [45]. Moreover, other reports indicate that there are considerable differences in conidiation and growth in response to pH, including values as low as 2.8, concluding that there is a *Trichoderma* species-specific pH effect [46,47]. The highest conidia production obtained here using 2.5 $g.L^{-1}$ of $(NH_4)NO_3$ is recommended for scaling experiments. Broadly, it appears that lowering the concentrations of $(NH_4)NO_3$ enhances conidia production.

Conclusions

In conclusion, three strains of *Trichoderma* with biocontrol capabilities were isolated form desert sandy soils. *T. asperellum* BP60 turned out to have the most effective control against *S. terrestris* on *in vitro* experiments. In addition, this study demonstrates that abundant conidia of *T. asperellum* can be produced in liquid media with the combination of V8, vinaze and $(NH_4)NO_3$.

Acknowledgements

The scholarship for Alvaro Montoya from CONACTY, México is fully acknowledged as well as support of Jose Jesus Bustamante, Owner of Bustamante-Parra y Asociados, located at San Luis Rio Colorado, Mexico.

References

1. INEGI (2009) Prontuario de información geográfica municipal de los Estados Unidos Mexicanos, San Luis Rio Colorado, Sonora. Clave geoestadística 26055. Instituto Nacional de Estadística y Geografía.
2. Avendaño B, Varela R (2010) La adopción de estándares en el sector hortícola de Baja California. Estud Front 11: 171-202.
3. SIAP (2015) Producción agrícola cierre 2013, Cebollín, riego y temporal.
4. de Gruyter J, Woudenberg JH, Aveskamp MM, Verkley GJ, Groenewald JZ, et al. (2010) Systematic reappraisal of species in Phoma section Paraphoma, Pyrenochaeta and Pl1europhoma. Mycologia 102: 1066-1081.
5. Schwartz HF (2011) Soil-borne diseases of onions. Colorado. Colorado State University. Fact Sheet 2.940.
6. Cavalcante RS, Lima HLS, Pinto GAS, Gava CAT, Rodrigues S (2008) Effect of moisture on Trichoderma conidia production on corn and wheat bran by solid state fermentation. Food Bioprocess Technol 1: 100-104.
7. Benítez T, Rincón AM, Limón MC, Codón AC (2004) Biocontrol mechanisms of Trichoderma strains. Int Microbiol 7: 249-260.
8. Vinale F, Sivasithamparam K, Ghisalberti EL, Marra R, Woo SL, Lorito M (2008) Trichoderma–plant–pathogen interactions. Soil Biol Biochem 40: 1-10.
9. Lo CT, Nelson EB, Hayes CK, Harman GE (1998) Ecological Studies of Transformed Trichoderma harzianum Strain 1295-22 in the Rhizosphere and on the Phylloplane of Creeping Bentgrass. Phytopathology 88: 129-136.

10. Lewis JA, Papavizas GC (1991) Biocontrol of plant diseases: the approach for tomorrow. Crop Prot 10: 95-105.

11. Howell CR (2003) Mechanisms employed by Trichoderma species in the biological control of plant diseases: the history and evolution of current concepts. Plant dis 87: 4-10.

12. Subash N, Meenakshisundaram M, Sasikumar C, Unnamalai N (2014) Mass cultivation of Trichoderma harzianum using agricultural waste as a substrate for the management of damping off disease and growth promotion in chilli plants (Capsicum annuum l). Int J Pharm Bio Sci 4: 1076-1082.

13. Papavizas GC, Dunn MT, Lewis JA, Beagle-Ristaino J (1984) Liquid fermentation technology for experimental production of biocontrol fungi. Phytopathology 74: 1171-1175.

14. Yang X, Chen L, Yong X, Shen Q (2011) Formulations can affect rhizosphere colonization and biocontrol efficiency of Trichoderma harzianum SQR-T037 against Fusarium wilt of cucumbers. Biol Fertil Soils 47: 239-248.

15. Harman GE, Jin X, Stasz TE, Peruzzotti G, Leopold AC, Taylor AG (1991) Production of conidial biomass of Trichoderma harzianum for biological control. Biol Control 1: 23-28.

16. Panahian GR, Rahnama K, Jafari M (2012) Mass production of Trichoderma spp. and application. Intl Res J Appl Basic Sci 3: 292-298.

17. Guzmán-Murillo MA, López-Bolaños CC, Ledesma-Verdejo T, Roldan-Libenson G, Cadena-Roa MA, et al. (2007) Effects of fertilizer-based culture media on the production of exocellular polysaccharides and cellular superoxide dismutase by Phaeodactylum tricornutum (Bohlin). J Appl Phycol 19: 33-41.

18. Bae JH, Hur SB (2011) Development of economical fertilizer-based media for mass culturing of Nannochloropsis oceanica. Fish Aquatic Sci 14: 317-322.

19. Watanabe N, Lewis JA, Papavizas GC (1987) Influence of nitrogen fertilizers on growth, spore production and germination, and biocontrol potential of Trichoderma and Gliocladium. J Phytopathology 120: 337-346.

20. Petrovski S, Tillett D (2012) Back to the kitchen: food-grade agar is a low-cost alternative to bacteriological agar. Anal Biochem 429: 140-141.

21. SMN (2014) Normales climatológicas para San Luis Rio Colorado, 1951-2010.

22. Hernandez LA (2010) Soils of Mexico. USDA-NRCS. Little Rock, AZ.

23. O'Donnell K, Cigelnik E, Nirenberg HI (1998) Molecular systematics and phylogeography of the Gibberella fujikuroi species complex. Mycologia 465-493.

24. White TJ, Bruns T, Lee S, Taylor J (1990) Amplification and direct sequencing of fungal ribosomal DNA for phylogenetics. pp: 315-322.

25. Dennis C, Webster J (1971) Antagonistic properties of species-groups of Trichoderma: I. Production of non-volatile antibiotics. Trans Br mycol Soc 57: 25-IN3.

26. Dennis C, Webster J (1971) Antagonistic properties of species-groups of Trichoderma: II. Production of volatile antibiotics. Trans Br mycol Soc 57: 41-IN4.

27. Royse DJ, Ries SM (1978) The influence of fungi isolated from peach twigs on the pathogenicity of Cytospora cincta. Phytopathology 68: 603-607.

28. Schwyn B, Neilands JB (1987) Universal chemical assay for the detection and determination of siderophores. Anal Biochem 160: 47-56.

29. Shen CR, Chen YS, Yang CJ, Chen JK, Liu CL (2010) Colloid chitin azure is a dispersible, low-cost substrate for chitinase measurements in a sensitive, fast, reproducible assay. J Biomol Screen 15: 213-217.

30. Sivan A, Elad Y, Chet I (1984) Biological control effects of a new isolate of Trichoderma harzianum on Pythium aphanidermatum. Phytopathology 74: 498-501.

31. Windham M, Elad Y, Baker R (1986) A mechanism for increased plan growth induced by Trichoderma spp. Phytopathology 76: 518-521.

32. Adiyaman T, Schisler DA, Slininger PJ, Sloan JM, Jackson MA, et al. (2011) Selection of biocontrol agents of pink rot based on efficacy and growth kinetics index rankings. Plant Dis. 95: 24-30.

33. Fatma FM (2003) Distribution of fungi in the sandy soil in Egyptian beaches. Pak J Biol Sci 6: 860-866.

34. Gomes DN, Cavalcanti MA, Fernandes MJ, Lima DM, Passavante JZ (2008) Filamentous fungi isolated from sand and water of "Bairro Novo" and "Casa Caiada" beaches, Olinda, Pernambuco, Brazil. Braz J Biol 68: 577-582.

35. Tondje PR, Roberts DP, Bon MC, Widmer T, Samuels GJ, et al. (2007) Isolation and identification of mycoparasitic isolates of Trichoderma asperellum with potential for suppression of black pod disease of cacao in Cameroon. Biol Control 43: 202-212.

36. Segarra G, Avilés M, Casanova E, Borrero C, Trillas I (2012) Effectiveness of biological control of Phytophthora capsici in pepper by Trichoderma asperellum strain T34. Phytopathol Mediterr 52: 77-83.

37. Da Silva AR, Steindorff AS, Ramada MHS, De Siqueira SJL, Ulhoa CJ (2012) Biochemical characterization of a 27 kDa 1,3-ß-D-glucanase from Trichoderma asperellum induced by cell wall of Rhizoctonia solani. Carbohydr Polym 87: 1219-1223.

38. Trillas MI, Casanova E, Cotxarrera L, Ordovás J, Borrero C, et al. (2006) Composts from agricultural waste and the Trichoderma asperellum strain T-34 suppress Rhizoctonia solani in cucumber seedlings. Biol Control 39: 32-38.

39. Sant D, Casanova E, Segarra G, Avilés M, Reis E, et al. (2010) Effect of Trichoderma asperellum strain T34 on Fusarium wilt and water usage in carnation grown on compost-based growth medium. Biol Control 53: 291-296.

40. Guigón-López C, Guerrero-Prieto V, Lanzuise S, Lorito M (2014) Enzyme activity of extracellular protein induced in Trichoderma asperellum and T. longibrachiatum by substrates based on Agaricus bisporus and Phymatotrichopsis omnivora. Fungal Biol 118: 211-221.

41. Chaudhari PJ, Shrivastava P, Khadse AC (2011) Substrate evaluation for mass cultivation of Trichoderma viride. Asiat J Biotechnol Resour 2: 441-446.

42. Rodríguez-León JA, Domenech F, León M, Méndez T, Rodríguez DE, et al. (1999) Production of spores of Trichoderma harzianum on sugar cane molasses and bagasse pith in solid state fermentation for biocontrol. Braz Arch Biol Technol 42: 1.

43. Marx IJ, van Wyk N, Smit S, Jacobson D, Viljoen-Bloom M, et al. (2013) Comparative secretome analysis of Trichoderma asperellum S4F8 and Trichoderma reesei Rut C30 during solid-state fermentation on sugarcane bagasse. Biotechnol Biofuels 6: 172.

44. Monga D (2001) Effect of carbon and nitrogen sources on spore germination, bio-mass production and antifungal metabolites by species of Trichoderma and Gliocladium. Indian Phytopath 54: 435-437.

45. Lewis JA, Papavizas GC (1983) Production of chlamydospores and conidia by Trichoderma spp. in liquid and solid growth media. Soil Biol Biochem 15: 351-357.

46. Gao L, Sun MH, Liu XZ, Che YS (2007) Effects of carbon concentration and carbon to nitrogen ratio on the growth and sporulation of several biocontrol fungi. Mycol Res 111: 87-92.

47. Steyaert JM, Weld RJ, Stewart A (2010) Ambient pH intrinsically influences Trichoderma conidiation and colony morphology. Fungal Biol 114: 198-208.

23

Synergistic Bioefficacy of Botanical Insecticides against *Zabrotes subfasciatus* (*Coleoptera:* Bruchidae) a Major Storage Pest of Common Bean

Tamiru A[1]*, Bayih T[2,3] and Chimdessa M[3]

[1]*International Centre of Insect Physiology and Ecology, Ethiopia*
[2]*Hawassa University, PO Box 05, Hawassa, Ethiopia*
[3]*Haramaya University, PO Box 138, Dire Dawa, Ethiopia*

Abstract

This experiment was conducted to determine possibilities of synergism among insecticidal plants against *Z. subfasciatus* with a view of augmenting potency and reducing dosage rates. Leaf and seed powders of five insecticidal plants, namely *Jatropha curcas* (L.), *Datura stramonium* (L.), *Chenopodium ambrosioides* (L.), *Schinus molle* (L.) and *Azadrachta indica* (A. Juss) were mixed to 1% and 2%w/w unitary and binary formulations. The synthetic insecticide primiphos methyl at the rate of 0.1/100 gm grain dust and untreated grains were used as positive and negative controls, respectively. Most binary formulation had better efficacy than their constituent unitary formulation especially at lower dosage rates. Synergistic combination of botanical powders resulted in highest adult mortality, F_1 progeny reduction and lowest weevil perforation index and weight loss comparable to chemical standard primiphos methyl. Among the botanical combinations, bean seeds treated with binary formulation of *C. ambrosioides* with *D. stramonium*, *J. curcas* and *S. molle* gave the best efficacy in controlling *Z. subfasciatus*.

Keywords: *Z. subfaciatus*; Binary formulation; Botanical insecticides; *P. vulgaris*; Synergism

Introduction

The common bean (*Phaseolus vulgaris* L.) is one of the important food and cash crops in eastern and southern Africa [1,2]. Pre-harvest and post-harvest damage by insect pests, inter alia, is a major limiting factor of bean production, especially in smallholder farming conditions, under which most beans are grown in the region. Stored beans suffer heavy losses in terms of both quality and quantity mostly by bean bruchids [1]. Common bean weevil, *Acanthoscelides obtectus* and Mexican bean weevil *Zabrotes subfasciatus* are the most important species of bruchids attacking stored beans, causing yield losses reaching up to 38% [3,4] Bruchids infestation damage quantity, quality and viability of bean seed [3]. The degree of loss depends on the storage period and storage conditions. Ref. [5] reported an average grain loss of 60% within 3-6 months of storage period due to bean bruchids.

To reduce storage losses due to insect pests, synthetic insecticides have been recommended. However, their use is limited under small scale farming condition due to high costs and infrequent supply [6,7]. Besides, indiscriminate use of insecticides may result in undesirable consequences such as resistance development by the pest, secondary pest outbreaks, wide spread environmental hazards and risk to spray operators [8,9]. For these reasons, development of other alternative control methods such as botanical insecticides have gained significant importance in bruchid management [4,10,11]. Use of botanical insecticides not only confers effective pesticidal effect against bruchids but also serves as ecologically sound and economically feasible control option with low health risks to consumers [8,12]. Different plant extracts may act synergistically to effectively inhibit pest growth and developments compared with a single constituent extract and development of pest resistance is less likely when used over time [13-15].

Even though encouraging efforts that have been made in the last 2-3 decades, to identify botanicals with better insecticidal potential for bruchid management [16-18] limited information is available in their synergistic potential, toxicology, optimal application and species specificity. Moreover, recommended rates were often high which created inconvenience in practical application of botanical. Hence, the current study was undertaken to examine the prospect of synergism among combinations of crude botanical formulations with the objective of enhancing effectiveness of constituent botanical in mixtures and reducing dosage rates. The botanical plants were chosen based on their local availability and their potential for bean bruchids control [17,19]. The insecticidal plants and parts used in this study are shown in Table 1.

Materials and Methods

Insect rearing

Adult bean bruchids (*Z. subfascitus*) were obtained from laboratory culture reared on disinfested common bean variety, Awash-1. The experimental insects were maintained under laboratory condition (27 ± 3°C, 60 ± 10% RH, 12L:12D) at Melkassa Agricultural Research Center (8°24'N; 39°21'E). The food medium (bean seeds) used for insect rearing was first disinfected by keeping the grains in the oven at 40°C for 4 hours and allowed to cool for 2 hrs before use [20]. Infestation was done by introducing 100 parental adults (1:1 sex ratio) in 1 L volume of glass jars containing 250 g of bean grains. The parental adults were sieved off 13 days after oviposition period and the grains were kept under laboratory condition until the emergence of F_1 progeny. New generations of adult bean bruchids (*Z. subfascitus*) obtained from this culture were used in the experiment.

***Corresponding author:** A Tamiru, International Centre of Insect Physiology and Ecology, PO Box 30772-00100, Nairobi, Kenya, E-mail: atamiru@icipe.org

No.	Scientific name	Common name	Parts used
1	*Azadirachta indica*	Neem tree	Seed
2	*Chenopodium ambrosioides*	Mexican tea	Leaf
3	*Datura stramonium*	Thorn-apple	Leaf
4	*Jatropha carcus*	Physic nut	seed
5	*Schinus molle*	Pepper tree	Seed

Table 1: List of botanical plants and parts to be used against Z. *subfascitus.*

Plant materials and treatment formulations

Fresh plant parts (leaves and seeds) of the botanical plants *J. carcus, S. molle* and *Datura stramonium* were collected from MARC and the surroundings. Whereas, plant materials from the other two insecticidal plants *C. ambrosioides* and *A. indica* were collected from natural habitat in Addis Ababa (9°1′48″N 38°44′24″E) and Worer Agricultural Research Center (9°20′ 27″ N and 40°10′ 53″E), respectively. The plant materials were air dried and crushed separately into fine powder using a pestle and mortar. The resultant powder was further sieved through a 0.25 mm mesh to obtain a fine dust. The powders were weighed into 0.5 and 1 gm samples and then mixed appropriately to constitute binary formulation at either 1 or 2% w/w admixture on 100 gm bean samples. The unitary formulations were weighed into 1 and 2 gm samples and admixed with 100 gm bean samples to represent dosage rates of 1 or 2% w/w respectively.

Toxicity assessment

Healthy disinfected common bean seeds (100 gm) treated with different unitary and binary formulations of botanical insecticide powders were placed in the 1 L volume glass jar. The glass jars tops were covered with nylon mesh to allow aeration and held in place with rubber bands. The effectiveness of the different treatments was assessed by introducing 8 pairs of 2-4 days old bruchids obtained from laboratory culture to the treated and untreated grains. The synthetic insecticide primiphos methyl at the rate of 0.1/100 gm grain dust and untreated grains were used as positive and negative controls, respectively. Percent insect mortality was calculated using Abbott's formula by counting number of dead insects in each jar 24 hrs, 48 hrs, 72 hrs, 96 hrs and 120 hrs after treatment application/ adult introduction. Adults were considered dead when no response was observed after probing them with forceps. At the end of each assessment, dead insects were removed. The experiment was arranged in completely randomized design (CRD) with three replications.

Abbot's formula: $Pt = \frac{Po-Pc}{100-Pc} x100$

Where P_t=percent (%) mortality; P_o=observed mortality; P_c=control mortality

Effect of powders on F_1 progeny

After toxicity assessment of plant powders, remaining *Z. subfascitus* adults on treated and untreated jars were kept for additional 10 days and were sieved and discarded (both live and dead). The infested jars were further maintained under laboratory condition (7 ± 3°C, 60 ± 10% RH, 12L: 12D) until adult emergence and effect of treatments on the F_1 progeny were assessed. To avoid overlapping generation, the number of F_1 progeny was counted upon emergence for a period of 45 days since the initial date of adult introduction. Percentage reduction in adult emergence or inhibition rate (% IR) was calculated using the following formula:

$$\% \text{ IR} = \frac{(Cn-Tn)}{Cn} x100$$

Where C_n=number of newly emerged insects in the untreated (control) jar

T_n=number of insects in the treated jar

Grain damage assessment

To determine grain damage level, samples of 100 grains were taken randomly from the treated and control jars. Both treated and untreated grains were assessed for extent of bruchids damage using exit-holes as a measure of damage to the grain. The number of damaged grains (with characteristic hole) and undamaged grains were counted and weighed. Percentage grains weight loss was calculated using the following formula.

$$\text{Weight loss (\%)} = \frac{(UNd)-(DNu)}{U(Nd+Nu)} x100$$

Where U=weight of undamaged grain; Nd=number of damaged grains; D=weight of damaged grain; N_u=number of undamaged grains.

Moreover, grains that are riddled with exit-holes were counted and the percentage damage (PD) and weevil perforation index (WPI) was calculated according to methods in Ref. [21,22] respectively.

PD=(total number of treated grains perforated/total number of grains) × 100

WPI=(% of treated grains perforated/% of control grains perforated+% of treated grains perforated) × 100

Germination test

Germination test was carried out by randomly picking 80 undamaged grains from each treatment jar. Then 20 grains from each treated and control groups were placed separately on a moistened filter paper in Petri dishes and kept at room temperature. Each treatment was replicated four times where healthy grains without botanical insecticide powder application were used as a control. The numbers of germinated grains were recorded starting from the first date of germination. Percent germination was computed using the following formula:

$$\text{Viability index (\%)} = \frac{NG}{TG} x100$$

Where NG=number of grains germinated and TG=total number of grains tested in each Petri dish.

Data analysis

All data were checked for normality before they were subjected to analysis. Data which lacked normality were transformed using appropriate transformations method. Data were analyzed with analysis of variance (ANOVA) using General Linear Model (GLM) in SAS software. Significant means were separated using Student-Newma Keuls (SNK) test.

Results

Effects of different botanical powder combinations on bruchids mortality

Results on adult mortality of *Z. subfasciatus* 24-120 hrs after application of unitary and binary formulations of different botanical powders at 1% w/w and 2% w/w dosage rates on common beans grain are shown in Table 2 and 3 and Figure 1. Significant difference ($P<0.001$) in adult mortality was observed among different treatments depending on type of botanicals and their combinations, dosage rates

Treatments	Adult mortality (% mean ± SE)					F-value	P-value
	24 hrs*	48 hrs	72 hrs	96 hrs	120 hrs		
A. indica	16.67 ± 2.08Ec**	47.92 ± 4.17Cb	64.58 ± 5.51Ca	70.83 ± 4.17Fa	80.28 ± 2.41Da	34.73	P<0.0001
C. ambrosioides	27.08 ± 4.17Dd	60.42 ± 2.08Bc	72.92 ± 2.08Bb	87.5 ± 3.61Ca	93.22 ± 3.85Ba	58.01	P<0.0001
D. stramonium	4.17 ± 2.08Gc	27.08 ± 2.08Eb	50 ± 3.61Ea	64.58 ± 4.17Ha	71.53 ± 7.22Fa	35.73	P<0.0001
J. curcas	22.92 ± 5.51Dc	47.92 ± 5.51Cb	68.75 ± 7.2Ca	79.17 ± 7.51Fa	87.83 ± 6.47Ca	14.21	P =0.0004
S. molle	14.58 ± 2.08Fc	35.42 ± 5.51Db	54.17 ± 4.17Da	62.5 ± 0.00Ia	71.81 ± 0.69Fa	40.13	P<0.0001
A. indica+C. ambrosioides	33.33 ± 2.08Cd	60.42 ± 4.17Bc	70.83 ± 2.08Cb	81.25 ± 0.00Ea	90.97 ± 0.28Ca	75.44	P<0.0001
A. indica+D. stramonium	29.17 ± 4.17Db	56.25 ± 3.61Ba	62.5 ± 3.61Ca	72.92 ± 5.51Fa	80.42 ± 3.97Da	17.18	P=0.0002
A. indica+J. curcas	31.25 ± 3.61Dc	54.17 ± 5.51Cb	70.83 ± 2.08Ca	81.25 ± 3.61Ea	88.75 ± 4.39Da	26.99	P<0.0001
A. indica+S. molle	33.33 ± 2.08Cd	60.42 ± 8.33Bc	72.92 ± 5.51Bb	83.33 ± 2.08Da	88.92 ± 4.58Ca	16.71	P=0.0002
C. ambrosioides+D. stramonium	35.42 ± 4.17Cd	64.58 ± 2.08Bc	75.0 ± 3.61Bb	89.58 ± 2.08Ba	93.44 ± 2.16Ba	57.44	P<0.0001
C. ambrosioides+J. curcas	22.92 ± 2.08Dd	47.92 ± 2.08Cc	75.0 ± 3.61Bb	87.5 ± 3.61Ca	94.36 ± 3.61Ba	82.84	P<0.0001
C. ambrosioides+S. molle	35.42 ± 5.51Cd	58.33 ± 2.08Bc	70.83 ± 5.51Cb	85.42 ± 5.51Ca	90.36 ± 3.61Ba	21.19	P<0.0001
D. stramonium+J. curcas	31.25 ± 3.61Dc	47.92 ± 5.51Cb	68.75 ± 6.25Ca	77.08 ± 2.08Fa	86.81 ± 1.81Da	22.52	P<0.0001
D. stramonium+S .molle	37.5 ± 3.61Bb	54.17 ± 2.08Ca	60.42 ± 5.51Ca	66.67 ± 5.51Ga	75.83 ± 6.31Ea	6.06	P=0.0097
J. curcas+S. molle	22.92 ± 4.17Dd	47.92 ± 4.17Cc	64.58 ± 5.51Cb	70.83 ± 8.33Fa	74.72 ± 3.37Da	15.45	P=0.0003
Primiphose methyl	87.5 ± 7.22Aa	91.67 ± 4.17Aa	100 ± 0.0Aa	100 ± 0.0Aa	97.92 ± 2.08Aa	2.12	P=0.1532
Control (untreated)	0.00 ± 0.00Ha	0.00 ± 0.00Fa	0.00 ± 0.00Fa	0.00 ± 0.00Ja	2.08 ± 2.08Ga	1.00	P=0.4516
F-value	23.85	20.11	20.48	25.93	28.53		
P-value	<0.0001	<0.0001	<0.0001	<0.0001	<0.0001		

*Hours after treatment application; **Means followed by the same letter (s) within a column (upper case letters) and within a row (lowercase letters) are not significantly different using Student-Newman-Keuls (SNK) test (P<0.05). Effectiveness of botanicals and their combinations was determined by computing percent insect mortality (Abbotts 1925) and comparing the mortality data by ANOVA using GLM procedure.

Table 2: Mortality (% mean ± SE) of adult *Z. subfasciatus* on common bean seeds admixed with unitary and binary formulations (2% w/w) of different botanical insecticide powders.

Treatments	F1 progeny	% Inhibition Rate (IR)	Weevil Perforation Index(WPI*)	% Weight Loss
A. indica	18.33 ± 0.88c**	58.89 ± 7.56d	37.36 ± 3.23b	1.33 ± 0.45a
C. ambrosioides	1.00 ± 1.0f	98.33 ± 1.67a	3.51 ± 3.51e	0.04 ± 0.04c
D. stramonium	22.33 ± 3.18b	48.52 ± 13.31e	32.77 ± 11.56b	1.00 ± 0.50b
J. curcas	14.00 ± 2.08d	69.26 ± 5.29c	18.80 ± 1.02d	0.95 ± 0.46b
S. molle	10.00 ± 1.73e	78.33 ± 3.53b	19.58 ± 2.50c	0.19 ± 0.18c
A. indica+C. ambrosioides	3.00 ± 1.73f	92.78 ± 3.89a	10.50 ± 5.44e	0.05 ± 0.05c
A. indica+D. stramonium	4.33 ± 2.19e	91.67 ± 4.19a	7.00 ± 3.76e	0.17 ± 0.13c
A. indica+J. curcas	3.33 ± 0.33f	92.59 ± 1.21a	10.45 ± 1.38e	0.18 ± 0.12c
A. indica+S. molle	7.33 ± 0.88e	83.15 ± 4.38b	18.90 ± 2.49d	1.13 ± 0.11b
C. ambrosioides+D. stramonium	1.33 ± 0.88f	97.59 ± 1.45a	3.04 ± 1.63e	0.00 ± 0.00c
C. ambrosioides+J. curcas	1.67 ± 0.33f	96.48 ± 0.49a	4.49 ± 1.23e	0.00 ± 0.00c
C. ambrosioides+S. molle	0.00 ± 0.00f	100.00 ± 0.00a	0.00 ± 0.00g	0.00 ± 0.00c
D. stramonium+J. curcas	7.67 ± 1.20e	83.15 ± 3.05b	8.63 ± 1.54e	0.18 ± 0.09c
D. stramonium+S. molle	8.00 ± 1.15e	82.59 ± 2.59b	12.17 ± 1.73e	0.35 ± 0.10c
J. curcas+S. molle	4.67 ± 1.20e	90.37 ± 1.03a	12.02 ± 0.84e	0.22 ± 0.09c
Primiphos methyl	0.33 ± 0.33f	99.26 ± 0.74a	1.19 ± 1.19f	0.04 ± 0.04c
Control (untreated)	47.00 ± 7.00a	0.00 ± 0.00f	50.00 ± 0.00a	1.5 ± 0.41a
F-value	28.23	31.00	14.29	4.89
P-value	<0.0001	<0.0001	<0.0001	<0.0001

Table 3: Mean number of F_1 progeny produced (mean ± SE), % inhibition rate (IR), weevil perforation index (WPI) and % weight loss caused by *Z. subfasciatus* on common bean seeds admixed with different unitary and binary botanical powder formulations at 1% w/w dosage rate.

and time after treatment application. Significantly higher *Z. subfasciatus* mortality was recorded under binary botanical formulations compared to unitary formulation at both dosage rates (1% w/w and 2% w/w). For example, mean bruchid mortality recorded 120 hrs after application of unitary formulation was 55% while for binary formulation it was 75%.

Adult *Z. subfaciatus* mortality due to botanical insecticide application was directly related to exposure time. Longer duration of exposure after treatment application resulted in significantly higher adult mortality and vice versa for both unitary and binary botanical formulations (P<0.05). For instance, mean adult mortality due to *C. ambrosioides+D. stramonium* treatment was only 35.42% after 24 hrs while the same botanical formulation caused 89.58% mortality after 96 hrs (4 days) (Table 3). Overall, lowest *Z. subfasciatus* mortality was recorded 24 hrs after treatment application where as the highest mortality was recorded 96 hrs after treatment application. In the present study, no significant difference in percent adult mortality was observed between 96 hrs and 120 hrs after treatment application except for binary formulation of *C. ambrosioides+J. curcas* at 1% w/w dosage rate.

Mortality effect of botanicals was dose dependent especially for

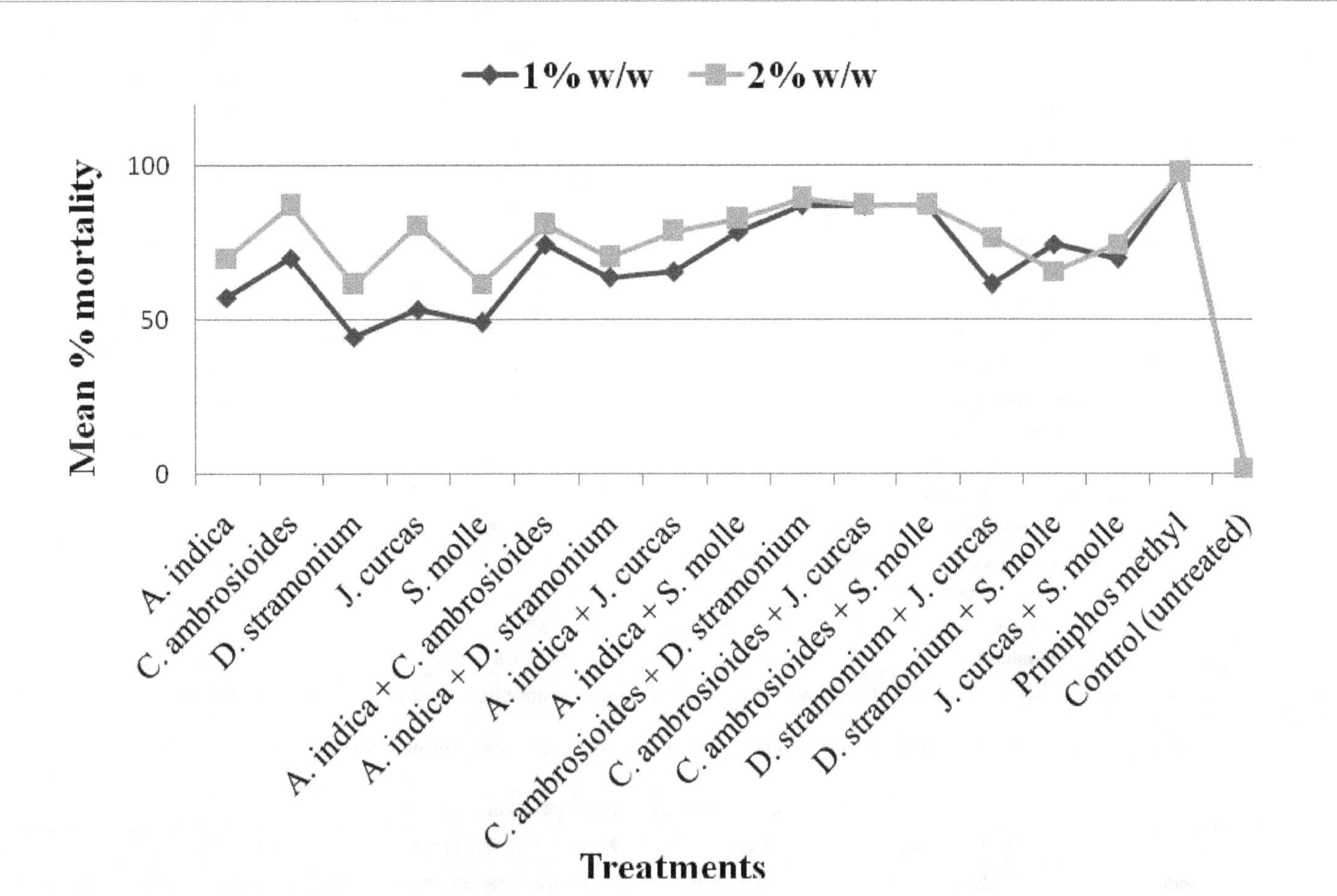

Figure 1: Cumulative mortality effect of unitary and binary formulations of botanical insecticide powders on adult *Z. subfasciatus*, applied at dosage rates of 1% w/w and 2% w/w.

Treatments	F1 progeny	% Inhibition Rate (IR)	Weevil Perforation Index(WPI*)	% Weight Loss
A. indica	8.67 ± 3.71b**	82.96 ± 4.86e	17.35 ± 4.63d	0.33 ± 0.18b
C. ambrosioides	0.33 ± 0.33b	99.26 ± 0.74a	2.42 ± 1.21i	0.14 ± 0.07b
D. stramonium	11.33 ± 2.33b	76.11 ± 2.00g	21.06 ± 1.31c	0.17 ± 0.05b
J. curcas	6.33 ± 3.33b	87.78 ± 4.75d	13.29 ± 4.79f	0.17 ± 0.17b
S. molle	9.33 ± 1.76b	78.89 ± 6.19f	31.41 ± 2.79b	0.77 ± 0.31b
A. indica+C. ambrosioides	2.33 ± 1.20b	95.37 ± 2.57b	4.29 ± 2.97i	0.18 ± 0.09b
A. indica+D. stramonium	5.33 ± 0.88b	87.96 ± 2.73d	19.35 ± 1.95d	0.46 ± 0.17b
A. indica+J. curcas	2.33 ± 0.88b	94.26 ± 2.80b	6.7 ± 2.16h	0.03 ± 0.03b
A. indica+S. molle	5.67 ± 1.45b	87.96 ± 2.91d	13.78 ± 2.31f	0.17 ± 0.09b
C. ambrosioides+D. stramonium	0.67 ± 0.33b	98.70 ± 0.67a	2.14 ± 1.09i	0.04 ± 0.04b
C. ambrosioides+J. curcas	0.00 ± 0.00b	100.00 ± 0.00a	0.00 ± 0.00k	0.00 ± 0.00b
C. ambrosioides+S. molle	0.00 ± 0.00b	100.00 ± 0.00a	0.00 ± 0.00k	0.00 ± 0.00b
D. stramonium+J. curcas	7.00 ± 1.00b	84.44 ± 2.94e	14.94 ± 3.65e	0.1 ± 0.09b
D. stramonium+S. molle	6.33 ± 0.88b	86.48 ± 0.19d	14.07 ± 3.83f	0.55 ± 0.23b
J. curcas+S. molle	4.33 ± 0.33b	90.56 ± 0.85c	11.61 ± 4.91g	0.22 ± 0.22b
Primiphos methyl	0.33 ± 0.33b	99.26 ± 0.74a	1.19 ± 1.19j	0.04 ± 0.04b
Control (untreated)	47.00 ± 7.00a	0.00 ± 0.00h	50.00 ± 0.00a	1.51 ± 0.41a
F-value	22.60	70.75	21.32	4.91
P-value	<0.0001	<0.0001	<0.0001	<0.0001

*WPI value above 50 indicate negative protectant ability; **Means followed by the same letter within a column are not significantly different using Student-Newman-Keuls (SNK) test (P<0.05). The data was analyzed by ANOVA using GLM procedure (SAS2002-2008)

Table 4: Mean number of F_1 progeny produced (mean ± SE), % inhibition rate (IR), weevil perforation index (WPI) and % weight loss caused by *Z. subfasciatus* on common bean grain admixed with different unitary and binary botanical powder formulations at 2% w/w dosage rate.

Treatments	Percent Germination (mean ± SE)	
	1% w/w	2% w/w
A. indica	96.67 ± 1.12a*	95.00 ± 2.46a
C. ambrosioides	98.75 ± 0.65a	97.92 ± 0.96a
D. stramonium	97.08 ± 1.14a	97.50 ± 0.75a
J. curcas	95.42 ± 1.14a	97.92 ± 0.74a
S. molle	98.75 ± 0.65a	97.08 ± 1.68a
A. indica+C. ambrosioides	96.67 ± 1.98a	96.67 ± 1.55a
A. indica+D. stramonium	95.00 ± 1.51a	96.67 ± 1.67a
A. indica+J. curcas	92.50 ± 1.90a	97.08 ± 1.79a
A. indica+S. molle	97.92 ± 0.94a	96.67 ± 1.12a
C. ambrosioides+D. stramonium	92.92 ± 1.99a	94.58 ± 2.42a
C. ambrosioides+J. curcas	97.50 ± 1.44a	99.17 ± 0.56a
C. ambrosioides+S. molle	96.25 ± 1.52a	97.50 ± 1.44a
D. stramonium+J. curcas	94.58 ± 1.68a	98.33 ± 0.94a
D. stramonium+S. molle	95.42 ± 1.44a	97.92 ± 1.14a
J. curcas+S. molle	93.33 ± 1.88a	95.83 ± 1.72a
Primiphos methyl	98.33 ± 0.71a	98.33 ± 0.71a
Control (untreated)	97.92 ± 0.96a	97.92 ± 0.96a
F-value	2.03	0.69
P-value	P=0.5308	P=0.7992

*Means followed by the same letter within a column are not significantly different using Student-Newman-Keuls (SNK) test (P<0.05). The data was analyzed by ANOVA using GLM procedure (SAS 2002-2008)

Table 5: Effect of unitary and binary botanical formulations treatment on percent germination (mean ± SE) of common bean seeds.

Treatments	*Z. subfasciatus* mortality (% mean ± SE)					F-value	P-value
	24 hrs*	48 hrs	72 hrs	96 hrs	120 hrs		
A. indica	20.83 ± 5.51Ec**	33.33 ± 9.08Eb	45.83 ± 9.08Hb	60.42 ± 2.08Fa	67.36 ± 5.14Ga	6.09	P=0.0095
C. ambrosioides	27.08 ± 9.08Dc	41.67 ± 7.51Dc	50.0 ± 7.22Fb	70.83 ± 4.16Da	79.28 ± 2.41Ea	8.29	P=0.0032
D. stramonium	12.5 ± 0.0Fc	31.25 ± 0.0Eb	43.75 ± 0.0Ha	47.92 ± 2.08Ha	54.58 ± 5.42Ja	31.57	P<0.0001
J. curcas	6.25 ± 3.61Gd	22.92 ± 4.17Fc	37.5 ± 3.61Ib	52.08 ± 4.17Ga	63.33 ± 1.82Ha	31.32	P<0.0001
S. molle	10.42 ± 2.08Gb	20.83 ± 5.51Fb	35.42 ± 2.08Ja	47.92 ± 4.17Ha	59.17 ± 2.92Ia	22.09	P<0.0001
A. indica+C. ambrosioides	29.17 ± 7.51Db	56.25 ± 9.55Ca	70.83 ± 9.08Ca	72.92 ± 7.51Da	79.72 ± 3.37Da	6.14	P=0.0092
A. indica+D. stramonium	12.5 ± 3.61Fc	39.58 ± 2.08Db	52.08 ± 4.17Fa	62.5 ± 3.61Fa	70.89 ± 2.50Fa	41.49	P<0.0001
A. indica+J. curcas	22.92 ± 2.08Db	43.75 ± 0.0Da	47.92 ± 2.08Ga	66.67 ± 7.51Ea	75.69 ± 9.98Fa	9.86	P=0.0017
A. indica+S. molle	25 ± 3.61Dd	43.75 ± 3.61Dc	60.42 ± 7.51Db	77.08 ± 9.08Ca	88.61 ± 7.91Ca	11.44	P=0.0009
C. ambrosioides+D. stramonium	27.08 ± 4.17Dd	56.25 ± 3.61Cc	72.92 ± 2.08Cb	85.42 ± 2.08Ba	87.22 ± 3.85Ba	57.63	P<0.0001
C. ambrosioides+J. curcas	37.5 ± 3.61Bb	45.83 ± 4.17Db	56.25 ± 6.25Eb	58.33 ± 4.17Fb	87.08 ± 6.67Ba	13.45	P=0.0005
C. ambrosioides+S. molle	35.42 ± 5.51Cc	62.5 ± 0.0Bb	77.08 ± 2.08Ba	87.5 ± 0.0Ba	92.22 ± 3.85Ba	48.07	P<0.0001
D. stramonium+J. curcas	20.83 ± 2.08Ee	39.58 ± 2.08Dd	50.0 ± 3.61Fc	58.33 ± 2.08Fb	61.67 ± 4.35Ga	30.25	P<0.0001
D. stramonium+S. molle	14.58 ± 4.17Fb	29.17 ± 9.08Fb	56.25 ± 3.61Ea	70.83 ± 2.08Da	74.58 ± 3.63Da	26.48	P<0.0001
J. curcas+S. molle	20.83 ± 4.17Ed	39.58 ± 7.51Dc	56.25 ± 6.25Eb	66.67 ± 7.51Ea	70 ± 8.32Ea	8.77	P=0.0026
Primiphos methyl	87.5 ± 7.22 Aa	91.67 ± 4.17Aa	100 ± 0.0Aa	100 ± 0.0Aa	97.92 ± 2.08Aa	2.12	P=0.1532
Control (untreated)	0.0 ± 0.0Ha	0.0 ± 0.0Ga	0.0 ± 0.0Ka	0.0 ± 0.0Ia	2.08 ± 0.08Ka	1.00	P=0.4516
F-value	16.85	13.66	17.79	22.39	18.93		
P-value	<0.0001	<0.0001	<0.0001	<0.0001	<0.0001		

* Hours after treatment application; **Means followed by the same letter (s) within a column (upper case letters) and within a row (lower case letters) are not significantly different using student Newman Keuls (SNK) test (P<0.05). Effectiveness of botanicals and their combinations was determined by computing percent insect mortality (Abbotts 1925) and comparing the mortality data by ANOVA using GLM procedure

Table 6: Mortality (% mean ± SE) of adult Z. *subfasciatus* on common bean seeds admixed with unitary and binary formulations (1%w/w) of different botanical insecticide powders.

unitary formulations. An increased *Z. subfaciatus* mortality was observed at higher doses for unitary formulation. For example, *Z. subfasciatus* mortality due to *J. curcas* application at 1%w/w dosage rate was 53.33% which increased to 80.83% at higher dosage rate (2%w/w). On the other hand, notable increase in adult mortality due to higher dose was not observed in most binary formulations. For instance, binary formulations of *C. ambrosioides, D. stramonium, J. curcas* and *S. molle* had more or less similar mortality effects at lower and higher application rates. An increase in adult mortality due to higher application rate in binary formulation was observed mainly during *J. curcas* combination with *A. indica* and *D. stramonium*.

Effects of botanical insecticides on weevil perforation index (WPI) and percent grains weight loss

Weevil Perforation Index (WPI) and % weight loss due to *Z. subfasciatus* on common bean seeds admixed with different application rates of unitary and binary botanical powder formulations is presented in Tables 4 and 5. All botanical insecticide formulations resulted in positive protective effect against damage by *Z. subfasciatus*, as the WPI values for all treatments were significantly less than 50. Generally, binary formulations showed better protectant ability compared to unitary formulation. For instance, the mean WPI for binary formulation at 1% w/w dosage rate was 8.72% whereas the mean WPI for unitary formulation was 22.41%. WPI was reduced with an increase dosage rate especially for unitary formulation. Binary formulations *D. stramonium+C. ambrosioides, C. ambrosioides+J. curcas* and *S. molle+C. ambrosioides* gave the best protection against *Z. subfasciatus* damage, with WPI value less than 5 at both test doses. This protective effect due to botanical formulation was on par with standard synthetic insecticide, primiphos methyl.

Weight loss due to *Z. subfasciatus* was significantly reduced ($P<0.0001$) after application of both unitary and binary botanical formulations compared to untreated control 7 weeks after infestation. The untreated bean grain had the highest weight loss due to damage by *Z. subfasciatus*. Overall, binary formulations had better effect in reducing weight loss compared to unitary formulation. For instance, mean weight loss after seed treatment with unitary formulation was 41% while that of binary formulation was 15% at 1%w/w dosage rate. Among the botanical treatments, the highest weight loss (65%) was recorded on beans grain treated with *A. indica* while the lowest weight loss due to *S. molle+C. ambrosioides* treatment at 1%w/w dosage rate. There was neither seed damage nor weight loss recorded on bean grains treated with binary formulations of *C. ambrosioides+J. curcas* and *S. molle+C. ambrosioides* at 2% w/w.

Effects of unitary and binary botanical formulations treatment on percent germination

Germination percent of common bean seeds treated with different unitary and binary botanical powder formulation is presented in Table 5. There was no significant ($P>0.05$) difference in the percent germination between disinfected common bean seeds treated with different botanical insecticide formulations and untreated control at both dosage rates. The percent germination of bean seed treated with different botanical powder formulations ranged between 92-99%, which was as good as untreated control, indicating botanical treatment didn't have effect on germination rate.

Discussion

Several studies have been carried on the potential of botanical insecticides in controlling insect pests including stored grain pests; however, only a few studies considered their synergistic combination. Results from the present study demonstrated synergistic potential of different botanical insecticide powders in controlling of bean bruchids (*Z. subfasciatus*) on stored beans (*P. vulgaris*). Overall, binary formulations had better effect in reducing damage compared to unitary formulation as assessed by different control parameters such as adult mortality (Table 6), F_1 progeny production, percent inhibition, and weevil perforation index (WPI) and weight loss. Moreover, combining more than one botanical insecticide which works synergistically will also make difficult for pests to develop resistance.

Significantly higher adult mortality was recorded in binary formulation compared to unitary formulation. For instance, adult mortality which ranged from 44.58-53.33% due to application unitary formulation of *D. stramonium, J. curcas* and *S. molle* increased to over 87% when combined with *C. ambrosioides*, even at lower dosage rate (1% w/w). This indicated combining different botanical insecticides will enhance their potency in controlling *Z. subfasciatus*. The current findings concur with previous reports which showed enhanced potency of botanicals in controlling pests when combined as binary formulations. For example, combination of *A. indica* with a pyrethroid resulted in a more effective management of silver white fly (*Bemisia argentifolii*), a major greenhouse pest of horticultural flowers [23]. Similarly, biological activities of botanical plants, tobacco (*Nicotiana tabacum*), Mexican marigold (*Tagetes minuta*), tephrosia *Tephrosia vogelli*, and *A. indica* were significantly enhanced in their binary formulation against common bean weevil, *A. obtectus* on stored beans reported reduced number of pests on cowpea plants and increased yield of grains as a result of synergistic activity of mixed botanical extract from herbal landraces.

Results from the present study demonstrated reduction in the application rates of constituent individual formulations without compromise in control efficacy due to enhanced potency in combining botanical insecticides. Higher bruchids mortality was achieved after seed treatment with binary formulation of *C. ambrosioides* with *D. stramonium, J. curcas* and *S. molle* at lower dosage rate which was on a par with the mortality recorded due to synthetic insecticide primiphos methyl. Besides, low dose binary formulations were more effective in controlling *Z. subfaciatus* than most of their constituent unitary botanical formulations at high dosage rate. In the current study, a binary formulation in which *C. ambrosioides* was included had the best mortality effect, often over 85%. Specifically, combination of *C. ambrosioides* with *D. stramonium, J. curcas* and *S. molle* were most potent as highest *Z. subfasciatus* mortality was achieved with very small quantity/proportion of botanicals applied. Besides, there was no significant difference in adult mortality due to these binary formulations at lower (1% w/w) and higher (2% w/w) dosage rates. Previous studies have demonstrated reduced application rates of synthetic insecticides due to increased potency of binary formulations.

Toxic effects of unitary and binary botanical formulations in the current study were directly related to exposure time of the pest to the treatments. Highest *Z. subfasciatus* mortality was recorded at longest exposure periods after botanical treatment and vice versa. It was also found out that mortality effect of botanical insecticides was dose dependent particularly for unitary formulations. An increased *Z. subfaciatus* mortality was observed at higher doses of unitary formulation. Interestingly, some binary formulations had more or less similar effects at both dosage rate, for example, binary formulation of *C. ambrosioides D. stramonium, J. curcas* and *S. molle*. Among individual formulations tested, lowest *Z. subfasciatus* mortality was recorded by *D. stramonium* at 1%w/w while the highest mortality was observed by *C. ambrosioides* at higher rate (2%w/w) application. The current findings are in agreement with the previous report by G/selase and Getu where mortality effect of botanicals was shown to be dose and exposure time dependent.

Overall, bean grain treatment with unitary and binary botanical formulations induced significant high reduction in F_1 progeny production by *Z. subfasciatus* compared to the untreated control. Besides, binary formulations showed better reduction in adult emergence compared to their unitary formulation. Bean grain treatment with binary formulations of *C. ambrosioides+A. indica, D. stramonium+C. ambrosioides, C. ambrosioides+J. curcas, S. molle+C. ambrosioides, A. indica+J. curcas* resulted in the highest reduction F_1 progeny produced. Moreover, there was no significant different in F_1 progeny reduction due to synthetic chemical primiphos methyl and the binary formulations. Significantly high reduction in F_1 progeny as a result of binary formulations application, demonstrated by none or below unitary adult emergent number, strongly suggested enhanced potency of different botanical combinations. Pest attack is population dependent where high pest populations build up lead to high infestation and damage, which in turn depends on number of emerging adults [24,25]. The highly toxic effects of the binary formulations against F_1 progeny in this study indicated the potential of synergists as an effective control option against *Z. subfasciatus.*

The synergistic effect of botanical insecticides in suppressing in F_1 progeny could be due to combined factors such as, increased adult mortality, ovicidal and larvicidal properties of botanical formulations and/or presence of chemicals that interfere with insect feeding [26,27]. Previous investigation on wheat treated with *A. indica* and *A. boonei* powder attributed suppression of F_1 generation of *S. zeamais* to high mortality of adult insects which disrupts mating and sexual communication as well as deterring females from laying eggs and affecting developmental stages of insects. Related studies showed botanical powder treatment act as oviposition-deterrent, inhibit oviposition by weakening adult bruchid to lay fewer eggs and kill the hatching larvae afterwards [28,29]. In related study reported neem seed kernel admixed to the groundnuts at the rate of 5% reduced the adult emergence of *C. serratus*. Even though synergistic combination of botanicals in insect suppression has not been widely examined, several studies revealed the potential of botanical insecticides in reducing F_1 progeny production on different insect pests.

Damage by *Z. subfasciatus* infestation was significantly reduced after treating bean grains with unitary and binary botanical formulations compared to untreated control. Binary formulations had better effect in reducing damage compared to their unitary formulations as judged by low to none weevil perforation index and weight loss. This has further confirmed that combining some botanicals as binary formulation will enhance their biological activity to effectively reduce damage by *Z. subfasciatus* on stored beans. Among the different botanical synergists, combination *C. ambrosioides* with *D. stramonium, J. curcas* and *S. molle* showed the most effective protecting ability against *Z. subfasciatus*. This was demonstrated by the least weevil perforation index and percent weight loss recorded after their application at lowest dosage rate (1% w/w). In addition to enhanced efficacy, botanical synergist discussed here have favorable toxicological properties such as rapid degradation, low residues and are safe for the consumer which make them preferred biopesticides in storage pest control.

Results from germination test demonstrated that all botanical powder formulations used to treat bean seeds against *Z. subfasciatus* didn't have negative effect in germination percents of the seeds at both dosage rates. Hence, bean seed for planting can be protected and kept viable from storage pest by treating them botanical formulations similar to the grain stored for food purposes. Though this is the first time to test combined effect botanical synergists on germination, previous study on seeds treated with unitary botanical formulation showed no significant effect on the germination rate. Our study results are in agreement with several reports which stated botanical insecticides which provided protection against storage pests didn't affect seed quality and viability [30-32]. In summary, the findings from the current study underscored synergistic combinations of botanicals enhance effective control of storage pests by optimizing potency of constituent botanicals while reducing dosage rates. Toxicity effect of these binary formulations was comparable with the standard chemical pesticide primiphos methyl. Besides, use of botanical insecticides has several comparative advantages over synthetic insecticides, which include low cost, in the context of small holder farmers, availability, reduced environmental pollution and minimal toxicity to humans and livestock [33]. The insecticidal properties of most of botanical plants studied here have been reported against different insect pests [34,35]. However, none of these studies considered synergistic potential of botanicals and similar insecticidal effects were reported at relatively high dosage rates and after longer exposure time. A new dimension of utilizing synergistic combination of different botanical formulations offers an excellent opportunity to increase the efficacy and reduce application rates of biopesticides in effort to successfully control storage pests of agricultural crops [36]. Hence, the authors recommend incorporation the information/ knowledge generated on synergistic combination of botanicals into regular biorational crop protection practice especially by resource limited small scale bean farmers.

Acknowledgements

The authors are grateful to Melkassa Research Center of Ethiopian Institute of Agricultural Research for providing necessary laboratory facilities for undertaking research. Technical assistance by Weinshet Belay in insect rearing is highly appreciated. BT received scholarship to undertake this research as part of his MSc study from Ethiopian Ministry of Education and Hawassa University. The funder had no role in study design, data collection and analysis, decision to publish, or preparation of the manuscript.

References

1. Abate T, Ampofo JKO (1996) Insect pests of common bean in Africa: Their ecology and management. Ann Rev Entomol 41: 45-75.
2. Schmale I, Wackers FL, Cardona C, Dorn S (2002) Field Infestation of Phaseolus vulgaris by Acanthoscelides obtectus (Coleoptera: Bruchidae), Parasitoid Abundance, and Consequences for Storage Pest Control. J Environ Entomol 31: 859-863.
3. Schoonhoven AV, Cardona C (1986) Main insect pests of stored beans and their control. Centre International de Agriculture Tropical.
4. Negasi F (1994) Studies on the economic importance and control of bean bruchids in haricot bean. Alemaya University.
5. Getu E, Ibrahim A, Iticha F (2003) Review of lowland pulse insect pest research in Ethiopia. Proceedings of grain legume workshop, Addis Ababa, Ethiopia.
6. Isman BM (2008) Perspective Botanical insecticides: for richer, for poorer. Pest Manag Sci 64: 8-11.
7. Kareru P, Rotich Z K, Maina EW (2013) Use of Botanicals and Safer Insecticides Designed in Controlling Insects: The African Case. Agricultural and Biological Sciences.
8. Williams P, Hammitt J (2001) Chemicals derived from natural sources are generally perceived to pose less environmental risks than synthetic pesticides. J Entomol Sci 45: 45-51.

9. Bruce TJA (2010) Tackling the threat to food security caused by crop pests in the new millennium. J Food Secu 2: 133-141.

10. Songa JM, Rono W (1998) Indigenous methods for bruchid beetle (Coleoptera: Bruchidae) control in stored beans (Phaseolus vulgaris L.). Inter J Pest Manag 44: 1-4.

11. Tadesse A (2008) Increasing Crop Production through Improved Plant Protection. Volume I. Plant Protection Society of Ethiopia (PPSE), Addis Ababa, Ethiopia.

12. Ileke KD, Oni MO (2011) Toxicity of some plant powders to maize weevil, Sitophilus zeamais (motschulsky) [Coleoptera: Curculiondae] on stored wheat grains (Triticumaestivum). Afric J Agric Res 6: 3043-3048.

13. Agona JA, Muyinza H (2003) Synergistic potential of different botanical combinations against bean brucids in storage. Afric Crop Sci Conf Proc 6: 216-219.

14. Oparaeke AM, Dike MC, Amatobi CI (2005) Evaluation of botanical mixtures for insect pests management on cowpea plants. J Agric Rural Devel Trop Subtrop 106: 41-48.

15. Rahman A, Talukder FA (2006) Bioefficacy of some plant derivatives that protect grain against the pulse beetle, Callosobruchus maculatus. J of Insec Sci 6: 3-4.

16. Hill J, Schoonhoven AV (1981) Effectiveness of vegetable oil fractions in controlling the Mexican bean weevil on stored beans. J Econ Entomol 74: 478-479.

17. Gselase A, Getu E (2009) Evaluation of botanical plants powders against Zabrotes subfasciatus (Boheman) (Coleoptera: Bruchidae) in stored haricot beans under laboratory condition. Afric J Agric Res 4: 1073-1079.

18. Ileke KD, Bulus DS (2012) Evaluation of contact toxicity and fumigant effect of some medicinal plant and pirimiphos methyl powders against cowpea bruchid, Callosobruchus maculatus (fab.) [Coleoptera: chrysomelidae] in stored cowpea seeds. J Agric Sci 4: 279-284.

19. Abate T, Negasi F, Ayalew G (1992) Botanicals in pest management Presented. CEE/EPC Joint Conference, Addis Ababa.

20. Jembere B (2002) Evaluation of the toxicity potentialof Milletia ferruginea (Hochest) Baker against Sitophilus zeamais (Motsch.). Inter J Pest Manag 48: 29-32.

21. Adedire CO, Ajayi TS (1996) Assessment of insecticidal properties of some plants as Grain protectants against the maize weevil, Sitophilus zeamais (Mots). Niger J Entomol 13: 93-101.

22. Fatope MO, Nuhu AM, Mann A, Takeda Y (1995) Cowpea weevil bioassay: a simple pre-screen for plants with grain protectant effect. Inter J Pest Manag 41: 84-86.

23. Greer L (2000) Greenhouse IPM: Sustainable Whitefly Control. Pest Management Technical Note.

24. Southwood TRE (1978) Ecological Methods. New York, USA.

25. Odum EP (1983) Basic Ecology. Holt-Saunders International Edition, Tokyo, Japan.

26. Ofuya TI (1990) Oviposition deterrence and ovicidal property of some plant powders against Callosobruchus maculates in stored cowpea (*Vigna unguiculata*) seeds. J Agric Sci 115: 343-345.

27. Omotoso OT (2005) Insecticidal and insect productivity reduction capacities of aloe vera and *bryophyllumpinnatum* on *tribolium castaneum* (herbst). Afr J Appl Zool Envi Biol 7: 95-100.

28. Mulatu B, Gebremedhin T (2000) Oviposition-deterrent and toxic effects of various botanicals on the Adzuki bean beetle, *Callosobruchus chinensis* L. Insec Sci Appl 20: 33-38.

29. El-Atta HA, Ahmed A (2002) Comparative effects of some botanicals for the control of the seed weevil *Caryedon serratus* Oilvier (Col., Bruchidae). J Appl Entomol 126: 577-583.

30. Onu I, Aliyu M (1995) Evaluation of powdered fruits of four peppers (*Capsicum* spp.) for the control of *Callosobruchus maculatus* (F) on stored cowpea seed. Inter J Pest Manag 41: 143-145.

31. Keita SM, Vincent C, Schmit JP, Arnason JT, Belanger A (2001) Efficacy of essential oil of Ocimum basilicum L. and O. gratissimum L. applied as an insecticidal fumigant and powder to control Callosobruchus maculatus (Fab.) (Coleoptera: Bruchidae). J Stor Prod Res 37: 339-349.

32. Dejen A (2002) Evaluation of some botanicals against maize weevil, *Sitophilus zeamais* Motsch. (Coleoptera: Curculionidae) on stored sorghum under laboratory condition at Sirinka. Pest Manag J Ethiop 6: 73-78.

33. Elhag EA (2000) Deterrent effects of some botanical products on oviposition of the cowpea bruchid *Callosobruchus maculatus* (F.) (Coleoptera: Bruchidae). Inter J Pest Manag 46: 109-113.

34. Asmanizar A, Djamin D, Indris AB (2011) Evaluation of jatropha carcus and annona muricata seed crude extracts against *Sitophilus zeamais* infesting stored rice. J Entomol 9: 13-22.

35. Abbott WS (1925) A method of computing the effectiveness of insecticides. J Econ Entomol 18: 265-267.

36. Su HCF (1991) Toxicity and repellency of chenopodium oil to four species of stored-product insects. J Entomol Sci 26: 178-182.

Potassium Fertilization and its Level on Wheat (*Triticum aestivum*) Yield in Shallow Depth Soils of Northern Ethiopia

Brhane H[1]*, Mamo T[2] and Teka K[3]

[1]*Mekelle Soil Research Center, Tigray Agricultural Research Institute, Mekelle, Ethiopia*
[2]*Agricultural Commercialization Cluster (ACC), Initiative and Ethiopian Soil Information System (EthioSIS), Agricultural Transformation Agency, Addis Ababa, Ethiopia*
[3]*Department of Land Resources Management and Environmental Protection, Mekelle University, Mekelle, Ethiopia*

Abstract

Many un-updated reports indicated that potassium was not deficient in Ethiopian soils. However, it was later proved that many Ethiopian soils are potassium deficient. Hence, the Ethiopian Soil Information System (EthioSIS) has initiated potassium fertilizer demonstrations in 2014 using K containing blended fertilizers in different parts of the country. But, there were no evidences about the K in the blended fertilizer is enough for wheat demand or not. Thus, a field experiment was conducted to evaluate the response of wheat to additional K rates on top of the K containing blended fertilizers. The experiments were laid out in Randomized Complete Block Design with 4 levels of potassium (0, 30, 60, 90, of K_2O kg/ha) replicated 3 times. Data on yield and yield components of wheat crop were collected and analysis of variance was done. Results depicted that plant height and harvest index were not significant. However, spike length, grain yield and straw yields of wheat were significantly affected by K application rates. Hence, the highest spike length was obtained at a rate of 90 kg/ha K_2O but the highest grain and straw yield of wheat were obtained at 30 kg/ha K_2O. Besides, the highest apparent K recovery and agronomic use efficiency were found at 30 kg K_2O/ha. Therefore, potassium fertilization is important and its level in the blended formula did not meet the wheat requirement in the study area.

Keywords: Potassium; Blended fertilizer; Wheat; Shallow soils; Enderta

Introduction

Wheat (*Triticum aestivum*) is an important food crop in Ethiopia but its mean yield is around 2.54 Mt/ha, well below experimental yields [1]. Low soil fertility is among the factors which limit wheat production in Ethiopia [2]. Usage of organic and mineral fertilizer is important in replacing the depleted soils, and to improve soil productivity and crop production in the country. In Ethiopia, many of the smallholder farmers have good awareness about the contribution of fertilizer for crop production, but nationally only 35% of farmers applies fertilizer on about 40% of the area under crop production [3].

Moreover, for the past several years, farmers in the country had used a limited type of mineral fertilizers such as blanket recommendation of nitrogen and phosphorus in the form of urea and DAP respectively as an input to increase crop production. However, studies in Ethiopia indicated that application of N and P fertilizers without considering other nutrients such as K, and micro-nutrients led to the depletion of other soil nutrients [4].

Potassium is among the macro nutrients which are taken up by plants in large amount. It plays significant roles in transportation of water, nutrients, nitrogen utilization, and stimulation of early growth and in insect and disease resistance [5]. Potassium is also important in the transportation of prepared food from the leaves to the rest of the plant parts, quality of seeds and fruits, strengthens the roots, stem and branches of plants and reduce lodging.

Potassium fertilization had shown yield improvement of crops in various areas of the world [6]. Research findings in India, Bangladesh and Iran indicated that Potassium fertilizer increased grain and straw yield of wheat at various rates [7-10].

Nowadays, efforts have been started to include K as fertilizer in Ethiopia through K containing blended fertilizers. On the other hand, there was no evidence whether the K in blended formula and the recommended rate was enough for wheat demand or not. Thus, this study was aimed at investigating the response of wheat to the recommended blended fertilizers and increasing rates of K fertilizer applied as KCl) in Enderta district, northern Ethiopia.

Materials and Methods

Study area

The study was carried out in Enderta district located in the southeastern zone of Tigray region, northern Ethiopia. Geographically, the area is located between 13°12'55" and 13°38'38" N latitudes and 39°16'43" and 39°48'08" longitudes. The average elevation of the district is about 2200 m above sea level [11]. Based on data collected from the nearest weather station (Mekelle airport), the annual rainfall of the ranges between 258 and 756 mm and mean annual temperature ranges between 11.5°C and 24.4°C.

Experimental design and procedures

The experiment consisted 4 levels of potassium (0, 30, 60 and 90 K_2O kg/ha) designed in a RCBD with three replications and applied as KCl on top of the initially recommended blended fertilizers (NPKSZn at 100 kg/ha) that contains 8 kg/ha of K in the form of K_2O.

On top of the blended fertilizers which contain 15.2% N, 48.8% Nitrogen was also added to satisfy N wheat requirements. The blended fertilizer was applied at planting, while the nitrogen and K fertilizers

***Corresponding author:** Hagos Brhane, Mekelle Soil Research Center, Tigray Agricultural Research Institute, Mekelle, Ethiopia, E-mail: hagos2015@yahoo.com

were applied twice in the crop growth stage that is 1/3 of the full dose at planting and the other 2/3 at the full tillering stage. Sowing was done manually at a seed rate of 150 kg/ha using manual row maker with a spacing of 0.20 m between rows.

The initial experimental field soils were analyzed for texture, pH, organic matter, cation exchange capacity (CEC), total nitrogen, available phosphorus and exchangeable K. The method used for soil physical and chemical analysis was: Soil pH, Organic carbon %, soil texture by hydrometer, available Phosphorus, total nitrogen by Kjeldhal method, Neutral Ammonium acetate method for cation exchange capacity and Exchangeable K^+ [12-17]. After maturity, wheat crop samples were collected and partitioned into grain and straw parts. The grain and straw samples were analyzed for nitrogen and potassium. Plant total nitrogen was analyzed using Kjeldhal method whereas potassium using dry ashing method. Data on plant height, spike length, grain yield, straw yield and harvest index were collected [16,18].

Moreover, apparent K recovery and K agronomic use efficiency were calculated by the formula developed [19].

$$\textit{Apparent K recovery (kg/kg)} = \left(\frac{Un - Uo}{n}\right)$$

Where; Un stands for nutrient uptake at 'n' rate of fertilizer and Uo stands for nutrient uptake at control (no fertilizer) and 'n' stands for fertilizer applied.

$$\textit{Agronomic K use efficiency(kg/kg)} = \left(\frac{Gn - Go}{n}\right)$$

Where Gn and Go stand for grain yield of fertilized plots at 'n' rates of fertilizer and grain yield of unfertilized plots, respectively, and 'n' stands for nutrient applied.

Data analysis

Analyses of variance (ANOVA) were carried out using Statistical Analysis Software (SAS) version 9. Whenever treatment effects were significant, mean separations were made using the least significant difference (LSD) test at the 5% level of probability.

Results and Discussion

Soil properties before planting

The physical and chemical properties of the soil before planting are indicated in Table 1. The soil of the study site is neutral in pH, medium in Organic carbon and total nitrogen, low in available P and high in CEC and Exchangeable K [15,17,20,21]. The medium organic carbon content might have contributed to the high level of CEC and medium total nitrogen because of mineralization in the study area.

Parameters	Value
pH Water (1:2.5)	7.32
Organic Carbon (%)	2.1
Total N (%)	0.22
P-Olsen(mg/kg)	2.64
Exchangeable K(cmol/kg)	0.69
CEC (cmol (+) kg^{-1})	35.8
% Sand	25
% Silt	31
% Clay	44
Textural class	Clay Loam

Table 1: Soil physio- chemical properties of the site before sowing.

Plant height, spike length, grain yield, straw yield and harvest index

Data presented in Tables 2 and 3 showed that the K fertilizer rate showed that average plant height and spike length had increased with increasing K application rates even though the trend was not consistent. The highest plant height was obtained from treatments receiving 60 kg/ha K_2O but it is statistically like other treatments. However, spike length, grain yield and straw yields of wheat significantly affected by K application rates. The highest spike length was obtained from plots receiving 90 kg/ha K_2O. In agreement with this result, higher K rate treatments had significantly longer spike length as compared to the check treatments (without potassium fertilizer) [22]. Grain and straw yields of wheat increased from 0 to 30 kg /ha K_2O and decreased from 30 to 90 kg/ha K_2O in the study site and the highest grain yield of wheat gained from plots receiving 30 kg/ha K_2O was significantly superior to all other treatments while in straw yield it is superior only to control. In line to this, it increased grain yield of wheat by 29.3%, 27.24% and 30.03% over treatments which received 0, 60 and 90 kg/ha K_2O respectively. On the other hand, the non-significant harvest index in the site might be due to the initial high level of soil exchangeable potassium and the mineralization of medium organic matter in the soil brought K, which is available to wheat plants requirement and increasing K level as fertilizer may be used as luxury consumption and contribute equal effects to grain and biological yield. This result agreed with the research findings that potassium effect on harvest index was not significant indicating approximately equal positive effects of potassium on seed and biological yield [23].

Generally, the significant grain yield obtained in this experiment with K application rate agrees with the research findings [10,22,24,25].

Treatment	Plant height (cm)	Spike length (cm)	Grain yield (kg/ha)	Straw yield (kg/ha)	Harvest Index
Control	41.7	4.3^{C}	633.5^{C}	1099.8^{B}	36.7
RBF (100 kg/ha)	52.3	5.2^{BC}	1076.3^{B}	2212.6^{A}	32.7
RBF+30 K_2O kg/ha	53.9	5.9^{BA}	1391.8^{A}	2519.3^{A}	35.5
RBF+60 K_2O kg/ha	54.7	6.0^{BA}	1093.8^{B}	2150.7^{A}	33.8
RBF+90 K_2O kg/ha	54.4	6.3^{A}	1070.4^{B}	2140.7^{A}	33.4
CV	9.88	9.47	13.07	12.87	6.47
P	0.06	0.012	0.002	0.002	0.26

Means followed by the same letter along columns are not significantly different. RBF: Recommended blended fertilizer (NPKSZn), CV: coefficient of variance, P: probability level

Table 2: Effect of potassium fertilizer rates on wheat plant height, spike length, grain yield, straw yield, and harvest index.

Treatment	Values	
K rates (K_2O kg/ha)	ARK (kg/kg)	AUEK (kg/kg)
30	0.52	25.28
60	0.25	7.67
90	0.15	4.85

ARK=Apparent recovery of potassium; AUEK=Agronomic use efficiency of potassium

Table 3: Effect of potassium fertilizer rate on apparent recovery and agronomic use efficiency.

Apparent recovery and agronomic K use efficiency

Potassium application rate influenced apparent potassium recovery and agronomic K use efficiency. Both recovery and agronomic K use efficiency consistently decreased with increasing potassium rates. Hence, the highest apparent K recovery and agronomic efficiency was obtained at 30 kg/ha K_2O.

Conclusions

A field experiment was done to evaluate the response of wheat to different potassium fertilizer rates and to determine the optimum rate of potassium for wheat on shallow soils of Eenderta district. Results from the experiment shown that the application of different potassium fertilizer rates significantly affected the yield of wheat in the study area. So, this falsifies the thought that potassium fertilization is unnecessary for Ethiopian soils and the level of potassium in the blended formula which was initially recommended by EthioSIS does not meet yield requirement of wheat since the yield of wheat has significantly responded from the additional K levels.

References

1. Report on Area Production for Crops (2007) Central Statistical Authority Agricultural Sample Survey.

2. Gebreselassie Y (2002) Selected chemical and physical characteristics of soils of Adet research center and its testing sites in North-western Ethiopia. Ethiopian Society of Soil Science.

3. Tekalign Mamo T, Saleem M (2001) Joint Vertisol Project as a Model for Agricultural Research and Development, Proceedings of the international symposium on Vertisol management, Ethiopia.

4. Bereket H, Stomph TJ, Hoffland E (2011) Teff (Eragrostis tef) production constraint on Vertisols in Ethiopia, farmers perceptions and evaluation of low soil zinc as yield-limiting factor. Soil Sci Plant Nutr 57: 587-596.

5. Lakudzala DD (2013) Potassium response in some Malawi soils. International Letters of Chemistry, Physics and Astronomy 8: 175-181.

6. Imran M, Gurmani ZA (2011) Role of macro and micro nutrients in the plant growth and development. Science Technology and Development, Pakistan.

7. Astatke A, Mamo T, Peden D, Diedhiou M (2004) Participatory on-farm conservation tillage trial in the Ethiopian highland Vertisols: The impact of potassium application on crop yields. Experimental Agriculture 40: 369-379.

8. Malek-Mohammadi M, Maleki A, Siaddat SA, Beigzade M (2013) The effect of zinc and potassium on the quality yield of wheat under drought stress conditions. International Journal of Agriculture and Crop Sciences 6: 1164.

9. Saha PK, Hossain ATMS, Miah MAM (2010) Effect of potassium application on wheat (*Triticum aestivum L.*) in old Himalayan piedmont plain. Bangladesh Journal of Agricultural Research 35: 207-216.

10. Tabatabaei SA, Shams S, Shakeri E, Mirjalili MR (2012) Effect of different levels of potassium sulphate on yield, yield components and protein content of wheat cultivars. Applied Mathematics in Engineering, Management and Technology 2: 119-123.

11. Gebre T, Kibru T, Tesfaye S, Taye G (2015) Analysis of Watershed Attributes for Water Resources Management Using GIS: The Case of Chelekot Micro-Watershed, Tigray, Ethiopia. Journal of Geographic Information System 7: 177.

12. Rhoades JD (1982) In: Methods of Soil Analysis. Part 2. 2nd edn. American Society of Agronomy, Madison, USA.

13. Walkley A, Black IA (1934) An examination of the Degtjareff method for determining soil organic matter and a proposed modification of the chromic acid titration method. Soil Science 37: 29-38.

14. Bouyoucos GJ (1962) Hydrometer method improved for making particle size analyses of soils. Agronomy Journal 54: 464-465.

15. Olsen SR (1954) Estimation of available phosphorus in soils by extraction with sodium bicarbonate. United States Department of Agriculture, Washington, USA.

16. Bremner JM, Mulvaney CS (1982) Nitrogen-total. Methods of soil analysis. Part 2, Chemical and microbiological properties, pp: 595-624.

17. Landon JR (1991) Booker Tropical Soil Manual: A Handbook for Soil Survey and Agricultural Land Evaluation in the Tropics and Sub-tropics. Longman Scientific and Technical, Essex, New York, USA, pp: 474.

18. Chapman HD (1965) Cation-exchange capacity. Methods of soil analysis. Part 2. Chemical and microbiological properties, pp: 891-901.

19. Fageria NK, Baligar VC (2003) Methodology for evaluation of lowland rice genotypes for nitrogen use efficiency. Journal of Plant Nutrition 26: 1315-1333.

20. Tadesse T, Haque I, Aduayi EA (1991) Soil, plant, water, fertilizer, animal manure and compost analysis manual. ILCA/PSD Working Document (ILCA).

21. Jones Jr JB (2002) Agronomic Handbook: Management of crops, soils and their fertility. CRC Press.

22. Maurya P, Kumar V, Maurya KK, Kumawat N, Kumar R, et al. (2014) Effect of potassium application on growth and yield of wheat varieties. The Bioscan 9: 1371-1373.

23. Zare M, Zadehbagheri M, Azarpanah A (2013) Influence of Potassium and Boron on Some Traits in Wheat. International Journal of Biotechnology 2: 141-153.

24. Tahir M, Tanveer A, Ali A, Ashraf M, Wasaya A (2008) Growth and yield response of two wheat (*Triticum aestivum L.*) varieties to different potassium levels. Pak J Life Soc Sci 6: 92-95.

25. Wassie H, Tekalign M (2013) The Effect of Potassium on the Yields of Potato and Wheat grown on the Acidic Soils of Chencha and Hagere Selam in Southern Ethiopia. International Potash Institute 35: 3-8.

Synergistic Effect of Seaweed Manure and *Bacillus* sp. on Growth and Biochemical Constituents of *Vigna radiata* L.

Lakshmanan Kasi Elumalai* and Ramasamy Rengasamy

Center for Advanced Studies in Botany, University of Madras, Guindy Campus, Chennai - 600025

Abstract

Seaweeds are one of the major marine renewable resources of the world. Seaweed fertilizers have often been more beneficial to the crop plants than the conventional chemical fertilizers. Seaweed meals provide nitrogen, phosphorus, potassium and salts besides some readily available micro elements to the plants. The seaweeds are also known to contain plant growth regulators such as cytokinin and auxin in appreciable quantities. In the present study, different physico-chemical characteristics of *Sargassum* manure obtained from the Seaweed Natural and Alginate Products Ltd, Ranipet, India have been assessed. Studies on the effect of different concentrations of *Sargassum* manure on the photosynthetic pigment composition, biochemical make-up and plant growth regulators of *Vigna radiata* L. were made for the *Sargassum* manure itself besides the same in combination with a Bacillus sp. added to the soil as five different combinations (0.1 %). Those plants applied with the *Sargassum* manure and the bacterium demonstrated superior growth in relation to the plants supplied with the *Sargassum* manure alone. Therefore, the use of the bacterium along with the *Sargassum* manure appeared to make this combination an efficient eco- friendly alternative to the conventional chemical fertilizers.

Keywords: *Sargassum manure*; *Bacillus*; Growth; Biochemical constituents; *Vigna radiata* L.; Plant Growth Regulators

Introduction

Two thirds of today's world population depends upon agriculture for livelihood. Sixty five percent of Indian population mostly depends on farming for livelihood. The abuse of pesticides or fungicides has been causing soil pollution besides exerting harmful effects in humans. Chemical fertilizers generally deteriorate soil quality there being disturb the homeostasis of the ecosystems, eventually leading to habitat loss.

Seaweeds constitute one of the important biotic components of the ocean and might serve as an alternative to inorganic fertilizers [1]. Seaweed fertilizers are often found to be more effective in promoting productivity rather than the chemical fertilizers [2]. Seaweed extract contains macro nutrients, trace elements, organic substances such as carbohydrate, amino acids and Plant Growth Regulators [3]. Use of organic manure in agriculture particularly, for seed treatment could be eco-friendly and cost effective. Seaweed extracts are marketed as fertilizer additives [4]. It has been reported that, seaweeds offer several benefits such as increased crop yield, improved growth, induced resistance to frost, fungal and insect resistance, reduced spider, aphid, nematode infestation and increased nutrient uptake from the soil [5]. Seaweed meals provide an approximately equivalent amount of N, less P, but more K and total salts and readily available micro elements compared to most animal manures [6]. Other than the macro and micro nutrients, seaweeds contain many plant growth regulators such as cytokinin, gibberellins and auxin [7-12]. The seaweed sludge does not degrade rapidly owing to the presence of certain complex substances like phenolic derivatives. The plant growth regulating and productive roles of a bacterium isolated from the organic manure was confirmed by culturing the same in a variety of media and their supernants were checked for the presence of bacterial secondary metabolites [13].

The aerobic spore forming *Bacillus* that constitutes the genus, *Bacillus*, majority of which are harmless saprophytes plays a significant role in the breakdown of complex polysaccharides and helps in the cycling of nutrients in nature. *Bacillus* that plays a major role in spore resistance, germination of seed apparently possesses enzymatic machinery that probably leads to entering into association with other organisms in the environment [14]. Owing to their genetic make-up, *Bacillus* spp. resist adverse conditions and can remain viable in soil for long periods and offer sustainable crop protection against pathogens.

More recently, it has been discovered that several microbes in the rhizhosphere produce Indole Acetic Acid (IAA) [15]. *Bacillus* populations also promote growth and health of crops by the action of plant hormones [16]. The strain, *Bacillus subtilis* AFI increased shoot length ranging between 4% and 15% and also raised the plant biomass by up to 15% [17]. *Bacillus cereus* QQ 308 produced growth enhancing substances in Chinese cabbage [18]. *Bacillus* sp. MRF produced IAA at a concentration of 3.71 μg / mL [19]. The rhizobacterial strains of *Bacillus pumilus* INR7 and *Bacillus subtilis* GBO3 when given as seed treatment and soil drenchment for cucumber were found to enhance plant growth gradually under green house conditions [20].

The main objective of this study is to evaluate the efficacy of *Sargassum manure* in combination with a *Bacillus* sp. isolated from the manure on plant growth and its influence on the biochemical composition of *Vigna radiata* L. under green house conditions.

Material and Methods

Collection of *Sargassum manure*.

The manure was collected after extraction of the alginate from *Sargassum* spp. from the Seaweed Natural and Alginate Products Ltd., Ranipet, India. The manure samples were stored in airtight containers and were later analyzed for different parameters.

***Corresponding author:** LK Elumalai, Centre for Advanced Studies in Botany, University of Madras, Guindy Campus, Chennai - 600 025, India, E-mail: elumalai25@gmail.com

The physico - chemical characteristics of *Sargassum manure*

The physico-chemical characters such as colour, odour, pH, macronutrients and micro elements were estimated by the methods as described by the American Public Health Association [21].

In the biochemical analysis total carbohydrate was estimated by the method as described by [22] total protein [23] total lipids [24] total phenol [25] and organic carbon by [26]. Humic acid was estimated by the method given by [27] Plant Growth Regulators [PGRs] such as, auxin, gibberellins and cytokinin were also measured [28].

Isolation of plant growth promoting bacteria

One gram of *Sargassum manure* was mixed with 10 mL sterile distilled nutrient agar plates. The plates were then incubated at 37°C for 24 h. The dominant bacterium from the culture plate was isolated and identified as a *Bacillus* sp.

Growth Study

The bacterial isolate was inoculated in nutrient broth and was incubated at 37°C for 24 h and growth was assessed by measuring the OD at 3h intervals at 600 nm.

Antagonistic activity of *Bacillus* sp. against phytopathogens

The test bacterium, *Bacillus* sp. was screened for the antagonistic activity against the phytopathogen, *Rhizoctonia solani*, on Potato Dextrose Agar (PDA) by dual culture technique.

Screening of the bacterial isolate for auxin production

The bacterial isolate was screened for IAA production [29] by inoculating it in the NB with and without tryptophan (1mg / mL) and incubated at 37°C for 7 days and analysed for auxin by following the method [30].

Sargassum manure and *Bacillus* sp.

Samples were prepared as follows: i] control, ii] unsterilized *Sargassum manure*, iii] sterilized *Sargassum manure*, iv] *Sargassum manure* inoculated with 25mL of 18 hr old *Bacillus* sp. v] Sterilized *Sargassum manure* inoculated with 25 mL of 18 h old *Bacillus* sp. and vi] 50 ml of 18 h old were chosen for the following study.

Test plants

The crop plant, *Vigna radiata* (Fabaceae) was chosen as the test plant for the present study. The seeds of the plant were procured from the local market. Seeds of uniform size, colour and weight were chosen for use in the experiments.

The experimental plants were raised in pots in a glass house. Six different experiments were conducted with four replicates. Seeds soaked in tap water overnight were used in further study. Experiments were conducted in earthen pots (15 cm dia., 13 cm depth and 14 cm height) filled with 700 g of the sterilized red soil.

Experimental

T1 - Control Plants without any set of treatment or application were irrigated with fifty mL of sterile water at 50 mL per day.

T2 *Sargassum manure* Fifty mL of 2% *Sargassum manure* (w/v) on 3rd, 6th, 9th day was applied.

T3 Autoclaved *Sargassum manure* - 50 mL of autoclaved 2% *Sargassum manure* (w/v) on 3rd, 6th, 9th day was applied.

T4 *Sargassum manure* with bacteria 50 mL of 2% non-autoclaved *Sargassum manure* (w/v) with *Bacillus* sp. twenty 5 mL of 18h old on 3rd, 6th, 9th day was applied.

T5 *Sargassum manure* autoclaved with bacterial culture Applied twenty five mL of 2 % autoclaved *Sargassum manure* (w/v) and *Bacillus* sp. twenty five mL of 18h old on 3rd, 6th, 9th day.

T6 - *Bacterial culture* of only fifty mL of 18 h old culture on *Bacillus* sp. on 3rd, 6th, 9th day was applied.

Observations were made on the test plants over a period of 10 days.

After the experiments, the plants were uprooted and analysed against the following morphometric parameters: shoot length (cm), root length (cm), fresh weight (g), dry weight (g), number of lateral roots (Nos.) and number of leaves (Nos). The plants were analyzed for the photosynthetic pigments Chl a, Chl b, total chlorophyll [31] and carotenoids [32]. The amounts of total carbohydrate, total protein, and total lipids were also estimated. The levels of Plant Growth Regulators (PGRs) such as, auxin, gibberellins and cytokinin were also measured [28].

Results

The manure looks brown in color with pH 9.0. Among the various nutrients, it has been found that the level of calcium (50 mg/g) was found to be maximum followed by nitrate (14 mg/g), magnesium (18.0 mg/g), sulphate (31.0 mg/g), potassium (8.0 mg/g) and total Kjeldhal nitrogen (54.64 mg/g) in the manure (Table1). Similarly, Organic substances like total carbohydrate (7.0 µg/mg) total protein (8.2 µg/mg), total lipids (1.5 µg/mg), and total phenol (42 µg/mg) were also quantified. Besides humic acid (35.4 %), total organic carbon (13.3 µg/g) occurring at a C: N ratios (1:5) were also detected in the manure (Table 2).

Growth promoters, auxin (135 µg/g), gibberellin (50 µg/g) and cytokinin (115 µg/g) were also recovered from the manure (Table 3).

The predominant bacterium that occurred in association with the

S.No.	Characters	Quality/Quantity
1	Appearance	Brown
2	Odour	Fermented Odour
3	pH 1g / 10mL 1g / 1000 mL	 9.80 **9.05**
4	Alkalinity Total (as $CaCO_3$) mg/g	96
5	Calcium (as Ca) mg /g	50
6	Magnesium (as mg) mg/g	18
7	Sodium (as Na) mg/g	105
8	Potassium (as K) mg/g	8
9	Iron (as Fe) mg/g	2.12
10	Manganese (as Mn) mg/g	0
11	Free Ammonia (as NH_3) mg/g	1.12
12	Nitrite (as NO_2) mg/g	0.06
13	Nitrate (as NO_3) mg/g	14
14	Chloride (as Cl) mg/g	90
15	Fluoride (as F) mg/g	0.36
16	Sulphate (as SO_4) mg/g	31
17	Phosphate (as PO_4)mg/g	1.71
18	Silica (as Si) mg/g	34.40
19	Total Kjeldhal Nitrogen mg (TKN) mg/g	52.64

Table 1: Physico-chemical characters of *Sargassum* manure.

S. No	Substances	µg / mg
1	Total carbohydrate	7.0
2	Total Protein	8.2
3	Total lipids	1.5
4	Humic acid (%)	35.4
5	Total carbon (%) dry weight*	13.3
6	C:N ratio	1:5
7	Total Phenol (µg/mg)	42.0

(* Total carbon on dry weight)

Table 2: Biochemical composition of *Sargassum* manure.

S.No	Plant Growth Regulators	µg / mg of. F. wt
1	Auxin	135
2	Gibberellin	50
3	Cytokinin	115

Table 3: Plant Growth Regulators of *Sargassum* manure.

Test	Results
Colony morphology	Pale yellow, irregular, smooth surface
Gram's staining	Positive
Cell Shape	Rod
Motility	Motile
Catalase	Postive
Oxidase	Postive
Indole Production	Postive
Methyl red	Postive
Voges Proskauer	Negative
Citrate Utilization	Postive
Urease	Postive
Triple Sugar Iron	Glucose fermented
Nitrate Reduction	Negative
Starch Hydrolysis	Postive
Protease	Postive

Table 4: Physicochemical characters of the Bacterial isolate.

manure was identified as *Bacillus* sp. based on its morphological and biochemical characteristics (Table 4).

The selected test plant, *Vigna radiata* L. was exposed to different combinations of manure and *Bacillus* sp. It was noted when treated with at 25 mL of 2% manure and 25 mL of 18h *Bacillus* sp. culture, the test plants have shown enhanced growth (Figure 1b) compared to control (Figurela) in terms of the shoot length (20.3 cm) (Figure 2), root length (5.4 cm) (Figure 3), fresh weight (0.87 g) (Figure 4), dry weight (0.12 g) (Figure 5), number of roots (27) (Figure 6) and number of leaves (31) (Figure 7 and Table 5).

On fresh weight basis, the addition of manure and *Bacillus* sp. culture had significantly enhanced the content of chl a, chl b, total chl. and total carotenoids in the test plants. Maximum concentration of Chl a (7.099 mg/g), Chl b (2.860 mg/g), total Chl (9.95 mg/g) (Figure 8) and total carotenoids (0.309 mg/g) (Figure 9), total carbohydrate (2.45 mg/g) (Figure 10), total protein (3.16 mg/g) (Figure 11) and total lipids (23.47 mg/g) (Figure 12) were realized from *Vigna radiata* L. treated with 25 mL of 0.1% manure and 25 mL of 18h *Bacillus* sp. culture (Table 6). It was noticed than all this parameter were higher than7%, 47.64%, 18.28%, 59.19%, 68.36%, 44.19% and 77.42% respectively in relation to the control. Plant growth regulators such as auxin, gibberellin and cytokinin were assessed in *Vigna radiata* L. Maximun auxin (94.33 µg/g), gibberellin (257.71 µg/g) and cytokinin (97.0 µg/g)

Figure 1a: Control - Treated with water.

Figure 1b: Treated - Treated *Sargassum* Manure With *Bacillus* Sp Culture.

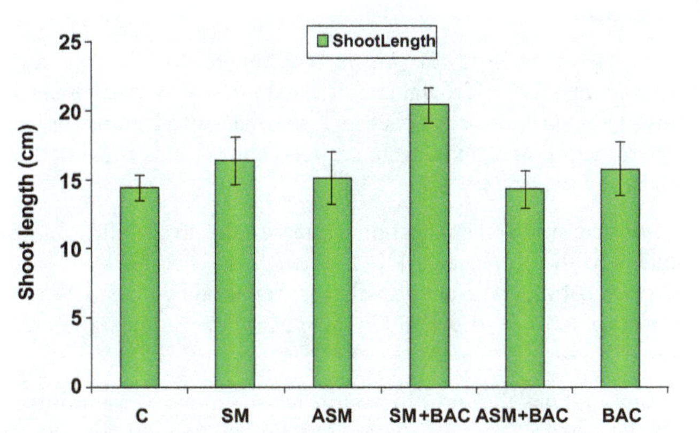

Figure 2: Effect of different proportions of *Sargassum* manure on shoot length of *Vigna radiata* L.
C- Control; SM – *Sargassum* Manure; ASM- Autoclaved *Sargassum* Manure; SM+BAC- *Sargassum* Manure + Bacterial culture; ASM+BAC- Autoclaved *Sargassum* Manure with Bacterial culture; BAC- Bacterial culture only.

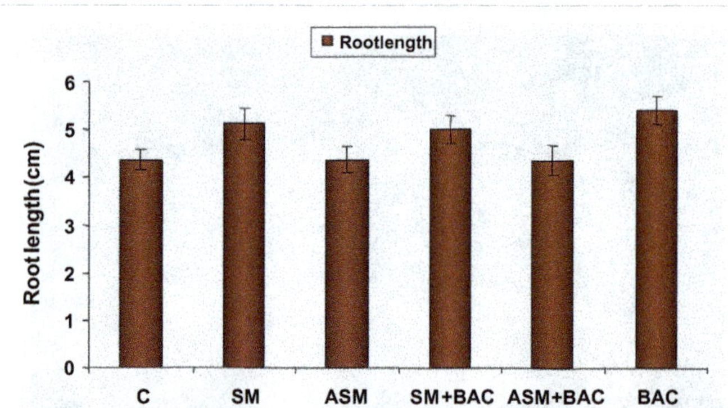

Figure 3: Effect of different proportions of *Sargassum* manure on root length of *Vigna radiata* L.
C- Control; SM – *Sargassum* Manure; ASM- Autoclaved *Sargassum* Manure; SM+BAC- *Sargassum* Manure + Bacterial culture; ASM+BAC- Autoclaved *Sargassum* Manure with Bacterial culture; BAC- Bacterial culture only.

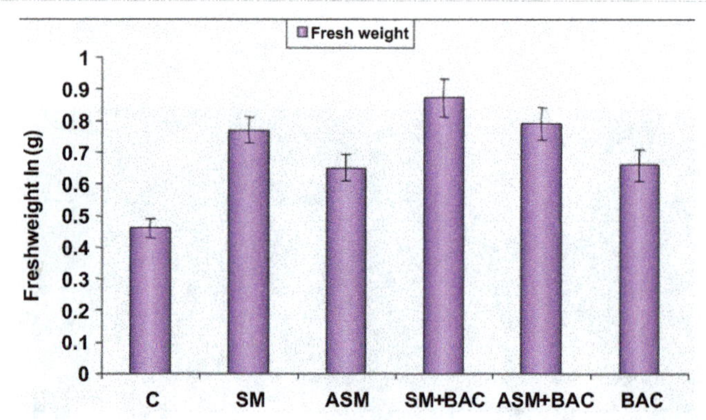

Figure 4: Effect of different proportions of *Sargassum* manure on fresh weight of *Vigna radiata* L.
C- Control; SM – *Sargassum* Manure; ASM- Autoclaved *Sargassum* Manure; SM+BAC- *Sargassum* Manure with Bacterial culture; ASM+BAC- Autoclaved *Sargassum* Manure with Bacterial culture; BAC- Bacterial culture only.

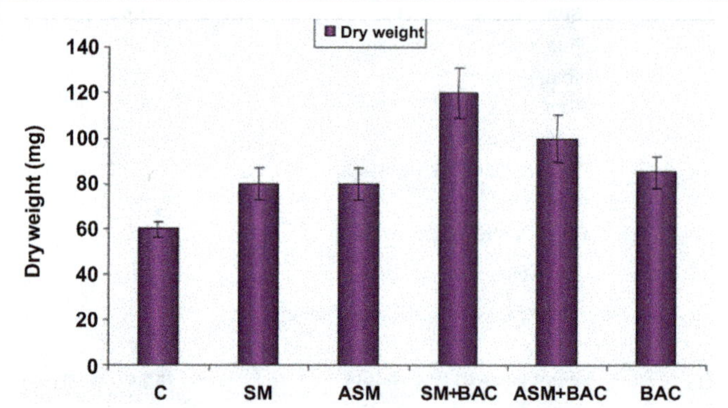

Figure 5: Effect of different proportions of *Sargassum* manure on dry weight of *Vigna radiata* L.
C- Control; SM – *Sargassum* Manure; ASM- Autoclaved *Sargassum* Manure; SM+BAC- *Sargassum* Manure with Bacterial culture; ASM+BAC- Autoclaved *Sargassum* Manure with Bacterial culture; BAC- Bacterial culture only.

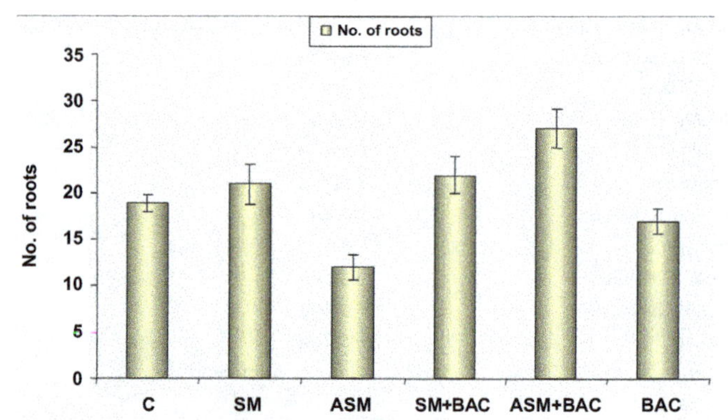

Figure 6: Effect of different proportions of *Sargassum* manure on number of lateral roots of *Vigna radiata* L.
C- Control; SM – *Sargassum* Manure; ASM- Autoclaved *Sargassum* Manure; SM+BAC- *Sargassum* Manure + Bacterial culture; ASM+BAC- Autoclaved *Sargassum* Manure with Bacterial culture; BAC- Bacterial culture only.

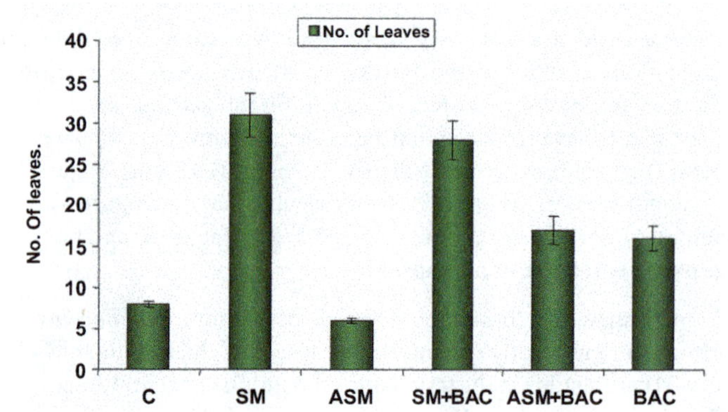

Figure 7: Effect of different proportions of *Sargassum* Manure on number of leaves of *Vigna radiata* L.
C- Control; SM – *Sargassum* Manure; ASM- Autoclaved *Sargassum* Manure; SM+BAC- *Sargassum* Manure with Bacterial culture; ASM+BAC- Autoclaved *Sargassum* Manure with Bacterial culture; BAC- Bacterial culture only.

were observed in the plants treated with 25 mL of 2% manure and the 18h *Bacillus* sp culture and the increase over the control was found to be to the tune of 78.73%, 57.0% and 53.0% respectively (Figure 13 and Table 7).

Discussion

The brown seaweed, *Sargassum* is commercially exploited for its principal polysaccharide, alginate. It is extensively used as emulsifier, gelling agent in food, cosmetics and pharmaceutical industries [33]. After extraction of alginate from seaweed, the waste is converted into *Sargassum manure.*

Seaweed manure application increases the availability of trace elements to the crop plants [4]. The beneficial effect of seaweeds on seed germination and plant growth were reported by [34,35]. Seaweed manure contains Fe, Cu, Zn, Co, Mo, Mn, Ni, vitamins and amino acids.

Application of the seaweed manure has raised the seed germination percentage even at lower concentration perhaps suggesting the presence of some plant growth promoting substances such as IAA and IBA, gibberellins (A and B) and cytokinin [36].

The growth enhancement potential of seaweed has been attributed to phenyl acetic acid [PAA] and micro and macro elements [37], vitamins and plant growth regulators such as gibberellins and cytokinin [39]. Maximum shoot length of 23.0 cm was observed in 10% Seaweed Liquid Fertilizer treated *Solanum* sp. seedlings [40]. The potential of *Bacillus* sp. to synthesis a wide range of metabolites with antibacterial

S. No	Parameter	Control	*Sargassum* manure	*Sargassum* manure + *Bacillus* sp.	Autoclaved *Sargassum* manure	Autoclaved *Sargassum* manure + *Bacillus* sp.	*Bacillus* sp.
1	Shoot length (cm)	14.37 ± 0.9	16.37 ± 1.7	20.37 ± 1.23	15.12 ± 1.9	16.25 ± 1.34	15.75 ± 1.96
2	Root length (cm)	4.37 ± 0.22	5.12 ± 0.32	5.42 ± 0.29	4.37 ± 0.29	4.37 ± 0.27	5.00 ± 0.31
3	Number of roots	19 ± 0.9	21 ± 2.2	19 ± 1.99	12 ± 1.32	27 ± 2.11	17 ± 1.38
4	Number of leaves	8.0 ± 0.34	31 ± 2.63	28 ± 2.36	6.0 ± 0.31	17 ± 2.63	16 ± 2.49
5	Fresh weight(g)	0.57 ± 0.03	0.77 ± 0.04	0.87 ± 0.06	0.65 ± 0.04	0.79 ± 0.05	0.66 ± 0.05
6	Dry weight (g)	0.06 ± 3.4	0.08 ± 7.2	0.12 ± 11.0	0.08 ± 6.9	0.10 ± 10.4	0.08 ± 6.8

Table 5: Effect of *Sargassum* manure and *Bacillus* sp. on the growth of *Vigna radiata* L.

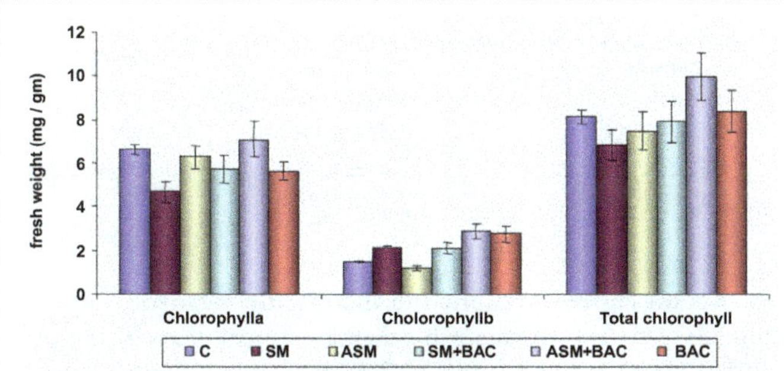

Figure 8: Effect of different proportions of *Sargassum* manure on photosynthetic pigments of *Vigna radiata* L.
C- Control; SM – *Sargassum* Manure; ASM- Autoclaved *Sargassum* Manure; SM+BAC- *Sargassum* Manure with Bacterial culture; ASM+BAC- Autoclaved *Sargassum* Manure with Bacterial culture; BAC- Bacterial culture only.

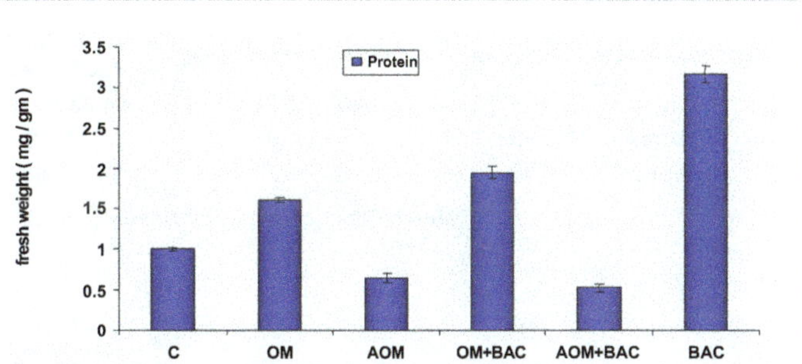

Figure 11: Effect of different proportions of *Sargassum* manure on total protein of *Vigna radiata* L.
C- Control; SM – *Sargassum* Manure; ASM- Autoclaved *Sargassum* Manure; SM+BAC- *Sargassum* Manure with Bacterial culture; ASM+BAC- Autoclaved *Sargassum* Manure with Bacterial culture; BAC- Bacterial culture only.

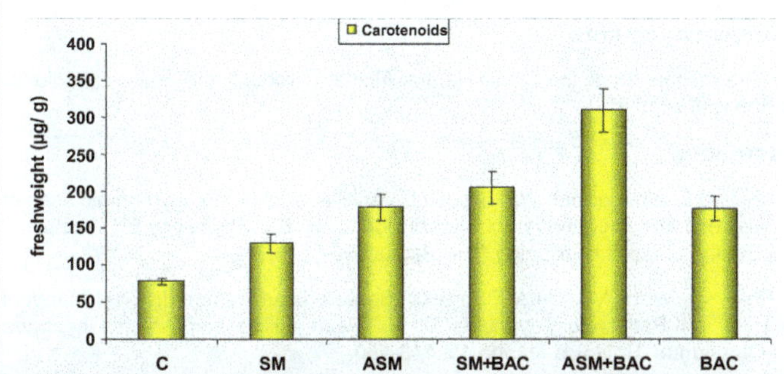

Figure 9: Effect of different proportions of *Sargassum* manure on carotenoids of *Vigna radiata* L.
C- Control; SM – *Sargassum* Manure; ASM- Autoclaved *Sargassum* Manure; SM+BAC- *Sargassum* Manure with Bacterial culture; ASM+BAC- Autoclaved *Sargassum* Manure with Bacterial culture; BAC- Bacterial culture only.

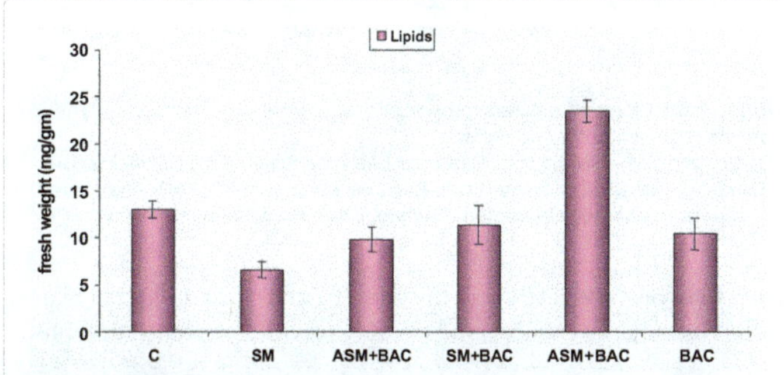

Figure 12: Effect of different proportions of *Sargassum* manure on total lipids of *Vigna radiata* L.
C- Control; SM – *Sargassum* Manure; ASM- Autoclaved *Sargassum* Manure; SM+BAC- *Sargassum* Manure with Bacterial culture; ASM+BAC- Autoclaved *Sargassum* Manure with Bacterial culture; BAC- Bacterial culture only.

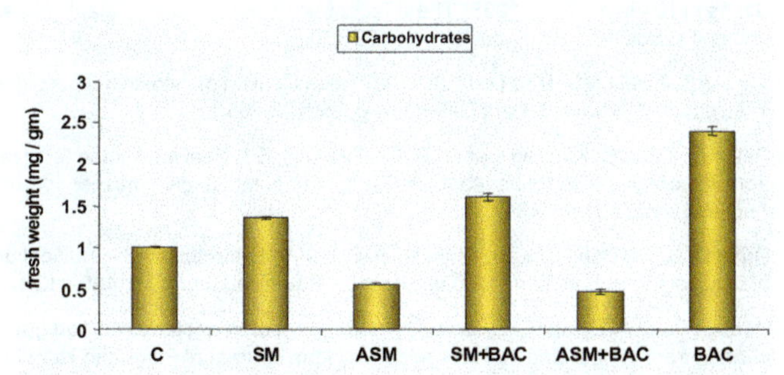

Figure 10: Effect of different proportions of *Sargassum* manure on total carbohydrate of *Vigna radiata* L.
C- Control; SM – *Sargassum* Manure; ASM- Autoclaved *Sargassum* Manure; SM+BAC- *Sargassum* Manure with Bacterial culture; ASM+BAC- Autoclaved *Sargassum* Manure with Bacterial culture; BAC- Bacterial culture only.

and antifungal activity has been extensively studied in medicine and industry and are well known for their ability to control plant disease when applied as biological control agents [41-43]. Numerous *Bacillus* strains have been implicated in activities that suppress pest and pathogen or otherwise promote plant growth. A number of these strains have already been developed commercially as biological fungicides, insecticides, and nematicides or plant growth promoters and their use in agriculture has been thoroughly reviewed [44-47]. Improvement in plant health and productivity are mediated by three different biological mechanisms namely, antagonism of resistant pathogens, promotion of host nutrition and growth and stimulation of plant host defenses.

In the present study, *Vigna radiata* L., showed maximum shoot length of (20.37 cm) and root length [5.42 cm] when it was treated with 2% *Sargassum manure* and *Bacillus* sp. culture. The maximum shoot length (33.10 cm/plant)root length (2.10 cm/plant) fresh weight (6.38 g/plant) dry weight (2.75 g/plant), number of lateral roots (39 numbers/

S. No	parameters	Control	*Sargassum* manure	*Sargassum* **manure + *Bacillus* sp.**	**Autoclaved *Sargassum* manure**	**Autoclaved *Sargassum* manure + *Bacillus* sp.**	*Bacillus* **sp.**
1	Chlorophyll a (mg/g)	6.640 ± 0.22	4.667 ± 0.5	7.099 ± 0.83	6.285 ± 0.54	5.703 ± 0.83	5.618 ± 0.43
2	Chlorophyll b (mg/g)	1.496 ± 0.01	2.157 ± 0.03	2.860 ± 0.34	1.211 ± 0.4	2.126 ± 0.34	2.749 ± 0.38
3	Total Chlorophyll (mg/ g)	8.138 ± 0.32	6.823 ± 0.71	9.958 ± 1.1	7.494 ± 0.89	7.828 ± 0.92	8.365 ± 0.95
4	Total Carbohydrate (mg/g)	1.0 ± 0.01	1.350 ± 0.02	2.450 ± 0.05	0.550 ± 0.01	0.450 ± 0.03	1.600 ± 0.04
5	Total Protein (mg/g)	1.0 ± 0.02	1.604 ± 0.03	3.160 ± 0.19	0.644 ± 0.06	0.520 ± 0.05	1.940 ± 0.08
6	Total Lipids (mg/g)	13.1 ± 0.9	6.570 ± 0.89	23.47 ± 2.17	9.820 ± 1.33	11.32 ± 0.09	10.44 ± 1.65
7	Carotenoids (mg / g. f wt)	0.0781 ± 4.2	0.129 ± 12.9	0.310 ± 29.03	0.1784 ±18.27	0.205 ± 21.99	0.176 ±16.42

Table 6: Effect of *Sargassum* manure and *Bacillus* sp. on biochemical constituents of *Vigna radiata* L.

S. No	parameters	Control	*Sargassum* manure	*Sargassum* **manure + *Bacillus* sp.**	**Autoclaved *Sargassum* manure**	**Autoclaved *Sargassum* manure + *Bacillus* sp.**	*Bacillus* **sp.**
1	Auxin	20 ± 0.8	14.65 ± 1.5	94.33 ± 9.44	14.54 ± 1.7	19.03 ± 1.9	24.91 ± 2.6
2	Gibberellins	110.28 ± 2.17	63.42 ± 1.9	257.71 ± 9.34	94.85 ± 4.28	114.28 ± 4.8	86.57 ± 3.1
3	Cytokinin	46.0 ± 7.22	21.80 ± 6.9	97 ± 23.84	40.2 ± 10.11	46.4 ± 11.8	31.80 ± 8.9

Table 7: Effect of *Sargassum* manure and *Bacillus* sp. on Plant Growth Regulators of *Vigna radiata* L.

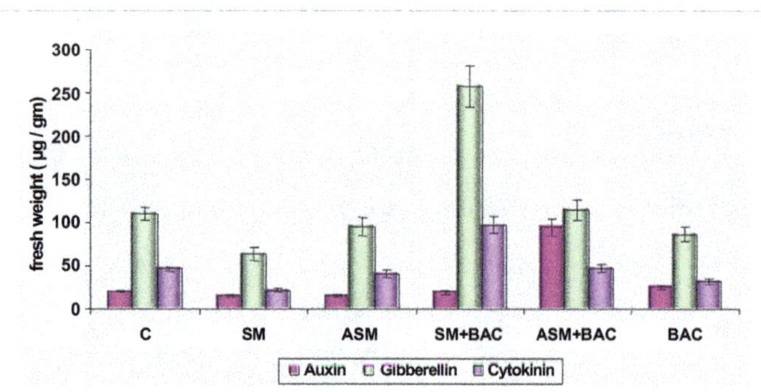

Figure 13: Effect of different proportions of *Sargassum* manure on Plant growth regulators of *Vigna radiata* L.
C- Control; SM – *Sargassum* Manure; ASM- Autoclaved *Sargassum* Manure; SM+BAC- *Sargassum* Manure with Bacterial culture; ASM+BAC- Autoclaved *Sargassum* Manure with Bacterial culture; BAC- Bacterial culture only.

plant) were observed at 20% in the case of *Turbinaria decurrens* plants. The lowest values were observed at 100% seaweed concentration [48].

In the present study, maximum fresh weight (0.87 g) and dry wt. (0.12 g) occurred with the plants treated with 2% *Sargassum manure* and *Bacillus* sp culture. Application of manure raised the contents of the photosynthetic pigments in *Vigna radiata*. Similar trends were observed in SLF treated crop plant, *Vigna unigiculata, Phaseolus radiata* [49], *Cymposis tetragonaloba* [50] and Zea *mays* [51]. The increase in the photosynthetic pigment contents were probably due to the presence of minerals such as Fe, Ni, Cu and Mg in the seaweeds [52] Iron, copper, and magnesium are essential elements which act as catalysts for the synthesis and efficient function of the chlorophyll [53]. The plant growth regulators from the seaweed may also have been responsible for the elevated synthesis of plant pigments [54]. Cytokinin inhibits degradation of chlorophyll, breakdown of protein molecules and aids in the increase of chl a, chl b in *Sorghum bicolor*. Seaweed extracts are known to promote seed germination and plant growth [55-57]. In the present study, elevated chl *a* and chl *b* contents were observed in the case of plants treated with manure. Maximum concentrations of chl *a*, chl *b* and total chlorophyll were observed in *Vigna radiata* L. treated with 25mL 0.1% of manure and 25 mL of *Bacillus* sp. culture .

The application of Seaweed Liquid Fertilizer (SLF) even at low concentration has been reported to increase the pigment content in Cucumber [58]. The enhanced biochemical constituents of the plant could be due to the increased availability of the essential elements available in the SLF [59].

Thus, the manure in combination with the culture of *Bacillus* sp. has been found to be promising for its growth promoting ability. Hence, this simple application of eco-friendly seaweed fertilizer to crop plants is recommended to the growers for better harvest.

Acknowledgements

The authors thank the SNAP Natural Alginate Products Ltd, Ranipet, India for the financial support.

References

1. Metting B, Zimmerman WJ, Crouch I, Van Staden J (1990) Agronomic uses of seaweed and microalgae. In: Introduction to Applied Phycology Ed. Akatsuka, I.SPB, Academic Publishing, The Hague 590-627.
2. Bokil KK, Metha VC, Datar DS (1972) Seaweeds as Manure III. Field Manurial Trials on Pennisetum typhoids S. H. (Pearl millet) and Arachis hypogea (Groundnut). Botanica Marina 15: 148-150.
3. Willams DC, Brain KR, Blunden G, Wildgoes PB (1981) Commercial seaweed extracts. Proc. 8th International Seaweeds Symposium, Bango, Marine science Laboratories, Menai Bridge 761-763
4. Booth E 1965 The Manurial value of Seaweed. Botanica Marina 8: 138-143.
5. Mooney PA, Van Staden J (1985) Effect of seaweed concentrate on the growth of wheat under conditions of water stress. S Afr J Sci 81: 632-633.
6. Simpson K, Hayes S F (1958) The effect of soil conditioners on plant growth and soil structure. J Sci Food Agric 9: 163-170.
7. Tay AAB, Palni LMS, Mac Leod JK (1987) Identification of cytokinin glucosides in a seaweed extract. J Plant Growth Regul 5: 133-138
8. Patier P, Yvin JC, Kloareg Lienart B, Rochas C (1993) Seaweed liquid fertilizer from Ascophyllum nodosum contains Elicitors of plants D-gluconases. Journal Applied Phycology 5: 343-350.
9. Duan D, Liu X, Pan P, Liu H, Chen N, et al. (1995) Extraction and identification of cytokinin from Laminaria japonica Aresch. Botanica Marina 38: 409-412.
10. Pěrez – Sanz A, Eymar E, Lucena JJ (1996) Effect of foliar sprays on Turf grass of an extract of peat and kelf amended with iron. Journal of Plant and Nutrition 19: 1179-1188.
11. Stirk WA, Van Staden J (1997) Isolation and identification of cytokinins in a new commercial seaweed product made from *Fucus serratus* L. Journal of Applied Phycology 9: 327-330.
12. Moller M, Smith ML (1998) The significance of the mineral component of

seaweed suspension on lettuce (*Lactuca sativa* L.) Seedling growth. Journal of Plant Physiology 153: 658-663.

13. Boyd KG, Mearns-Spragg A, Burgess JG (1999) Screening of marine bacteria for the production of microbial repellents using a spectrophotometric chemotaxis assay. Marine Biotechnology 1: 359-363.
14. Driks A (2002) Protein of the spore coat and coat. In *Bacillus subtilis* and its closest relatives, American Society for microbiology (Washington) 527-536.
15. Patten CL, Glick BR (1996) Bacterial biosynthesis of indole-3-acetic acid. Can J Microbiol 42: 207-220
16. Priest F (1993) Systematics and ecology of *Bacillus*. In: *Bacillus subtilis* and Other Gram- Positive Bacteria (Sonenshein, AL, Hoch, JA and Losick R Eds.) American Society for Microbiology, Washington, DC. 3-16
17. Manjula K, Podile AR (2001) Chitin supplemented formulations improve biocontrol and plant growth promoting efficiency of *Bacillus subtilis* AF1. Can J Microbiol 47: 618- 625.
18. Wen-Teish Chang, Yu-Chung Chen, Chia-Ling Jao (2007) Antifungal activity and enhancement of plant growth by *Bacillus cereus* grown on shellfish chitin wastes Bioresour Technol 98: 1224-1230.
19. Paul SD, Nongkynrih (1996) Enzyme activity - a diagnostic tool for toxicity/ deficiency of trace elements in rice plant. Indain Journal of Plant Physiology 1: 225- 257.
20. Raupach GS, Kloepper JW (1998) Mixtures of plant growth-promoting rhizobacteria enhance biological control of multiple cucumber pathogens. Phytopathology 88: 1158-1164.
21. APHA, AWWA and WPCF 1995 Standards Methods for Examination of water and waste water analysis. 19th ed., Washington.
22. Dubios M, Gilles KA, Hamilton TK, Rebers PA, Smith F (1956) Calorimetric method for determination of sugars and related substance. Annuals of Chemistry 28: 350-356.
23. Bradford MM (1976) A rapid and sensitive method for the quantification of microgram qualities of protein utilizing the principle of protein dye binding. Anal Biochem 72: 248-254.
24. Folch J, Less M, Stoare Stanley GH (1957) A simple methods for the isolation and purification of total lipids from animal tissues. J Bio Chem 226: 497-508.
25. Waterman PG, Mole S (1994) Analysis of phenolic plant metabolites. Blackwell Scientific, London, UK.
26. SK el Wakeel, Riley JP (1957) The determination of organic carbon in marinemuds. J Cons Int Explor Mer 22: 180-183
27. Welte (1952) Determination of humic acid. Anjew chemistry 67: 153.
28. Unyayar S, Faith Topcuoglu, Ali Unyaya (1996) A modified method for extraction & identification of IAA, GA3, ABA & Zeatin produced by *P.chrysoporium* ME446 Bulg. Journal of Plant Physiology 22: 105-110.
29. Loper JE, and Schroth MN (1986) Influence of bacterial sources of indole -3-acetiec acid on root elongation of sugar beet. phytopathology and Biochemistry 76: 386-389.
30. Hirte W (1961) Glyzerin-Pepton-Agar, ein vorteilhafter Nährboden für bodenbakteriologische Arbeiten. Zbl Bakteriology 114: 141-146.
31. Mackinney G (1941) Absorption of light by chlorophyll solutions. J Biol Chem 140: 315-322.
32. Wellburn AR, Lichtenthaler H (1984) Formulae and program to determine total carotenoids and Chlorophylls a and b of leaf extracts in different solvents. In Advances in Photosynthesis Research, Vol 2. C. Sybesma, editor. Martinus Nijhoff, The Hague, The Netherlands. 9-12.
33. Joseph T, Baker OBE (1984) Seaweeds in pharmaceutical studies and applications. Hydrobiologia 116/117: 29- 40.
34. Rajeshwari M, Lakshmanan KK, Chitra AS (1983) Effect of seaweed on tomato In : Proc. of National Seminar on the Production Technology of Tomato and Chillies, TNAU, Coimbatore 87-89.
35. Kannan L, Tamil Selvan C (1990) Effect of seaweed manures on *Vigna radiatus* L.(Fabaceae) In: *Perspective in phycology* (Prof. MOP Iyenger Centenary Celebration Volume) VN Today and Tomorrow's Printers and Publishers, New Delhi, India. 427- 430.
36. Challen SB, Hemingway JC (1965) Growth of higher plants in response to feeding with Seaweed extracts. Proc. 5th Indian Seaweed Symposium.
37. Taylor IEP, Wilkinson AJ (1977) The Occurrence of gibberellins - like substance in algae. Phycologia 16: 37- 42.
38. el-Sheekh MM, el-Saied A el-D (1999) Effect of seaweed extracts on seed germination, seedling growth and some metabolic processes of Fabae beans (*Vicia faba* L) Cytobios 101: 23-35.
39. Rajalakshmi MC, Sudha SPS, Sebasthiar S, Sudha N, Raju S (2008) Growth promoting effect of commercial Seaweed Liquid Fertilizer on *Solanum melongena* and *Mucuna pruriens*. Seaweed Research and Utilization 30: 183-187.
40. McKeen CD, Reilly CC, Pusey PL (1986) Production and Partial Characterization of Antifungal Substances Antagonistic to *Monilinia fructicola* from *Bacillus subtilis*. Phytopathology 76: 136-139.
41. Silo-suh LA, Lethbridge BJ, Raffel SJ, He H, Clardy J, et al. (1994) Biological activities of two fungistatic anti-biotics produced by *Bacillus cereus* UW85. Appl Environ Microbiol 60: 2023-2030.
42. Leifort C, Sigee DC, Epton HAS, Stanley R, Knight C (1992) Isolation of bacterial antagonistic to post harvest fungal disease of cold stored *Brassica* spp. Phytoparas 20:158-163.
43. Siddiqui ZA, Mahmood I (1999) Role of bacteria in the management of plant parasitic Nematodes: a review Bioresource Technology 69: 167-179.
44. Paulitz TC, Blanger RR (2001) Biological control in greenhouse systems. Annual Review of Phytopathology 39: 103-133.
45. McSpadden Gardener B, Weller DM (2001) Changes in populations of rhizosphere bacteria associated with take-all disease of wheat. Appl Environ Microbiol 67: 4414-4425.
46. Jacobson CB, Pasternak JJ, Glick BR (1994) Partial purification and characterization of 1-amino- cyclopropane-1-carboxylate deaminase from the plant growth promoting rhizobacterium Pseudomonas putida GR 12–2. Can J Microbiol 40: 1019–1025.
47. Sivasankari S, Chandrasekaran M, Kannathasan K, Venkatesalu V (2006) Studies in the biochemical constitiuents of *Vigna radiate* L. Treated with seaweed liquid fertilizer. Seaweed Resesrch and Utilization 28: 151-158.
48. Lingakumar K, Jayaprakash R, Manimuthu C, Haribaskar A (2002) *Gracillaria edulis* an effective alternative source as growth regulators for legumm crops. Seaweeds Research Utilization 24: 117-123.
49. Thirumal Thangam R, Maria Victorial Rani S, Peter marian M (2003) Effect of seaweed liquid fertilizers on the growth and biochemical constituents of *Cyamopsis tetragonaloba* (L) Taub. Seaweed Research Utilization 25: 99 -103.
50. Asir selin kumar R, Edwin James J, Saravana Babu S (2004) Comparative studies on the impact of seaweeds and seagrass liquid fertilizer on the chlorophyll content of *Zea mays*. Seaweeds Research Utilization 26: 167 - 170.
51. Ramavat B K, Dosh YA, Parekh RG, Chauhan VD (1986) Concentration of poly valent metals by seaweeds from Okha Coast. Phykos 25: 44-50.
52. Paul SD, Nongkynrih (1996) Enzyme activity - a diagnostic tool for toxicity/ deficiency of trace elements in rice plant. Indian Journal of Plant Physiology 1: 225- 257.
53. Blunden G, Wildgoose PB (1977) The effect of aqueous seaweed extract and kinetin on potato yields. Journal of Science Food Agriculture 28: 121- 125.
54. Bhosle NB, Dhargalkar VK, Untawale AG (1975) Effect of seaweed extracts on the growth of *Phaseolus vulgaris* L (*Fabaceae*) Indian Journal of Marine Science 4: 208-210.
55. Venkataraman Kumar V, Mohan VR, Murugeswari R, Muthuswamy M (1993) Seaweed Research Utilization 16: 23-27.
56. Whaphan CA, Blunden G, Jenkins T, Hankins SD (1993) Significant of betaines in the increased Chlorophyll content of plant treated with seaweed extract. Journal of Applied Phycology 5: 231-234.
57. Kannan L, Tamil Selvan C (1990) Effect of seaweed manures on *Vigna radiates* L. (Fabaceae) In: *Perspective in phycology* (Prof. M.O.P. Iyenger Centenary Celebration Volume) (ed.) Raja Rao, VN Today and Tomorrow's Printers and Publishers, New Delhi, India. 427-430.
58. Raj Kumar Immanuel S Subramanian SK (1999) Effect of fresh extracts and seaweed liquid fertilizers on some cereals and millets 21: 91-94.
59. Anantharaj M, V Venkatesalu (2001) Effect of seaweed liquid fertilizer on *Vigna catajung* seaweed Research and Utilization 23: 33-39.

Phosphate Solubilizing Fungi Isolated and Characterized from Teff Rhizosphere Soil Collected from North Showa and Gojam, Ethiopia

Birhanu Gizaw*, Zerihun Tsegay, Genene Tefera, Endegena Aynalem, Misganaw Wassie and Endeshaw Abatneh

Microbial Biodiversity Directorate, Ethiopian Biodiversity Institute,Addis Ababa, Ethiopia

Abstract

Phosphorus (P) is one of the major bio elements limiting agricultural production. Phosphate solubilizing fungi play a noteworthy role in increasing the bioavailability of soil phosphates for plants. The present study was aimed at isolating and characterizing phosphate solubilizing fungi from teff rhizosphere soil. Fungi were identified using lactophenol cotton blue staining confirmation and Biolog micro station. Fungi isolates were screened and transferred to Biolog universal yeast agar media. Pure yeast cells and filamentous fungi were suspended in sterile water and filamentous fungi (FF) inoculum fluid at 49 ± 2 and 75 ± 2 turbidity measured by Biolog turbidimeter respectively. 100 µ-L transferred from each suspension into 96 wells of the biolog yeast micro Plate and filamentous fungi microplate tagged with different carbon source and incubated at 26°C for 24 to 72 h and read by the micro station reader at a single wavelength of 590 nm, results were recorded and processed for identification by micro log3 software ver. 4.20.05. Biolog micro station produce 24 fungi read results. Filamentous fungi ≤ 0.5 similarity index (62.5%), yeast ≥ 0.5 similarity index (25%), yeast ≤ 0.5 similarity index (12.5%). The identified fungi were evaluated for phosphate solubilization by the pikovskaya's agar (PVK) selective media. Seven species were positive in phosphate solubilizing ability. *Trichosporon beigelii B, Phichia norvegensis, Cryptococcus albidus var aerius, Candida etchellsii, Cryptococcus albidus var albidus, Rhodotrula aurantiaca A, Rhodotorula aurantiaca B Cryptococcus luteolus, Cryptococcus albidus var diffluens, Neosartorya fisheri var. Fischeri, Cryptococcus terreus A, Candida montana, Penicillium purpurogenum var. Rubrisclerotium, Yeast isolate GTRWS18, GTS9B, GTS7C.* At 15 days' incubation, the solubilizing index ranges 1.2-5.3. The *Trichosporon beigelii B, Phichia norvegensis, Cryptococcus albidus var aerius* were superior in phosphate solubilization with 5.3, 3.35, 3.2 solubilizing index respectively. Therefore, these species can be candidated and exploited after further evaluation as bio fertilizers for teff productivity.

Keywords: Biolog microorganisms; Microstation; Phosphorus; Rhizosphere; Soil solubilization; Teff

Introduction

Phosphorus (P) is the second important nutrient after nitrogen that affects plant growth and metabolism processes making up about 0.2% of plant dry weight [1]. Phosphorus contributes remarkably to photosynthesis, energy and sugar production, nucleic acid synthesis, and promotes N_2 fixation in legumes [2]. In plants, phosphorous increases the strength of cereal straw, promotes flower formation and fruit production, stimulates root development and essential for seed formation [3]. It also plays a role in root development, stalk and stem strength, flower and seed formation, maturity and production, crop quality and resistance to plant diseases [4]. Mobility of phosphate ions in the soil is very low due to their high retention in soil. In 1986 Stevenson and Holford reported that the recovery rate of P fertilizer by plants is only about 10-30% [5]. The remaining 70-90% is accumulated in soil or in the form of immobile that is bound by Al or Fe in acid soils, or Ca and Mg in alkaline soils [6,7]. Phosphorus is highly insoluble and unavailable to plants. It must be converted into soluble form. Phosphate solubilizing microorganisms can play an important role in dissolving both of fertilizer phosphorus and bound phosphorus in the soil that is environmentally friendly and sustainable [8]. Several groups of microorganisms including fungi, bacteria and actinomycetes are known as efficient fixed P solubilizes [9]. Fungi are the important components of soil microbes typically constituting more of the soil biomass than bacteria, depending on soil depth and nutrient conditions. Fungi have been reported to have greater ability to solubilize insoluble phosphate than bacteria [10]. A wide range of soil fungi are reported to solubilize insoluble phosphorous such as *Aspergillus niger* and *Penicillium sp.* which are the most common fungi capable of phosphate solubilization [11]. Exploration of phosphate solubilizing microorganisms has been conducted by many researchers from soils, mangrove and rhizosphere [1,12-17] respectively. From such explorations, various types of phosphate solubilizing microorganisms have been successfully identified. In last few decades a large array of rhizosphere bacteria and fungi including species of *Penicillium, Pseudomonas, Azospirillum, Azotobacter, Klebsiella, Enterobacter, Alcaligenes, Arthrobacter, Burkholderia, Bacillus, Rhizobium* and *Serratia* have reported to enhance plant growth [18]. Many fungal species can solubilize rock phosphate, aluminum phosphate and tricalcium phosphate, such as *Aspergillus niger, Aspergillus tubingensis, Aspergillus fumigatus, Aspergillus terreus, Aspergillus awamor, Penicillium italicum, Penicillium radicum, Penicillium rugulosum, Fusarium oxysporum, Curvularia lunata, Humicola sp., Sclerotium rolfsii, Pythium sp., Aerothecium sp., Phoma sp., Cladosporium sp, Rhizoctonia sp., Rhizoctonia solani, Cunninghamella spp., Rhodotorula sp., Candida sp., Schwanniomyces occidentalis, Oideodendron sp., Pseudonymnoascus sp.* [11,19-22]. The soil yeasts *Candida tropicalis, Geotrichum candidum, Geotrichum capitatum, Rhodotorula minuta* and *Rhodotorula rubra* solubilized insoluble phosphate reported by Al falih.

***Corresponding author:** Birhanu Gizaw, Microbial Biodiversity Directorate, Ethiopian Biodiversity Institute, Addis Ababa, Ethiopia
E-mail: mgbirhanu@ibc.gov.et (or) gizachewbirhan@gmail.com

Yeast belonging to genus *Saccharomyces, Hansenula, Klockera, Rhodotorula* and *Debaryomyces spp.* were phosphate solubilizing yeast [23]. The principal mechanism for many soil fungi and bacteria can solubilize inorganic phosphate into soluble forms through the process of acidification, chelation, exchange reactions and production of organic acids by Han. Acid phosphatases play a major role in the mineralization of organic phosphorus in soil phosphate solubilization effect is mainly through the reaction between organic acids excreted from organic matters with phosphate binders such as Al, Fe, and Ca, or Mg to form stable organic chelates to free the bound phosphate ion [12,24]. Phosphorus deficiency is the most important problem of Ethiopian soil and more than 70-75% of highland soils are characterized by phosphorus deficiency. The deficiency is very severe in the acidic soils of the southern, southwestern and western regions. Areas Al^{3+} and Fe^{3+} are totally incriminated with phosphorus fixation [25]. Around 70% of Ethiopian vertisol have available phosphorus below 5 ppm, which is very low for supporting good plant growth and fixation in vertisols is related more to calcium, which is the predominant cation in all profiles than Al^{3+} and Fe^{3+} [26]. Vertisols are dark, montmorillonite-rich clay soils with characteristic shrinking and swelling properties. They have high clay content (>30% to at least 50 cm depth from the surface) and when dry they show cracks of at least 1 cm wide and 50 cm deep. They have high calcium and magnesium contents (FAO) [27]. Teff [*Eragrostistef* (Zucc.) Trotter] is the major indigenous cereal crop of Ethiopia, where it was originated and diversified. It is a highly demanded and a staple food grain for majority of the Ethiopian people. In a country of over 80 million people, teff accounts for about 15% of all calories consumed in Ethiopia [28]. The teff grain is ground to flour which is mainly used for making popular pancake-like local bread called injera and sometimes for making porridge. The grain is also used to make local alcoholic drinks, called tela and katikala. Teff straw, besides being the most appreciated feed for cattle [29]. Teff is the only cultivated of all 300 *Eragrostis spp.* Its agro ecological adaptability has resulted in its cultivation as an important crop in 10 of 18 agro ecological zones of the country. Teff adapted to a wide range of environments and cultivated under diverse agro-climatic conditions [29]. It can be grown in altitudes ranging from near sea level to 3000 ms, but the best performance occurs between 1100 and 2950 m.a.s.l [30]. Annual rainfall of 750-850 millimeter (mm), growing season rainfall of 450-550 mm and a temperature range of 10°C- 27°C. A very good result can also be obtained at an altitude range of 1700-2200 m and growing-season rainfall of 300 mm [31]. The crop performs well in both water logged vertisol in the highlands as well as water-stressed areas in the semi-arid regions throughout the country and consequently it is preferred over other grain crops such as maize or barley [32]. Teff production and productivity have been far below the potential. Currently the average national productivity is estimated to be less than 0.5 ton per ha. This is very low compared to other cereals such as wheat and sorghum grown in the region. Lower grain yield is mainly attributed to low soil fertility, especially nitrogen (N) and phosphorus (P) deficiencies and weed control practices [31]. Declining soil fertility because of continuous cropping without replenishing soil nutrients, continues application of phosphate fertilizer and soil erosions is the major factors that reducing production and productivity of the crop in Ethiopia. Higher grain yield of teff was recorded by applying inorganic fertilizers. However chemical fertilizers are neither easily available nor affordable for many poor Ethiopian farmers and not environmentally friendly. Such economic considerations necessitate for an alternative less expensive and environmentally friendly agricultural technologies to improve yield and quality of grain. Screening and characterization of phosphate solubilizing microorganisms are important for proper utilization of their beneficial effects to increase crop production and sustain agricultural productivity of the country without contaminating environments. In Ethiopia, only few studies on teff root-associated microorganisms have been undertaken. The effect of phosphate solubilizing fungus on growth and yield of teff was studied, inoculation of teff by vascular arbuscular mycorrhizal (VAM) and other phosphate solubilizing fungi that gives great results on teff yield improvement. So, these previous research works that tell us bio fertilizers made by plant growth promoting rhizobacteria (PGPR) are better indicative to improve teff production and productivity to the significant level. This study was aimed to isolate, identify and evaluating of phosphate solubilizing fungi from teff rhizosphere soil collected from North Showa and Gojam farm land and selecting superior solubilizing fungi that will be candidated for bio fertilizer after further evaluation for agricultural productivity.

Materials and Methods

Study area

The study was conducted in North Showa and Gojam in selected districts, particularly in Bichena, Bahirdar zuria, Huletejunaesae, Debrwork, Dejen. Tarmaber, kewot, Sidebirna wayu, Moretna giru, North showa the elevation ranges 1100-3009 meters above sea level. Geographic coordinates Latitude: 9°46'8.4", Longitude: 39°40'4.8". The zone is located approximately average 200 km far from Addis Ababa. Gojam Zone is bordered on the south by the Oromia Region, on the west by west Gojjam, on the north by south Gondar, and on the east by south Wollo; the bend of the Abay Riverdefines the Zone's northern, eastern and southern boundaries.10°31'44.7"N and 37°51'10.2"E. West Gojjam (Mirab Gojjam) is a Zone in the Amhara Region of Ethiopia. West Gojjam isborded by North Gondar, on the north by Lake Tana, and the Abay River which separates it from the South Gondar, and on the east by east Gojam. Coordinates: Latitude: 10.97379 North, Longitude: 37.46814 East. Gojam at Average altitude, 1788 m.a.s.l. (Figure 1).

Sample collection

One hundred fifty teff farmland site were selected based on five teff varieties, two soil types and 200 m difference within 1200-2200 m.a.s.l altitude from the districts of North Showa and Gojam. One hundred fifty rhizosphere soil and 150 teff root samples were collected through drillings at 5, 10, and 15 cm depth (Figure 2). Approximately 15 g of soil were taken from each depth of sampling point and a total of 45 g composite soil per sampling farmland were stored in sterile sample tube and icebox during April 08-28/2016 and transported microbial directorate laboratory in Ethiopian biodiversity institute to Addis Ababa and kept in +4°C until processed (Figure 2).

Screening and isolation of fungi from teff rhizospher soil

Soil samples were clustered according to altitude, soil type and teff varieties and merged into 50 composite samples. From each soil samples 1 g was taken and diluted in distilled water serially up to 10^{-6} mL. About 0.1 mL inoculum sample was transferred by swab through and streaked by nichrom loop on extract peptone dextrose agar media (YPDA), Rose Bengal agar, potato dextrose agar. Primary cultures were incubated for 26°C in digital incubator for 48 h. Isolates were subculture twice until pure colony obtained for morphological identification. A single yeast colony and pure filamentous fungi was streaked to Biolog universal yeast agar (BUY agar plate, (60 g / 1 L) and incubated for 48 h at 26°C for yeast and filamentous fungi micro plate (YT/FF Microplate) inoculum preparation. The yeast and filamentous fungi were identified

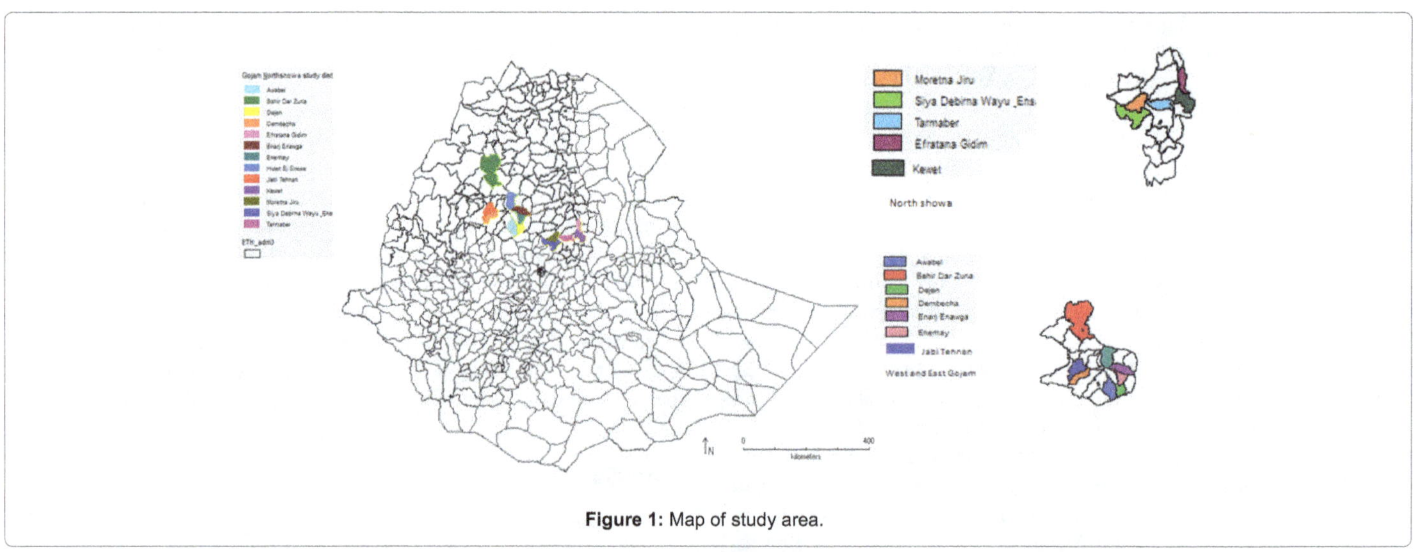

Figure 1: Map of study area.

Figure 2: Activities during teff rhizospher soil collection from Gojam and Northshowa.

according to the Biolog micro station reading and procedure.

Screening and isolation of fungi from teff rhizoplan soil

Fifty composite Teff root sample were washed with sterile distilled water and the root wash solution was kept in test tube. About 0.1 mL inoculum sample was transferred by swab and streaked by nichrom loop on yeast extract peptone dextrose agar media (YPDA), Rose Bengal agar, potato dextrose agar. Primary cultures were incubated for 26°C in digital incubator for 48 h. Isolates were subculture twice until pure colony obtained for morphological identification. A single yeast colony and pure filamentous fungi was streaked to Biolog universal yeast agar (BUY agar plate, (60 g/1 L) and incubated for 48 h at 26°C for yeast and filamentous fungi micro plate (YT/FF Microplate) inoculum preparation.

Teff root endophyte identification

Surface sterilization on teff root was carried out using 97% ethanol alcohol and 5% hypochlorate. Teff root were first washed by distilled water until all soil particle release and soaked in 97% ethanol alcohol for 10 minutes. Disinfected teff root again soaked in 5% solution Hypochlorite solution for 10 minutes then again washed by distilled water for six times. 1 cm root is cut and macerated by mortar and pestle until mix together. Each of the samples is transferred into 50 mL distilled water and mixed, shacked by vortex. About 0.1 mL inoculum sample was transferred by swab and streaked by nichrom loop on yeast extract peptone dextrose agar media (YPDA), Rose Bengal agar, potato dextrose agar. Primary cultures were incubated for 26°C in digital incubator for 48 h. Isolates were subculture twice until pure colony obtained for morphological identification. A single yeast colony and pure filamentous fungi was streaked to Biolog universal yeast agar (BUY agar plate, (60 g / 1 L) and incubated for 48 h at 26°C for yeast and filamentous fungi micro plate (YT/FF Microplate) inoculum preparation.

Colony morphology identification for phosphate solubilizing fungi

The colony morphology of the isolated fungi was examined after growth on yeast extract peptone dextrose agar media and Biolog universal yeast agar media at 26°C for 48 h and its colony morphology, form, size, elevation, margin/edge, colony color was observed using hand lens as well as its percentage frequency were recorded.

Identification of yeast from teff rhizosphere, rhizoplane soil and root endophyte using biolog micro station

Pure yeast isolates after grown on yeast extract potato dextrose agar were transferred to biolog universal growth agar and incubated at 26°C for 48 h. Pure colony of yeast suspension were prepared in 9 mL sterile distilled water and adjusted to 47 ± 2T using biolog turbidiameter.100 µL of inoculum was dispensed using digital pipettor to each of 96 wells of yeast microplate(YT) tagged with different carbon source and incubated at 26°C 24-72 h. The YT micro plate measures both metabolic reactions as well as turbidity growth to produce identifications. YT micro Plate is configured with both oxidation tests and assimilation tests. The first 3 rows of the panel (rows A - C) contain carbon source oxidation tests using tetrazolium violet as a colorimetric indicator of oxidation. The next five rows of the panel (rows D - H) contain carbon source assimilation tests. Results from these tests are scored turbid metrically. The last row of the panel (row H) has wells that contain 2 carbon sources. These wells test for the co-utilization of various carbon sources with D-xylose. YT micro plate was read by the micro station reader at 24 h, 48 h, and 72 h at a single wavelength of 590 nm. The Biolog software micro log3 ver. 4.20.05 compared the results obtained with the test strain to the database and provided identification based on distance value of match and separation score produces similarity index value and probability for species identification (Biolog) (Figure 3).

Identification of filamentous fungi from teff rhizosphere soil using biolog micro station

Filamentous fungi isolated and screened on rose bengal agar and potato dextrose agar were stained by lactophenol cotton blue in order to confirm to which genera the fungi is belonged to then pure filamentous fungi were transferred into Biolog universal growth agar media and incubated at 26°C for 48 h. Pure sporulate filamentous fungi suspension were prepared using 15 mL filamentous fungi inoculum fluid and adjusted to 75 ± 2T using Biolog turbidiameter. 100 µ-L of inoculum was dispensed using digital pipettor to each of 96 wells of filamentous fungi microplate (FF) tagged with different carbon source and incubated at 26°C, 24-240 h. After incubation, the FF micro plate measures both metabolic reactions as well as turbidity growth to produce identifications. Filamentous fungi micro plate (FF) was read by the micro station reader at 24 h, 48 h, and 72 h at a single wavelength of 590 nm. The Biolog software micro log3 ver. 4.20.05 compared the results obtained with the test strain to the database and provided identification based on distance value of match and separation score produces similarity index value and probability for species identification (Figure 3).

Identification of phosphate solubilizing microorganisms from teff rhizoshere soil

Fungal isolate identified by Biolog micro station were tested for their phosphate solubilizing ability. Pure fungi colonies were collected using a needle nose and spot at 4 quadrants on sterile solid Pikovskaya media having composition of 2.5 g $Ca_3(PO_4)$, 0.5 g $(NH_4)_2SO_4$, 0.2 NaCl, 0.1 g $MgSO_4.7H_2O$, 0.2 g KCl, 10 g glucose, 0.5 g of yeast extract, 20 g agar, small amounts of $MnSO_4$ and $FeSO_4$, and 1000 mL distilled water [33]. $Ca_3(PO_4)_2$ was used as a source of phosphate. Observations were made until the formation of a clear zone around the colonies of fungi that indicated the occurrence of phosphate dissolution. At 5 days' intervals solubilization index (SI) was measured using following formula. Fungi that formed the fastest clear areas with the greatest

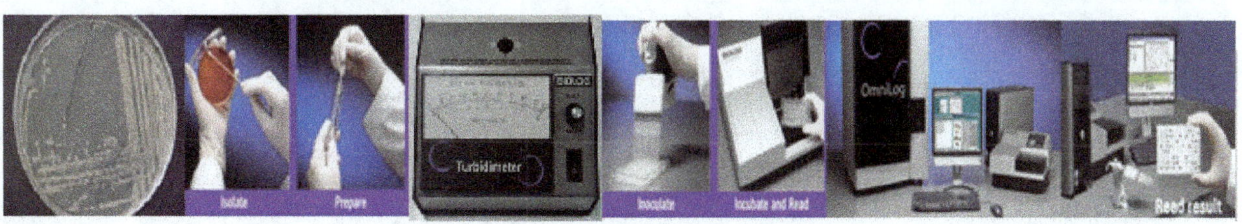

Figure 3: Biolog micro station identification steps.

diameter indicate the most superior phosphate solubilizing fungi.

SI=Colony diameter + Halozone diameter/Colony diameter

Statistical analysis

The data analysis involved various descriptive statistics such as means and percentages frequency. STATA ver.13 was used for phosphate solubilization index data analysis.

Results

Percentage frequency of fungal species isolated from teff rhizosphere soil

A total of 290 fungal colonies were grown and counted on different growth media and identified pure colonies having similar morphology were clustered in order to detect percentage frequencies of the microorganisms. Sixty one percent were filamentous fungi and 39% were non-filamentous fungi. From filamentous fungi, *Aspergillus* species were dominant (32%), *Penicillium* species (30%), *Fusarium* species (15%). *Trichoderma* (14%), *Colletotrichum* (9%). From yeast species *Cryptococcus* and *Rhodotrula* were dominant. The phosphate solubilizer fungi isolates were also identified based on their colony morphology that is pigmentation, shape, size, texture, elevation and margin, the following table will summarize (Table 1).

Identification of filamentous fungi species using lactophenol cotton blue staining (LPCB) and biolog micro station

Representative filamentous fungal isolates from clustered group were stained using lacto phenol cotton blue to confirm to which genera filamentous fungi belonged to and read by biolog micro station where equivalent to molecular method. The result revealed that 15 filamentous fungi species associated teff rhizosphere soil. Both lacto phenol cotton blue staining result and biolog microstation read showed that a filamentous fungi ≤ 0.5 similarity index (62.5%) *Colletotrichum lindemuthianum, Emericella quadrilineata, Fusarium melanochlorum, Aspergillus brevipes, Fusarium juruanum, Trichoderma piluiliferum, Fusarium avenaceum, Penicillium vulpinum, Neosartorya fischeri var. Fischeri, Fusarium udum, Hypocrea pseudokoningii, Trichoderma cittrinoviride, Trichoderma aureoviride, Penicillium purpurogenum var. rubrisclerotium, Penicillium pinophilum* (Table 2).

Identification of yeast species using biolog micro station

Biolog microstion read at 24, 48 and 72 YT microplate incubation result revealed that yeast ≥ 0.5 similarity index value identify *Rhodotorula aurantiaca A, Candida etchellsii, Kluyveromyces delphensis, Cryptococcus luteolus, Cryptococcus albidus var aerius, Cryptococcus terreus A, Rhodotorula aurantiaca, B, Phichia norvegensis, Zygoascus hellenicus, Candida Montana, Cryptococcus albidus var diffluens* (Table 3).

Phosphate solubilization test

A total of 36 funguses were read by biolog micro station from teff rhizospher soil and tentatively evaluated for their phosphate solubilization efficiency on Pikovskaya's agar selective media. Among all 19 isolates were positive for phosphate solubilization (Table 4), they produced the largest halos around their colony approximately 1.2-5.3 cm within 15 days of incubation. *Trichosporon beigelii B* showed superior solubilization index (PSI) 5.3, followed by 3.35, 3.2 *Phichia norvegensis, Cryptococcus albidus var aerius* respectively the smaller solubilization index recorded 1.2 by yeast isolate GTS7C (Table 4 and Figure 4).

Discussion

Phosphorus deficiencies are wide spread on soil throughout the world and one of the limiting factors for crop productivity. Phosphorus fertilizers represent major cost for agricultural production. Many bacteria, fungi and a few actinomycetes are potential solubilizes of bound phosphates in soil thus playing an important role making it available to plants in the soluble form [34-37]. Solubilization of insoluble phosphorus by microorganisms was reported by Pikovskaya [38]. During the last two decades' knowledge on phosphate solubilizing microorganisms increased significantly [39]. In this study, a total of 290 yeast isolates were screened from teff rhizosphere soil collected from North showa and Gojam, Ethiopia and 19 yeasts were read by Microstation. Sixteen yeast have got full species identification (ID) and 3 with no species ID (Table 2). All yeast species evaluated for their phosphate solubilization ability on Pikovskaya (PVK) selective media. Among all 16 yeast species and 3 isolates were positive for phosphate solubilization. *Phichia norvegensis, Cryptococcus albidus var albidus, Cryptococcus luteollus, Rhodotrula aurantiaca B, Cryptococcus albidus var diffluens, Candida etchellsii, Cryptococus terrus A, Cryptococcus albidus* var aerius, *Rhodotrula aurantiaca A, Trichosporon beigelii B, Cryptococcus luteolus, Rhodotrula aurantiaca A, Penicillium purpurogenum var. rubrisclerotium, Neosartorya fisheri var. fischeri, Candida montana, Zygo ascus hellenicus.* Yeast isolate, GTRWS1, GTS9B reported yeast belonging to genus *Saccharomyces, Hansenula, Klockera, Rhodotorula* and *Debaryomyces spp* were phosphate solubilizing yeast [23]. The soil yeasts *Candida tropicalis,*

P-solubilizing fungi	Shape	Elevation	Size	Marigin	Surface texture	Color
Trichosporon beigelii B	Irregular	Flat	Large	Lobate	Concentric	White yellow
Rhodotrula aurantiaca A	Rhound	Flat	Large	Undulate	Radiate	White
Penicillium purpurogenum Var.rubrisclerotium	Circular	Umbonate	Large	Filamentous	Radiate	Gray
Neosartoryafisheri var. fischeri	Circular	Umbonate	Large	Filamentous	Rugose	Olive green
Cryptococcus luteolus	Irregular	Flat	Lobate	Obate	Radiate	Yellow
Zygoascus hellenicus	Irregular	Flat	Lobate	Obate	Radiate	Yellow
Candid montana	Round	Flat	Large	Smooth	Radiate	White pink
Cryptococcus albidus var aerius	Irregular	Flat	Large	Lobate	Concentric	Yellow
Cryptococcus terreus A	Irregular	Flat	Large	Undulate	Smooth	White
Cryptococcus albidus var albidus	Entire	Pulvinate	Large	Entire	Radiate	White yellow
Rhodotorula aurantiaca B	Irregular	Raised	Large	Undulate	Concentric	White
Phichia norvegensis	Circular	Flat	Large	Entire	Concentric	White yellow
Candida etchellsii	Circular	Flat	Medium	Entire	Concentric	Yellow white
Cryptococcus albidus var diffluens	Entire	Flat	Large	Erose	Concentric	Yellow white

Table 1: Colony morphology for phosphate solubilizer fungi.

	Fungus species	LPCB Staining result	Probability	Similarity	Distance	Teff farm land districts
1	*Colletotrichum lindemuthianum (Saccardo & Mangus) Briosi*	+	-	0.001	32.96	Ejersa Qubete
2	*Emericella quadrilineata (Thom & Raper) C.R. Benjamin*	+	-	0.001	29.89	Kewot (Worentele)
3	*Fusarium melanochlorum (Caspary) Sacc.*	+	-	0.001	32.37	Efratana Gidm (Karalgoma)
4	*Aspergillus brevipes G. Sm.*	+	-	0.002	28.24	Tarma Ber (Asfachew)
5	*Fusarium juruanum*	+	-	0.000	48.72	Tarma Ber (Armania)
6	*Trichoderma piluliferum Webster & Rifai*	+	-	0.000	48.48	Efratana Gidm (Karalgoma)
7	*Fusarium avenaceum s.sp.nurragi Summerell & L.W. Burgess*	+	-	0.002	27.20	Kewot (Korebta)
8	*Penicillium vulpinum (Cooke & Massee) Seifert & Samson*	+	-	0.003	25.98	Tarma Ber (Chira Meda)
9	*Neosartorya fischeri var. Fischeri (Wehmer) Malloch & Cain*	+	-	0.001	32.18	Ejersa Qubete
10	*Fusarium udum E.Butler*	+	-	0.000	39.36	Tarma Ber (Chira Meda)
11	*Hypocrea pseudokoningii*	+	-	0.000	33.5	Tarma Ber(Chira Meda)
12	*Trichoderma cittrinoviride Bissett BGA*	+	-	0.000	46.16	Efratana Gidm(Karalgoma)
13	*Trichoderma aureoviride Rifai*	+	-	0.000	39.81	Efratana Gidm(Karalgoma)
14	*Penicillium purpurogenum var. Rubrisclerotium Thom*	+	-	0.004	24.46	Ejersa Qubete
15	*Penicillium pinophilum Hedge. BGB*	+	-	0.002	26.74	Mendida (Moyesilasie)

Table 2: Biolog micro station filamentous fungi identification result read.

	Fungus species	Probability	Similarity	Distance	Fungi isolated from North showa	Fungi isolated from Gojam
1	*Rhodotorula aurantiaca A*	100/100/100	0.604/0.533/0.584	6.1/7.77/6.48	Kewot (Abay Atir)/ Kewot (Worentele)	Bichena Gotera Kebele
2	*Candida etchellsii*	74/78	0.5450.658/	4.01/2.34	Deneba (Dacho)	Awabel, Enebi chifri
3	*Kluyveromyces delphensis*	77	0.533	4.75	Kewot (Worentele)	-
4	*Cryptococcus luteolus*	-/-	0.728/0.659	4.13/3.19	Tarmaber (ChiraMeda)	Dejen Zemetin Kebele
5	*Cryptococcus albidus var aerius*	-/100	0.226/0.6	7.22/14.34	Mendida (Moyeselase)	Jabitehnan
6	*Trichosporon beigelii B*	99	0.523	7.51	Mendida (Moyeselase)	-
7	*Cryptococcus albidus var aerius*	100	0.542	7.22	-	Hulet eju enese,Debre Gubae Kebele
8	*Cryptococcus terreus A*	99	0.605	6.03	-	Hulet eju enese,Debre Gubae Kebele
9	*Cryptococcus albidus var albidus*	93	0.598	5.49	-	Jehabitenan,Jiga Yelimdar
10	*Rhodotorula aurantiaca B*	86	0.588	4.86	-	Hulet eju enese,Debre Gubae Kebele
11	*Phichia norvegensis*	82	0.520	5.61	-	Hulet eju enese, Debre Gubae Kebele
12	*Zygoascus hellenicus*	-	0.216	14.1	Tarmaber (ChiraMeda)	
13	*Candida montana*	-	0.47	3	Kewot (Abay Atir)	
14	*Cryptococcus albidus var diffluens*	-	0.358	7.81		Hulet eju enese, Debre Gubae Kebele

Table 3: Biolog micro station yeast identification result read.

Geotrichum candidum, Geotrichum capitatum, Rhodotorula minuta and *Rhodotorula rubra* solubilized insoluble phosphate reported by Al falith. Woyessa and Assefa reported bacteria isolated from teff rhizosphere soil from agricultural fields of Alemgena and Bushoftu Ethiopia, isolates teff rhizosphere contains a diverse flora of microorganisms. The genera were *Pseudomonas, Chryseomonas, Burkholderia, Bacillus, Brevibacillus, Stenotrophomonas and Aeromonas.* These 4 species *Bacillus subtilis, Burkholderia cepacia, Pseudomonas fluorescens, Bacillus coagulans* were superior phosphate solubilizer bacteria. However many rhizospheric bacteria and fungi isolated from different crop rhizosphere soil, there is little information regarding teff rhizophere yeast and potential phosphate solubilizer yeast. This study will confirm that there are a diverse teff rhizosphere yeast and superior phosphate solubilizer isolated from North showa and Gojam teff farm land (Tables 2 and 3). The yeast species *Rhodotrula aurantiaca A* are phosphate solubilizer fungi species discovered in this study are also similar with the work of [19,40]. In this study phosphate solubilization index (PSI) were measured within 5 days' intervals for 15 days and they showed 1.2 -3.35 PSI clear zone diameter over colony diameter ratio (Table 3). Narsian reported yeast belonging to genus Saccharomyces *Hansenula, Klockera, Rhodotorula* and *Debaryomyces* exhibited highest SI (1.33-1.50). The study by Yasser et al. phosphate solubilization index recorded 1.05- 1.45. *A. japonicas* (SI=1.45), *A. niger* (SI=1.12), *Penicillium expansum* (SI=1.20), *Penicillium funiculosum* (SI=1.40), *Penicillium variable* (SI=1.13), *Penicillium purpuragenum*(SI=1.30). In this study the largest solubilization index recorded by *Phichia norvegensis* (SI. 3.35), *Cryptococcus albidus var aerius* (SI. 3.2), *Candida etchellsii* (SI 2.9). The smallest solubilization index recorded by *Cryptococcus terrus A* (PSI, 1.72) (Figure 3 and Table 3). According to De freitas et al., good phosphate solubilizers produce halos around their colonies with diameters higher than 1.5 cm. Most efficient phosphate solubilizer on Pikovskaya's agar plates with PSI=3.29. Whereas among fungi *P. canescens* showed highest solubilizing index [10]. Phosphate solubilization index (PSI) values up to 2.4 have been recorded for *Aspergillus niger*, with values of 3.1 for *Penicillium italicum* and 3.0 for *Paecilomyces lilacinus* [40-42]. Fungal strains isolated from sugarcane and sugar beet rhizosphere showed SI in range of 1.13 to 1.59 reported PSI of the fungal strains

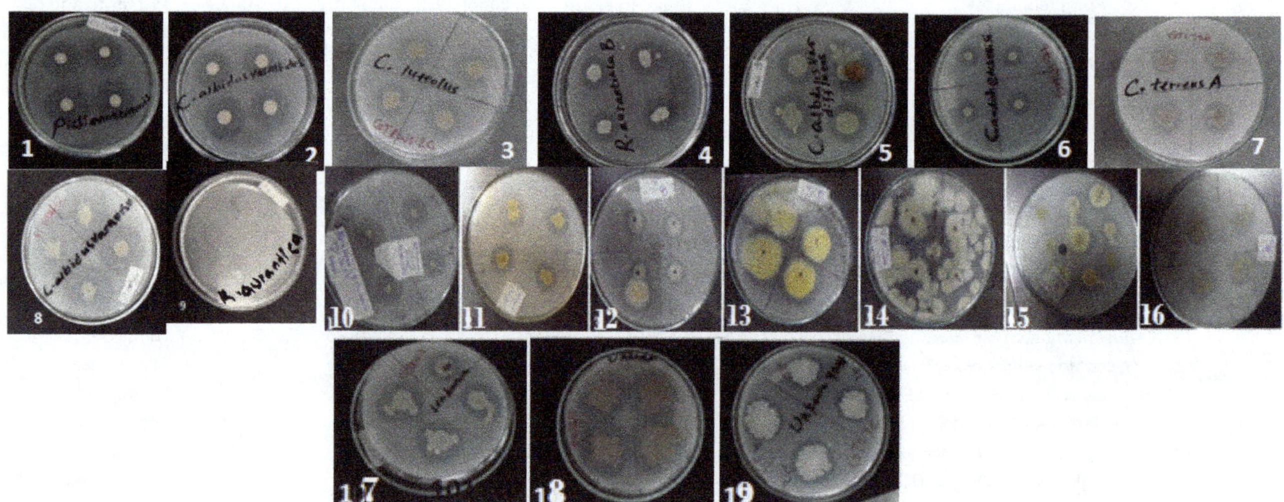

Figure 4: Phosphate solubilizing yeast on Pikovskaya's agar media. (1) *Phichia norvegensis* (2) *Cryptococcus albidus var albidus* (3) *Cryptococcus luteollus* (4) *Rhodotrula aurantiaca B* (5) *Cryptococcus albidus var diffluens* (6) *Candida etchellsii* (7) *Cryptococus terrus A* (8) *Cryptococcus albidus* var aerius (9) *Rhodotrula aurantiaca A* (10) *Trichosporon beigelii B,* (11) *Cryptococcus luteolus,* (12) *Rhodotrula aurantiaca A* (13) *Penicillium purpurogenum var.rubrisclerotium,* (14) *Neosartorya fisheri var. fischeri,* (15) *Candida montana,* (16) *Zygo ascus hellenicus* (17) Yeast isolate GTRWS18 (18) Yeast isolate GTS9B (19) Yeast isolate GTS7C.

	Fungus species isolated from teff rhizosphere soil	Phosphate solubilization index (PSI)		
		5th day	10th day	15th day
	Trichosporon beigelii B	3.8	4.3	5.3
1	*Phichia norvegensis*	2.51	3	3.35
2	*Cryptococcus albidus var aerius*	1.32	2.51	3.2
3	*Candida etchellsii*	1.76	2.54	2.90
4	*Cryptococcus albidus var albidus*	2.49	2.57	2.9
5	*Rhodotrula aurantiacaA*	1.16	1.8	2.6
6	*Rhodotorula aurantiaca B*	1.55	2	2.24
7	*Cryptococcus luteolus*	1.80	1.82	2.22
8	*Cryptococcus albidus var diffluens*	1.54	1.67	1.9
9	*Neosartorya fisheri var. fischeri*	1.53		1.88
10	*Cryptococcus terreus A*	1.34	1.44	1.72
11	*Candida montana*	1.2	1.3	1.6
12	*Penicillium purpurogenum var.rubrisclerotium thom*	1.318	1.4	1.5
13	*Yeast isolatGTRWS18*	1.4	1.5	1.54
14	*Yeast isolatGTS9B*	1.2	1.33	1.49
15	*Yeast isolatGTS7C*	0.9	1.1	1.2

Table 4: Phosphate solubilization index (PSI).

isolated from maize rhizosphere that ranged from 1.53 to 1.80. In this study new phosphate solubilizer yeast *Trichosporon beigelii B, Phichia norvegensis, Cryptococcus albidus var aerius* identified from teff rizhosphere soil with superior solubilization index (PSI) 5.3, 3.35 and 3.2 respectively in 15 days incubation. Therefore, these strains can be candidated and exploited as bio fertilizers through further evaluation and optimization test to increase agricultural productivity of teff crop.

Conclusion

Twenty nine fungi isolated from teff rhizosphere soil using lactophenol cotton blue staining and biolog microstation identification system where equivalent to molecular techniques and the dominant species were filamentous fungi. Seven fungi species *Trichosporon beigelii B, Phichia norvegensis, Cryptococcus albidus var aerius, Candida etchellsii, Cryptococcus albidus var albidus, Rhodotrula aurantiacaA, Rhodotorula aurantiaca B, Cryptococcus luteolus, Cryptococcus albidus var diffluens, Neosartorya fisheri var. Fischeri, Cryptococcus terreusA, Candida montana, Penicillium purpurogenum var. Rubrisclerotium, yeast isolate GTRWS18, GTS9B, GTS7C* were positive for phosphate solubilization efficiency. *Trichosporon beigelii B* was the superior among the isolated fungi in solubilizing index 5.3 followed by *Rhodotrula aurantiaca A* 2.9 and good candidate after further evaluation on *invitro* test, green house and field trials as bio fertilizer. The rise in the cost of chemical fertilizer, the lack of fertilizer industries in developing countries and the growing environmental issue and biodiversity loss using chemical fertilizer timely important concern using alternative ecofriendly bio fertilizer to increase yield and productivity of teff crop.

Recommendation

The beneficial effects of plant growth promoting microorganisms(PGPM) have not been exloited well. In the past, some microbial inoculants prepared from *Rhizobium* for leguminous crops, *Azotobacter* and *Azospirillium* for cereal crops and *Frankia* for tree crops have been used as nitrogen providers in many developed and developing countries. However enormous interest increase in research in recent years in PGPM such as nitrogen fixer, phosphate solubilizer, pathogen suppressor. There is no well-organized microbial inoculant industry for bio fertilizer production especially for phosphate solubilizer and there is no link with researcher working on microbial bio fertilizer in Ethiopia, therefore agricultural research institute, microbiologist, soil scientist agronomist, and stockholders in general must work together in depth on structural and functional diversity of PGPM and selecting superior biofertilizer, biopesticide, biostimulant to increase crop yield and prductivity. Further research should be continued with selecting efficient phosphate solubilizer microorganism (PSM) isolates. These may be used for inoculum production and their inoculation effect on the plant growth must be studied *in vitro*, green house and field trials.

Acknowledgements

It gives me a great pleasure to acknowledge Dr. Genene Tefera for his unreserved guidance and encouragement and support in providing and facilitating the necessary equipment. And extremely greatful to acknowledge North show zonal and district agriculture office and leader of all area in the research area who helped me guiding the study site and finally goes to Ethiopian biodiversity institute, microbial directorate for every budget grant to carry out this study and its research team for their un reserved support at laboratory and field work especially Misganaw Wasie,, Endeshaw Abatneh, Endegena Aynalem, as part of research group in tireless effort in teff rhizospher soil sample collection. Lastly, I acknowledged Woyenshet Lule for her kindly support especially laboratory chemicals facilitation.

References

1. Widawati S, Suliasih D (2006) Augmentation of potential phosphate solubilizing bacteria (PSB) stimulate growth of green mustard (Brasica caventis Oed) in marginal soil. J Biodiversity 7: 10-14.
2. Saber KL, Nahla AD, Chedly A (2005) Effect of P on nodule formation and N fixation in bean. Agron Sustain Dev 25: 389-393.
3. Sharma S, Vijay K, Tripathi RB (2011) Isolation of Phosphate Solubilizing Microorganism (PSMs) from soil. J Microbiololgy Biotechnology Research 1: 90-95.
4. Richardson AE (2007) Making microorganisms mobilize soil phosphorus. In: Velazquez E, Rodriguez-Barrueco C (eds.), First International Meeting on Microbial Phosphate Solubilization, pp: 85-90.
5. Stevenson FJ (1986) Cycles of soil carbon, nitrogen, phosphorus, sulphur and micronutrients. Wiley, New York, USA, p: 201.
6. Prochnow LI, Fernando J, Quispe S, Artur E, Francisco B, et al. (2006) Effectiveness of phosphate fertilizers of different water solubility's in relation to soil phosphorus adsorption 65: 1333-1340.
7. Yang M, Ding G, Shi L, Feng J, Xu F, et al. (2010) Quantitative trait loci for root morphology in response to low phosphorus stress in Brassica napus. Theor Appl Genet 121: 181-193.
8. Khan MS, Zaidi A, Wani PA (2007) Role of phosphate-solubilizing microorganisms in sustainable agriculture A review. Agronomy and Sustainable Development 27: 29-43.
9. Sundara B, Natarajan V, Hari K (2002) Influence of phosphorus solubilizing bacteria on the changes in soil available phosphorus and sugar cane and sugar yields. Field Crops Research 77: 43-49.
10. Nahas E (1996) Factors determining rock phosphate solubilization by microorganisms isolated from soil. World J Microbiol Biotech 12: 567-572.
11. Whitelaw MA, Harden JT, Helyar RT(1999) Phosphate solubilization in solution culture by the soil fungus Penicillum radicum. Soil Biol Biochemi 31: 655-665.
12. Gupta M, Kiran S, Gulati A, Singh B, Tewari R (2012) Isolation and identification of phosphates olubilizing bacteria able to enhance the growth andaloin-A biosynthesis of Aloe barbadensis miller. Microbiological Research 167: 358-363.
13. Vazquez P, Holguin G, Puente ME, Lopez CA, Bashan Y (2000) Phosphate solubilizing microorganisms associated with the rhizosphere of mangroves in a semiarid coastal lagoon. Biology and Fertility of Soils 30: 460-468.
14. Holguin G, Vazquez P, Bashan Y (2001) The role of sediment microorganisms in the Productivity, conservation and rehabilitation of mangrove ecosystems: an overview. Biol Fertil Soils 33: 265-278.
15. Chung H, Park M, Madhaiyan M, Seshadri S, Song J, et al. (2005) Isolation and characterization of phosphate solubilizing bacteria from the rhizosphere of crop plants of Korea. Soil Biol and Biochem 37: 1970-1974.
16. Poonguzhali S, Madhaiyan M, Sa T (2008) Isolation and identification of phosphate solubilizing bacteria from Chinese cabbage and their effect on growth and phosphorus utilization of plants. J Microbiol Biotechnol 18: 773-777.
17. Oliveira CA, Alves VMC, Marriel IE, Gomes EA, Scotti MR, et al. (2009) Phosphate solubilizing microorganisms isolated from rhizosphere of maize cultivated in an oxisol of the Brazilian Cerrado Biome. Soil Biol and Biochem 41: 1782-1787.
18. Kloepper JW, Lifshitz R, Zablotowicz RM (1989) Free-living bacterial inoculum for enhancing crop productivity. Trends in Biotechnol 7: 39-43.
19. Isbelia RL, Bernier RR, Simard P, Tanguay G, Antoun H (1999) Characteristics of phosphate solubilization by an isolate of a tropical Penicillium rugulosum and two UVinduced mutants. FEMS Microbiol Ecol 28: 291-295.
20. Sparks LD (1999) Advances in Agronomy: V 69. Academic Press, p: 12.
21. Didiek GS, Sugiarto Y (2000) Bioactivation of poorly soluble phosphate rocks with a phosphorus solubilizing fungus. Soil Sci Soc Am J 64: 927-932.
22. Helen JBP, Graeme K, Ritz D, Fordyce A, Geoffrey GM (2002) Solubilization of calcium phosphate as a consequence of carbon translocation by *Rhizoctonia solani*. FEMS Microbiol Ecol 40: 65-71.
23. Varsha N, Ahmed Abu Samaha SM, Patel HH (2010) Rock phosphate dissolution by specific yeast. Indian J Microbiol 50: 57-62.
24. Arcand MM, Schneider KD (2006) Plant-and microbial-based mechanisms to improve the agronomic effectiveness of phosphate rock: a review. Anais da Academia Brasileira de Ciencias 78: 791-807.
25. Sertsu S, Ali A (1983) Phosphorus sorption characteristics of some Ethiopian soil. Soil Eth Agric Sci 34: 28-407.
26. Mamo T, Haque I, Kamara CS (1988) Phosphorus status of some Ethiopian high land vertisoles in sub-saharan Africa. J E S 26: 232-252.
27. Food and Agricultural Organization (1988) Master Land Use Plan, Ethiopia Range/Livestock Consultancy Report prepared for the Government of the People's Democratic Republic of Ethiopia. Technical Report, AG/ETH/82/010, Food and Agricultural Organization (FAO), Rome, p: 94.
28. Bekabil F, Befekadu B, Rupert S, Tareke B (2011) Strengthening the Teff Value Chain in Ethiopia. Ethiopian Agricultural Transformation Agency.
29. Ketema S (1997) Teff [Eragrostis tef (Zucc.) Trotter] Promoting the conservation and use of underutilized and neglected crops. International plant genetics resources institute, Rome, Italy.
30. Hailu T, Seyfu K (2000) Production and importance of tef in Ethiopia Agriculture. In: Tefera H, Getachew B, Mark S (eds.), Narrowing the Rift: Tef research and Development Proceedings of the international Tef Genetics and improvement, pp: 16-19.
31. Ketema S (1993) Phenotypic variations in tef (Eragrostis tef) germplasm morphological and agronomic traits. A catalon Technical Manual No. 6. Institute of Agricultural Research. Addis Ababa, Ethiopia.
32. Zeleke G (2000) Landscape dynamics and soil erosion process modeling in the north-western Ethiopian highlands, African Studies Series A 16, Geographica Bernensia, Berne, Switzerland.
33. Rao NSS (1982) Phosphate solubilization by soil microorganisms. In: Rao NS (ed.), Advanced in Agricultural Microbiology, New Delhi. Oxford and IBH Publishing Co.
34. Halder AK, Mishra AK, Chakarbarthy PK (1991) Solubilization of inorganic phosphate by Bradyrhizobium. Ind J Exp Biol 29: 28-31.
35. Abd-alla MH (1994) Phosphatases and the utilization of organic phosphorus by Rhizobium leguminosarum biovar viceae. Lett Appl Microbiol 18: 294-296.
36. Whitelaw MA (2000) Growth promotion of plants inoculated with phosphate solubilizing fungi. Adv Agron 69: 99-151.
37. Goldstein AH (1986) Bacterial solubilization of mineral phosphates: historical perspectives and future prospects. Am J Altern Agric 1: 51-57.

38. Pikovskaya RI (1948) Mobilization of phosphorus in soil connection with the vital activity of some microbial species. Microbiologiya 17: 362-367.

39. Richardson AE (2007) Making microorganisms mobilize soil phosphorus. In: Velazquez E, Rodriguez-Barrueco C (eds.), First International Meeting on Microbial Phosphate Solubilization. pp: 85-90.

40. Manal MY, Ahmad SM, Osama NM, Siada HN (2014) Solubilization of inorganic phosphate by phosphate solubilizing fungi isolated from Egyptian soils. J Biol Earth Sci 4: 83-90.

41. El-Azouni IM (2008) Effect of phosphate solubilizing fungi on growth and nutrient uptake of soybean (*Glycine max L*) plants. J Appl Sci Res 4: 592-598.

42. Hernandez-Leal T, Carrion G, Heredia G (2011) Solubilizacion in vitro de fosfatos por una cepa de Paecilomyces lilacinus (Thom) Samsom. Agrociencia 45: 881-892.

27

Toward the Quantification of Confrontation (Dual Culture) Test: A Case Study on the Biological control of *Pythium aphanidermatum* with *Trichoderma asperelloides*

Babak S Pakdaman[1]*, Ebrahim Mohammadi Goltapeh[2]*, Bahram Mohammad Soltani[3], Ali Asghar Talebi[4], Mohsen Nadepoor[2], Joanna S Kruszewska[5], Sebastian Piłsyk[5], Sabrina Sarrocco[6] and Giovanni Vannacci[6]

[1]*Department of Phytomedicine, RaminAgricultural and Natural Resources University, Ahwaz, Iran*
[2]*Department of Plant Pathology, Agricultural Faculty, Tarbiat Modares University, Tehran, Iran*
[3]*Department of Biological Sciences, Department of Genetics,Tarbiat Modares University, Tehran, Iran*
[4]*Department ofAgricultural Entomology, Agricultural Faculty, Tarbiat Modares University, Tehran, Iran*
[5]*Laboratory of Fungal Glycobiology, Division of Genetics, Institute of Biochemistry and Biophysics, Warsaw, Poland*
[6]*Department ofTree Science, Entomology and Plant Pathology "Giovanni Scaramuzzi", Faculty ofAgriculture, University of Pisa, Pisa, Italy*

Abstract

Trichoderma species are well-appreciated filamentous fungi applied in agriculture for biological control and biofertilization purposes. One of the primary steps to screen for potent biological control isolates and/to study the effect of gene(s) on the biological control of phytopathogenic fungi and oomycetes is to perform confrontation or dual culture tests. Despite the comprehensiveness of the test, it still suffers from the lack of a reliable methodology for the mathematical data collection and statistical analyses. Here, different aspects of data collection are critically studied and new parameters are introduced. With this method the statistical comparison of different fungal biological control isolates and their counterpart phytopathogenic fungi as well as oomycetes becomes feasible. In the mean time, with this new approach, it becomes possible to statistically analyze the effect of different factors included in the medium of interaction. And with those interested in genetic studies on this type of interactions, the results of this study indicate the probable feasibility of statistical analysis of the possible impact of genetic manipulation and transformation of fungal biological control agents on the biological control of phytopathogenic fungi and oomycetes, and vice versa.

Keywords: Confrontation test; Dual culture; Control; *Trichoderma*; *Pythium*

Introduction

With the increase of human knowledge of and experience with the hazardous impacts of agricultural pesticides on the environment and human health, integrated pest management has changed to a choice of more attraction [1]. Biological control is a non-ignorable part of such a management system in modern agriculture [2]. Fungal biological control agents are among the most successful tools applied in the combat against plant pathogenic fungi [3-5] and *Trichoderma*-based fungicides are regarded as the most commonly used fungicides [6]. Some strains have the ability to reduce the severity of plant diseases by inhibiting plant pathogens, mainly in the soil or on plant roots, through their high antagonistic and mycoparasitic potential [7]. Effective *Trichoderma* isolates can play important roles in sustainable agriculture. These isolates would be able to (i) control phytopathogenic fungi [8], (ii) control plant nematodes [9-12], (iii) increase plant growth, development, and yield [13-22], (iv) induce plant systemic resistance of induced systemic resistance [22,23] acquired systemic resistance [24] against plant pests and diseases (v) increase plant resistance to abiological stresses [25,26], (vi) remediate polluted agricultural soils [7,27,28], (vii) improve soil environment [29-32], (viii) potentially control insect pests [33,34] and weeds [35-37].

Interestingly there are some data indicating the higher chance for biological control as *Trichoderma* isolates can be applied some while after given herbicides that are also of their effects against important oomycetous, ascomycetous, as well as basidiomycetous pathogens, although to this end field studies are required [38]. Such positive effects if also obtained under field conditions would be of peculiar practicality in integrated pest management systems as the dead bodies of the chemically controlled weeds could be inoculated and colonized by *Trichoderma* biological control isolates prior to the re-establishment of fungal pathogens and/re-start of the bioactivities of the herbicide-imposed suppressed fungal pathogens. It is clear that *Trichoderma* fungi can effectively colonize dead plant materials thanks to their strong potential of the secretion of cellulases and other glucanases while some humidity is available during routine practice of soil preparation before cultivation. In the meantime and as implied in (*vi*), *Trichoderma* isolates as mycoremediants are able to remediate herbicide-treated soils-an effect of high importance if non-selective soil herbicides are applied. The colonized weed plant bodies would help *Trichoderma* isolates establishment and survival in soil [39,40]. Considering the beneficial roles *Trichoderma* isolates can play in eco-friendly agricultural systems, it would be very economic to search for the effective domestic isolates and use them in integrated pest management programs. After the isolation of numerous numbers of such isolates, their effectiveness in

***Corresponding author:** Babak Pakdaman, Department of Phytomedicine, Ramin Agricultural and Natural Resources University, Ahwaz, Iran,
E-mail: bpakdaman@yahoo.com

Ebrahim Mohammadi Goltapeh, Department of Plant Pathology, Agricultural Faculty, Tarbiat Modares University, Tehran, Iran, E-mail: emgoltapeh@modares.ac.ir

biological control programs has to be tested in the first screening step to reduce the number in a way biased for more antagonistic ones. Also with the attempts to generate genetic transformants more effective in biological control of phytopathogenic fungi (for scientific purposes or practical uses wherever it is legally allowed), or to understand the effect/role of the given genes on/in the process of biological control, the necessity of a quantitative method is increasingly felt. One of the conventionally applied screening tests is the confrontation test (also known as dual culture test) first introduced by Weindling [41]. The test is a comprehensive experiment that exhibits overall antagonistic potential of a fungal biological control agent, and can be applied after preliminary fast screening tests. The important constriction in the use of dual culture test lies in the collection of the experimental data. If pathogen growth inhibition is considered in the time of a pathogen full growth in check cultures, the full antagonistic potential of the fungal biological control agent will not probably be understood because full overgrowth of the biological control agent will often take a time period more than that required for the full growth of pathogen in check cultures. On the other hand, the growth inhibition percentage will be dependent on the plate diameter. If pathogen growth inhibition is to be considered in the time of confrontational contact between two opposed colonies [42], there will be a drawback with most of pathogens which their growth are temporarily promoted in the presence of fungal biological control agents. Another disadvantage of the confrontation test is the ignorance of time, so the speed of biological control activity is missed in pathogen growth inhibition percentage calculations. Vice versa, some researchers have only considered time period for full growth of a fungal biological control agent over a pathogen forgetting pathogen growth inhibition in the biological control. Some authors emphasize on the use of the temporal period that two colonies need in order to come in contact in dual culture test as well as pathogen growth inhibition percentage as applied in the conventional dual culture method. However, the time period required for coming in contact will not include the period of overgrowth on a pathogen colony. Here we attempt to develop a new methodology for data collection and analysis. Also, we will use the methodology for a comparative study on the biological control potentials of two *T. asperelloides* isolates against *P. aphanidermatum* at two different temperatures.

Materials and Methods

Fungal isolates

Two *Trichoderma* isolates T13 and T92 were isolated from the rhizosphere soil of tarragon (*Artemisia dracunculus* L.) in Hokmabad, Tabriz, Iran. These were selected out from more than 100 *Trichoderma* isolates based on the results obtained with some preliminary studies including different *in vitro* tests performed to evaluate their *in vitro* activities against various plant pathogens (fungi such as *Fusarium* spp., *Macrophomina phaseolina*, *Rhizoctonia solani*, and the oomycetous plant pathogen *Pythium aphanidermatum*), and to assess their abilities to grow on glycerol and mannitol, at different temperatures, and to grow on a layer of a long chain alkane. The specification of both isolates as *T. asperelloides* was performed at Vienna University of Technology, Vienna, Austria. An isolate of *Pythium aphanidermatum* was obtained from the collection of fungi available at the laboratory of Plant Pathology, Agricultural Faculty, Tarbiat Modares University, Tehran, Iran.

Confrontation tests to study the effect of temperature on the biological control

Malt extract agar plates in disposable plastic Petri dishes were used for dual inoculations with 5mm discs from 4 day old cultures grown at 25°C under dark conditions. The discs were put in a 5mm distance from the edge, opposed and on the same diagonal line. Three plates were considered for each fungus-fungus interaction so that the biological control potential of each *Trichoderma* isolate against *Pythium aphanidermatum*, an oomycetous soil born plant pathogen was studied. Two sets of each interaction were synchronously prepared and incubated in two incubators adjusted to 15°C and 25°C under dark conditions. Four parameters were determined per interaction: (*i*) days of the period after inoculation till two colonies came in contact (C); (*ii*) days of the period after inoculation till the fungal biological control agent fully grew over a pathogen colony (Z); (*iii*) days of the period after inoculation till the fungal biological control agent fully grew on the plate (M); and finally (*iv*) the radial distance (in mm) of pathogen colony growth between the edge of the inoculation disc and the marginal point of the colony located on the presumed diagonal line connecting centers of two discs in the plate (P). A parameter for pathogen resistance (R) to the biological control agent was defined based on the periods required for the full growth of a fungal biological control agent in the presence (Z) and absence (M) of pathogen. R was defined as the ratio (without any dimension) below:

$$R = \frac{Z(\text{in day})}{M(\text{in day})}$$

As the success of biological control decreases with the increase of pathogen resistance to the fungal biological control fungus and the increase of pathogen colony growth, therefore a new index was defined combining temporal parameters and pathogen growth parameter. Pakdaman's biological control index (PBCI) was calculated following the equation below. The dimension of the index was determined as L^{-1} with the unit of mm^{-1}.

$$\text{PBCI}(\text{in mm}^{-1}) = \frac{M(\text{in day})}{Z(\text{in day})\cdot P(\text{in mm})}$$

Statistical analysis of data

Data analysis was performed based on the experimental design of complete random blocks taking advantage of SAS 9.1.3 portable for Windows, SAS institute Inc. The means of parameters as well as indices were compared based on Tukey's studentized range (HSD) tests (α=0.05). Also, the correlation among parameters was checked based on Pearson correlation coefficients using the above-mentioned software.

Results

Analysis of data indicated that there were highly significant differences in the time period required for contact between pathogen and fungal biological control colonies (C) in the treatments (P<0.0001).

Comparison of mean C values through Tukey's studentized range (HSD) test (α=0.05) indicated that both *T. asperelloides* isolates had statistically equal speed in growth toward *P. aphanidermatum* colonies, however, reduction of temperature from 25°C to 15°C caused the increase of C values from 2 to 4 days (Table 1).

Similarly the differences in Z values of the treatments were also found very meaningful (P<0.0001). While two *T. asperelloides* isolates could grow over *P. aphanidermatum* colonies after an equal length of time period at 25°C (4 days), T92 could dominate the oomycetous pathogen sooner than T13 at 15°C (Table 2). Furthermore, while both biological control isolates had equal C values at 15°C, they had significantly different Z values under the same conditions.

Treatment	Mean C value	Tukey grouping
15°C Pa-T1	4	A
15°C Pa-T92	4	A
25°C Pa-T1	2	B
25°C Pa-T92	2	B

Table 1: Tukey's studentized range test-based comparison (α=0.05) of average C values (in day) resulted from two different *Trichoderma asperelloides* isolates (T13 and T92) interactions with *Pythium aphanidermatum* (Pa) at two temperatures.

Treatment	Mean Z value	Tukey grouping
15°C Pa-T13	12	A
15°C Pa-T92	11	B
25°C Pa-T13	4	C
25°C Pa-T92	4	C

Table 2: Tukey's studentized range test-based comparison (α=0.05) of average Z values (in day) resulted from two different *Trichoderma asperelloides* isolates (T13 and T92) interactions with *Pythium aphanidermatum* (Pa) at two temperatures.

Treatment	Mean M value	Tukey grouping
15°C Pa-T13	12	A
15°C Pa-T92	12	A
25°C Pa-T13	4	B
25°C Pa-T92	4	B

Table 3: Tukey's studentized range test-based comparison (α=0.05) of average M values (in day) resulted from two different *Trichoderma asperelloides* isolates (T13 and T92) interactions with *Pythium aphanidermatum* (Pa) at two temperatures.

Treatment	Mean P value	Tukey grouping
25°C Pa-T13	47.6667	A
25°C Pa-T92	46.3333	AB
15°C Pa-T13	45.3333	AB
15°C Pa-T92	44.3333	B

Table 4: Tukey's studentized range test-based comparison (α=0.05) of average P values (in mm) resulted from two different *Trichoderma asperelloides* isolates (T13 and T92) interactions with *Pythium aphanidermatum* (Pa) at two temperatures.

Treatment	Mean R value	Tukey grouping
15°C Pa-T13	1.000	A
25°C Pa-T92	1.000	A
25°C Pa-T13	1.000	A
15°C Pa-T92	0.917	B

Table 5: Tukey's studentized range test-based comparison (α=0.05) of average R values (without unit) resulted from two different *Trichoderma asperelloides* isolates (T13 and T92) interactions with *Pythium aphanidermatum* (Pa) at two temperatures.

Treatment	Mean PBCI value	Tukey grouping
15°C Pa-T92	0.0235333	A
25°C Pa-T92	0.0223667	B
15°C Pa-T13	0.0220333	BC
25°C Pa-T13	0.0210000	C

Table 6: Tukey's studentized range test-based comparison (α=0.05) of average PBCI values (in mm-1) resulted from two different *Trichoderma asperelloides* isolates (T13 and T92) interactions with *Pythium aphanidermatum* (Pa) at two temperatures.

Other temporal parameter (M) was also found to be of very notably different values in different temperature-*Trichoderma* isolate combinations (P<0.0001). The effect of temperature on the fungal biological control agent full growth in check cultures was very meaningful (P<0.0001). Comparison of average M values showed both isolates of *T. asperelloides* were of equal M values at a given temperature and under equal conditions of the experiment but with the fall of temperature from 25°C down to 15°C, the average M value raised three times (Table 3).

Notable differences were also found in the radial growth of *P. aphanidermatum* in the presence of *T. asperelloides* isolates T13 and T92 (P=0.0181), where two temperatures did not have any meaningful impact on pathogen growth in the interaction with biological control isolates which were of highly significant impact on the growth of *P. aphanidermatum* (P=0.0090). *P. aphanidermatum* grew more in the presence of biological control isolates at 25°C compared to 15°C (Table 4).

With the parameter R, *P. aphanidermatum* exhibited highly significant differences in its resistance against *T. asperelloides* in different treatments (P<0.0001). The effect of temperature on the resistance was very meaningful (P<0.0001). The R values obtained in interactions to *Trichoderma* isolates were also highly different (P<0.0001). R values (calculated as the ratio of Z/M) equal to 1 indicate the equal facility of biological control isolate growth in the presence of the pathogenic fungus and in the absence of it on a rich substrate like malt extract agar. R values less than 1 indicate more facility of a biological control isolate in the presence of a pathogenic fungus compared to its growth on a medium and as a result of growth promoting effect of a pathogen. Based on the mean R values shown in Table 5, *P. aphanidermatum* could be regarded a so-easy-target for the *T. asperelloides* isolates, T13 and T92. *P. aphanidermatum* had a growth arising impact on T92 at 15°C.

Pakdaman's biological control indices (PBCIs) were statistically highly different (P=0.0008), varying highly significantly as the result of temperature impact (P=0.0025) and *T. asperelloides* isolate effect (P=0.0123).

Comparison of average PBCI values indicated that T92 was more effective than T13 in the biological control of *P. aphanidermatum*, and the success of biological control was higher at 15°C (Table 6).

Study of the possibility of correlations between C, Z, R, and P values in one hand and PBCI in other hand revealed no significant correlation between PBCI and each of the parameters respectively as C (R^2=0.56561^{ns}, P=0.0553), Z (R^2=0.51192^{ns}, P=0.0889), M (R^2=0.56561^{ns}, P=0.0553), and P (R^2=0.0838^{ns}, P=0.0838). However, a highly significant negative correlation was found between the parameter R and PBCI (R^2=-0.77186**, P=0.0033). No correlation was found between P and other parameters including C (R^2=-0.12309^{ns}, P=0.7031), Z (R^2=-0.14706^{ns}, P=0.6483), M (R^2=-0.12309^{ns}, P=0.7031) and R (R^2=-0.14213^{ns}, P=0.6595).

Discussion

The quantification of the confrontation or dual culture test in order to get access to analyzable data was an inescapable task in the comparison of fungal biological control isolates. Without any mathematical data in hand, precise analyses would be impossible. Here a new index for biological control was introduced based on the direct radial growth of the pathogen in the presence of a biological control agent (P), and the parameter for the resistance of a pathogen against a biological control agent applied. In as much as like what we observed with the interaction of two *T. asperelloides* isolates with *P. aphanidermatum* it was possible to have equal numbers for Z and M values, then resistance parameter (R) was defined as the ratio of Z/M. Pakdaman's biological control index was defined as 1/P·R or M/P·Z. The index unit was determined as mm^{-1}. Time has been considered inside the concept of R also a representative of phytopathogenic fungus resistance against biological control mechanism(s). Because of usually enfaced transitory promoting impact of the biological control agents on the growth of phytopathogenic fungi in one side, and the ignorance of full process of biological control and other problems in the calculation method conventionally applied in confrontation tests (explained in introduction), development of a new methodology for this important test was necessary. Considering the parameter C, this parameter seemed incapable in the revealing of the differences in the whole potentials of fungal biological control isolates. In contrast to C, another parameter Z comprehending whole of fungus-fungus interaction period reflected existent variation in the biological control potentials of isolates. C is only a part of the whole Z. Other point that is better to explain here is that the growth of pathogen at the time before C is not a suitable criterion. This is because of the promoting impact of *Trichoderma* isolates on the growth of plant pathogenic fungi as well oomycetes then followed with the invasion by *Trichoderma* isolates. Another point is that here in this method, no control was considered for the pathogen. The reason for this was that in most reactions pathogens were of very fast growth and could fill their own plates earlier than the end of the dual culture experiments and this can lead to error.

Using Pakdaman's biological control index it was possible to get the mathematical data required for statistical analyses and Tukey's studentized range (HSD) test was preferred to other tests because of its proper sensitivity in grouping of data means. The new methodology was found easy to perform and simple to understand. Using the new methodology it would be feasible to compare the performance of several biological control species and or isolates under various conditions and confronted with different phytopathogenic fungi.

Comparing the results from Tukey grouping based on C, Z, and P values with the results of PBCI, it is illustrated no of former three parameters can represent the level of biological control. Also, synchronous consideration of C and P did not end to the same result obtained based on PBCI. As C-based grouping could not show the differences between two biological control isolates, then Z was preferentially applied instead, and the formula for PBCI was developed using two temporal parameters to form a combined parameter, R representative of pathogen resistance under peculiar conditions. R values more than 1 indicate pathogen resistance to biological control mechanism(s) applied by a control agent in confrontation with a pathogen. R value equal to 1 indicates absence of any resistance by a pathogen. R values less than 1 indicate that a pathogen not only does not resist, but also promotes the growth and development of a biological control agent. With our study, *P. aphanidermatum* either did not show any resistance against *T. asperelloides* isolates, or it promoted the growth and development of biological control isolates. However, such a promoting impact was only observed with the isolate T92 and only at 15°C. This promotion might be because of the compensational provision of one or more growth promoting factors (such as vitamins, amino acids, etc) for a biological control isolate under special eco-physiological conditions. Additionally, such a promoting effect might be originally induced by peculiar biological control isolates under specific conditions.

Based on PBCI values obtained in this study, biological control of *P. aphanidermatum* with T92 isolate at 15°C would lead to more successful control. This finding is in agreement with thermophily of *P. aphanidermatum* and mesophily of *T. asperelloides*. Our findings indicate that PBCI can be regarded as a proper criterion useful in confrontation tests. Indeed, there is no need to change the formulas when other plant pathogens are studied and the introduced formulae have successfully been applied with other plant pathogens such as *F. moniliforme*, *M. phaseolina*, and *R. solani*.

Acknowledgement

The authors dedicate their intimate thanks to Mrs. Dr. Monika Komon-Zelazowska, Mrs. Dr. Irina S. Druzhinina and Mr. Prof. Dr. Christian Peter Kubicek from Research Area Gene Technology and Applied Biochemistry, Institute of Chemical Engineering, Vienna University of Technology, Vienna, Austria, who kindly helped through molecular identification of Trichoderma isolates applied in this research without any expectations.

References

1. Monte E (2001) Understanding *Trichoderma*: between biotechnology and microbial ecology. Int Microbiol 4: 1-4.
2. Seidl V, Huemer B, Seiboth B, Kubicek CP (2005) A complete survey of *Trichoderma* chitinases reveals three distinct subgroups of family 18 chitinases. FEBS J 272: 5923-5939.
3. Burges HD (1998) Formulation of microbial pesticides. Kluwer Academic Publishers, Dordrecht, Netherlands.
4. Butt TM, Goettel MS, Papierok B (1999) Directory of specialists involved in the development of fungi as biocontrol agents. Colin Butt Design & Print, Warley, West Midlands, USA.
5. Butt TM, Jackson CW, Magan N (2001) Fungal biological control agents: Progress, Problems and Potential. Wallingford, CABI International, Oxon, UK
6. Verma M, Brar SK, Tyagi RD, Surampalli RY, Valéro JR (2007) Antagonistic fungi, *Trichoderma* spp.: panoply of biological control. Biochem Eng J 37: 1-20.
7. Viterbo A, Horwitz BA (2010) Cellular and molecular biology of filamentous fungi: Mycoparasitism. American Society for Microbiology, Washington, USA 42: 676–693.
8. Harman GE, Kubicek CP (1998) *Trichoderma* and *Gliocladium.* Taylor & Francis, London, UK. 278
9. Sahebani N, Hadavi N (2008) Biological control of the root-knot nematode *Meloidogyne javanica* by *Trichoderma harzianum*. Soil Biol Biochem 40: 2016-2020.
10. Sharon E, Bar-Eyal M, Chet I, Herrera-Estrella A, Kleifeld O, et al. (2001) Biocontrol of the root-knot nematode Meloidogyne javanica by *Trichoderma harzianum*. Phytopathology 91: 687-693.
11. Sharon E, Chet I, Viterbo A, Bar-Eyal M, Nagan H, Samuels GJ (2007) Parasitism of *Trichoderma* on *Meloidogyne javanica* and role of the gelatinous matrix. Eur J Plant Pathol 118: 247-258.
12. Suarez B, Rey M, Castillo P, Monte E, Llobell A (2004) Isolation and characterization of PRA1, a trypsin-like protease from the biocontrol agent *Trichoderma harzianum* CECT 2413 displaying nematicidal activity. Appl Microbiol Biotechnol 65: 46-55.
13. Bailey BA, Bae H, Strem MD, Roberts DP, Thomas SE, et al. (2006) Fungal and plant gene expression during the colonization of cacao seedlings by endophytic isolates of four *Trichoderma* species. Planta 224: 1449-1464.
14. Baker R (1991) The rhizosphere and plant growth: Induction of rhizosphere competence in the biocontrol fungus *Trichoderma*. (D. L. Keister, and P. B. Cregan edn). 221-228
15. Chang YC, Chang YC, Baker R, Kleifeld O, Chet I (1986) Increased growth of plants in presence of the biological control agent *Trichoderma harzianum*. Plant Disease 70: 145-148.
16. Ghahfarokhy MR, Goltapeh EM, Purjam E, Pakdaman BS, Modarres Sanavy SAM, et al. (2011) Potential of mycorrhiza-like fungi and *Trichoderma* species in biocontrol of Take-all Disease of wheat under greenhouse condition. J Agricultural Technology 7: 185-195.
17. Harman GE (2001) Microbial tools to improve crop performance and profitability and to control plant diseases. (D. S. Tzeng, and J. W. Huang edn). Proceeding of international symposium on biological control of plant diseases for the new century-mode of action and application technology, NCHU, Taichung, Taiwan. 71-81.
18. Hohmann P, Jones E E, Hill RA, Stewart A (2011) Understanding *Trichoderma* in the root system of *Pinus radiata*: associations between rhizosphere colonisation and growth promotion for commercially grown seedlings. Fungal Biol 115: 759-767.
19. Inbar J, Abramsky M, Cohen D, Chet I (1994) Plant growth enhancement and disease control by *Trichoderma harzianum* in vegetable seedlings grown under commercial conditions. Eur J Plant Pathol 100: 337-346.

20. Lo CT, Lin CY (2002) Screening strains of *Trichoderma* spp. for plant growth enhancement in Taiwan. Plant Pathol Bull 11: 215-220.

21. Lo CT, Nelson EB, Harman GE (1997) Improved biocontrol efficacy of *Trichoderma harzianum* 1295-22 for controlling foliar phases of turf diseases by spray applications. Plant Disease 81: 1132-1138.

22. Yedidia I, Benhamou N, Chet I (1999) Induction of defense responses in cucumber plants (*Cucumis sativus* L.) by the biocontrol agent *Trichoderma harzianum*. Appl Environ Microbiol 65: 1061-1070.

23. Korolev N, Rav David D, Elad Y (2008) The role of phytohormones in basal resistance and *Trichoderma* induced systemic resistance to Botrytis cinerea in Arabidopsis thaliana. BioControl 53: 667-683.

24. Martinez C, Blanc F, Le Claire E, Besnard O, Nicole M, et al. (2001) Salicylic acid and ethylene pathways are differentially activated in melon cotyledons by active or heat-denatured cellulase from *Trichoderma longibrachiatum*. Plant Physiol 127: 334-344.

25. Donoso EP, Bustamante RO, Caru M, Niemeyer HM (2008) Water deficit as a driver of the mutualistic relationship between the fungus *Trichoderma harzianum* and two wheat genotypes. Appl Environ Microbiol 74: 1412-1417.

26. Bae H, Sicher RC, Kim MS, Kim SH, Strem MD, et al. (2009) The beneficial endophyte *Trichoderma hamatum* isolate DIS 219b promotes growth and delays the onset of the drought response in *Theobroma cacao*. J Exp Bot 60: 3279-3295.

27. Ezzi MI, Lynch JM (2002) Cyanide catabolizing enzymes in *Trichoderma* spp. Enzyme Microb Tech 31: 1041-1047.

28. Harman GE, Howell CR, Viterbo A, Chet I, Lorito M (2004) *Trichoderma* species- opportunistic, avirulent plant symbionts. Nat Rev Microbiol 2: 43-56.

29. Altomare C, Norvell WA, Björkman T, Harman GE (1999) Solubilization of phosphates and micronutrients by the plant growth promoting and biocontrol fungus *Trichoderma harzianum* Rifai 1295-22. Appl Environ Microbiol 65: 2926-2933.

30. Iskandar NL, Mohd Zainudin NAI, Tan SG (2011) Tolerance and biosorption of copper (Cu) and lead (Pb) by filamentous fungi isolated from a freshwater ecosystem. J Environ Sci 23: 824-830.

31. Kacprzak M, Malina G (2005) The tolerance and Zn^{2+}, Ba^{2+} and Fe^{3+} accumulation by *Trichoderma atroviride* and *Mortierella exigua* isolated from contaminated soil. Can J Soil Sci 85: 283-290.

32. Krantz-Rülcker C, Allard B, Schnürer J (1993) Interactions between a soil fungus, *Trichoderma harzianum*, and IIb metals- adsorption to mycelium and production of complexing metabolites. BioMetals 6: 223-230.

33. Ganassi S, Moretti A, Stornelli C, Fratello B, Pagliai AMB, et al. (2000) Effect of Fusarium, *Paecilomyces* and *Trichoderma* formulations against aphid *Schizaphis graminum*. Mycopathologia 151: 131-138.

34. Shakeri J, Foster HA (2007) Proteolytic activity and antibiotic production by *Trichoderma harzianum* in relation to pathogenicity to insects. Enzyme Microb Tech 40: 961-968.

35. Evans HC (1998) The safe use of fungi for biological control of weeds. Phytoprotection 79: 67-74.

36. Heraux FMG, Hallett SG, Ragothama KG, Weller SC (2005) Composted chicken manure as a medium for the production and delivery of *Trichoderma virens* for weed control. HortScience 40: 1394-1397.

37. Heraux FMG, Hallett SG, Weller SC (2005) Combining *Trichoderma virens*-inoculated compost and a rye cover crop for weed control in transplanted vegetables. Biol Control 34: 21-26.

38. Pakdaman BS, Khabbaz H, Goltapeh EM, Afshari HA (2002) *In vitro* studies on the effects of sugar beet field prevalent herbicides on the beneficial and deleterious fungal species. Pakistan J Plant Pathol 1: 23-24.

39. Pakdaman BS, Goltapeh EM (2006) An *in vitro* study on the possibility of rapeseed white stem rot disease control through the application of prevalent herbicides and *Trichoderma* species. Pak J Biol Sci 10:7-12.

40. Pakdaman BS, Mohammadi Goltapeh E, Sepehrifar R, Pouriesa M, Rahimi Fard M, et al. (2007) Cellular membranes as the sites for the antifungal activity of the herbicide sethoxydim. Pak J Biol Sci 10: 2480-2484.

41. Weindling R (1932) *Trichoderma lignorum* a parasite of other soil fungi. Phytopathology 22: 837-845.

42. Sarrocco S, Matarese F, Moretti A, Haidukowski M, Vannacci G (2012) DON on wheat crop residues: effects on mycobiota as a source of potential antagonists of *Fusarium culmorum*. Phytopathologia Mediterranea 51: 225-235.

Phosphorus Modeling in Tile Drained Agricultural Systems Using APEX

Francesconi W[1]*, Williams CO[2], Smith DR[3], Williams JR[4], Jeong J[4]

[1]*International Center for Tropical Agriculture (CIAT), USA*
[2]*USDA-NRCS National Soil Survey Center, Lincoln, NE, USA*
[3]*USDA-ARS Grassland, Soil and Water Research Laboratory, USA*
[4]*Blackland Research an Extension Center, Texas A&M University System, Temple, TX, USA*

Abstract

Phosphorus (P) losses through tile drained systems in agricultural landscapes may be causing persistent eutrophication problems observed in surface water. The purpose of this paper is to evaluate the state of the science in the Agricultural Policy/Environmental extender (APEX) model related to surface and tile P transport. This was accomplished using data from a monitored corn-soybean rotation field in the St. Joseph River watershed, IN. The estimation of soluble phosphorus (SP) in surface runoff and tile flow in APEX includes a user defined linear (based on GLEAMS) and nonlinear (Langmuir) sorption option. The results suggest that the inclusion of the Langmuir isotherm improved (18%) SP sorption estimates in surface runoff during the corn year only when P inputs were added, whereas the linear method was more appropriate during the soybean year when no fertilizers were applied. Similarly, SP estimates in tile flow were improved (30%) when using the Langmuir option during the corn year, though the overall model performance predicting this variable were very poor. Modeling improvements of P partitioning processes in APEX can help predict more realistic outputs. Yet to achieve this in tile flow, water percolation processes need to be improved to reflect preferential flow conditions often found in long-term no-till fields and in soils with high clay content. Greater accuracy in the estimation of the effect of artificial drainage systems, common in the US Midwest, should result in the improved evaluation of agricultural conservation practices in order to examine strategies that could reduce P losses for water quality purposes.

Keywords: Adsorption; Hydrological modeling; No-till; Water quality

Introduction

Phosphorus (P) from non-point sources such as agricultural lands can have a major environmental impact on the quality of receiving waters. In artificially drained agricultural fields, P losses have been considered a fraction of that in the surface, and these account for a small amount of the fertilizer application [1]. However, small P concentrations (between 0.03 to 0.06 mg/L) in water bodies can result in algal blooms and subsequently in eutrophication and hypoxic conditions [2]. The management of P on agricultural lands with hydrological pathways to sensitive receiving surface water bodies is therefore fundamentally important. Models that estimate the effects of agricultural conservation practices on water quantity and quality are increasingly important tools for short- and long-term assessments [3-5]. Due to the time and financial resources needed to adequately monitor P transport to receiving waters, simulation models have served as a valuable management tool. One such tool is the Agricultural Policy/ Environmental extender Model (APEX) [6]. APEX is an extension of the Environmental Policy Integrated Climate (EPIC) model which was developed to assess the impact of erosion productivity [6,7], and later expanded to allow for the simulation of many agricultural management processes for field-sized areas, up to 100 ha [4,8]. The major components of EPIC are weather, hydrology, sedimentation, nutrient cycling, pesticide fate, crop growth, tillage, economics, and plant competition. APEX was developed to extend EPIC functions to whole farms and small watersheds and to include routing of nutrients, pesticides, water and sediment across landscapes (e.g., fields or subareas), through shallow groundwater, and into channel systems to a watershed outlet [9]. APEX has been used to assess the effectiveness of conservation practices and is one of few models capable of simulating the routing of chemical pollutants and water at the field scale [10]. Because APEX is able to consistently model various land management strategies at scales ranging from field to farm to small watersheds, it was adopted by USDA NRCS for the Conservation Effects Assessment Program (CEAP) national assessment [11]. Recent updates to the APEX model include the simulation of soluble phosphorus (SP) transport processes from the surface to the tile drains and the inclusion of the Langmuir equation. The APEX model, similar to other models, utilizes a simplified P cycle model development by Jones et al. and Sharpley et al. [12,13] to describe soil P transformations. In APEX, following fertilizer applications, the model divides the P content into inorganic and organic pools. The inorganic pool is divided into labile, active, and stable pools. Phosphorus applications as inorganic or organic fertilizer are assumed to be in the soluble form and contribute to labile P, making the nutrient available for plant uptake. The rapid equilibrium between labile P and active P, and consequently the slower equilibrium between active P and stable P, are calculated as a function of the chemical and adsorption properties of the soil. At equilibrium, the labile P pool is proportional to the active P pool. In APEX, the labile P pool is partitioned between the solid and solution phases. Limitations of the model include the linear relationship between labile and active P pools as well as between active and stable P pools. These relationships lead to a linear relationship between solution and solid phase P pools and therefore do not reflect the nonlinear relationships of soil P sorption. Although the linear relationship may be suitable at low P soil solution concentrations, it may underestimate solution P concentrations at higher soil P concentrations which may lead to an under prediction of P in surface runoff and leaching. Applications of APEX for simulating tile drainage dynamics have been limited, likely as a result of the simplistic way in which tile flow is simulated in the

***Corresponding author:** Wendy Francesconi, International Center for Tropical Agriculture, USA, E-mail: w.francesconi@cgiar.org

model. Tile flow in APEX is a function of lateral subsurface flow and the time required for the drainage system to reduce plant stress. Although limited in use, [14] Gassman et al. evaluated APEX to simulate tile flow in the Upper Maquoketa River Watershed in northeast Iowa, and computed a satisfactory R^2 value of 0.70 for average monthly tile flow. Tile flow has also been satisfactorily calibrated using the EPIC model [15,16]. Soluble P transport to tile is a function of percolation and soil properties in a given soil layer. As the water in the upper soil layer is divided to simulate runoff, evaporation, and infiltration, SP percolates to the underlying soil layer. The SP concentration percolated from the layer above is included in the SP concentration estimated for the layer in question, which is then used for estimating the losses that occur through evaporation and plant uptake. The SP nutrient transport and extraction occurs at each soil layer until it reaches the artificial drainage layer, where lateral flow results in tile flow and SP losses. The depth for each layer depends on the soil characteristics for the areas studied. Along with P modeling in tile flow, APEX was also modified to include a nonlinear P sorption algorithm, adapted from the Langmuir equation [17,18]. Inorganic fertilizer becomes readily available for leaching from the top soil layer when rainfall induced infiltration occurs. Meanwhile, it takes time for organic P to mineralize and become available for leaching. Both organic and inorganic P are subject to loss in surface runoff (soluble P and sediment-bound organic P). When inorganic P is added to soil, it has been shown that labile P and P sorption decreases in concentration [19,20]. This decline is nonlinear indicating that the P losses reach a fairly constant value. In APEX, the inorganic P flow rate did not reflect the nonlinear decrease in labile P and P sorption concentrations over time. Futhermore, the partitioning coefficient reflected linearity between the solution and the solid P phases, which is not typical of P sorption in some soils [21]. To overcome this limitation, the Langmuir adsorption isotherm was incorporated to adequately simulate environmental conditions not currently included in APEX. The inclusion of the Langmuir modification will influence the partitioning of P between soluble and adsorbed in the labile pool. It does not influence the rate of P flux to the active pool nor the partitioning between active and labile. APEX is an integral part of the evaluation of USDA conservation programs. This paper reviews the conceptual approach in APEX for estimating P losses in an artificially drained agricultural field. To help provide testing of these features the present study 1) examined the use of APEX for estimating tile flow and P losses through tile, and 2) evaluated the nonlinear P estimation (Langmuir) option as an effective method for estimating SP in surface runoff and tile flow.

Materials and Methods

General hydrology and phosphorus modeling in APEX

Hydrology: The water hydrology component in APEX is routed through channels and flood plains using a daily time step average flow, or a short interval complete flood routing method [6]. The watershed is divided into subareas (homogeneous hydrologic land-use units, HLU), and each subarea contains channel specifications. Water flow is computed from the most distant subarea to the watershed outlet. The average water flow at the outlet is a function of the water inflow volume, the area of the watershed, the frequency and duration of rainfall, and the time of concentration of the watershed above the reach. The time of concentration is dependent on the channel flow length and the average channel velocity, which is estimated using Manning's equation [6].

Tile Drainage in APEX: Applications of APEX to simulate tile drainage dynamics have been limited. Artificial drainage systems in APEX are simulated by modifying the natural lateral subsurface flow routines [6]. Tile in APEX is simulated by indicating the depth of the drainage system and the time required (days) for the drainage system to reduce plant stress. Storage routing in APEX allows percolation of soil moisture from a soil layer when the soil water content exceeds field capacity. In APEX, surface layer is defined by a 10 mm top soil layer. Subsurface layers are typical soil layers having different properties such as texture, permeability, organic C contents, and etc. The soil layers definition and properties are often loaded from public soil databases such as STATSGO or SSURGO. The top soil layer is the hypothetical layer and its property is usually the same as the second layer, which are drawn from the database. The depth of soil layers vary by location and soil types. Water drains from one layer to the layer below as a function of storage and saturated conductivity:

$$SWC_1 = (SWC_0 - FC)^{*} \exp^{\left(\frac{-24}{TT_V}\right)} + FC \quad (1)$$

Where SWC_1 and SWC_0 are the soil water contents at the end and start time interval (24 hours), FC is the field capacity in mm, and TT_V is the vertical travel time through a soil layer in hours. Travel time through a soil layer is calculated as:

$$TT_V = \frac{(PO - FC)}{SC} \quad (2)$$

Where PO is porosity in mm and SC is saturated conductivity in mm h^{-1}. The lateral subsurface flow rate (Q_H) is estimated in mm d^{-1} by partitioning the changes in soil moisture storage between vertical and horizontal flow.

$$Q_H = (SWC - FC) * \frac{X_V \cdot X_{VH}}{X_V + X_H} \quad (3)$$

where

$$X_V = 1 - \exp^{(-24/TT_V)}$$

$$X_H = 1 - \exp^{(-24/TT_H)}$$

$$X_{VH} = 1 - \exp^{(-24/TT_V)+(-24/TT_H)}$$

Where TT_H is the horizontal travel time h. Horizontal flow is partitioned into subsurface flow to the adjacent subarea and or outflow (tile) and quick return flow. Drainage is simulated by indicating the depth of the drainage system and the time required for the drainage system to reduce plant stress. The drainage time replaces the horizontal travel time (TT_H) in Equation 3 for the layer containing the system.

Phosphorus dynamics: Specific to phosphorus dynamics, APEX has two components, an organic and an inorganic P. Organic P is divided into fresh residue and stable organic P pools. Soil inorganic P is divided into active P, stable P, and labile P pools. Labile P is plant-available P that is extracted by anion exchange resin [22] (Sharpley et al.) and represents soluble and weakly sorbed P. Studies have indicated that after an inorganic fertilizer application, labile P concentrations decrease rapidly (several days to weeks) followed by a much slower decrease in labile P which may continue for years [23,24] (Indiati et al. and Paulter and Sims). To account for the initial fast and subsequent slow decrease in labile P, APEX assumes that the labile mineral pool is in rapid equilibrium with the active mineral pool and the active mineral pool is in slow equilibrium with the stable mineral pool. The P sorption coefficient governs the equilibration between the mineral labile and active P pools. This coefficient represents the fraction of fertilizer P extracted by anion exchange resin after an incubation period 6 months [25] and is represented as:

$$F_{MP} = P_{LAB} - P_{ACT} * \frac{PSC}{(1 - PSC)} \quad (4)$$

F_{MP} is the mineral P flow rate between mineral labile and active P pools, P_{LAB} is labile mineral P, P_{ACT} is active mineral P, and PSC is the P sorption coefficient, which is a function of physical and chemical soil properties and is described by [25] Jones et al. At equilibrium, the stable mineral P pool is four times the size of the active mineral P pool. A detailed description of the soil P model in APEX may be found in Jones et al. [3,12], Sharpley et al. and Williams et al.

In addition to the P model presented by [25] Jones et al., APEX estimates solution phase P based on the concept of partitioning mineral labile P into the solution and solid phases as described by [26] Knisel and is expressed as,

$$K_D = \frac{P_{SOL}}{P_L} \tag{5}$$

Where K_D is a partitioning coefficient, P_{SOL} is the concentration of P in the solid phase, and P_L is the concentration of P in the solution phase. The default value for the partitioning coefficient is set at 100 [21].

Description of new features introduced to APEX: Though the above relationships (linear) may be suitable at low soil P concentrations, it could likely underestimate solution P at higher soil P concentrations such as in cases where there is an application of animal waste or fertilizer application in excess of plant uptake requirements. Due to this limitation, the Langmuir isotherm [18] was added to APEX to account for large soil P concentrations. At equilibrium, the Langmuir isotherm model is described as:

$$C_S = S_{\max} * K_D * \frac{C_L}{(1 + K_D * C_L)} \tag{6}$$

Where C_S is the soluble P concentration sorbed in the soil, C_L is the soluble P concentration in the liquid, and K_D is a partitioning coefficient, which in APEX is the concentration of the labile P in the solid phase divided by that of the solution phase. K_D is an adsorption constant related to the binding energy at equilibrium ranging from 1 to 20.

$$S_{\max} = 1000 * \frac{Clay}{Clay + \exp^{(3.3519 - 0.027 * Clay)}} \tag{7}$$

where S_{max} is the maximum P sorption capacity of the soil and Clay is the percentage of clay in soil layer 2. More information for this equation can be found in [27] Salton. To determine soluble P in solution we rearranged to solve for C_L given C_S,

$$C_L = \frac{C_S}{(K_D * (S\max - Cs))} \tag{8}$$

As the C_S approaches S_{max}, the C_L increases. In the current APEX model, transport of soluble P in runoff is estimated as:

$$QP_L = 0.01 * P_{LAB} * \frac{Q}{K_D} \tag{9}$$

where QP_L is the mass (kg ha^{-1}) of soluble P lost in runoff volume, Q (mm) and P_{LAB} is the concentration (g t^{-1}) of labile P in soil layer 1. The inclusion of the Langmuir isotherm estimates soluble P in runoff (QP_{LAN}) as:

$$QP_{LAN} = 0.01 * C_L * Q \tag{10}$$

The tillage component in APEX mixes P within the top layer which is then made available for plant uptake from the root zone soil solution. Routines in APEX were added to reflect labile P in subsurface drainage and are a modification of the GLEAMS [28] (Leonard et al.) leaching component. Phosphorus leaching is expressed as a function of time, concentration, and flow rate through a soil layer using the equation

$$SP = SP_0 * \exp^{\left(\frac{-QT}{0.01*ST+0.1*K_D*BD}\right)} \tag{11}$$

where SP_0 is the initial P in solution in the soil layer (g ha^{-1}), SP is the amount of P that remains after the amount of flow (QT) passes through a soil layer, ST is the initial water storage in mm, and BD is the bulk density. The amount of P leached by the amount of water QT is obtained by subtracting SP from SP_0 using the equation

$$P_L = SP_0 * \left(1 - \exp^{\left(\frac{-QT}{0.01*ST+0.1*K_D*BD}\right)}\right) \tag{12}$$

where P_L is the amount of P leached by QT.

Case study-cedar creek watershed: Monitoring Site- The model simulations presented is based on edge-of-field measurements at a long-term USDA-ARS-National Soil Erosion Research Laboratory (NSERL) site in Cedar Creek sub watershed at the St. Joseph River watershed. The field is located in the northeast region of the state of Indiana (Figure 1), and has an area of 1.7 ha [29]. The predominant soil series is Glynwood silt loam, and the topography is relatively flat with a 3% slope. Corn and soybean production have been managed under no-till practice for the past 25 years. A major soil limitation for row crop production in the field as well as for the Cedar Creek sub watershed is the somewhat poorly to poor drainage. As a result, tile drainage systems are placed approximately 1 meter below the soil surface to reduce saturation. The St. Joseph River watershed has been monitored by the NSERL as part of the Conservation Effects Assessment Project (CEAP) Watershed Assessment Study [30]. Surface runoff, sediment and P (total (TP) and soluble (SP)) have been collected since 2003, though data inconsistencies occured during the first few years. Tile flow and soluble P in tile (SP-tile) have been measured starting in 2008. Surface and tile discharge were collected at 10 minute intervals using a modified ISCO automated water sampler [31], while agricultural management practices were recorded by the land managers. Meteorological data (precipitation, air temperature, solar radiation, etc.) were recorded at the field site and complemented with the data collected by a nearby weather station located in Garret, IN.

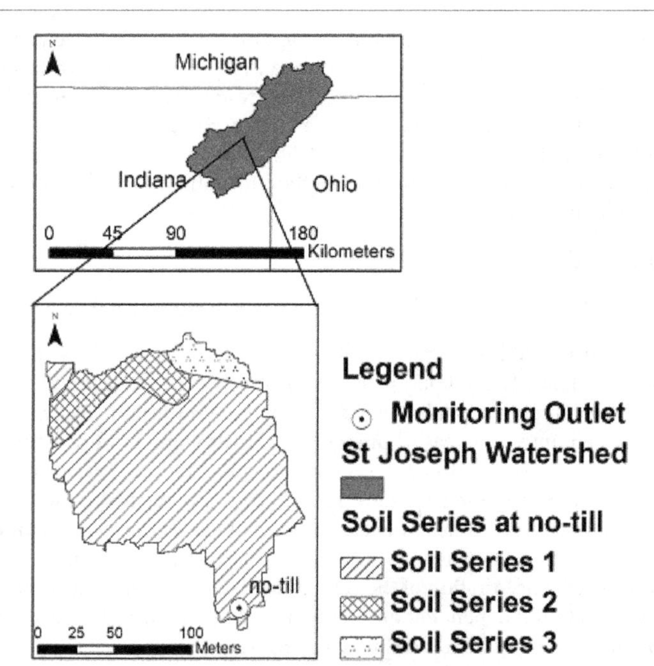

Figure 1: St. Joseph River Watershed and NSERL monitoring field site in Indiana. Watershed delineation at No-Till field depicting soil series distribution.

Modeling parameterization: APEX model Ver. 0806 [32] was used for the present study to evaluate the inclusion of Langmuir and P leaching subroutines. The initial calibration and validation of the model for the no-till site was done using an older version of APEX 0806 [33] and can be found in Francesconi et al. [32]. Due to the inclusion of more accurate input data such as fertilization dates, and upgrades to the code estimating erosion driven by runoff (Jaehak Jeong personal conversation), the recalibration of the model was conducted. The changes in the sediments values estimated using the upgraded APEX version, affected the estimation of P values as well. The subsequent recalibration and revalidation of the variables of interest was required for the present analysis. In order to recalibrate the model, several parameters including those that influence soluble P sorption capacity (PARM 8 and 96) were modified. The parameters targeted for recalibration to improve the model's predictive capability and to test the P sorption options are listed in Table 1. Once the model was recalibrated, the P sorption linear or nonlinear equations were selected by choosing the linear (LBP = 0) or Langmuir (LBP =1) adsorption isotherm options in the command file. Tile flow estimates were also improved by removing an outlier value. A zero tile flow value was reported by NSERL's database on May 25, 2010. However, there was a precipitation event of 48 mm that day. Given the patterns observed for tile flow compared to the amount of precipitation and antecedent soil moisture in the soil, the May 25 data point was assumed to be an error. By removing this data point the calibration of tile flow was improved from R^2 = 0.42 and NS = 0.40 to R^2 = 0.52 and NS = 0.45 in 2010. Model efficiency for both tile flow and surface runoff was considered satisfactory at the daily time scale (NS > 0.40). Furthermore, sediment and P in surface runoff had acceptable statistical scores (NS > 0.40) according to the standards described by Moriasi et al. [34] for a daily time step. Overall, the recalibration of APEX reasonably simulated hydrology, erosion, and P transport in surface runoff (Table 2).

Results and Discussion

Soluble P in surface runoff

The inclusion of the Langmuir isotherm equation and parameter (K_{LAN}) in APEX proved to be effective at improving the estimation of P at the no-till site during the corn planting year. Compared to the calibration values produced by the linear equation, the Langmuir sorption option resulted in better calibration scores estimating TP and SP in surface runoff during 2010 (Table 2). While R^2 values remained the same, NS scores improved by 18% for both TP and SP. In 2010, 4.5 Mg of manure were applied to the field prior to corn planting. In typical agricultural settings, that the same management hardly repeats every year. From the modeling standpoint, these varying inputs give dynamic and more realistic modeling environment to APEX, allowing for users to test model processes like P loads in tile drainage more realistically. This is analogous to weather input, for example, rainfall in year 1 is different from year 2 and temperature is also different day by day. If correctly implemented, changing management should be a positive point to introduce when it comes to model evaluation. The incorporation of an option in APEX that includes a maximum sorption capacity, which is based on the fraction of clay content, sets the upper limits for P binding capacity to soil particles in the top soil layer. Hence, the estimation of P takes into account the soil's characteristics to establish a nonlinear relation between the concentration found in solution and that bound to the soil phase, which is a relationship that has been observed experimentally [35] (Sharpley). Given the nonlinear characteristics of the Langmuir equation compared to the linear method, its application may be considered when P content in the soil is relatively high. The work by Rossi et al. [21] validated the use of Langmuir to adequately simulate higher P concentration in watershed nutrient analysis using the Soil and Water Assessment Tool (SWAT). Under the high fertilizer application conditions observed at the no-till site for the year 2010, the nonlinear adsorption method would be more appropriate for

No.	Parameter	Parameter Description	Range	Value Assigned	
				Initial	Final
1	PARM 47	RUSLE C factor coefficient	0.5 – 1.5	1.5	0.5
2	PARM 18	Sediment routing exponent	1.0 – 1.5	1.5	2.0
3	PARM 76	Standing dead fall rate	0.0001 – 0.1	0.001	0.005
4	PARM 18	Sediment routing exponent	1.0 – 1.5	1.5	2.0
5	PARM 58	P Enrichment ratio exponent for routing	0.3 – 0.9	0.24	0.35
6	PARM 57	P Enrichment ratio coefficient or routing	0.05 – 0.20	0.78	0.05
7	PARM 59	P Upward movement by evaporation coefficient	1 – 20	0.5	1.0
8	PARM 8	Soluble P adsorption coefficient	10 – 20	20	16
9	PARM 62	Manure erosion equation coefficient	0.1 – 0.5	0.1	0.3
10	LBP	Soluble phosphorus runoff estimate equation	0 or 1	0	1
11	PARM 96	Soluble P leaching K_D value*	1 – 15	10	1.0

*Parameter modified to calibrate soluble P in tile.

Table 1: Model recalibration for utilizing APEX0806 (February, 2014) version with new routing capabilities for SP-tile and Langmuir equation.

Variable	Linear				Langmuir			
	2010		2011		2010		2011	
	R^2	NS	R^2	NS	R^2	NS	R^2	NS
Runoff	0.90	0.84	0.84	0.79	0.90	0.84	0.84	0.79
Sediment	0.96	0.96	0.96	0.86	0.96	0.96	0.96	0.86
Total P	0.95	0.50	0.74	0.38	0.95	0.68	0.73	0.36
SP	0.93	0.75	0.58	0.36	0.93	0.92	0.58	-15
Tile Flow	0.52	0.45	0.37	0.37	0.52	0.45	0.37	0.37
SP-Tile	0.17	-1.83	NA	NA	0.16	-1.28	NA	NA

NA = Data not available.

Table 2: APEX R^2 and Nash Sutcliffe Efficiency (NS) recalibration (2010) and revalidation (2011) values.

estimating P losses. In addition to the fertilizer rates and the soil characteristics, other aspects of crop management may have influenced the effectiveness of the P adsorption method used in 2010. At the no-till site, the application of the manure was conducted using a surface-spread application technique. The manure was applied in mid-February while the ground was still frozen prior to planting the corn, and the type of manure used was chicken litter. Even though the nutrient composition of poultry manure varies depending on several factors (such as species, feed ratio, type of litter), P content is generally high compared to cattle manure (at about 30 – 50% that of nitrogen) [36] (Nicholson et al.). The application of chicken litter in fields that may be considered to have nutrient rich soils can result in the overapplication of P and nutrient transport to surface water [37] (Sharpley et al.). Furthermore, the application of manure fertilizers without incorporation into the soil would most likely result in nutrient losses if high precipitation events occur before crop planting takes place. Given the placement, the source, and the timing of the fertilizer application at the no-till site, high P losses could be expected. Thus a nonlinear adsorption option would better simulate P losses as was observed when using the Langmuir option in 2010. The Langmuir option was not advantageous when estimating P values under the low P conditions in 2011. On the contrary, the use of the Langmuir option made the calibration of TP and SP in runoff worse (Table 2). According to the management at the no-till site, no fertilizer applications were added during the soybeans year in 2011. The P applied during 2010 through the chicken manure application would have been taken-up by the corn crop roots, extracted in runoff, moved into the soil profile, transformed into other P pools, or degraded [25,28] (Jones et al. and Leonard et al.). The amount of labile P available for extraction into runoff in 2011 would have been greatly reduced from the topsoil layer, thus making the linear nutrient adsorption method more appropriate during the soybean year. The reduction of P in 2011 is not only observed by the monitoring data, but was also simulated in APEX (Figure 2). When using Langmuir, the P losses estimated by the

			Observed	Linear	Langmuir
SP (kg/ha)	2010	Storm Mean	0.030	0.044	0.031
		Total Annual	0.360	0.524	0.371
	2011	Storm Mean	0.009	0.007	0.002
		Total Annual	0.121	0.096	0.029
SP-Tile (kg/ha)	2010	Storm Mean	0.005	0.044	0.005
		Total Annual	0.063	0.532	0.060
	2011*	Storm Mean	N/A	N/A	N/A
		Total Annual	N/A	N/A	N/A

*N/A data was not available for SP-Tile this year.

Table 3: Observed and APEX values for SP in surface runoff and SP-tile using the linear and Langmuir adsorption options for the corn (2010) and soybean (2011) years. Storm mean and total annual values were estimated for storm events during the growing season.

model were lower compared to those observed and estimated by the linear option (Table 3). While R^2 scores remained the similar for both TP and SP, NS values were slightly reduced (2%) for TP, and resulted in an error value (NS = -15) for SP. The negative NS score indicates that the observed mean would be a better predictor than the model for estimating SP in runoff [38] (Nash and Sutcliffe).

Soluble P in tile

Soluble P in tile is a recent nutrient routing capability in APEX. As per the modification of the subsurface lateral flow to simulate the presence of an artificial drainage system, the model has been updated to predict nutrients in tile flow. Compared to the observed data in 2010, the model results predicting SP-in tile were very poor (R^2 = 0.17 NS = -1.87). Given that no monitoring data was available for 2011, no modeling results were available for that year. The poor SP-tile prediction values in 2010 can be partly explained by the relatively low tile flow calibration scores (R^2 = 0.52 and NS = 0.45). Given that the estimation of nutrients in tile depends on the accuracy of water percolation process in the soil profile, low tile flow calibration values will result in even lower nutrient prediction. Currently, tile flow modeling in APEX offers a practical, but simplified approach to measuring artificial drainage systems in agricultural fields. The predicted values provide a broad estimate of drainage flow that can be used to ensure the reasonable water balance estimation by the model during calibration, and/or to provide a rough prediction of nutrient losses through tile flow. In addition to P, APEX has also been modified to simulate N losses in tile flow. Model performance estimating this nutrient however, have been more successful. Francesconi et al. reported NS values of 0.27 for APEX estimations of N in tile flow in a corn-soybean rotation [32]. Similar to P, N losses through drainage flow in APEX are determined by estimating the change in N concentration at the soil depth where the tile is present [21]. The simulation of N in tile in APEX is derived from the EPIC model. Both EPIC and APEX incorporate various N cycle processes (i.e., nitrification, denitrification, fixation, transformation), which provide a more complex simulation of its transformation and transport to the tile compared to P. The values estimated by Chung et al. [16] using EPIC indicate a satisfactory model performance (R^2 = 0.52 - 0.62 and NS = 0.43 - 0.54) when comparing monthly average N losses via tile in a corn-soybean system. Furthermore, the average monthly estimations by Gassman et al. and Saleh et al. [1,39] using APEX or a combination of APEX-SWAT, resulted in R^2 values of0.63 and 0.74 (respectively). Yet, R^2 values and monthly averages are less rigorous evaluation methods than comparisons at a daily time step. Even though N routines in APEX utilize a more comprehensive approach than P, which has led to better estimations in tile drainage, N simulations have also been considered a simplified representation of this nutrient cycling process in APEX

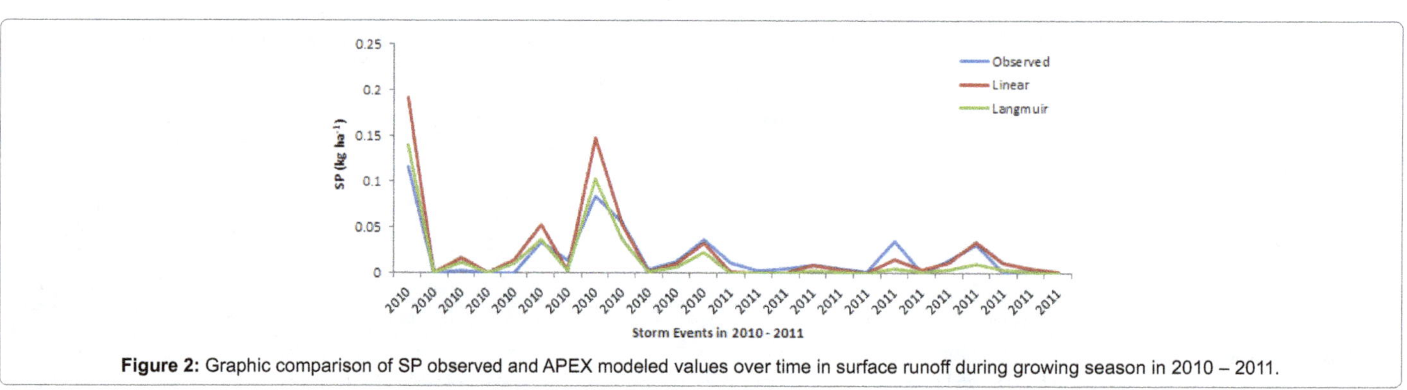

Figure 2: Graphic comparison of SP observed and APEX modeled values over time in surface runoff during growing season in 2010 – 2011.

(Wang et al.) [33]. Phosphorous transfer processes from the surface through the soil profile is less understood than that of N. The tendency of P to bind to soil particles makes it mostly immobile and resistant to leaching [40] (Holford). Hence, little consideration had been given to P losses through tile as an environmental pollution problem in agriculture. In light of the persistent eutrophication processes observed, and the accumulating evidence showing P losses through artificially drained agricultural systems, attention has been brought to potential P leaching and transport mechanisms [1,24,41-43] (Breeuwsma et al., Beauchemin et al., Sims et al., Smith et al. and Gentry et al.). The observed results for P in tile flow suggest that the mechanisms by which P moves through the soil profile could be primarily associated to water percolation and preferential flow (macropore) processes. While preferential flow is not currently simulated in APEX, there is a vertical crack component that could be used for this purpose [44] (Steglich and Williams). However, this feature was not utilized in the present analysis as the relatively satisfactory tile-flow values were obtained prior to examining the P-tile model output. In the absence of a preferential flow modeling component, and disregarding the user-defined vertical soil crack characteristic, our results did not show an adequate simulation of P transport process in APEX. Given the high clay content in the St. Joseph River watershed [45], the soils are susceptible to shrink-swell conditions leading to cracking and macropores under dry conditions [46]. Macropores can also be caused by soil fauna and decomposing plant materials. The potential presence of macropores in the Midwest has been validated as an important contribution pathway of nutrients and pesticides to tile drainage systems and subsequently to ditches, streams, and lakes [47]. Furthermore, the long-term no-till management (more than 25 years) at the simulated site could result in a large number of biopores in the soil [23]. When comparing surface runoff and tile flow discharge at the field site, peak values tend to occur almost simultaneously; which is indicative of the quick water transfer from the surface to the tile [32]. In contrast, peak discharge in the tile from water percolation through the soil matrix is expected to occur following surface runoff [46]. Given that the P losses in the tile at the no-till site were associated with storm events, the main water transfer mechanism from the surface to the tile may be due to the presence of macropores. Hence, the absence of a modeling feature specific to preferential flow in APEX may be reflected in the relatively low tile flow, through satisfactory calibration score, and subsequently in the very poor SP-tile calibration results. In addition, the hydrology routine (storage-routing concept) in APEX may be less accurate than a more physically based model using the Richard's equation when estimating water percolation through the soil matrix. Regardless of the water flow path, one could argue that the satisfactory calibration of tile flow should provide reasonable SP-tile output values. Yet, estimations of SP in tile flow in APEX are subjected to the extraction and plant uptake at each layer, which are dependent on soil characteristics. After the model was recalibrated for tile flow as well as for surface runoff, TP, and SP, the values predicted for SP-tile using the linear option for P were two orders of magnitude higher than those observed. Using the default setting for PARM 96, which modifies SP-tile, the initial outputs resulted in extremely poor calibration values (R^2 = 0.0 and NS = -18.0). After modifying the user defined value for SP leaching (K_D) value (PARM 96 = 1) to comply with the lower SP observed in the tile, some improvements were observed (R^2 = 0.16 and NS = -1.83) (Figure 3). While the calibration of tile flow can be done by modifying a few parameters (time lapse, concentration, and tile depth), it is unlikely that improvements in the prediction of this variable would result in the satisfactory estimation of SP-tile values in the present study. Given the limited dataset and the low P loads being measured, it is important to acknowledge that uncertainty in the data collection and analysis will influence the

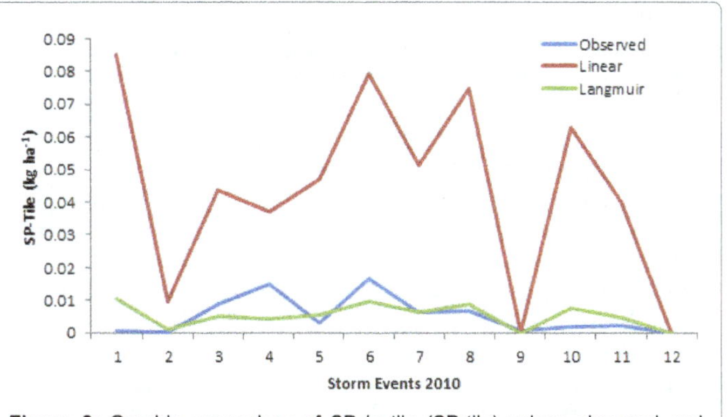

Figure 3: Graphic comparison of SP in tile (SP-tile) values observed and predicted in APEX during growing season in 2010.

modeling results (Kavetski et al.) [14]. In addition to this potential source of error, it becomes apparent that a preferential flow component would be essential in the estimation of tile flow, as well as the need to better understand and model P dynamics to improve SP-tile predictions. While the calibration values for SP-tile were unsatisfactory using the linear method, the Langmuir option provided a reasonable approximation for the P in tile values. Despite the still very poor calibration scores using the Langmuir P adsorption option, the nonlinear model followed the trends of the observed data and in most instances it under predicted the mass of P transported through the tile (Figure 3). The results indicate that storm mean and total values during the growing season in 2010, were similar between the observed and APEX (Table 3). While the observed storm mean value was 0.005 kg ha^{-1} and the total annual value was 0.063 kg ha^{-1}, the APEX predicted storm mean was 0.005 kg ha^{-1} and the total annual value was 0.065 kg ha^{-1}. Given the high SP concentration in water resulting from the chicken litter application in 2010, high SP values can be expected to move from the surface to the tile layer. The effective use of the nonlinear adsorption option at the surface corresponds with the more effective use of the same method at the subsurface for estimating SP-tile. These results are promising. Even though APEX modeling of SP losses in tile will need to be further improved, prediction outputs from the current model can serve as a starting point for the generalized comparison between practices and their effect on reducing nutrient transport through artificial drainage.

Conclusion

Provided the satisfactory calibration of APEX for most of the variables of interest, SP-tile modeling still needs improvement. Phosphorus estimations for tile use the same analytical approach as in surface runoff. Even though the model takes into consideration the soil characteristics, as well as other important processes in the fate of P such as plant uptake, a better understanding of the main transfer mechanisms and how they interact is necessary. On the other hand, the monitoring data may have not been sufficient to adequately evaluate SP-tile outputs in APEX. A larger dataset would provide greater confidence. Despite the restrictions of the measured data and the model limitations predicting SP-tile, APEX may be on the right track simulating P dynamics in the soil. The inclusion of the Langmuir adsorption isotherm provides evidence of the model's plasticity to adapt to environmental conditions in order to better simulate P transfer and transformation processes. Among the existing process-bases models, APEX is capable of broadly simulating tile flow and nutrient transport in tile, and can be utilized to roughly assess a variety of agricultural practices, which otherwise could not be reproducible experimentally due to cost and time constraints.

References

1. Beauchemin S, Simard RR, Cluis D (1998) Forms and concentrations of phosphorus in drainage water of twenty-seven tile drained soils. J Environ Qua 27: 721-728.
2. Walker WW, Havens KE (1995) Relating algal bloom frequencies to phosphorus concentrations in Lake Okeechobee. Lake and Reserv Manage 11: 77-83.
3. Lowrance R, Altier LS, Williams RG, Inamdar SP, Sheridan JM et al. (2000) REMM: The riparian ecosystem management model. J Soil Water Conserv 55: 27-34.
4. Sharpley AN, JR Williams (1990) EPIC-Erosion/Productivity Impact Calculator: 1. Model Documentation. US Department of Agriculture Technical Bulletin No. 1768.
5. Williams JR, Arnold JG, Kiniry JR, Gassman PW, Green CH (2008). History of model development at Temple, Texas. Hydro Sci J 53: 948-960.
6. William JR, Izaurralde RC (2000) The APEX model. BRC Report No. 006. Research Center, Temple, Texas, USA.
7. Williams JR, Jones CA, Dyke PT (1984). A modeling approach to determining the relationship between erosion and soil productivity. Trans ASAE 27: 129-144.
8. Williams JR (1995) The EPIC model. In: Computer Models of Watershed Hydrology, Water Resources Publications, Highlands Ranch, Colorado, USA.
9. Smith DR, King KW, Johnson L, Francesconi W, Richards P et al. (2014) Surface runoff and tile drainage of phosphorus in the Midwestern. US J Environ Qual 44: 495-502.
10. Wang X, Potter SR, Williams JR, Atwood JD, Pitts T (2006) Sensitivity analysis of APEX for national assessment. Trans ASABE 49: 679-688.
11. Jones CA, Cole CV, Sharpley AN, Williams JR (1984) A simplified soil and plant phosphorus model: I. Documentation. Soil Sci Soc Am J 48: 800-805.
12. Sharpley AN, Jones CA, Gray C, Cole CV (1984) A simplified soil and plant phosphorus model: II. Prediction of labile, organic, and sorbed phosphorus. Soil Sci Soc Am J 48: 805-809.
13. Gassman PW, Osei E, Saleh A, Rodecap J, Norvell S (2006) Alternative practices for sediment and nutrient loss control on livestock farms. Agric Ecosys Environ 117: 135-144.
14. Chung, SW, Gassman PW, Gu R, RS Kanwa (2002) Evaluation of EPIC for assessing tile flow and nitrogen losses for alternative agricultural management systems. Trans ASAE 45: 1135-1146.
15. Chung SW, Gassman PW, Huggins DR, Randall GW (2001) EPIC tile flow and nitrate loss predictions for three Minnesota Cropping Systems. J Environ Qual 30: 822-830.
16. Langmuir I (1918) The adsorption of gases on plane surfaces of glass, mica and platinum. Journal of the American Chemical society 40: 1361-1403.
17. Olsen SR, Watanabe FS (1957) A Method to Determine a Phosphorus Adsorption Maximum of Soils as Measured by the Langmuir Isotherm. Soil Sci Soc Am Proc 21: 144-149.
18. Pautler MC, Sims JT (2000) Relationships between soil test phosphorus, soluble phosphorus, and phosphorus saturation in Delaware soils. Soil Sci Soc Am J 64: 765-773.
19. Vadas PA, Krogstad T, Sharpley AN (2006) Modeling phosphorus transfer between labile and non-labile soil pools: updating the EPIC model. Soil Sci Soc Am J 70: 736-743.
20. Rossi CG, Heil DM, Bonumà NB, Williams JR (2012) Evaluation of the Langmuir model in the Soil and Water Assessment Tool for a high soil phosphorus condition. Environ Mod Soft 38: 40-49.
21. Vanden BAJ, Protz R, Tomlin AD (1999) Changes in pore structure in a no-till chronosequence of silt loam soils, southern Ontario. Can J Soil Sci 79: 149-160.
22. Sims JT, Simard RR, Joern BC (1998) Phosphorus losses in agricultural drainage: Historical perspective and current research. J Environ Qual 27: 277–293.
23. Knisel WG (1980) CREAMS: A field-scale model for chemicals, runoff, and erosion from agricultural management systems. U.S. Department of Agriculture, Science and Education Administration, Conservation Research Report No. 26.
24. Salton NA, Wayne ES (2009) Arkansas Soil Fertility Studies -2008, Arkansas Agricultural Experiment Station, Division of Agriculture, University of Arkansas System, Research Series 569.
25. Leonard RA, Knisel WG, Still DA (1987) GLEAMS: Groundwater Loading Effects of Agricultural Management Systems. Trans ASAE 30: 1403-1418.
26. Richardson CW, Bucks DA, Sadler EJ (2008) The Conservation Effects Assessment Project benchmark watersheds: synthesis of preliminary findings. J Soil Water Conserv 63: 590-604.
27. Flanagan DC, Huang C, Pappas EA, Smith DR, Heathman GC (2008) Assessing conservation effects on water quality in the St. Joseph River watershed. Proc Agro Environ 1-12.
28. Francesconi W, Smith DR, Heathman GC, Wang X, Williams CO (2014) Monitoring and APEX modeling of no-till and reduced-till in tile drained agricultural landscapes for water quality. ASABE 57(3): 777-789.
29. Wang X, Williams JR, Gassman PW, Baffaut C, Izaurralde RC et al. (2012) EPIC and APEX: Model use, calibration, and validation. Trans ASABE 55: 1447-1462.
30. Moriasi DN, Arnorld JG, Van LMW, Bingner RL, Harmel RD et al. (2007) Model evaluation guidelines for systematic quantification of accuracy in watershed simulations. Am Soc Agric Biol Eng 50: 885-900.
31. Sharpley AN (1982) A prediction of the water extractable phosphorus content of soil following a phosphorus addition. J Environ Qual 11: 166-170.
32. Nicholson FA, Chambers BJ, Smith KA (1996) Nutrient composition of poultry manures in England and Wales. Biores Techn 58: 279-284.
33. Sharpley AN, Chapra SC, Wedepohl R, Sims JT, Daniel TC et al. (1994) Managing agricultural phosphorus for protection of surface waters: Issues and options. J Environ Qual 23: 437-451.
34. Nash JE, Sutcliffe JV (1970) River flow forecasting through conceptual models: Part I. A discussion of principles. J Hydro 10: 282-290.
35. Gassman PW, Williams JR, Wang S, Saleh A, Osei E et al. (2010) The Agricultural Policy Environmental Extender (APEX) model: An emerging tool for landscape and watershed environmental analyses. Trans ASABE 53: 711-740.
36. Saleh A, Gassman PW, Abraham J, Rodecap J (2003) Application of SWAT and APEX models for Upper Maquoketa River watershed in northeast Iowa. ASAE Paper No. 032063.
37. Holford ICR 1997. Soil phosphorus: its measurement, and its uptake by plants. Aust J Soil Res 35: 227-239.
38. Breeuwsma A, Reijerink JGA, Schoumans OF 1(995) Impact of manure on accumulation and leaching of phosphate in areas of intensive livestock farming. Animal waste and the land-water interface. Lewis Public, CRC, New York pp: 239-249.
39. Smith DR, Haggard BE, Warnemuende EA, Huang C (2005) Sediment Phosphorus dynamics for three tile fed drainage in Northeast Indiana. Agri Water Manage 71: 19-32.
40. Gentry LE, David MB, Royer TV, Mitchell CA, Starks KM (2007) Phosphorus transport pathways to streams in tile-drained agricultural watershed. J Environ Qual 36: 408-415.
41. Steglich E, Williams JR (2013) Agricultural Policy/Extension extender: User manual version 0806. Blackland Research and Extension Center.
42. Richards, RP, Calhoum FG, Matisoff G (2002) The Lake Erie agricultural systems for environmental quality project: an introduction. J Environ Qual 31: 6-16.
43. Riedel M, Selegean J, Dahl T (2010) Sediment budget development for the Great Lakes region. 2nd Joint Federal Interagency Conference, Las Vegas, NV.
44. Stone WW, Wilson JT (2006) Preferential flow estimates to an agricultural tile drain with implications for glyphosate transport. J Environ Qual 35: 1825-1835.
45. Algoazany AS, Kalita PK, Czapar GF, Mitchell JK (2007) Phosphorus transport through subsurface drainage and surface runoff from a flat watershed in east central Illinois, USA. J Environ Qual 36: 681-693.
46. Shipitalo MJ, Dick WA, Edwards WM (2000) Conservation tillage and macropore factors that affect water movement and the fate of chemicals. Soil Tillage Res 53: 167-183.
47. Kavetski D, Kuczera G, Franks SW (2006) Bayesian analysis of input uncertainty in hydrological modelling: 1. Theory, Water Resources Research, 42 Article W03407.

Life-Cycle Assessment of Neonicotinoid Pesticides

Natalia TG[1] and Robert MH[2]*

[1]*CAPES Foundation, Ministry of Education of Brazil, Brazil*
[2]*Sustainable Futures Institute, Michigan Technological University, Houghton, MI, USA*

Abstract

Neonicotinoid pesticides have been an increasing focus of the environmental community, due to their potential impacts on bee populations and other important insects. The goal of this study was to develop a life-cycle assessment (LCA) approach that could be used to quantify the environmental impacts of two common neonicotinoid pesticides, Imidacloprid and Thiamethoxam. In order to develop the LCA study, an equivalent scenario was created for each pesticide that incorporated data on the production of each pesticide, followed by transportation to a model farm site in Brazil and application with an in-furrow pesticide application system. Data sources for the materials and energy used, combined with resulting emissions to air, water, and soil, were gathered from peer-reviewed literature, government reports, life-cycle inventory databases, and other sources. The SimaPro LCA modeling platform was used to assess the impacts of each life cycle on Human Health and Ecosystem Quality, according to the Impact 2002+ method. Results indicate that important differences exist between the pesticide life cycles, with Thiamethoxam resulting in lower LCA impacts in both impact areas and most mid-point categories under study. Pesticide production impacts varied by over an order of magnitude between the Imidacloprid and Thiamethoxam, while pesticide transport was determined to be a negligible source of environmental impact in both systems. Pesticide application activity using tractors was a larger contributor to Human Health and Ecosystem Quality impacts than the ultimate effect of the pesticide emission to the environment, which should be an area of further study to confirm this finding with the pesticides in question and also to focus impact on the application systems as a potential method of reducing environmental impact, in addition to pesticide toxicity.

Keywords: Neonicotinoid; Pesticide; Imidacloprid; Thiamethoxam

Introduction

Insecticides are a class of pesticides used to kill, harm, or repel different species of insects. They act in different ways in organisms based on their active ingredients. For instance, corn plantations commonly use insecticides that have organophosphates and carbamate as the active ingredient, which acts on the enzyme acetylcholinesterase within an insect nervous system. In many cases, these standard insecticide products are being phased out for a new class of insecticide known as neonicotinoids, which use nicotine as the active ingredient. Neonicotinoid compounds interact with nicotinic acetylcholine receptors (nAChR) of the central nervous systems of insects [1]. Nicotine acts in an insect's system in the same way that it acts in the human body. However, neonicotinoids are more toxic for invertebrates than they are to mammals, birds and other higher organisms. Neonicotinoids became popular because of their high water solubility, which makes their soil application travel through the entire plant. Nowadays, neonicotinoids are one of the most widely used class of insecticides for controlling sucking insects and soil insects. In 2004, the worldwide annual usage of neonicotinoids was approximately 11-15% of the total insecticides in the market [2]. Different generations of neonicotinoids have been created over time. They have the same principle of action in the nervous system; however, the specific active ingredients are different. The first generation of this pesticide class used was 1-(6-chloro-1,3-thiazol-5-ylmethyl)-1,3,5-oxadiazinan-4-ylidenene(nitro)amine, known as Imidacloprid. It was first registered for use in the United States by the United States Environmental Protection Agency in 1994. It is the most widely used generation of neonicotinoids, and there are several hundred Imidacloprid-based products for sale in U.S. In 2006, the worldwide sales of those products were near $1.6 billion [3]. In 2009, Imidacloprid was applied to hundreds of thousands of acres in California, one of the most used pesticides that time [4]. Some commercial examples of this class of insecticide are Gaucho® (seed treatment), Admire® (soil application), Provado® (foliar application), Merit® (turf and ornamental use) and Premise® (termite control). Most are produced by Bayer Company [5]. The second generation of neonicotinoids is known as Thiamethoxam. The most active ingredient is 1-(6-chloro-1,3-thiazol-5-ylmethyl)-1,3,5-oxadiazinan-4-ylidenene(nitro)amine. It was first approved in 1999 for use as an antimicrobial wood preservative and as a pesticide. The main products of this generation are produced by Syngenta Company, including Platinum®, Actara®, Centric®, Cruiser®. Flagship® and Helix® among others [6]. Those products were introduced in the US market in 2001 [5]. Even though neonicotinoids have different effects in mammals and insects, they are a source of large concerns in the world. Many countries in the EU and around the world have banned the use of these chemicals. This is primarily because some classes of neonicotinoids have been demonstrated to be quite toxic for bees and other beneficial insects. Many studies show that bee disorders are being caused by the contamination of this type of insecticide in plant nectars that feed bees [7]. These beneficial insects are responsible for more than $15 billion in crop production in United States annually [8]. Many studies are being developed to analyze the actual risk of neonicotinoids in those insects. Also, neonicotinoids can persist in the soil for years, so it may contaminate other plants and non-target species over time. Pesticides may contaminate water, soil, fish, and other living

***Corresponding author:** Robert MH, Sustainable Futures Institute, Michigan Technological University, Houghton, MI, USA, E-mail:rhandler@mtu.edu

species [9,10]. One important study method used to assess the potential environmental impacts of products and systems is Life Cycle Assessment (LCA). This technique has been applied to many different products throughout many different industries, but the generally accepted best practices for LCA consist of (1) rigorously defining the goal, scope, and system boundary of the study, (2) compiling an inventory of the important flows of materials and energy throughout the product life cycle system boundary, (3) assessing the environmental impacts of this inventory data with a transparent and replicable impact assessment method, and (4) interpreting data to define new study conditions or suggest improvements to the product life cycle [11]. Ideally, an LCA study should be cradle-to-grave, examining environmental impacts throughout the product life cycle, from extraction of raw materials, to product assembly, use, and finally to disposal of wastes. In this article, a life cycle assessment method is used to quantify the environmental impacts of production, transport, and usage of the active ingredients in two common neonicotinoids, Imidacloprid and Thiamethoxam. SimaPro 8.0 LCA software was used in conjunction with the Ecoinvent life cycle inventory database [12] to construct product life cycles for both pesticide active ingredients and assess the impacts, in terms of regards of human health and ecosystem quality. Additional discussions regarding the toxicity and persistence in the environmental after their application, and how this would impact the environmental impacts of these products, are also offered based on a literature review of their use in typical cultivation systems.

Study Methods

Goal and scope

The goal of this study is to determine the environmental impacts of two different generations of neonicotinoids, Imidacloprid and Thiamethoxam, resulting from their production, transport, and ultimate application in the environment. The environmental impacts being assessed are terrestrial ecotoxicity, aquatic ecotoxicity, and human health impacts. The functional unit used in this experiment is 1 kg of insecticide applied to land. The system boundary includes production and transport of materials and energy used in the production, transport, and application stages of each pesticides life cycle. Figure 1 displays the system boundary along with key inputs that are described in more detail below. In each scenario for the two different pesticides, the application stage is assumed to take place on a farm in the State of Ceará, Brazil, as a common point of reference.

Life cycle inventory

To accomplish this life cycle assessment study, information on the inputs of materials and energy related to the full product life cycles was collected from a variety of peer-reviewed publications, government reports, and other sources. The following sections summarize the key inputs of materials and energy that have been considered in each stage of the insecticide life cycle, along with key assumptions relating to the development of this input data.

Production stage: Imidacloprid production was summarized in a 2006 article originating from the Institut fur Lebensmittelchemie in Germany [13], and this information is utilized in this study to model the production of this pesticide. There are alternate routes to production [14], but few studies include as much detail on the synthesis as the Schippers and Schwack work [13]. The synthesis is divided in five steps with an overall yield of 10%. In the beginning, ammonia is used to react with coumalic acid methyl ester to form the hydroxyl nicotinic acid. This is the cheapest way to start the production of the insecticide. All products and their amounts are specified in Table 1, with an estimated production of 1.01 grams of Imidacloprid. These amounts were scaled up proportionally to the reported yield in order to model a production of 1 kg in the life cycle model. In the synthesis procedure, some reactions need heating and cooling as part of the process. In order to estimate the energy usage required for these steps in a commercial application, an assumption was made that heat can be provided or removed from the system via a heat exchange system with an efficiency of 80%, using water as the exchange fluid. Some reagents utilized in this synthesis method were not available in the Ecoinvent database, and in these cases similar products were used or the creation of the required reagents was modeled using required inputs when present within the Ecoinvent database. These assumptions have been noted in Table 1. Reagents used just to wash or extract a chemical were assumed to require 50 mL per washing step in this production process, unless otherwise indicated in Table 1. Thiamethoxam synthesis was modeled according to the route described in a paper from Syngenta Crop Protection researchers [15]. The materials and energy requirements for synthesis of 220 g of Thiamethoxam at 98% purity are presented in Table 2, and are scaled up to a production of 1 kg with our LCA model to be consistent with the functional unit of this study. Similar assumptions are used in this production unit operation as were used for Imidacloprid, namely that heat exchangers can or remove heat from the system while operating at an efficiency of 80%, and the specific heat of water is used to represent the solution heat capacity and working heat exchange fluid. The main compound used in this production system is dimethyl carbonate. It is used in the production in different steps, however in Table 2 it is shown in a single amount.

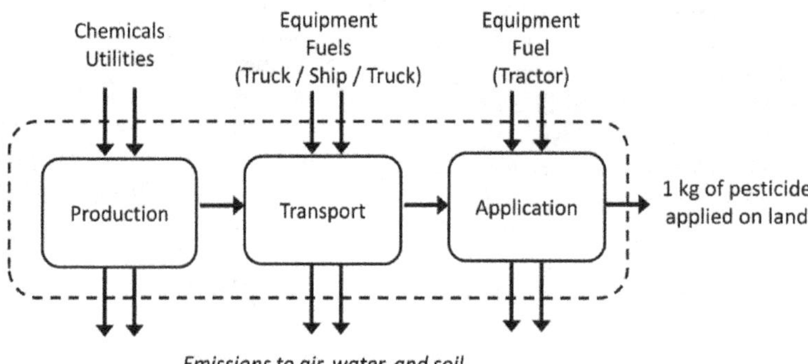

Figure 1: General description of LCA system boundary for both pesticide scenarios under study. Inputs at each stage are normalized to ultimate application of 1 kg of the pesticide to agricultural land. Emissions to air, water, and soil are used to calculate environmental impact according to common LCA methods.

Item	Amount	Comments
		Chemicals
Coumalic Acid methyl ester	4 g	Coumalic acid methyl ester is synthesized from malic acid, using fuming H_2SO_4 and absolute methanol. This compound was created in Ecoinvent database according to Smith and Wiley [16].
Ammonia	10 g	
Sodium hydroxide (21%)	14.7 g	Density=1.222 g/mL
Hydrochloric acid	12.2 g	Required amount necessary until pH of 4.5. Assumed equivalent volume to NaOH. Density=1.017 g/mL
Phosphorus pentachloride	3.56 g	
Phosphorus oxychloride	2.80 g	Utilized Phosphoryl chloride in Ecoinvent database. Density=1.645 g/mL
Sodium borohydride	700 mg	Utilized sodium tetrahydridoborate in Ecoinvent database.
Ethanol	150 mL	Density=0.789 g/mL, mass=118.35 g.
Sodium hydroxide (0.01N)	1.5 mL	Density=1.011 g/mL, mass=1.52 g.
Sodium chloride	50 mL	Assumed amount to treat the residue.
Diethyl ether	50 mL	Assumed amount to extract the organic phase.
Anhydrous sodium sulfate	0.94 g	Assumption: used for drying and can contain $7H_2O$ when fully hydrated, 1 mol of sodium sulfate is needed to remove 10 mols of water in each mol of product
Thionyl chloride	0.79 g	Assumption: In chloroform, and 1 mol of thionyl chloride (118.97 g/mol) is necessary to react with 1 mol of N-(6-chloropyridin-3-yl)methanol (143.57 g/mol) that has 0.95g grams.
Potassium carbonate	3.9 g	
Acetone	100 mL	
Ethylenediamine	0.462 g	Synthesis was created in SimaPro. 60.10 g/mol. 1 mol necessary to produce 1 mol of 2-nitroiminoimidazoline.
$(NH_2)_2CNNO_2$	0.799 g	Used in the production of 1 gram of 2-nitroiminoimidazoline (130.11 g/mol). Ecoinvent reference: nitro compound.
Ethyl acetate	100 mL	Assumption of the amount to extract the solution (3 times).
		Energy[a]
Heat to 48°C	1.69 kJ	Used the total mass (14 g), starting at room temperature
Heat to 100°C	7.81 kJ	Assumptions: Heated from 48°C. The total mass is the sum of all compounds in the solution at the moment.
Heat to 75°C	2.28 kJ	Assumptions: The mass of the previous product (2.33 g), and the both phosphorous compounds are used as the total mass. Starting at room temperature.
Heat to 120°C	2.05 kJ	Assumptions: The same total mass was used as previous. Heated from 75°C.
Cooling to 4°C	14.22 kJ	Assumptions: The total mass is the previous product+all compounds added. Initial temperature=room temperature.

a: In all calculations of required energy at each heating/cooling phase, the specific heat of water was assumed, along with a heat exchanger efficiency of 80%.

Table 1: Material and Energy Requirements for 1.01 g of Imidacloprid Production.

Item	Amount	Comments
		Chemicals
Dimethyl carbonate (DMC)	1050 g	Assumption: Created in SimaPro using Ecoinvent data for reagents of carbon monoxide, oxygen, and methanol according to Tundo and Selva [17]. This amount of DMC is used in different parts of the procedure.
3-methyl-4-nitroimino-perhydro-1,3,5-oxadiazine	184 g	Assumption: Used generic Diazine Compound present within Ecoinvent database as substitute.
2-chloro-5-chloromethylthiazole	168 g	Assumption: Created in Ecoinvent database using information present in Decker [18]
Tetramethylammonium hydroxide pentahydrate	4 g	Assumption: Created in Ecoinvent database using information present in Walker [19]
Potassium carbonate	242 g	
Water	900 g	Deionized water
Hydrochloric acid	260 g	
		Energy[a]
Heat to 65°C	157.54 kJ	Assumptions: Mass: sum of all compounds for the first mixture. Initially at room temperature.
Heat at 62°C to 68°C	18.73 kJ	Assumptions: The sum of all compounds mass in the mixture are used in the calculation. Initial temperature=62°C.
Cool to 47°C	65.55 kJ	Assumptions: When 99% of product is formed, the mixture is cooled. Total mass of reagents used as mass.
Heat to 65°C	111.04 kJ	Assumptions: The mass used was the sum of a possible product of 200 g+the water and hydrochloric acid. Initial temperature is 45°C.
Heat to 65°C	125.7 kJ	Assumptions: The mass used was the final mass after the heating. The initial temperature was room temperature after waiting for phase separation.
Cool at 65°C to 5°C	188.55 kJ	Assumptions: The mass used was the final mass after the heating.
Cool at 5°C	1 kWh	Electricity used to keep the mixture temperature at 5°C for 1 hour, 1 kW power requirement
Vacuum at 70°C	176.77 kJ	Assumptions: the mass used was the sum of a possible product of 200 g+the water and 300 g of DMC. Initial temperature is room temp.

a: In all calculations of required energy at each heating/cooling phase, the specific heat of water was assumed, along with a heat exchanger efficiency of 80%.

Table 2: Material and Energy Requirements for 220 g of Thiamethoxam Production.

	Imidacloprid	Thiamethoxam	Comments
Production Facility	Grundhof, Schleswig-Holstein State, Germany	Corlim, Goa State, India	Thiamethoxam produced at Syngenta Agro Chemicals facility [20] Imidacloprid is primarily produced in Germany, with minor amounts produced at other locations [21,22]
Distance to Port Facility	17 km Flensburg	32 km Mormugao	Truck transport assumed in Ecoinvent (16-32 t lorry)
Ocean Shipping Distance	6043 nm	9481 nm	Distance between ports determined via http://ports.com, long-distance transoceanic ship transport assumed in Ecoinvent (Distances presented in nautical miles)
Distance to Agricultural Site	150 km	150 km	Truck transport assumed in Ecoinvent (16-32 t lorry)

Table 3: Key inputs and assumptions for Transportation stage of pesticide life cycles.

Transportation stage: Inputs and assumptions for the transportation stage can be seen in Table 3, where 1 kg of the pesticide was moved according to the following multi-modal transportation scheme. Each of the neonicotinoid pesticides are produced in commercial quantities at locations far away from the farm field in Ceará where the pesticides are used in this scenario, which reflects the global supply chains that many chemicals currently have. For each pesticide, the predominant production facility was assumed according to available public information. Transportation from the production facilities in India (Imidacloprid) and Germany (Thiamethoxam) to the nearest local commercial scale port facility was accomplished with truck transportation profiles available in the Ecoinvent database, followed by shipping from the ports in the respective countries to the city of Fortaleza. Final truck transportation to a farm field in central Ceará was assumed to be equivalent in each scenario.

Application stage: In the application stage, this LCA study attempts to quantify the impacts associated with the actual application activity required to spread the pesticides on a farm field, along with the resultant impacts of that soil application. The use of machinery to apply insecticides to the field is included in the study. The most common way to apply neonicotinoids in the environment is using the in-furrow method with tractors, which is faster and safer for workers. This is possible because both pesticides area able to translocate throughout the plants from the roots. AmTide Imidacloprid 2F, a commercial product of Imidacloprid, is applied in a maximum rate in of 0.35 kg / ha in potatoes and other tuber-based agricultural cultivation systems [23]. Thiamethoxam needs to be applied in a rate of 0.14 kg/ha of soil [24] in for cultivation of potatoes and other tubers. Using a consistent tractor fuel usage rate of 2.6 L diesel/ha of soil [25], it is possible to calculate the tractor fuel requirements for spreading 1 kg of each pesticide. To account for the other inputs of materials and energy (tractor production, equipment use) and other engine emissions, a pesticide application process profile in Ecoinvent was modified to include the appropriate level of fuel consumption and fuel combustion per hectare as indicated in Ref. [25], and used to model the overall application process. After being applied in the environment, pesticides generally move into soil and water at variable rates, where they may interact with several different non-target species, and this is a subject of much recent research interest. The half-life of Imidacloprid in soil has been reported in different conditions. On average, it is 130 days, but it can increase with the pH or absence of light [26]. For Thiamethoxam, the half-life is 47-54 days in presence of light [27]. It means that both pesticides persist in the soil for a long time so it can interact with other plants or living species in the soil. The study of hydrolysis of those insecticides is a useful method to analyze their persistence and potential contamination of water bodies. After application, neonicotinoids can reach either lakes or groundwater. Sunlight makes the hydrolysis faster, however Imidacloprid can persist in the water for more than 30 days in pH 7 [26] and Thiamethoxam can be stable at pH 5 and persist for 580 days in pH 7 [27]. As insecticides are persistent in the soil and water, they can interact with different living species. Neonicotinoids act in different ways in mammals and insects, but they can be toxic to humans and mammals in general. Many studies show the lethal dose in rats or rabbits as model species. Once in the environment, Imidacloprid and Thiamethoxam can get in human bodies from different routes, such as oral, dermal, and inhalation, with potentially different toxic effects from different routes of exposure. Using rats as an example, the lethal dose for 50% of exposed animals (LD_{50}, a common toxicity metric) via oral exposure is 450 mg/kg for Imidacloprid [28]. For Thiamethoxam, the LD_{50} is 100 mg/kg [29]. They are classified as moderately toxic. Some signs of toxicity with Imidacloprid in humans are drowsiness, dizziness, vomiting, disorientation, increased heart and respiratory rates, and fever. However, when neonicotinoids contaminate foods, they generally do not represent a risk for human health due to the low dosage and relatively rapid excretion. In regards to carcinogenicity, Imidacloprid has been classified into Group E, meaning there is no evidence of carcinogenicity in studies with rats [30]. However, a study of human lymphocytes exposed to greater than 5200 μg/ml of Imidacloprid demonstrated a slight increase in chromosome abnormalities *in vitro* [31]. Studies for Thiamethoxam show that it does not have a carcinogen risk for humans and rats too [32]. In bees or other insects, the impacts can be more serious. Since bees are more similar to the target insects of neonicotinoids, the lethal dose for them is smaller. Because of it, many disorders are being caused in different cultivations around the world. Pesticides can affect bees though direct contact or though ingestion of pollen or nectar. The lethal dose needed to kill 50% of bees is 0.0037 μg/bee of Imidacloprid via oral and 0.0179 μg/bee via contact [33]. For Thiamethoxam, the LD_{50} for honey bees is 0.005 μg/bee via oral and 0.024 μg/bee via contact [34]. Thus, both classes of insecticides are classified as highly toxic to honey bees. Since Imidacloprid and Thiamethoxam are not present within the set of pesticides available within the current Ecoinvent database or the environmental impact assessment methods available to quantify the ecosystem or human health impacts, a similar insecticide available within the LCA modeling platform was used as a proxy. Relative toxicity values between the three pesticides were then used to simulate the appropriate exposure impact in land, water, and air. Fenvalerate is a pyrethrin insecticide, and is considered moderately toxic to mammals and highly toxic to insects, including bees. The LD_{50} for Fenvalerate in rats is 451 mg/kg, which is very similar to Imidacloprid and 4.5-fold higher than Thiamethoxam. The assumption used for this study followed the guidance of available rat exposure data, that 1 kg of Imidacloprid emission to the environment could be reasonably represented by 1 kg of Fenvalerate emissions, while 4.5 kg of Fenvalerate emissions would have to be used to represent 1 kg of Thiamethoxam emissions to the environment. Fenvalerate is classified as highly toxic to bees with a reported LD_{50} of 0.017 μg / bee, similar to the neonicotinoid pesticides [35].

Impact assessment

To analyze different environmental impacts, the SimaPro LCA tool includes several established impact assessment methods that characterize different aspects of environmental impact. The method used in this experiment is Impact 2002+. This methodology proposes a feasible implementation a combined midpoint/damage approach, linking all types of life cycle inventory results via 14 midpoint categories to 4 damage categories [36], which is explained in greater detail in other resources. For this study, the two endpoint damage categories being assessed are ecosystem quality and human health impacts. Ecosystem quality is quantified by combining multiple midpoint indicators and is expressed in terms of Potentially Disappeared Fraction (PDF) of species over a certain area per year, in this case PDF/m^2/yr. Human Health impacts are also quantified by combining several midpoint indicators, and are cumulatively expressed in terms of Disability-adjusted Life Years (DALY), accounting for human health in terms of changing mortality and morbidity related to environmental impacts.

Results and Discussion

A summary of endpoint results for the life cycle Human Health and Ecosystem Quality impacts of each pesticide can be seen in Table 4. Results are separated into each stage of the life cycle, and several of the most impactful items in each stage are shown as well. Figures 2 and 3 display the results of the midpoint indicators that contribute to the ultimate quantification of the two endpoint categories. Because each of the midpoint categories are assessed using different units, a normalization procedure was performed in order to display them on the same Figures with each other. Life cycle results for Imidacloprid in each of the midpoint categories were normalized to a value of 1.0, to remove the influence of different units. Life cycle results for Thiamethoxam were also normalized on the same basis, in order to facilitate comparison between the two pesticides across a range of midpoint categories.

Imidacloprid life cycle

The production stage of the imidacloprid life cycle is a significant source of environmental impacts, especially concerning Human Health, where roughly 90% of the life cycle impacts occur (7.11×10^{-4}

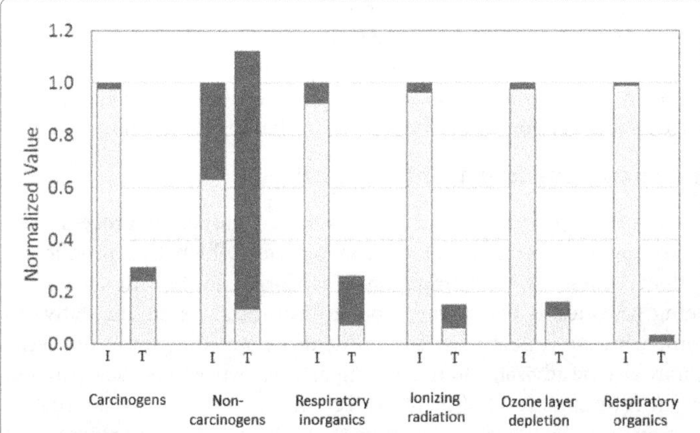

Figure 2: Midpoint indicator results for categories that contribute to Human Health endpoint impact metric, using a functional unit of 1 kg pesticide. Data for different midpoint indicators are measured according to different units, but here they have been normalized by normalized the life cycle impact value of Imidacloprid ("I" columns) to a value of 1.0. Thiamethoxam ("T" columns) values for each midpoint category were normalized according to the Imidacloprid scale, to illustrate the comparison between the two pesticides. In all columns, light colored bars represent impacts for the Production stage and dark colored bars represent impacts from the Application stage. The Transportation stage is too small to be observable in all categories.

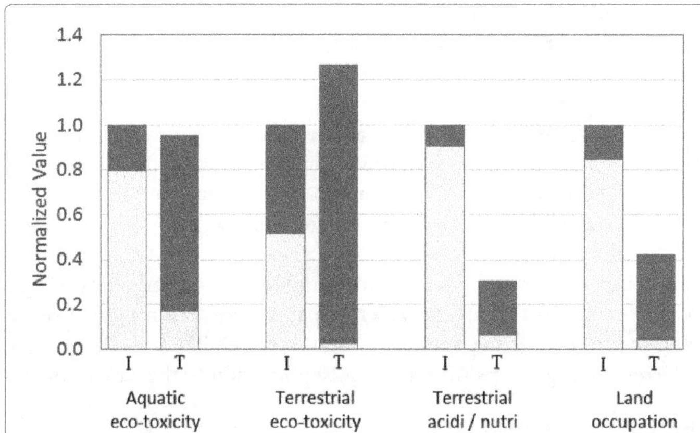

Figure 3: Midpoint indicator results for categories that contribute to Ecosystem Quality endpoint impact metric, using a functional unit of 1 kg pesticide. Data for different midpoint indicators are measured according to different units, but here they have been normalized by normalized the life cycle impact value of Imidacloprid ("I" columns) to a value of 1.0. Thiamethoxam ("T" columns) values for each midpoint category were normalized according to the Imidacloprid scale, to illustrate the comparison between the two pesticides. In all columns, light colored bars represent impacts for the Production stage and dark colored bars represent impacts from the Application stage. The Transportation stage is too small to be observable in all categories.

	Imidacloprid		Thiamethoxam	
Stage	Human Health (DALY)	Ecosystem Quality (PDF / m^2 / yr)	Human Health (DALY)	Ecosystem Quality (PDF / m^2 / yr)
Production	**7.11 x10^{-4}**	**98.21**	**7.80 x10^{-5}**	**5.70**
	ethyl acetate 2.03 x10^{-4}	*ethyl acetate 34.0*	*coumlic acid 4.35 x10^{-5}*	*diazine 2.08*
	diethyl ether 1.78 x10^{-4}	*diethyl ether 25.7*	*diazine 1.55 x10^{-5}*	*coumalic acid 2.04*
	acetone 1.01 x10^{-4}	*NaCl 12.2*	*dimethyl carbonate 7.95 x10^{-6}*	*dimethyl carbonate 0.71*
Transport	**1.72 x10^{-7}**	**0.02**	**2.59 x10^{-7}**	**0.03**
Application	**6.37 x10^{-5}**	**72.0**	**1.61 x10^{-4}**	**184.0**
	tractor use 6.23 x10^{-5}	*tractor use 69.5*	*tractor use 1.55 x10^{-4}*	*tractor use 173*
	pesticide emission 0.14 x10^{-5}	*pesticide emission 2.5*	*pesticide emission 0.06 x10^{-4}*	*pesticide emission 9.0*
Total	**7.75 x10^{-4}**	**170.24**	**2.40 x10^{-4}**	**189.73**

Table 4: Life cycle endpoint results for Human Health and Ecosystem quality impacts (1 kg basis).

DALY). These large impacts at the production stage stem from the reliance of pesticide production on organic chemicals such as ethyl acetate or acetone to serve as solvents or washing agents for various parts of the production process. In addition to these organic agents, the NaCl used in the production stage also contributes strongly to the overall Ecosystem Quality damage of the life cycle. The small amounts of heating and cooling required for the various steps of the production stage did not contribute a significant fraction of the overall impact in either endpoint category, less than 1% of the total environmental impact in each category. Transportation of Imidacloprid from its production location in Germany to the farm field in Brazil contributes a negligible amount of impact to both the Human Health and Ecosystem Quality impact metrics. Despite the long voyage, transportation of these products is done on a large scale where 1 kg of pesticide constitutes a trivial portion of the payload in every transport step. Pesticide application is an important part of the life cycle, contributing ~8% of the overall Human Health impact, and 42% of the Ecosystem Quality impact. Interestingly, in both of these impact metrics, the contribution of the tractor use and associated infrastructure for the application of the pesticide was ~30-40X more impactful than the resulting environmental impacts associated with emission of the pesticide into the environment, as modeled by our fenvalerate proxy in the SimaPro modeling tool. Midpoint indicators from the LCA study are presented below in Figures 2 and 3 for indicators contributing to the overall Human Health and Ecosystem Quality results, respectively. In regards to the Human Health impact of Imidacloprid, the production stage was responsible for over 95% of the impacts in 5 of 6 midpoint categories, the only exception being exposure to Non-carcinogenic air pollutants. For the midpoint indicators leading to the Ecosystem Quality result, the impacts were also primarily attributable to the production stage. For Terrestrial ecotoxicity, however, impacts were split roughly evenly between production and application stages, which illustrate the impact on terrestrial organisms from pesticide application to the landscape.

Thiamethoxam life cycle

The LCA results for the Thiamethoxam life cycle (Table 4) illustrate some important differences from Imidacloprid. The production stage of the Thiamethoxam life cycle results in Human Health impacts that are roughly 10X smaller than the Imidacloprid system and Ecosystem Quality impacts that are roughly 20X smaller. The items responsible for the largest impact in the Thiamethoxam life cycle in both impact categories are actual regents or building blocks of reagents in the production system (coumalic acid), as opposed to the organic solvents and washing agents in the Imidacloprid system. Heating and cooling requirements were also shown to contribute less than 1% of the environmental impacts associated with the production stage, similar to the Imidacloprid case. Also similarly, the transportation stage for moving Thiamethoxam from India to Brazil was negligible in the overall product life cycle, compared to impacts associated with production or application. In the application stage, the Thiamethoxam system used more tractor transport and associated materials and energy to spread the pesticide, because the reported dosage per unit land area was lower. Most of the environmental impact at the application stage was again due to the usage of the tractor for pesticide application, as opposed to the impacts associated with Thiamethoxam exposure on the landscape, as modeled by the fenvelarate proxy compound in this study. Overall, the Thiamethoxam life cycle had Human Health impacts that were 70% lower than Imidacloprid (2.40×10^{-4} DALY vs. 7.75×10^{-4} DALY). This general result can be seen in most of the midpoint indicators that contribute to the Human Health impact (Figure 2), where the Thiamethoxam life cycle impacts are generally less than 25% of the comparable midpoint indicator score for Imidacloprid. Non-carcinogenic air pollution is the one midpoint indicator that does not follow this trend, with a higher score for Thiamethoxam that is primarily due to emissions from the production stage, which makes sense due to the higher tractor use assumed for this scenario in our 1 kg functional unit comparison, and engine exhaust is an important source of these air emissions in this life cycle. Total Ecosystem Quality impacts are 11% higher for the Thiamethoxam (189.73 vs. 170.24 PDF / m^2 / yr). In the comparison of midpoint indicators contributing to the Ecosystem Quality impact score (Figure 3), very clear differences are observed between the two pesticides. In the Thiamethoxam life cycle, the production stage is the primary contributor to all Ecosystem-related midpoint indicators, and for Aquatic and Terrestrial Ecotoxicity, the total impacts are equivalent or higher than Imidacloprid.

Alternative functional unit comparison

When conducting an LCA study of a product or process, it is important to keep in mind the function for which that product or system is intended. For a product such as a pesticide, the ultimate service being provided is plant protection from insects. As an alternative to comparing equivalent masses of pesticide over a comparable life cycle, it may also be advisable to make comparisons between pesticides based on an equivalent level of plant protection. To illustrate this alternative comparison, the life cycle scenarios for both pesticides are represented in Table 5 on the basis of 1 ha of field protection, using the same input data and assumptions that have been outlined previously in the article. Because the reported dosage rate for Imidacloprid (the first-generation neonicotinoid pesticide) is 2.48X higher than Thiamethoxam, less Thiamethoxam is needed to provide the same level of plant protection. When the life cycles for each pesticide are normalized to 1 ha of plant protection, the inherent advantage of this lower pesticide requirement is illustrated more clearly. On a 1 ha equivalent basis, the Thiamethoxam life cycle is 87% lower in Human Health impact and 45% lower in Ecosystem Quality, compared to the Imidacloprid life cycle. As opposed to the data presented in Table 4, in this alternative LCA scenario the impacts associated with pesticide application are nearly equivalent, because each case assumes tractor usage for 1 ha of pesticide application, with minor differences due to toxicity impacts of the pesticides after application. Many of the same general points about the two pesticide

	Imidacloprid		Thiamethoxam	
Stage	**Human Health (DALY)**	**Ecosystem Quality (PDF / m^2 / yr)**	**Human Health (DALY)**	**Ecosystem Quality (PDF / m^2 / yr)**
Production	2.47×10^{-4}	34.1	1.09×10^{-5}	0.798
Transport	5.96×10^{-8}	0.008	3.63×10^{-8}	0.004
Application	2.21×10^{-5}	25.0	2.26×10^{-5}	25.8
Total	2.69×10^{-4}	59.1	3.36×10^{-5}	26.6
Percent Reduction from Imidacloprid			87%	45%

Table 5: Life cycle endpoint results for Human Health and Ecosystem quality impacts (1 ha basis).

life cycles made in the preceding sections about individual life cycle stages and the important factors within each stage are still true in this alternative LCA scenario, however, such as the importance of individual ingredients in each pesticide production stage, the minor contribution of pesticide transport, and the importance of the tractor usage in comparison to effects due to emissions of pesticide onto the landscape in contributing to the overall life cycle impacts calculated here.

Conclusion

This study represents an illustration of a method for comparing environmental impacts of pesticides across their life cycles of production, transportation, and application. This LCA method has been applied to 2 neonicotinoid pesticides, using the best available data, to compare the environmental impacts of each life cycle and offer guidance on potential areas for improvement, both for future studies and the ultimate pesticide life cycle designs. In this study, Thiamethoxam appears to offer considerable advantages over the first generation Imidacloprid pesticide, on the basis of Human Health and Ecosystem Quality. Production data for commercial products made by private industry is always difficult to acquire for the purposes of public LCA studies, but due to the importance of the production stage in this LCA, efforts should be made to continually improve the understanding of how these pesticides are made on a commercial scale, in order to reduce the uncertainly associated with modeling this stage. Once the commercial processes are better understood, more guidance can be offered in terms of how best to reduce environmental impacts of concern through eliminating certain synthesis routes, reducing use of impactful chemicals and solvents, and other approaches. The life cycles of these pesticides are likely to involve global supply chains that link production and consumption locations, but initial efforts at modeling life cycles that involve significant transportation steps reveal that these transport modes appear to influence the overall life cycle in a minimal fashion. Pesticide application should be performed as efficiently as possible, and the ability to use less pesticide when protecting a given quantity of agricultural land can have positive repercussions throughout the supply chain, as less material needs to be produced, transported, and applied to yield the same impact. Although the neonicotinoid pesticides under study here have come under considerable scrutiny for their potential impacts on bee populations, the initial attempts here to model what the Human Health and Ecosystem Quality impacts would be for pesticide application seem to indicate that the impacts associated with operating the tractor have a larger ultimate impact than the actual emissions of the pesticide itself onto the land. This toxicity modeling was performed by relating the available common toxicity metrics for the neonicotinoid pesticide and a third proxy pesticide in the SimaPro LCA modeling platform. This approach has been described in detail, but additional toxicity data for the neonicotinoid pesticides should ultimately be verified by multiple research teams and made publicly available within life cycle inventory databases and modeling platforms to facilitate the study of these environmental impacts. In addition to focusing on the toxic impact of the pesticides as they are released in the environment, this initial study suggests that the environmental impacts associated with tractor usage during pesticide application are actually a larger source of environmental harm in multiple impact categories, and this should be one large focus of the goal of achieving life-cycle reductions in the impact of pesticide use in agricultural systems.

References

1. Yamamoto I, Casida JE (1999) Nicotinoid insecticides and the nicotinic acetylcholine receptor. p: 300.
2. Tomizawa M, Casida JE (2005) Neonicotinoid insecticide toxicology: mechanisms of selective action. Annu Rev Pharmacol Toxicol 45: 247-268.
3. Jeschke P, Nauen R (2008) Neonicotinoids-from zero to hero in insecticide chemistry. Pest Manag Sci 64: 1084-1098.
4. California Department of Pesticide Regulation (CA DPR) (2010) Summary of Pesticide Use Report Data 2009.
5. Fisher DL, Chalmers A (2007) Neonicotinoid Insecticides and Honey Bees: Technical Answers to FAQs. Ecotoxicology Section, Bayer CropScience LP.
6. Syngenta Crop Protection Group (2005) Thiamethoxam Envirofacts.
7. Hopwood J, Vaughan M, Shepherd M, Biddinger D, Mader E, et al. (2012) Are Neonicotinoids Killing Bees? A Review of Research into the Effects of Neonicotinoid Insecticides on Bees, with Recommendations for Action. The Xerces Society for Invertebrate Conservation, Portland.
8. Morse RA, Calderone NW (2000) The value of honey bees as pollinators of US crops in 2000. Bee culture 3: 1-15.
9. Hladik ML, Kolpin DW, Kuivila KM (2014) Widespread occurrence of neonicotinoid insecticides in streams in a high corn and soybean producing region, USA. Environ Pollut 193: 189-196.
10. Huseth AS, Groves RL (2014) Environmental fate of soil applied neonicotinoid insecticides in an irrigated potato agroecosystem. PLoS One 9: e97081.
11. International Organization for Standardization (ISO) (2006) 14044: environmental management—life cycle assessment—requirements and guidelines.
12. Adler PR, De Grosso SJ, Parton WJ (2007) Life-cycle assessment of net greenhouse-gas flux for bioenergy cropping systems. Ecol Appl 17: 675-691.
13. Amtide Imidacloprid 2F, Insecticide Informational Label. Amtide, LLC.
14. Weidema BP, Bauer C, Hischier R, Mutel C, Nemecek T, et al. (2013) Overview and methodology: Data quality guideline for the ecoinvent database version 3. Swiss Centre for Life Cycle Inventories.
15. Schippers N, Schwack W (2006) Synthesis of the ^{15}N-labelled insecticide imidacloprid. Journal of Labelled Compounds and Radiopharmaceuticals 3: 305-310.
16. Moriya K, Shibuya K, Hattori Y, Tsuboi SI, Shiokawa K, et al. (1992) 1-(6-Chloronicotinyl)-2-nitroimino-imidazolidines and related compounds as potential new insecticides. Bioscience, biotechnology, and biochemistry 56: 364-365.
17. Maienfisch P (2006) Synthesis and properties of thiamethoxam and related compounds. Zeitschrift für Naturforschung B 61: 353-359.
18. Krohn J, Hellpointner E (2002) Environmental fate of Imidacloprid. Pflanzenschutz-Nachr 55: 1-25.
19. Kollman W, Segawa R (1995) Interim Report Of The Pesticide Chemistry Database 1995. California Department of Pesticide Regulation.
20. Karmakar R, Singh SB, Kulshrestha G (2009) Kinetics and mechanism of the hydrolysis of thiamethoxam. J Environ Sci Health B 44: 435-441.
21. Meister RT (1995) Farm Chemicals Handbook. Meister Publishing Company. pp: 120-125.
22. Food and Agriculture Organization (2014) United Nations. FAO specifications and evaluations for agricultural pesticides: Thiamethoxam.
23. US Environmental Protection Agency (1995) Pesticide Tolerance and Raw Agricultural commodities: Imidacloprid. 40 CFR Part 180 Section 472.
24. World Health Organization (WHO) (2004) Toxicological Evaluations: Imidacloprid. International Programme on Chemical Safety, Roland Solecki.
25. Green T, Toghill A, Lee R, Waechter F, Weber E, et al. (2005) Thiamethoxam induced mouse liver tumors and their relevance to humans part 1: Mode of action studies in the mouse. Toxicological Sciences 86: 36-47.
26. European Food Safety Authority (EFSA) (2013) Conclusion on the peer review of the pesticide risk assessment for bees for the active substance thiamethoxam. EFSA J 11: 3067.
27. Laurino D, Manino A, Patetta A, Porporato M (2013) Toxicity of neonicotinoid insecticides on different honey bee genotypes. Bulletin of Insectology 66: 119-126.

28. US Environmental Protection Agency (EPA) (1991) Asana: Honey Bee Data Memorandum. Office of Pesticides and Toxic Substances.

29. Jolliet O, Margni M, Charles R, Humbert S, Payet J, et al. (2003) IMPACT 2002+: a new life cycle impact assessment methodology. The International Journal of Life Cycle Assessment 8: 324-330.

30. Decker M (2001) US Patent No. 6,214,998. Washington, DC: US Patent and Trademark Office.

31. Mathiesen K, Goldenburg S (2016) US government says widely used pesticide could harm honeybees. The Guardian.

32. Smith NR, WileyRH (1963) Isodehydroacetic acid and ethyl isodehydroacetate. Organic Syntheses. pp: 76-76.

33. Syngenta (2016) Company History.

34. The Economic Times (2004) Bayer Crop invests Rs 15 cr in insecticide plant via JV.

35. Tundo P, Selva M (2002) The chemistry of dimethyl carbonate. Acc Chem Res 35: 706-716.

36. Walker J, Johnston J (1905) C—Tetramethylammonium hydroxide. Journal of the Chemical Society, Transactions 87: 955-961.

Some Physical and Chemical Properties of Bio-fertilizers

Ramy Hamouda[1]*, Adel Bahnasawy[1], Samir Ali[1] and El-Shahat Ramadan[2]

[1]*Agricultural Engineering Department-Faculty of Agriculture-Benha University 13736, Egypt*
[2]*Microbiology Department-Faculty of Agriculture-Ain Shams University, Egypt*

Abstract

An experiment was conducted to study the effect of fermentation conditions on the properties of bio fertilizer from cow manure. The results indicated that the lowest value of bulk density was 946.63 kg/m³ at temperature of 50°C, 300 rpm agitation speed and 5 liter/min, ventilation rate while the highest value (983.17 kg/m³) obtained at temperature 30°C, 500 rpm agitation speed and 1 liter/min ventilation rate. The lowest moisture content was 79.87% at temperature of 30°C, agitation speed of 200 rpm and ventilation rate of 1 liter/min while the highest value was 83.19% at temperature of 50°C, agitation of 500 rpm and ventilation rate of 5 liter/min. The electrical conductivity increased from 11.6 ds m^{-1} at the start of fermentation to 35.07 dS m^{-1} at the end of fermentation period depending on the treatments under study. The pH decreased from 8.13 at the start of fermentation to 6.77 at the end of fermentation period. The total solid of the biofert decreased from 21.2% at the start of fermentation to 16.81% at the end of fermentation period, where the lowest value was 16.81% at temperature of 50°C, agitation speed of 500 rpm and ventilation rate of 5 liter/min and the highest value (20.13%) recorded at temperature of 30°C, agitation speed of 200rpm and ventilation rate of 1 liter/min. The lowest TN% was 0.41% at 50°C fermentation temperature, agitation speed of 500 rpm and ventilation rate of 1 liter/min while, the highest value (1.18%) obtained at temperature of 35°C, agitation speed of 300rpm and ventilation rate of 5 liter/min. The O.M% decreased from 34.2% at the start of fermentation to 10.97% at the end of fermentation period. Regarding the microbial changes, all treatments showed disappearance of pathogenic microorganisms at temperature of 50°C, at all agitation speeds and ventilation rates.

Keywords: Dairy manure; Biofert; fermentation; Agitation; Pathogenic; Ventilation

Introduction

Animal slurry is widely used as a fertilizer in organic farms. Dairy cattle typically produce between 42 kg and 64 kg (depending on body weight) of manure per day, so if they are housed for 50% of the year that corresponds to 7.6-11.6 tonnes per cow. In many developing nations, animal faeces have been composted and used to fertilize farm fields [1,2]. Many factors, including the type and concentration of substrate, temperature, moisture, pH, etc., may affect the performance of the anaerobic digestion process in the bioreactor [3,4]. The anaerobic digestion of organic waste is also an environmentally useful technology. [5] described the benefits of this process to reduce environmental pollution in two main ways: the sealed environment of the process prevents exit of methane into the atmosphere, while burning of the methane will release carbon-neutral carbon dioxide (no net effect on atmospheric carbon dioxide and other greenhouse gases). On the other hand, the anaerobic process has some disadvantages such as long retention times and low removal efficiencies of organic compounds [6]. Consequently, various physical, chemical and enzymatic pre-treatments are required to increase substrate solubility and accelerate the biodegradation rate of solid organic waste [7,8]. In the described manner of treating the liquid manure its temperature is augmented; our findings show that in the summer time that it is possible to supply the air from outside the barn to the aerator pump whereas in winter it is recommendable to supply warmed air from the barn interior. In this way the liquid manure heating is accelerated and at the same time the barn microclimate, from which the bad smell is removed, is improved. If during aeration the temperature of the liquid manure rises to 25-30°C, germination of the weed plant seeds, coming from the ingredients of the feed meal through the animals' digestive tract into the liquid manure, is reduced and a considerable number of parasites and disease-causers are destroyed. It has been found out that the fly larvae and the rodents, which are regular companions in the liquid manure storages; do not have optimal conditions for the procreation after a short time of the liquid manure aeration and homogenization. The main objective of this work was to study the possibility of using dairy manure in Biofert production to eliminate the pollution effect and contribute in agricultural bio fertilizer sacristy problem. To achieve this goal, the fermentation temperature, and agitation speed and ventilation rate as the most important factors affecting the physical, chemical and microbiological properties the Biofert were studied.

Materials and Methods

Materials used

Dairy manure: This waste was obtained from animal farms at Sekem, Sharkia Governorate, Egypt. The main components of this waste was: Total nitrogen (TN) of 1.42%; Phosphorus (P), of 0.012% ; Potassium (K), of 0.015%; Organic Matter (OM), of 34.2% and C/N, of 13.97 ; Moisture, of 78.8% pH, of 8.13 and Electrical Conductivity (EC), 11.6 dSm^{-1}(9280 ppm).

Measurements and instrumentation

Scanning thermometer was used to measure temperature (model, Digi-Sense 69202-30 measuring range from-250 to1800°C ±

***Corresponding author:** Ramy Hamouda, Agricultural Engineering Department–Faculty of Agriculture–Benha University 13736, Egypt,
E-mail: ramyya7oby@yahoo.com

0.1%,USA). Dissolved oxygen was measured by a dissolved oxygen meter (model, HI9143 measuring range from 0.01 to 300% O_2 ± 1.5% full scale % O_2, Italy). The pH was measured by the pH meter (model, ORION230A measuring range from 2 to 19.999 ± 0.005, USA). Electrical conductivity was measured by EC meter, ICM model ORION 105 measuring range from 0 to 199.990 0.1, USA). Ammonia (NH_3) as nitrite (NO_2) and nitrate (NO3) was measured by kjeldahl digestion (model, Vapodest measuring range from 0.1 mgN to 200gN ± 1.5% full scale % N,Germany). The total phosphorus was measured by the spectrophotometer (model, 6320 D measuring range from 0.1 to 1000 Concentration ± 1 nmλ,UK). The total potassium was measured by the flame photometer (model, Jenway PFP7 measuring range from 0 to 10 ± 0.2 ppm,USA). Voltage (volt) and current (ampere) were measure by the Avometer (model, DT266 clamp meter measuring range from 200 to 1000A, China).

BioFlo 110 fermentor: A 20 L batch fermentor was used for the production of biofert from dairy manure. The experimental set-up (Figure 1) consisted of the fermenter, the air supply and the computer based data acquisition and control system. The fermenter and all accessories were chemically sterilized using 2% potassium metabisulfite solution and then washed with hot water several times before starting the experiment in order to remove any chemical traces. The reactor was then filled with 20 L of dairy manure. .the reactor was operated at air flow rate 1, 2 and 5 L/min and mixing speed of 200, 300 and 500 rpm. The dissolved oxygen and temperature of the reactor were monitored continuously. The experiment was devoted to study physical, chemical and microbiological properties of nutrient solution (Biofert) as affected by temperature with (30°C, 35°C, and 50°C), aeration rate (1, 2, and 5 L/min) and agitation rotation speed (200, 300, and 500 rpm). A total of 81 runs including 3 replicates were conducted.

Biofert analysis

Physical and chemical analysis: Bulk density was calculated as a ratio between dry weights of the sample (g) to its volume (cm^3), according to [8]. Electrical conductivity measurements were run in 1:1 Biofert water extracts according to [9], using EC meter, ICM model 71150. Total nitrogen was determined Kjeldahl digestion method as described by [10]. Soluble nitrogen forms were determined according to the method described by [11]. Total phosphorus was determined using spectrophotometer method [12]. Total potassium was determined using flame photometric method [13]. Microbiological analysis was performed according to [14].

Results and Discussion

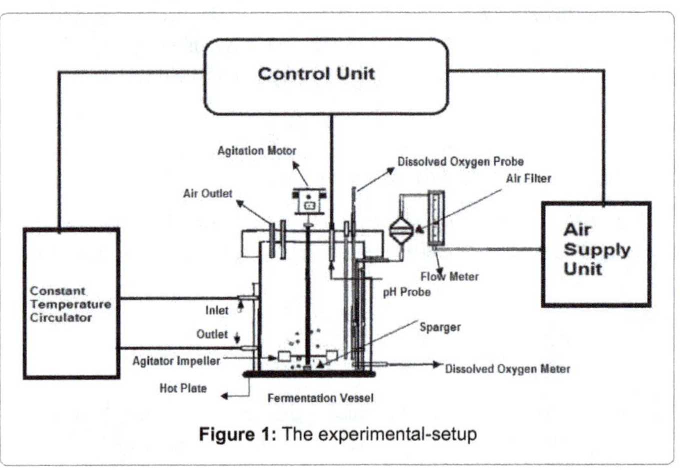

Figure 1: The experimental-setup

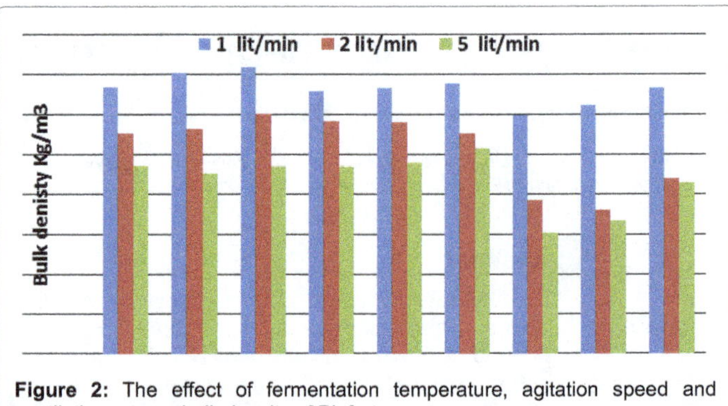

Figure 2: The effect of fermentation temperature, agitation speed and ventilation rate on bulk density of Biofert.

This study was carried out to investigate the properties of the biofert that produced under different condition of fermentation temperatures, agitation speeds, and ventilation rates. Physical, chemical and microbiological properties were studied. These properties are: moisture content, bulk density, total solid, electrical conductivity, hydrogen ion concentration, total nitrogen, organic matter, phosphorus, potassium, and microbial changes.

Physical properties

Biofert bulk density: Figure 2 shows the bulk density (BD) of the biofert as affected by the fermentation temperature, agitation speed and ventilation rate. The results indicated that the BD decreased with increasing the ventilation rate at different agitation speeds and fermentation temperature, where, it decreased from 970.60 to 951.04 kg/m^3. On the other hands, the BD increased with increasing the agitation speed, where, it increased from 966.41 to 970.28 $kg/m^3$with changing the agitation speed from 200 to 500 rpm at 30°C, from 967.19 to 962.83 kg/m^3 at 35°C, and from 946.32 to 953.66 kg/m^3 at 50°C . Regarding the effect of the fermentation temperature, it could be seen that, the BD of the biofert where they were 967.80, 967.59, 955.95 at fermentation temperature 30°C, 35°C and 50°C, respectively. The BD decreased with increasing the ventilation rate, where, it decreased from 970.60 to 951.04 kg/m^3. Multiple regressions analysis was carried out to find a relation between fermentation temperature, agitation speed, ventilation rate and the biofert Bulk density, the best form obtained was as follows:

$$BD = 993.99 - 0.75T + 0.02A - 3.11V \quad R^2 = 0.82 \tag{1}$$

Where:-

BD is the bulk density, kg m^3

T is the fermentation temperature, °C

A is the agitation speed, rpm

V is the ventilation rate, L / min

Biofert moisture content: Figure 3 shows the effect of fermentation temperature, T (30°C, 35°C and 50°C), agitation speed, A (200, 300, and 500 rpm) and ventilation rate, V (1, 2, and 5 L/min), on the moisture content (MC) of the resultant biofert. It could be seen that the lowest moisture content (79.87%) was obtained at 30°C fermentation temperature, 1 L/min ventilation rate and 200 rpm agitation speed, meanwhile, the highest MC of the biofert (83.11%) was recorded at 5 L/min ventilation rate and 500 rpm agitation speed. At 35°C fermentation temperature, the lowest moisture content (80.12%) of Biofert recorded at 1 L/min ventilation rate and 200 rpm agitation speed, meanwhile, the

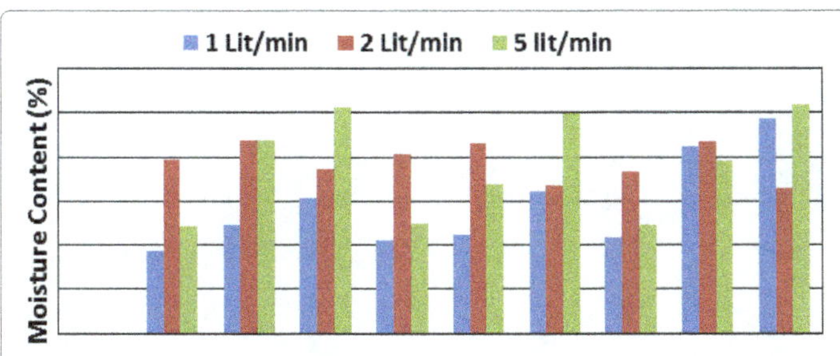

Figure 3: The effect of fermentation temperature, agitation speed and ventilation rate on the moisture content of the Biofert.

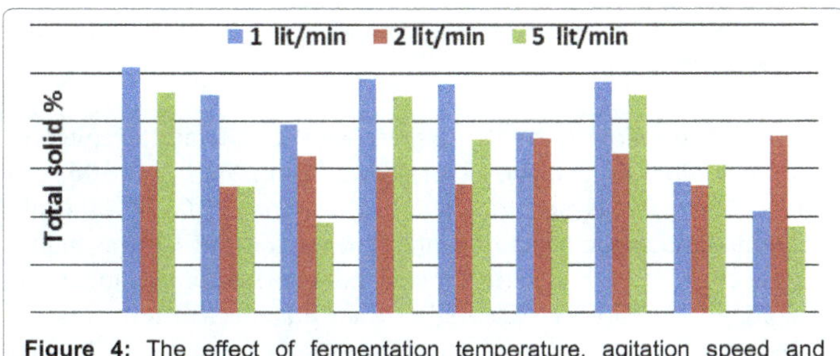

Figure 4: The effect of fermentation temperature, agitation speed and ventilation rate on the total solid of Biofert.

highest MC of the Biofert (83.01%) was recorded at 5 L/min ventilation rate and 500 rpm agitation speed. At 50°C fermentation temperature, the lowest moisture content (80.18%) of Biofert recorded at 1 L/min ventilation rate and 200 rpm agitation speed, meanwhile, the highest MC of the Biofert (83.19%) was recorded at 5 L/min ventilation rate and 500 rpm agitation speed.

Biofert total solid%: Figure 4 shows the effect of fermentation temperature (T), agitation speed (A) and ventilation rate (V) on the total solid (TSS) of the biofert. The results showed that the TSS varied slightly from 18.20 to 18.64% as fermentation temperature changed from 30-50°C. TSS varied from 17.55 to 19.26% as the agitation speed changed from 200 to 500 rpm. On the other hand, TSS ranged from 18.10 to 19.08% as the ventilation rate changed from 1 L/min to 5 L/min.

Chemical properties

Biofert electrical conductivity (EC): Figure 5 shows the effect of fermentation temperature (T), agitation speed (A) and ventilation rate (V) on the EC of the biofert. It could be seen that the biofert EC increased with increasing all treatments under study (T, A and V). The average EC increased from 29.54 to 33.97 ds/m. as the fermentation temperature from 30°C to 50°C, on the other hand, the EC increased from 28.33 to 30.96 ds/m as the agitation speed increased from 200 to 500 rpm at 30°C, 31.91 to 33.60 ds/m at 35°C and from 33.11 to 34.61 ds/m at 50°C for the same previous range of the agitation speed (200 to 500 rpm). These results could be attributed to with increasing these factors (T, A and V) caused increasing in the water evaporation while result in increasing of the salinity of the biofert solution. It means that the mineralization process of organic wastes increased gradually which led to release cations and anions showing the highest peak at the end of fermentation [15]. The best fit for the relationship between the electrical conductivity, temperature, agitation speed and ventilation rat was as follows:

$$EC = 21.02 + 0.19T + 0.006A + 0.64V \quad R^2 = 0.75 \tag{2}$$

Where, EC is the electrical conductivity dsm-1 (ds/m=800 ppm)

Biofert hydrogen ion concentration (pH): Figure 6 shows the effect of fermentation temperature (T), agitation speed (A) and ventilation rate (V) on the pH of the biofert. The results indicate that the pH ranged from 4.79 to 5.48 as the fermentation temperature changed from 30°C to 50°C and from 4.61 to 5.79, as the agitation speed changed from 200 to 500 rpm. The pH ranged from 4.95 to 5.29 as the ventilation rate varied from 1 L/min to 5 L/min.

Changes in biofert total nitrogen (TN)%: Figure 7 shows the effect of the fermentation temperature, agitation speed and ventilation rate on the TN% of the biofert. It could be seen that the TN% increased slightly with the agitation speed, where, it ranged from 0.61 to 1.06% as the agitation speed changed from 200 to 500 rpm .On the other hand, TN was affected by the ventilation rate, where it increased from 0.68 to 1.00% as ventilation rate increased from 1 to 5 L/min. TN% ranged

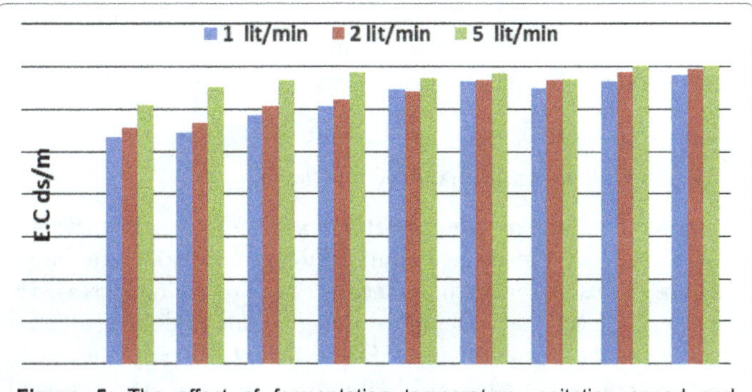

Figure 5: The effect of fermentation temperature, agitation speed and Ventilation rate on the electrical conductivity of Biofert.

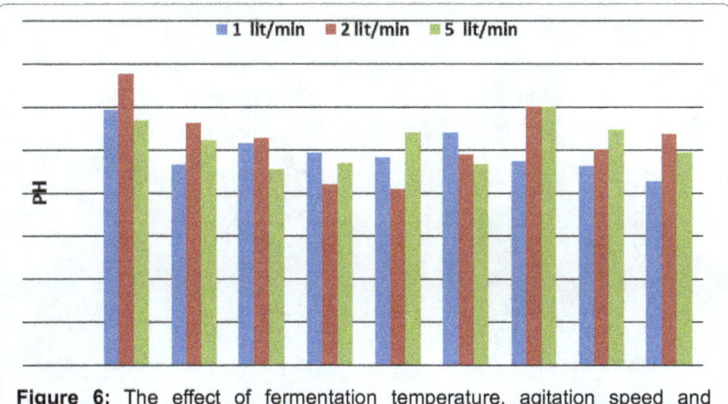

Figure 6: The effect of fermentation temperature, agitation speed and Ventilation rate on pH of Biofert.

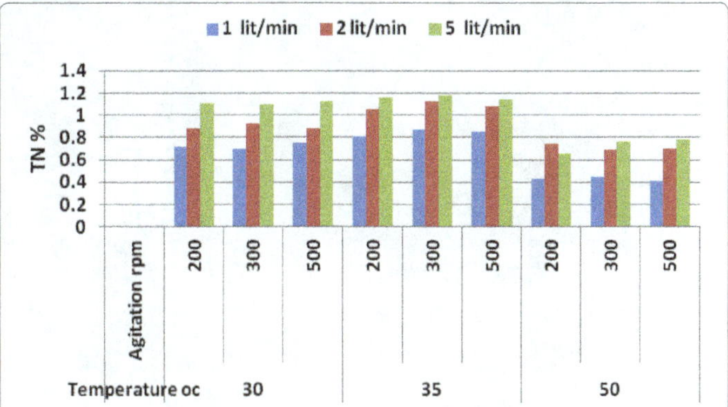

Figure 7: The effect of fermentation temperature, agitation speed and Ventilation rate on total nitrogen (TN)% Biofert.

from 0.63 to 1.04% as the fermentation temperature ranged from 30°C -50°C. The best fit for the relationship between the total nitrogen, Temperature, agitation and ventilation rat was as fallowed equation:

$$TN = 1.30 - 0.017T + 4.87A + 0.072V \quad R^2 = 0.72 \tag{3}$$

Where, TN is total nitrogen (TN), %.

Changes in biofert organic matter (O.M)%: Figure 8 shows the effect of O.M% of the biofert as affected by the fermentation temperature, agitation speed and ventilation rate. It could be seen that the O.M% ranged from 11.53-16.07% depending on the agitation speed. On the other hand, the OM changed slightly as the ventilation rate varied, where it changed from 13.57 to 13.84% where the VR varied from 1 to 5 L/min. Regarding the effect of fermentation temperature, it was found that the OM decreased from 15.01 to12.64% as the fermentation temperature increased from 30 to 50°C. Regression analysis was carried out to find a relationship between the organic matter, temperature, agitation and ventilation rat, and the most appropriate form is shown as follows:

$$OM = 18.47 - 0.06T - 0.007A + 0.29V \quad R^2 = 0.77 \tag{4}$$

Where, OM is organic matter (O.M), %.

Changes in biofert phosphorus: Figure 9 shows the effect of fermentation temperature, agitation speed and ventilation rate on the total phosphorus content of the biofert. It seems that the T.P increased by increasing the fermentation temperature, agitation speed and ventilation rate under study, where it increased from 145.59 to 275.70 ppm as the ventilation rate increased from 1 L/min to 5 L/min , on the other hand,

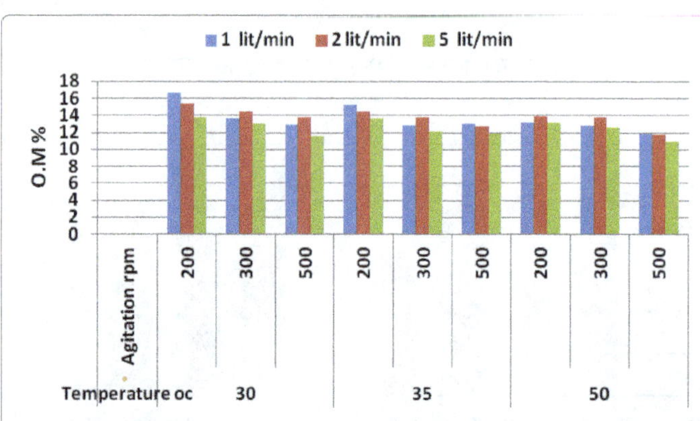

Figure 8: The effect of fermentation temperature, agitation speed and Ventilation rate on organic matter (O.M)% Biofert.

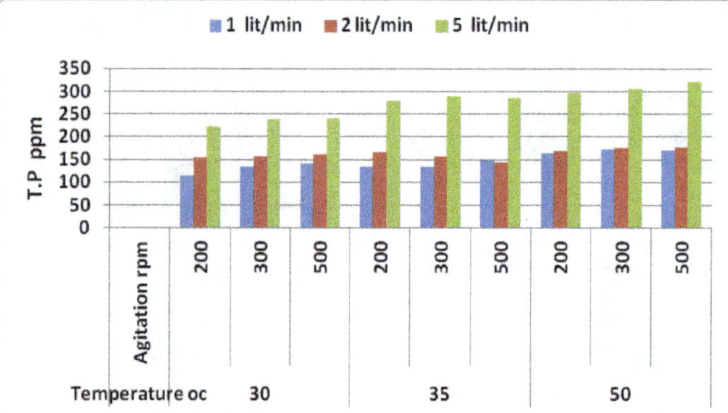

Figure 9: The effect of fermentation temperature, agitation speed and ventilation rate on the total phosphorus Biofert.

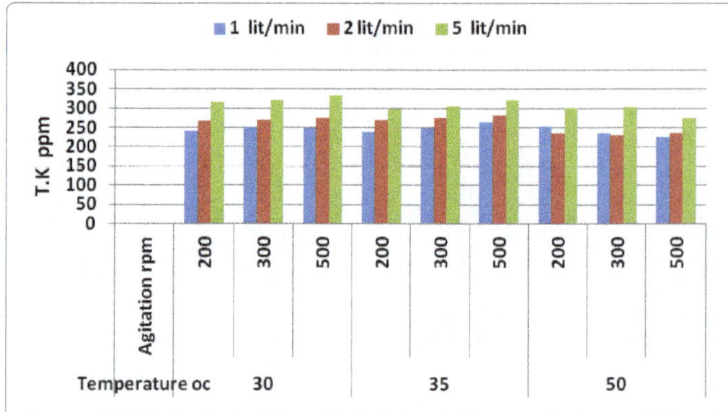

Figure 10: The effect of fermentation temperature, agitation speed and Ventilation rate on the total potassium Biofert.

it increased from 163.44 to 223.33 ppm with increasing the agitation speed from 200-500 rpm. The T.P increased from 173.81 to 216.96 ppm with increasing the fermentation temperature from 30°C-50°C.It could be conducted that the effect of ventilation rate on the T.P was higher than the effect of both agitation speed and fermentation temperature. But agitation speed effect was higher than the effect of fermentation temperature. The decreases of phosphorus concentration in the first stage of composting could be due to the microbial immobilization of available phosphorus [10], while the increase of phosphorus during the further stages of composting (mesophilic) could be due to either the microbial mineralization of organic phosphorus or the chelation of unavailable phosphorus with the organic acids, that found during the microbial decomposition of organic wastes. These data are in agreement with those of [16,17]. The best fit for the relationship between the total phosphorus, Temperature, agitation and ventilation rat was as fallowed equation:

$$TP = 15.88 + 2.03T + 0.03A + 33.75V \quad R^2 = 0.94 \tag{5}$$

Where, TP is Total phosphorus, ppm

Changes in biofert potassium: Figure 10 shows the effect of fermentation temperature, agitation speed and ventilation rate on the total potassium (T.K) content of the biofert. It could be seems that the T.K increased from244.63 to 295.30 ppm as the ventilation rate varied from 1 to 5 L/min. It increased from 274.89 to 286.44 ppm, 268.89 to 289.11 ppm, and 238.11 to 245.00 ppm at 30°C, 35°C and 50°C, respectively as the agitation speed increased from 200-500 rpm. Concerning the effect of fermentation temperature, it was found that the T.K decreased from 280.74 to241.19 ppm as the temperature increased from 30°C to 50°C. These data are in harmony with those of Singh and Sharma (2002) who found that the significant increase in potassium concentration by the end fermentation period could be due to the mineralization of organic matter. The best fit for the relationship between the total phosphorus, Temperature, agitation and ventilation rat was as followed equation:

$$Tk = 275.07 - 1.33T + 0.015A + 15.785V \quad R^2 = 0.89 \tag{6}$$

Where, Tk is Total potassium, ppm.

Microbial changes: Figures 11a-11c shows the microbial load of tested human pathogens (E.coli, Salmonella sp and Shigella sp) during biofert process as affected by ventilation, agitation and temperature. Results clearly depicted that a sharp decrease in the counts of all tested human pathogens with the increase of ventilation up to 5 L/min .The reduction percentages of all tested were ranged from 88.18 to 99.82%. It means that the ventilation levels had a deleterious effect on the proliferation of human pathogens. On the other hand the increase of

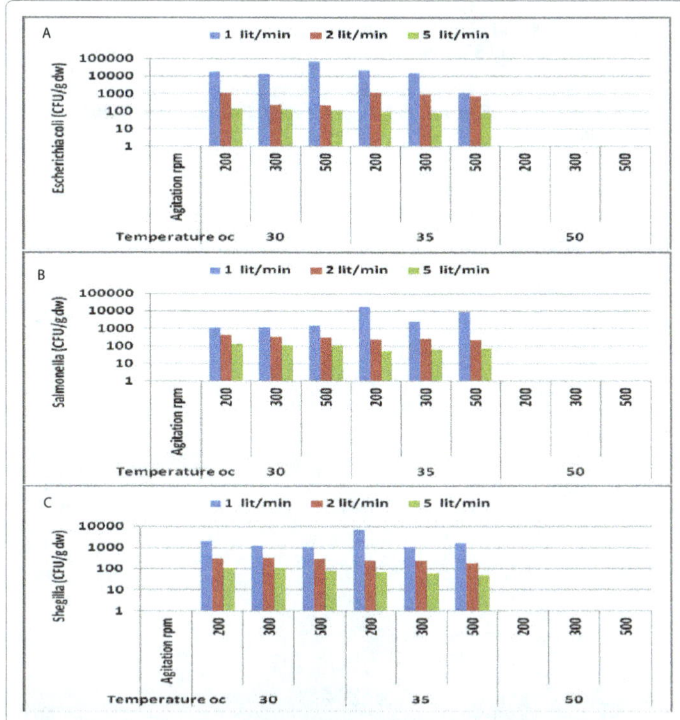

Figure 11a,11b,11c: The effect of fermentation temperature, agitation speed and Ventilation rate on Total account of Pathogenic (CFU/gDW) on biofert.

dissolved oxygen led to increase the biological activity of saprophytic microorganisms and consequently increase their metabolites especially antagonistic agents and antibiotics. The two levels of temperatures being 30°C and 35°C did not show a distinct difference whereas 50°C completely destroyed all human pathogens during biofert process. Agitation also exhibited the same trend of both tested temperature. This result is in line with [14] who observed that the disappearance of pathogens could be explained on the basis that when a beneficial microbe fills an ecological niche that would otherwise be exploited by a pathogen. For example, a beneficial organism may out-compete a pathogen for energy, nutrients, or "living space," thereby decreasing the survival of the pathogen.

Conclusions

This study was carried out to investigate the most important factors affecting the biofert production such as temperature, agitation speed, and aeration rate to obtain the proper factors for optimum production of the nutrient solution. The most important results could be summarized as follows:-

The results indicated that the lowest value of bulk density was 946.63 kg/m^3 at temperature of 50°C, 300 rpm agitation speed and 5 liter/min, ventilation rate while the highest value (983.17 kg/m^3) obtained at temperature 30°C, 500rpm agitation speed and 1 liter/min ventilation rate.

The lowest moisture content was 79.87% at temperature of 30°C, agitation speed of 200rpm and ventilation rate of 1 liter/min while the highest value was 83.19% at temperature of 50°C, agitation of 500rpm and ventilation rate of 5 liter/min.

The electrical conductivity increased from 11.6 ds m^{-1} at the start of fermentation to 35.07 ds m^{-1} at the end of fermentation period depending on the treatments under study. The pH decreased from 8.13 at the start of fermentation to 6.77 at the end of fermentation period.

The total solid of the biofert decreased from 21.2% at the start of fermentation to 16.81% at the end of fermentation period, where the lowest value was 16.81% at temperature of 50°C, agitation speed of 500 rpm and ventilation rate of 5 liter/min and the highest value (20.13%) recorded at temperature of 30°C, agitation speed of 200rpm and ventilation rate of 1 liter/min.

The lowest TN% was 0.41% at 50°C fermentation temperature, agitation speed of 500 rpm and ventilation rate of 1 liter/ min while, the highest value (1.18%) obtained at temperature of 35°C, agitation speed of 300 rpm and ventilation rate of 5 liter/min.

The O.M% decreased from 34.2% at the start of fermentation to 10.97% at the end of fermentation period. The lowest value of O.M % was 10.97% at temperature of 50°C, agitation speed of 500rpm and ventilation rate of 5 liter/min while the highest value was 16.63% at temperature of 30°C, agitation speed of 200rpm and ventilation rate of 1 liter/min.

The total phosphorus and potassium increased by increasing the fermentation temperature from 120 to 322 ppm, and from 150 to 334.33 ppm, respectively by the end of fermentation period.

Regarding the microbial changes, all treatments showed disappearance of pathogenic microorganisms at temperature of 50°C, at all agitation speeds and ventilation rates.

References

1. Ogbeide SE, Aisien FA (2000) Biogas from Cassava peelings, Afr. J Environ. Stud (1-12): 42-47.
2. Audu TO, Aisien FA, Eyawo EO (2003) Biogas from Municipal Solid Waste, NJEM, 4: 26-30.
3. Behera SK, Park JM, Kim KH, Park H (2010). Methane production from food waste leachate in laboratory-scale simulated landfill. Waste Manage. 30: 1502-1508.
4. Jeong E, Kim H, Nam J, Shin H (2010) Enhancement of bioenergy production and effluent quality by integrating optimized acidification with submerged anaerobic membrane bioreactor. Bioresour. Technol. 101, 1873-2976.
5. Ward AJ, Hobbs PJ, Holliman PJ, Jones DL (2008) Optimization of the anaerobic digestion of agricultural resources. Bioresour. Technol. 99, 7928-7940.Water Sci. Technol. 53: 187-194.
6. Park C, Lee C, Kim S, Chen Y, Chase HA (2005) Upgrading of anaerobic digestion by incorporating two different hydrolysis processes. J. Biosci. Bioeng. 100: 164-167.
7. Torres ML, de Llorens MC (2008) Effect of alkaline pretreatment on anaerobic digestion of solid wastes. Waste Manage. 28: 2229-2234
8. Clark ER, Harman JP, Forster JR (1985) Production of metabolic and waste products by intensively farmed rainbow trout, Salmo gairdneri Richaedson. Journal of Fish Biology, 27: 381-393.
9. Richards LA (1954) Diagnosis and Improvement of Saline and Alkali Soil. U.S.Dept. Agric., Handbook, No. 60 Gov. print off.
10. Jackson ML (1973) Soil Chemical Analysis, Prentice-Hall Englewood cliffs, New Jersey.
11. Page I, Ott CR, Pottle DS, Cocci AA, Landine RC(1999) Anaerobic-aerobic treatment of dairy wastewater: a pilot study. In: Proceedings of the 1999 31st mid-Atlantic industrial and hazardous waste conference; p. 69-78.
12. Olsen SR, Cole CV, Watanabe FS, Dean LA (1954) Estimation of available phosphorus in soils by extraction with sodium bicarbonate. US Dept. Agric., Circular No. 939, P. 19.
13. Chapman HD, Pratt FP (1961) Methods of analysis for soil, plants and water. Univ. of California,Div.of Agric. Sci.

14. Brinton WF, Evans ML, Brinton RB, Droffner M L (2001) A Standardized Dewar Test for Evaluation of Compost Self-Heating. BioCycle Report 1-16.

15. Raviv M, Chen Y, Inbar Y(1987) Peat and Peat substitutes as growth media for container grown plants-A review. In: The rol of organic matter in modern agriculture.(Chen,Y.and Avnimelech,Y.(Ed)) Martinus Nijhoff, the Hegue 257-87.

16. Elvira C, Sampedro L, Benitez E, Nogales R (1998) Vermicomposting of sludge from paper mill and dairy industries with Eisenia Andrei: A pilot- scale study. Bioresource Technol 63: 205-211.

17. Singh M, Sharma SD (2002) Bioefficacy of azafenidin for weed management in citrus. Weed Sci. Soc. Amer. Annual Meeting. February 10-13, 2002. Reno, NV, Abstr. 21, p. 7.

Natural Occurrence of Potential Fungal Biopesticide *Nomuraea Rileyi* (Farlow) Samson Associated with Agriculture Fields of Tamil Nadu, India and it's Compatibility with Metallic Nanoparticles

Karthick Raja Namasivayam*, R S Arvind Bharani and Moinuddin Raqib Ansari
Department of Biotechnology, Sathyabama University, Chennai,Tamil Nadu, India

Abstract

In the present study, natural occurrence of major fungal biopesticide *Nomuraea rileyi* (Farlow) Samson associated with agricultural field soil in an area around TamilNadu, India was studied adapting culture dependent method. Agricultural field soil samples were collected from ten different sites. Soil dilution method was used to isolate *N. rileyi*. A total of 123 isolates of *N.rileyi* were obtained. Among the 10 sampling sites, *Nomurea rileyi* was isolated from 4 sites belonging to Hasthampatty (Salem) Rajakoil (Vellore) Vadavali (Coimbatore), Pullarakottai (Viruthunagar), High frequency of fungal occurrence was recorded in vadavali (60%) followed by Pullarakottai (17%), Hasthampatty (13%,) Rajakoil (10%). Non *Nomurae rileyi* strains belong to *Beauveria* sp and *Metarhizium* sp were also isolated. Fungal occurrance was highly influenced by soil physico-chemical parameters. Effect of metallic nanoparticles such as silver, copper and the respective nanoparticles coated with chitosan on the post treatment persistance of *N. rileyi* was also studied.Distinct effcet on the growth of *N. rileyi* was recorded in copper nanoparticles with high concentration.

Keywords: *Nomuraea rileyi*; Natural occurrence; Agricultural field soil; Compatibility; Metallic nanoparticles

Introduction

Among the various microsymbions, fungi (mycorhizal and pathogens) is the major component of soil. Soil is also the major source of entomopathogenic fungi and the isolation of these fungi involves soil sampling since that is their natural habitat [1,2]. Entomopathogenic fungi are distributed in a wide range of habitats including aquatic forest, agricultural, pasture, desert, and urban habitats [3-5]. Their ability to regulate insect populations has been studied in tropical and temperate habitats [6-8]. Soil is considered an excellent environmental shelter for entomopathogenic fungi since it is protected from UV radiation and other adverse abiotic and biotic influences [9]. Fungal entomopathogens in the genera *Beauveria, Conidiobolus, Metarhizium* and *Isaria* (*Paecilomyces*) are commonly found in soil [9]. Occurrence of the entomopathogenic fungi was highly influenced by various factors geographical location, climatic conditions, habitat type, cropping system, and soil properties, as well as the effects of biotic factors [5,10,11]. It is increasingly recognized that the biodiversity of agro ecosystems delivers significant services, such as biological control of pests, to agricultural production [8]. The contribution of the entomopathogenic component of this biodiversity to the regulation of pest populations has often been ignored [12] and when it has been acknowledged, it has usually been discussed if the introduction of exotic strains of fungi, or the augmentation of endemic strains, is an appropriate biocontrol strategy [13]. Among the several existing entomogenous fungi, *Nomuraea rileyi* is a cosmopolitan species infecting many noctuids such as *Helicoverpa armigera, Spodoptera litura, Trichoplusia ni, Anticarsia gemmatalis, Pseudoplusia* includes and has a potential for development into mycoinsecticide and occurs in soils of various agro ecosystem.

Nanotechnology is significant for the comprehension, use, and control of matter at magnitudes of a minute scale, approaching atomic levels, with which to manufacture new substances, instruments, and frameworks. At present, silver and various metallic nanoparticles are in great use in the medicinal, pharmaceutical, agricultural industry and in water purification [14]. Increasing numbers of commercial products, from cosmetics to medicine, incorporate manufactured nano materials (MNMs) that can be accidentally or incidentally released to the environment and adversely affect the various biotic components in the ecosystem.

Gold, silver, and copper have been used mostly for the synthesis of stable dispersions of nanoparticles, which are useful in areas such as photography, catalysis, biological labelling, photonics, optoelectronics and surface-enhanced Raman scattering (SERS) detection. Concern over the potentially harmful effects of such nanoparticles has stimulated the advent of nanotoxicology as a unique and significant research discipline. However, the majority of the published nanotoxicology articles have focused on mammalian cytotoxicity or impacts to animals and bacteria, and only a few studies have considered the toxicity of nanoparticles to plants and other non target organism [15]. In the present study, the toxic effect of metallic nanoparticles on *Nomuraea rileyi* was discussed.

Materials and Methods

Sampling Sources

Agricultural field soil was collected from the different parts of Tamil Nadu belonging to Nanmangalam (Kanchipuram), Hasthampatty (Salem), Rajakoil (Vellore), Veeranam Thiruvanamalai,Mattuthanvani (Madurai), Vadavalli (Coimbatore), Pullarakottai (Viruthunagar), Pullarakottai Road (Viruthunagar) Pondicherry, Cunnur (Nilgiris).

***Corresponding author:** Karthick Raja Namasivayam, Department of Biotechnology, Sathyabama University, Chennai, Tamil Nadu, India,
E-mail: biologiask@gmail.com

For each soil sampling site, 1.5 to 2 kg soil was collected from 5 points randomly and mixed to obtain homogenous sample, the homogenized soil samples were kept in sterile polythene bags and brought to the laboratory.

Physicochemical of soil analysis

Before microbial analysis, soil aggregates were broken by hands, trays with soil were kept open until moisture was at equilibrium [2]. Soil texture pH electrical conductivity organic matter nitrate, phosphorous, potassium, calcium, magnesium sulphur, sodium, zinc, iron, copper were determined for all soil collected. These measurements were determined in national agro foundation at Taramani Tamil Nadu, India.

Isolation of *N. rileyi*

Soil dilution method was adopted for natural occurrence of *N. Rileyi.* 1 gm of homogenized sample was suspended in 99 ml of sterilized distilled water mixed well and serially diluted. 1 ml of aliquots was transferred to sterile petriplates, 20 ml of sterile molten CTC (chloramphenicol, Thiabendazole, cycloheximide media)+PDA Agar media consisting of potato dextrose agar supplemented with 0.5 gm/l Chloramphenicol 0.01 gm, Thiabendazole 0.25 gm, Cyclohexamide was added, allowed to solidify. The seeded plates were incubated at 25°C in an incubator for 3-7 days. Fungal colonies were isolated after the incubation period, respective fungal colonies were purified and the pure culture was stored on CTC media agar slant. Identification of fungal culture was determined by morphological characteristics and microscopic examination of the spores by lactophenol cotton blue staining.

Evaluation of toxic effect of metallic nanoparticles on the post treatment resistance of *N.rileyi*

In the present study, chitosan coated metallic nanoparticles such as silver; copper was selected to evaluate toxic effect.

Synthesis of silver and copper nanoparticles: Silver nanoparticles were synthesized by chemical reduction of 1 mM silver nitrate with 1 mM sodium borohydride as reducing agent. Synthesis of silver nanoparticles was confirmed by the conversion of the reaction mixture into brown colour and further characterization of the synthesized silver nanoparticles was carried out with determination of Plasmon absorption maxima with UV-VIS spectroscopy and particle morphology with electron microscopy (SEM). Similarly copper nanoparticles were synthesized by chemical reduction of 1 mM copper sulphate (Sigma, analytical grade) with 1 mM sodium borohydride (Sigma, analytical grade) as reducing agent. Synthesis of copper nanoparticles was confirmed by the conversion of the reaction mixture into green colour and further characterization of the synthesized nanoparticles was carried out with determination of Plasmon absorption maxima with UV-Vis spectroscopy and particle morphology with Scanning Electron Microscopy (TEM) and energy dispersive X-ray spectroscopy. Synthesized nanoparticles were purified by successive centrifugation by 10,000 rpm and the collected pellets were washed thrice with deionised water, the washed suspension thus obtained was freeze dried.

Characterization

Determination of Plasmon absorption maxima of the reaction mixture with UV-VIS spectra is the primary confirmation of the synthesis of nanoparticles. UV VIS absorption spectrum was carried out with Thermo scientific spectrascan UV 2700 spectrophotometer operating in the transmission mode. Scanning Electron Microscopy (SEM) images were recorded by using Carl zeiss subra (Germany scanning electron microscope equipped with an Energy-dispersive Spectrum (EDS) capability.

Synthesis and characterization of chitosan coated nanoparticles

In a typical procedure of chitosan stabilized silver nanoparticles, 5 ml of 0.1 M silver nitrate, 1 ml of tri sodium citrate of 0.1 M and 1 ml of 0.1 M sodium borohydride, 10 ml of a solution containing chitosan (6.92 mg mL^{-1}) was mixed and kept under magnetic stirrer for 3 hours to obtain homogeneous solution. The homogenous thus obtained was transferred to the screw cap vial and incubated for 12 h at 95°C. The colour of the solution changed from colourless to light yellow, finally to yellowish brown which primarily confirmed the coating of chitosan with silver. Similarly, chitosan coated copper nanoparticles were synthesized by reduction of 10 ml of 0.1 M copper sulphate, 2.5 ml of 0.1 M tri sodium citrate, 1 ml of 0.1 M sodium borohydride 10 ml of a solution containing chitosan (6.92 mg mL^{-1}) separately were mixed and kept under magnetic stirrer for 3 hours to obtain homogeneous solution The homogenous thus obtained was transferred to the screw cap vial and incubated for 12 hours at 95°C. The colour of the solution changed from colourless to light yellow, and finally to yellowish brown which primarily confirmed the stabilization of chitosan with copper. After the preliminary confirmation, the reaction mixture was freeze dried and characterized with SEM, FTIR and EDAX.

Pot assay

Fertile loam soil was collected from the garden of Sathyabama University in sterile polythene bag, kept in ice box brought to the laboratory. Homogenized soil was sterilized by autoclaving at 121°C for 15 minutes by autoclaving. Plastic pot of diameter of 16 cm was filled with 1.5 kg of sterilized soil and inoculated with 1 ml of *N. rileyi* spore suspension, (1.0×10^8 spores/ml) mixed well. The different concentration of nanoparticles was added into each pots already inoculated with spore suspension of *N. rileyi*. The pots were closed with cheese cloth inoculation at 25°C. Every 10 days interval, 1 gm of soil sample from each Nanoparticles treated pot was taken in a sterile Petri plate, serially diluted and plated on sterile CTC media. Plates were incubated at 25°C for 3-7 days.

Effects of nanoparticles on spore germination of *N. rileyi*

Spore germination inhibition was carried out by microscopic method to determine the effect of Nanoparticles on spore germination of *N. rileyi*. In this method one drop of sterile CTC media was added to clean microscopic slide, a drop of fungal spore suspension previously incubated with different concentration of Nanoparticles was added, mixed well by sterile tooth pick. The slides were then transferred to the sterile petriplates lined with sterile filter paper and a thin layer of cotton. The plates were incubated at 25°C. The slides were examined for spore germination under microscope. Spore germination inhibition (%) was calculated by following formula:

Spore Germination Inhibition (%) = {(No of spore germination in control)-No. of Spores germination in treatment)} * 100 No of spore germination in treatment

Result and Discussion

Among the 10 sampling sites, *Nomurea rileyi* was isolated from 4 sites belonging to Hasthampatty (Salem), Rajakoil (Vellore), Vadavali (Coimbatore), Pullarakottai (Virruthagar), High frequency of distribution was recorded in vadavali (60%) followed by Pullarakottai (17%), Hasthampatty (13%) and Rajakoil (10%) (Figure 1). Non *N. rileyi* such as *Beauveria* sp. and *Metarhizium* sp. were also isolated associated with *N. rileyi* in the sampling sites. Soil physico-chemical parameters highly influenced the natural occurrence of *Nomurea rileyi. Nomurea rileyi* isolated from respective soil samples reveals high organic matter, available nitrogen and phosphorous (Tables 1 and 2).This may favour the viability of the fungal spore and thus improved the natural occurrence of *Nomurea rileyi.* Moreover the presence of other trace element may affect the fungal distribution. High content of available zinc may inhibited the occurrence of *Nomurea rileyi. N. rileyi* was not obtained from any soil sample consists of more availbale zinc, available iron,available copper and available mangnese. Electrical conductivity and pH did not effect the natural occurrence of *Nomurea rileyi.* Natural occurrance of entomopathogenic fungi associated with soil was carrried out in various parts of the world [2,16,17].

Natural occurrence of the entomopathogenic fungi is highly influenced by various physico chemical parameters of the soil. The effect of soil factors (organic matter, clay, sand, silt content, and pH) and geographical location (latitude, longitude and altitude) on the occurrence of entomopathogenic fungi has been reported. They reported that frequency of occurrence was found to be more in soils contain high organic matter and total nitrogen content. Moreover, soil pH and geographical distribution did not affect the occurrence. In the present study, soils revealed the occurrence *N. rileyi* showed more organic matter and the soil pH did not cause any distinct effect.

Evaluation of toxic effect of nanoparticles on post treatment presence of *N. rileyi* Synthesis and characterization of metallic nanoparticles

Silver nanoparticles synthesis adopting chemical reduction was primarily confirmed by colour change of the reaction mixture from pale yellow to brown which clearly indicates formation (Figure 2a). Synthesized silver nanoparticles characterized by UV-Vis spectroscopy which reveals a strong broad surface Plasmon peak located at 420 nm (Figure 2b). Particle morphology size and shape with Transmission electron microscopy reveals spherical particles with the size of 19-21 nm (Figure 3a). In EDAX, strong signals from the silver particles were observed (42.44% in mass), while weaker signals from C, O, Al and S atoms are also recorded which confirmed silver nanoparticles (Figure 3b). CuNPs synthesized by chemical reduction method with copper sulphate reduced with sodium borohydride as a reducing agent adapting one phase synthesis. CuNPs were formed immediately after the addition of sodium borohydride to the copper sulphate solution and the synthesis was primarily confirmed by the colour change of the reaction mixture into green colour (Figure 4a).

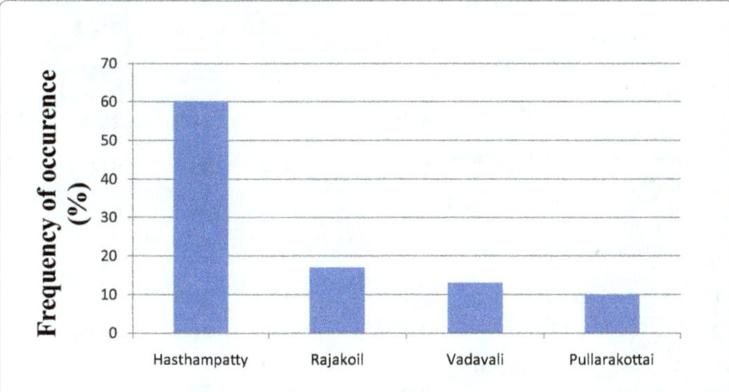

Figure 1: Frequency of occurrence of N.rileyi associated with agricultural field soil

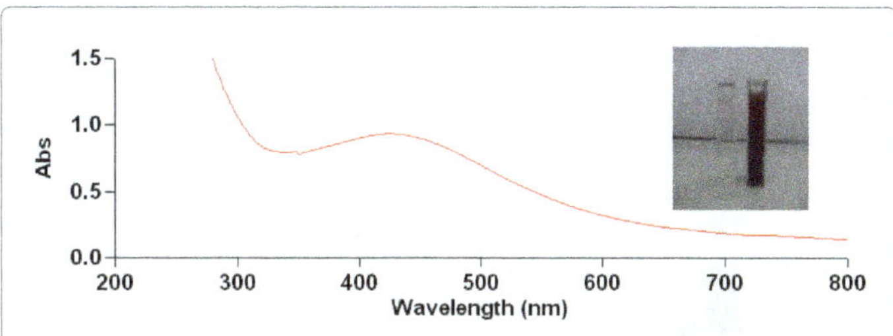

Figure 2: Characterization of synthesized silver nanoparticles (a) reaction mixture, (b) UV-VIS absorption spectra of the synthesized nanoparticles.

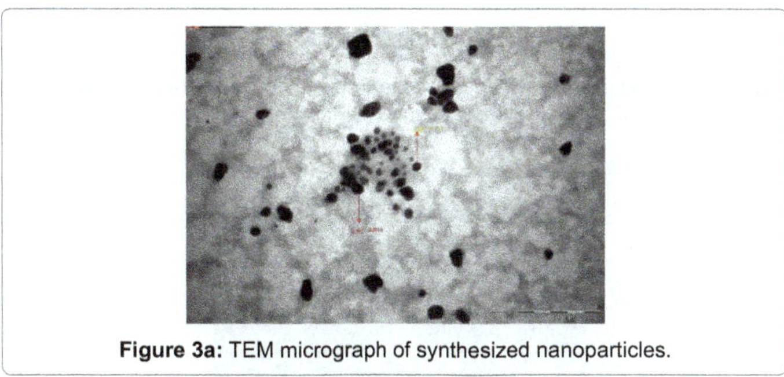

Figure 3a: TEM micrograph of synthesized nanoparticles.

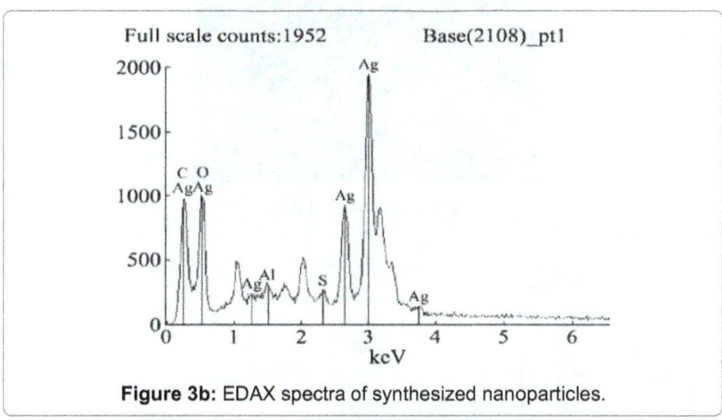

Figure 3b: EDAX spectra of synthesized nanoparticles.

Characterization of this synthesized particle was further studied by the Plasmon absorption maxima at 610 nm with UV-VIS Spectrophotometer which can be attributed to the plasma resonance absorption of non-oxidized the copper particles (Figure 4b). Size and shape with SEM which reveals 132 nm (Figure 5a) presence of elemental copper as strong peak was confirmed by EDAX (Figure 5b). Chitosan stabilized respective nanoparticles were synthesized by chemical reduction of respective metal salt precursor with nontoxic and biocompatible polymer chitosan which primarily confirmed by FTIR, SEM and EDAX. When the FTIR spectrum of free and stabilized nanoparticles was compared, it was found that almost the all the absorbed peaks were modified upon coating with chitosan. FTIR spectra of chitosan coated silver and copper nanoparticles are presented in Figure 6a and 6b.

The IR spectra of the chitosan capped Nano silver shows prominent peaks at 3788 cm^{-1}, 3427.4005 cm^{-1} corresponding to O – H stretching,

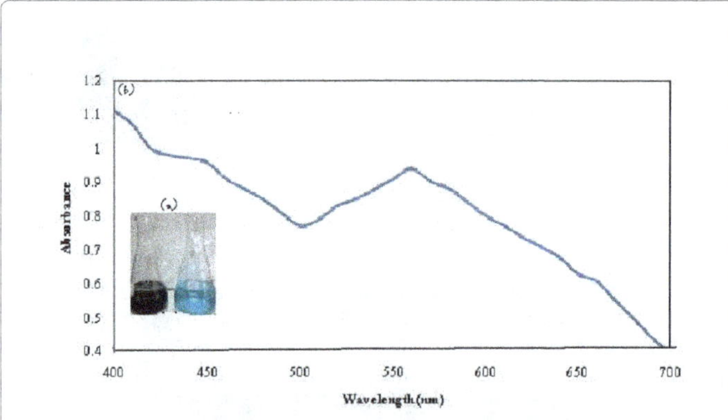

Figure 4: (a) Synthesized Copper nanoparticles Reaction mixture (b) UV-VIS absorption spectra of the synthesized nanoparticles.

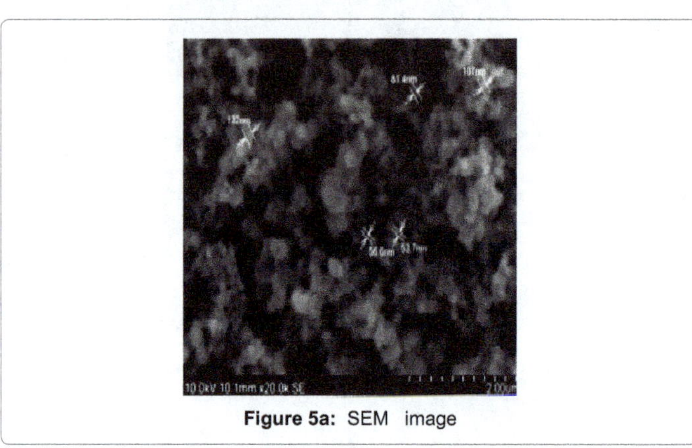

Figure 5a: SEM image

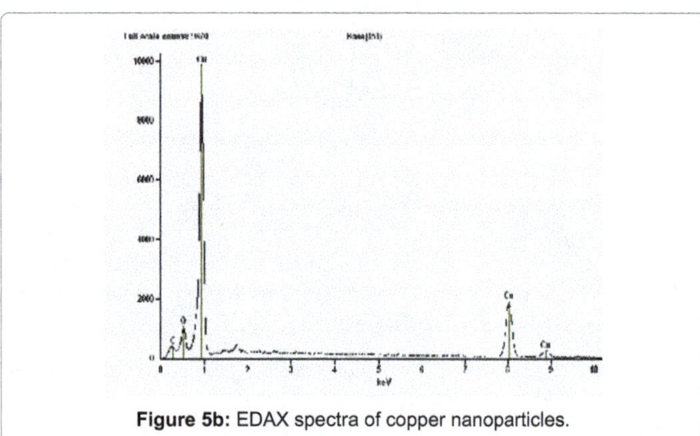

Figure 5b: EDAX spectra of copper nanoparticles.

strong polymerization, at 2928.1733 cm^{-1} for aliphatic C – H stretching. Peaks at approximately 2369.0387 cm^{-1}, 2345.3187 cm^{-1} represent N – H stretching vibration. Peaks at 1637.5845 cm^{-1} and 1389.4649 cm^{-1} represent N – H bending and a peak at 1026.6957 represent C – N vibration in aliphatic compounds. A is also observed at 617.7611 cm^{-1} showing the presence of inorganic metal ions (silver ions). The IR spectra of the Chitosan capped Nano copper shows prominent peaks at 3432.35 cm^{-1} corresponding to strong polymerization. Peaks at approximately 2044.52 315 cm^{-1} represent N – H stretching vibration. Peaks at 1633.03 cm^{-1} represent N – H bending and a peak at 1384.37 cm^{-1} represents O – H bending and C – H stretching.

A peak is also observed at 1091.66 cm^{-1} showing C – O – C stretching in polysaccharide of chitosan. Finally the presence of inorganic metal ions (copper ions) is shown by a peak at 463.13 cm^{-1}. The SEM analyzer built in with and EDAX analyzer allows a quantitative deduction on localization of elements in the nano specimens Scanning electron microscopy (SEM) study of chitosan stabilized copper nanoparticles reveals. The uniform spherical smooth morphology. within the size range of 101.78 nanometer and electron dense thin chitosan coating shell of diameter 3-5 nanometer (Figure 7a) Such size distribution analysis primarily confirms that the particles are well dispersed and less aggregated The EDAX images illustrated the presence of large amounts of C, O, N (Figure 7b). SEM and EDAX analysis of chitosan stabilized copper nanoparticles shows the spherical particles with electron dense thin chitosan coating shell of 50-65 nm diameter (Figure 8a) EDAX analysis further confirms the high amount of C, O, N and Cu (Figure 8b).

Toxic effect

Distinct effect on post treatment or viability on *N.rileyi* was not observed in silver nanoparticles treatment. In free silver nanoparticles treatment, 24×10^3, 21×10^3, 19×10^3, $19\times10^{\ 3}$ and 18×10^3 CFU/g of *N. rileyi* was recorded in the respective concentration. Similar finding was reported in chitosan coated silver nanoparticles treatment. Colony count was not recorded in free copper nanoparticles at the concentration of 1000, 750µg /ml and similar effect was recorded in chitosan coated copper nanoparticles (Table 3). Spore germination was not inhibited in all tested concentration of both free silver and chitosan coated silver nanoparticles treatment (Table 4).

Significant inhibition in spore germination (P>0.05) was recorded in all the concentration of free and copper nanoparticles. Nanotechnology is currently employed as a tool to explore the darkest avenues of medical sciences in several ways like imaging, sensing, targeted drug delivery, gene delivery systems and artificial implants [18]. In present situation, silver and various metallic nanoparticles nanoparticles are in great use in the medicinal, pharmaceutical, agricultural industry and in water purification [14].

Increasing numbers of commercial products, from cosmetics to medicine, incorporate manufactured nanomaterials (MNMs) that can be accidentally or incidentally released to the environment. Concern over the potentially harmful effects of such nanoparticles has stimulated

S.no	Parameters	Hasthampatty	Rajakoil	Vadavali	Pullarakottai
1	pH	7.95	7.50	7.80	7.95
2	Electrical conductivity(ms/cm)	0.600	0.865	0.425	0.418
3	Organic matter (%)	2.33	2.57	2.74	0.78
4	Nitrate nitrogen (ppm)	24.9	20.1	1606	27.1
5	Available phosphorous(ppm)	237.7	388.0	31.6	314.6
6	Potassium exchangeable k(ppm)	93	314	246	234
7	Calcium exchangeable (ppm)	1932	2277	2401	2579
8	Magnesium exchangeable (ppm)	511	477	356	644
9	Sulphur available s as so4 ((ppm)	49.3	58.5	25.4	83.1
10	Sodium exchangeable Na((ppm)	302	329	142	585
11	Zinc available Zn (ppm)	2.15	2.22	0.68	1.55
12	Manganese available Mn (ppm)	4.72	5.25	21.57	10.88
13	Iron available Fe (ppm)	1.36	0.79	1.24	1.18
14	Copper available	1.84	0.45	1.91	1.83

Table1: Physico chemical parameters of soil samples collected from different agricultural fields of Tamil Nadu

S.no	Parameters	Nanmangalam	Veeranam	Pullarakottai Road	Mattuthavani	Pondicherry	Cunnor
1	Ph	7.2	7.30	8.10	5.40	6.64	5.75
2	Electrical conductivity(ms/cm)	0.995	0.885	0.715	0.076	0.168	0.040
3	Organic matter (%)	0.44	0.47	0.60	0.13	1.07	0.80
4	Nitrate nitrogen (ppm)	17.4	29.1	5.9	7.7	9.1	5.7
5	Available phosphorous(ppm)	189.6	293.5	7.9	56.1	347.0	8.6
6	Potassium exchangeable k(ppm)	58	219	104	43	59	37
7	Calcium exchangeable (ppm)	1728	2249	3101	1162	1914	418
8	Magnesium exchangeable (ppm)	412	463	674	327	262	134
9	Sulphur available s as so4 ((ppm)	67.0	60.9	3.5	3.9	11.3	4.4
10	Sodium exchangeable Na((ppm)	226	315	122	104	149	135
11	Zinc available Zn (ppm)	4.00	2.47	0.26	1.29	5.00	0.26
12	Manganese available Mn (ppm)	5.74	5.95	7.53	6.67	38.23	6.37
13	Iron available Fe (ppm)	11.24	0.94	0.87	224.00	26.51	22.39
14	Copper available	2.55	0.78	1.37	4.88	2.79	2.12

Table 2 : Physico chemical parameters of soil samples collected from different agricultural fields of Tamil Nadu.

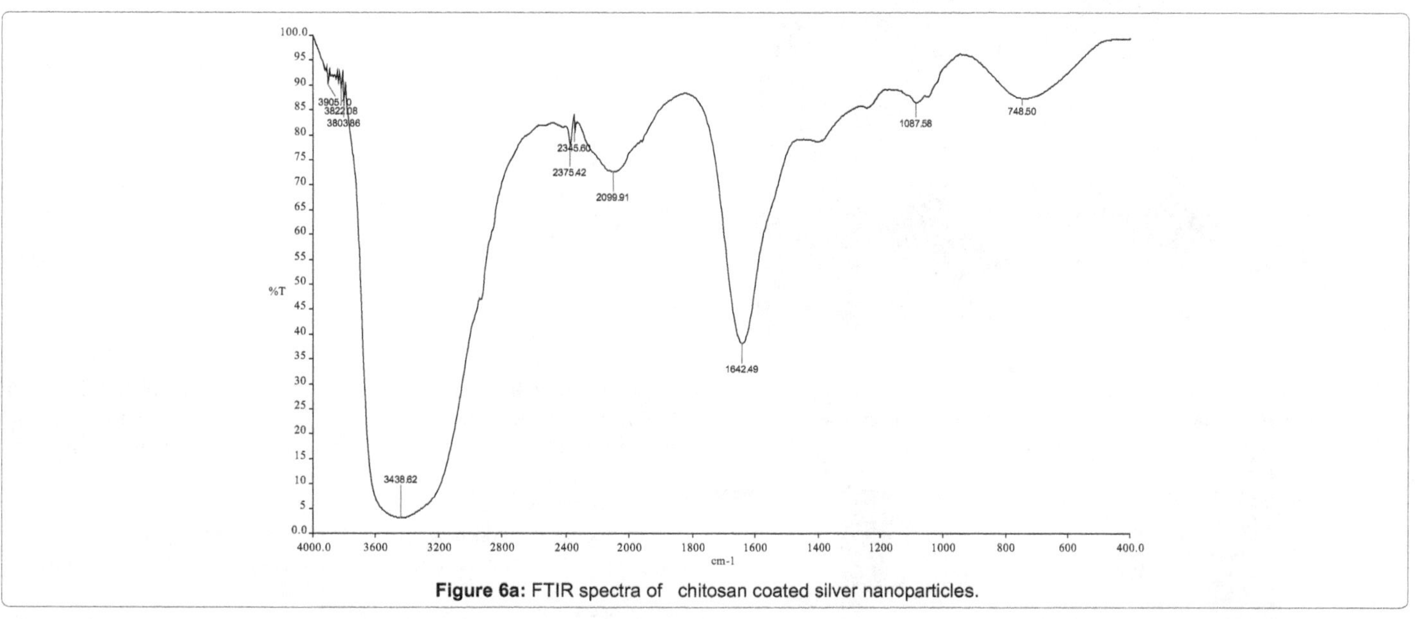

Figure 6a: FTIR spectra of chitosan coated silver nanoparticles.

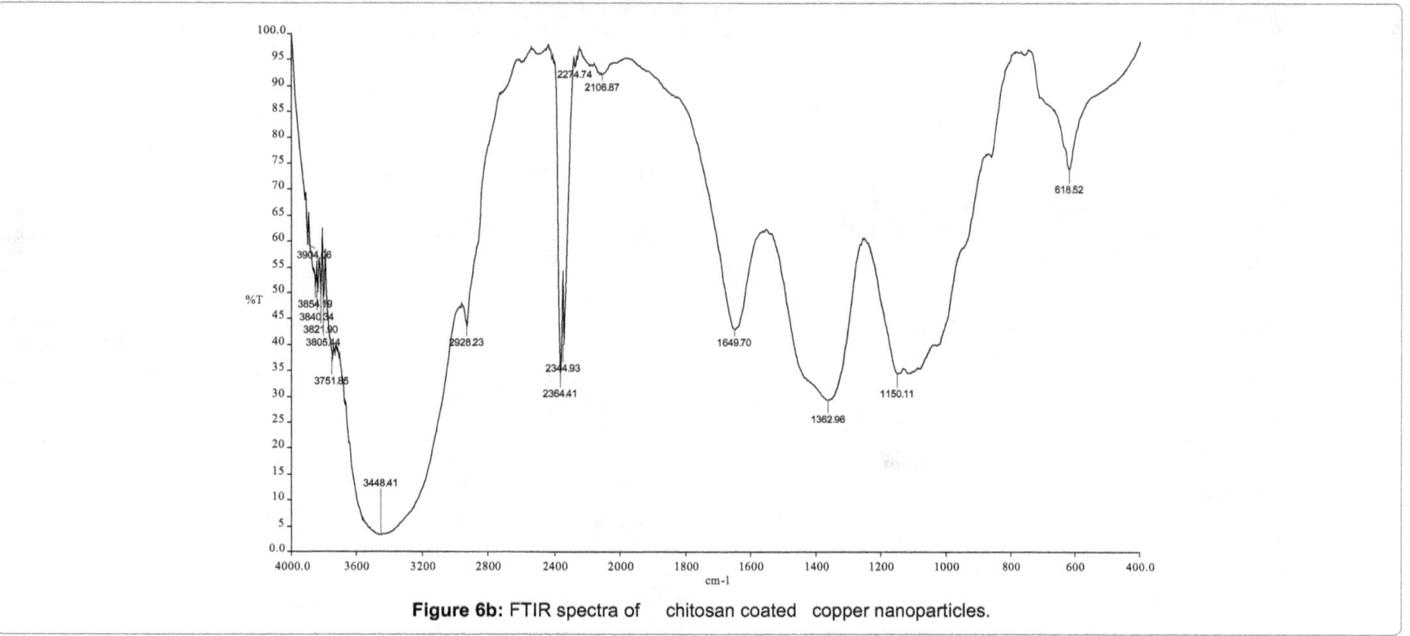

Figure 6b: FTIR spectra of chitosan coated copper nanoparticles.

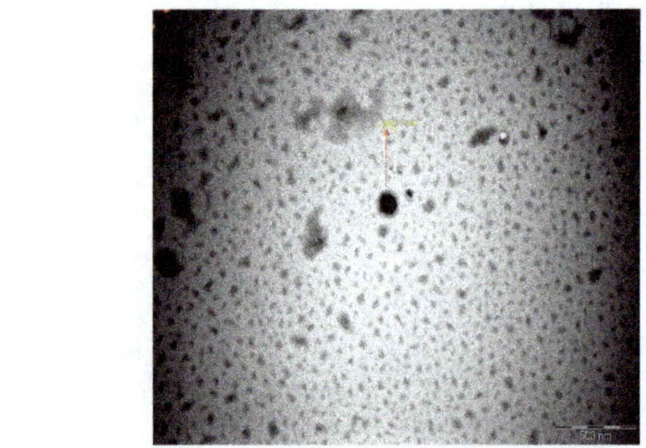

Figure 7a: SEM image of chitosan coated silver nanoparticle.

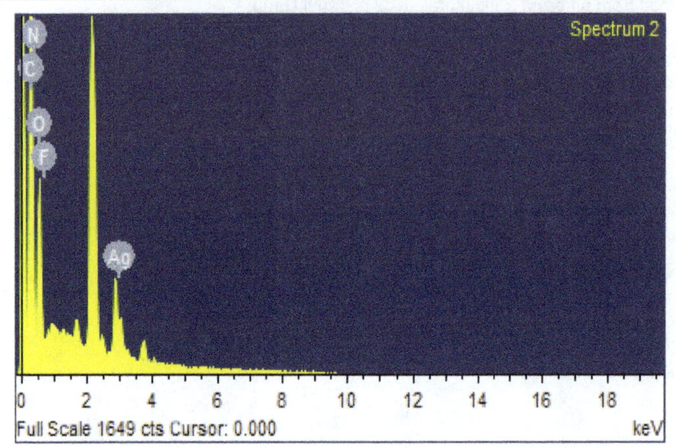

Figure 7b: EDAX image of chitosan coated silver nanoparticles.

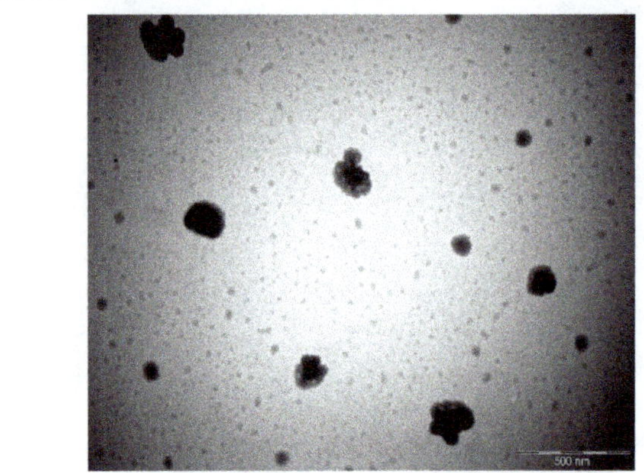

Figure 8a: Chitosan coated copper nanoparticles.

S.No	Treatment	Colony count (CFU/g)				
		100	250	500	750	1000
1	Control	24X10³				
2	Free silver nanoparticles(AgNps)	24X10³	21X10³	19 X10³	19X10³	18X10 ³
3	Chitosan coated silver nanoparticles (CS-AgNps)	23X10³	21X10³	19X10³	19X10³	18X10³
4	Free copper silver nanoparticles (CuNps)	11X10²	11X10²	7X10¹	0.0	0 .0
5	Chitosan coated copper nanoparticles (CS-CuNps)	10X10²	8X10²	7X10¹	0.0	0 .0

Table 3: Effect of nanoparticles on the post treatment persistence of *N.rileyi*.

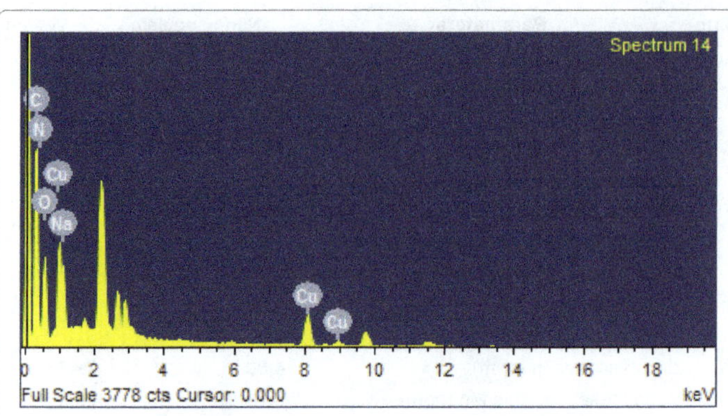

Figure 8b: EDAX image of chitosan coated copper nanoparticles

S.No	Treatment	Germination inhibition(%)				
		100	250	500	750	1000
1	Control	0.0				
2	Free silver nanoparticles(AgNps)	0.0	0.0	0.0	0.0	0.0
3	Chitosan coated silver nanoparticles (CS-AgNps)	0.0	0.0	0.0	0.0	0.0
4	Free copper silver nanoparticles (CuNps)	10.0	25.0	65.0	75.0	90.0
5	Chitosan coated copper nanoparticles (CS-CuNps)	20.0	45.0	70.0	80.0	98.0

Table 4: Effect of nanoparticles on the spore germination inhibition (%) of *N.rileyi*

the advent of nanotoxicology as a unique and significant research discipline. However, the majority of the published nanotoxicology articles have focused on mammalian cytotoxicity or impacts to animals and bacteria, and only a few studies have considered the toxicity of MNMs to plants and other non target organism [15]. This is the first report of studying compatibility of N.rileyi with nanoparticles. Further study under field trail will be used to understand the non target effect.

References

1. Gams W (1992) The analysis of saprophytic microfungi with special reference to soil fungi. Fungi in vegetation science. W Winderhoff. Klower Acad Publication, Netherlands 183-223.
2. Asensio L , Carbonell T , Jimenez L , Liorca L (2003) Entomopathogenic fungi in soils from Alicante province. Spanish Journal of Agriculture Research 3: 37-45.
3. Sánchez-Peña SR (1990) some insect- and spider-pathogenic fungi from Mexico with data on their host range. Florida Entomologist 73: 517-522.
4. Lacey LA, Fransen JJ, Carruthers R (1996) Global distribution of naturally occurring fungi of Bemisia, their biologies and use as biological control agents. Taxonomy, biology, damage, control and management Andover: Intercept 401-433.
5. Chandler D, Hay D, Reid AP (1997) Sampling and occurrence of entomopathogenic fungi and nematodes in UK soils. Appl Soil Ecol 5: 133-141.
6. Evans CH (1982) Entomogenous fungi in tropical forest ecosystems: an appraisal. Ecological Entomology 7: 47-60.
7. Subinprasert S (1987) Natural enemies and their impact on overwintering codling moth populations (*Laspeyresia pomonella* L.) (Lep., Tortricidae) in South Sweden. Journal of Applied Entomology 103: 46-55.
8. Meyling NV, Eilenberg J (2007) Ecology of the entomopathogenic fungi *Beauveria bassiana* and *Metarhizium anisopliae* in temperate agro ecosystems: Potential for conservation biological control. Biological Control 43: 145-155.
9. Keller S, Zimmermann G (1989) Mycopathogens of soil insects. Insect-Fungus Interactions. Academic Press 239-270.

10. Bruck DJ (2004) Natural occurrence of entomopathogens in Pacific Northwest nursery soils and their virulence to the black vine weevil, *Otiorhynchus sulcatus* (F.) (Coleoptera: Curculionidae). Journal of Environmental Entomology 33: 1335-1343.

11. Quesada-Moraga E, Navas-Cortés JA, Maranhao EA, Ortiz-Urquiza A, Santiago-Alvarez C (2007) Factors affecting the occurrence and distribution of entomopathogenic fungi in natural and cultivated soils. Mycol Res 111: 947-966.

12. Gurr GM, Wratten SD, Luna JM (2003) Multi-function agricultural biodiversity: pest Management and other benefits. Basic and Applied Ecology 4: 107-116.

13. Carruthers RI, Onsager JA. (1993) Perspective on the use of exotic natural enemies for biological control of pest grasshoppers (Orthoptera: Acrididae). Environmental Entomology 22: 885-903.

14. Ganesh, S, Karthick Raja Namasivayam S (2012) Biofilm Inhibitory Effect of Chemogenic Nano Zerovalent Iron against Biofilm of Clinical Isolate of *Staphylococcus aureus*. Asian Journal of Chemistry 24: 5533-5535.

15. Lee WM, An YJ, Yoon H, Kweon HS (2008) Toxicity and bioavailability of copper nanoparticles to the terrestrial plants mung bean (Phaseolus radiatus) and wheat (Triticum aestivum): plant agar test for water-insoluble nanoparticles. Environ Toxicol Chem 27: 1915-1921.

16. Roberts DW, St Leger RJ (2004) Metarhizium spp., cosmopolitan insect-pathogenic fungi: mycological aspects. Adv Appl Microbiol 54: 1-70.

17. Sun B, Lu X (2008) Occurrence and diversity of insect associated fungi in natural soils in China. Appl Soil Biol 39: 100-108.

18. Gleiter H (2000) Nanostructured materials: basic concepts and microstructure. Acta Materialia 48: 1-29.

Toxic and Biochemical Effects of Some Bioinsecticides and Igrs on American Bollworm, *Helicoverpa armigera* (h ü b.) (noctuidae: lepidoptera) in Cotton Fields

Hatem Mohamed Al-shannaf[1], Hala Mohamed Mead[1], Al-Kazafy Hassan Sabry[2*]

[1]*Plant Protection Research Institute, Agriculture Research Centre, Dokki, Giza, Egypt*
[2]*Pests and Plant Protection Dep. National Research Centre, Dokki, Giza, Egypt*

Abstract

Field experiments were carried out to evaluate efficiency of three bioinsecticides (Dipel DF, Protecto and Bioranza) and two insect growth regulators (IGRs) (chlorfluazuron and pyriproxyfen) against larvae of American bollworm, *Helicoverpa armigera* (ABW) and their side effects on some common predators in Egyptian cotton fields during 2009 and 2010 at Aga region, Dakahlia Governorate, Egypt. Results indicated that chlorfluazuron showed the highest initial reduction (75.00 and 80.6%); residual mean (83.75 and 79.45%) and annual mean (80.83 and 79.83%) on *H. armigera* during the two successive seasons, respectively. Moreover, chlorfluazuron was the most toxic and gave the highest reduction in predator numbers recorded (20.70, 23.20 and 22.37%) in the 2009 season and (23.30, 20.90 and 21.70%) in the 2010 season at the initial, residual and annual means, respectively.

Chlorfluazuron, pyriproxyfen and Dipel DF gave the lowest significant decrease in the activity of amylase enzyme (61.86 and 59.86% relative to control), invertase enzyme (75.28 and 80.13%) and trehalase enzyme (73.64 and 83.74%), respectively after 3 and 7 days post treatment. Insect growth regulators (chlorfluazuron and pyriproxyfen) caused highly significant increases in the activity of chitinase enzyme (130 and 122.6% 141.89 and 131.64%, respectively) at both interval times, respectively.

Keywords: *Helicoverpa armigera*; biopesticides; insect growth regulators (IGRs); enzyme

Introduction

The American bollworm (ABW), *Helicoverpa armigera* (Hüb.), Lepidoptera: Noctuidae, is one of the most important economic insect pests in Egypt [1]. The larvae of this pest feed on a wide range of the economically important crops including cotton, corn, tomato, sunflower, legumes, tobacco, and several cucurbitous and citrus crops [2]. In India, where *H. armigera* commonly destroys more than half the yield crop, losses were estimated at over $300 million per annum [3]. Field failure resulting from *H. armigera* resistance to pyrethroids has been reported worldwide by many authors [4]. Because of their economic advantages and low toxicity to mammals and to some predators [5] much effort has been directed towards developing management aimed at using biopesticides and insect growth regulators in control programs.

Metabolism of carbohydrate hydrolyzing enzymes that play a principal role in digestion and utilization of carbohydrate by insect [6] is controlled mainly by amylase, invertase and trehalase enzymes. Trehalose is one of the most important storage carbohydrates that is present in almost all forms of life except mammals. Trehalose is split into glucose units by trehalase enzyme. Amylase enzyme is required to digest carbohydrates (polysaccharides) into smaller units (disaccharides), and eventually into even smaller units (monosaccharides) such as glucose. Amylase enzyme plays a key role in plant defense toward pests and pathogens [7] which cause severe damage to field crops and stored grains [8]. Invertase enzyme hydrolyzes sucrose, forming fructose and glucose.

Chitin is a structural component of the cuticle and peritrophic membrane in the mid-gut of insects, and strict regulation of its metabolism is essential for the normal growth of insects. Chitenase is among a group of proteins that insects use to digest this structural polysaccharide in their exoskeleton ad gut linings during the molting process [9].

The present study was proposed to evaluate the effect of three bioinsecticides and two IGRs against *H. armigera* as well as their side effects to some common predators. Additionally, this research was proposed to elucidate some biochemical relationships among treatments and activities of some enzymes in *H. armigera*.

Materials and Methods

Tested compounds

Insect growth regulators (IGRs)

-Benzoylurea, Chlorfluazuron (Atabron® 5 % EC) 1- (2, 6, -difluorobenzoyl 3 - [4 (chloro – 5- trifluoromathyl-2-pyridyloxy) 3, 5,-dichlorophenyl] urea. used at rate of 400 ml/ feddan. Basic product of Syngenta Agro, Switzerland (local manufacture: Syngenta Agro, Dokki, Giza, Egypt)

-Juvenile hormone mimic, pyriproxyfen (Admiral®) 2-[1-methyl-2-(4-phenoxyphenoxy) ethoxy] pyridine, used at rate of 200 ml/ feddan. Basic product of Sumitomo chemical Co., Tokyo, Japan.

***Corresponding author:** Al-kazafy Hassan Sabry, Pests and Plant Protection Department, National Research Centre, Dokki, Giza, Egypt, E mail: kazafyhassan@yahoo.com

Bio-insecticides compounds

Bacterial insecticides: Dipel DF®, *Bacillus thuringiensis* subsp. *Kurstaki* (32, 258 Potency I.U. / mg) WP used at rate of 200 g/feddan (feddan = 4200 m^2). Basic product: Valent Biosciences Corporation, Libertyville, USA.

-Protecto®, *Bacillus thuringiensis* var. *Kurstaki*, (32000 I.U. /mg) WP used at rate of 300 g/feddan. Basic product: Insect pathogens unite, Plant Protection Research Institute, Agriculture Research Centre, Dokki, Giza, Egypt.

Fungus: Bioranza®, *Metarehizum aneasopliea* Sorok. Bioranza 10% WP (32x10^6 spores /ml) at rate of 200 g/feddan. Insect pathogens unite, Plant Protection Research Institute, Agriculture Research Centre, Dokki, Giza, Egypt.

Field trial

Field experiments were carried out at Aga region, Dakahlia Governorate, Egypt during two consecutive cotton growing seasons of 2009 and 2010. The experiment area of one feddan (4200 m^2) was divided into 6 equal randomized blocks (one for either treatment plus one for control). Each block was divided 4 experimental plots as replicates (175 m^2 for each) cultivated with the Egyptian cotton variety, Giza 86. Cotton plants were treated once with each treatment at 18th and 21st June during the both seasons, respectively. A plastic curtains used as borders between treatments during spray to avoid drift.

Samples

Weekly 20 cotton plants were investigated from the first of June until the mean numbers of ABW larvae reached 3 larvae / sample (0.15 larvae/ plant) [10] and then cotton plants were treated. The tested insecticides were applied at the recommended field rate, while control plots were sprayed with water only. Treatment plots were sprayed using a knapsack motor sprayer, 20 liters in capacity and using 200 -liter volume of insecticidal solution (insecticide + water as solvent) per feddan.

Directly pre treatment count was made visually on cotton plant (including leaves, stem, squares and bolls) for each treatment. Post treatment counts were recorded after 3, 7 and 10 days. The common predators: green lacewing, *Chrysoperla carnea*; lady beetles, *Coccinella* spp.; anthocoride bugs, *Orious* spp.; staphylinid beetle, *Peaderus alfierii* and True spiders found on cotton plants were counted and recorded at the same time. The efficiency of tested treatments was measured as a percentage of reduction in population density of American bollworm and some common predators using the Henderson and Tilton equation [11].

Biochemical assay

Preparation of samples: The present experiment was designed to study the changes in the activities of carbohydrate hydrolyzing enzymes and chitinase enzyme after treatment the field strain larvae of *H. armigera* with the tested biopesticides and insect growth regulators as compared to untreated larvae (control). The samples in the last season (2010) were studied only because this study not concerned on the inheritance of enzymes or follow the recipe in insect resistance. The biochemical analysis processes were carried out after 10 days because chlorfluazuron (IGR) has a latent effect and it was distinct after 10 days.

The preparation of samples involved the use of four healthy American bollworms larvae from each replicates (16 larvae/ each treatment and control) after 3 and 7 days of treatment with all tested compounds and control. The field populations of ABW were collected from cotton bolls during growing season 2010 and transferred to the laboratory in paper bags then placed in clean jars. Larvae were homogenized in distilled water (1 larvae/1 ml) using a Teflon homogenizer surrounded with jacket of crushed ice for 3 minutes. The homogenate was centrifuged at 3500 r.p.m for 10 minutes at 5°C to remove the haemocytes. The samples were divided into small portions and kept in a deep freezer at (- 20°C) until required [12]

Enzymes measurements

The methods used to determine the activities of carbohydrates hydrolyzing enzymes (amylase, trehalase and invertase) in digesting sucrose, trehalose and starch, respectively, were determined according to Ishaaya and Swiriski [13]. The free aldehedic group of glucose after starch, trehalose and sucrose digestion was determined using 3,5 dinitrosalicylic acid reagent.

Chitinase activity was determined using 3, 5-dinitrosalicylic acid reagent to determine the free aldehydic groups of hexoamines liberated on chitin digestion according to the method described by Ishaaya and Casida [14]

Statistical analysis

The significance of enzyme activities was determined by analysis of variance (ANOVA). The significance of various treatments was evaluated by Duncan's multiple range test ($p < 0.05$) [15]. Data were subjected to statistical analyses using a software package CoStat® Statistical Software [16] a product of Cohort Software, Monterey, California.

Results

Effect of different compounds on American bollworm larvae

Data in Table 1 indicate that the initial reduction percentages of the ABW Larvae in 2009 were 75.0, 17.5, 17.5, 11.4 and 10.0% after three days of treatment with chlorfluazuron, pyriproxyfen, Protecto, Dipel DF and Bioranza, respectively. After 7 days of application residual mean of reduction percentages were 83.3, 20.0, 13.3, 20.0 and 9.1% with chlorfluazuron, pyriproxyfen, Bioranza, Protecto and Dipel DF, respectively. In addition, chlorfluazuron caused the highest annual mean of reduction percentage 80.8%, while Dipel DF recorded the lowest one, 11.5%. During the 2010 season, the initial reduction percentage of ABW larvae after three days of treatment were 80.6, 15.8, 15.8, 10.2, and 10.2% for chlorfluazuron, pyriproxyfen, Dipel DF, Protecto and Bioranza in Table 1, respectively. The highest residual mean was 79.8% recorded with the chlorfluazuron followed descendingly by 17.9, 16.6, 13.9 and 12.0% recorded with Bioranza, pyriproxyfen, Protecto and Dipel DF, respectively. Chlorfluazuron caused the highest annual mean of reduction percentage (79.8%) against the ABW larvae, whereas Dipel DF recorded the lowest one (12.0%).

The statistical analysis shows that there are significant differences between chlorfluazuron and other pesticides, while no significant differences among Dipel DF, Protecto, Bioranza and pyriproxyfen. The LSD value is 10.5.

Effect of different compounds on the common predators

Data in Table 2 show that, the side effects of the compounds on the common predators found on cotton plants during both seasons in cotton field. The initial reduction percentages of common predators were 20.7 and 23.3% 10.2 and 11.5% 10.2 and 11.5% 4.6 and 11.1% and 2.8 and 5.6% for chlorfluazuron, pyriproxyfen, Protecto, Dipel DF and

Treatments	2009 Season									2010 Season								
	Pre count ±SE*	Mean number of *H. armigera* larvae ± SE			Reduction %					Pre count ±SE	Mean number of *H. armigera* larvae ± SE			Reduction %				
					Initial	Residual			An-nual Mean					Initial	Residual			Annual Mean
		3 days	7 days	10 days	3 days	7 days	10 days	Mean			3 days	7 days	10 days	3 days	7 days	10 days	Mean	
Dipel DF	11.0±0.4	13.0±0.4	15.0±.08	15.0±0.8	11.4	9.1	13.9	11.5	11.5[b]	14.0±0.8	14.0±0.8	17.0±0.8	16.0±0.9	15.8	11.7	8.6	10.2	12.0[b]
Protecto	10.0±0.9	11.0±0.9	12.0±0.9	13.0±0.9	17.5	20.0	17.90	18.9	18.5[b]	15.0±0.9	16.0±0.4	16.0±0.9	17.0±0.9	10.2	22.4	9.3	15.9	13.9[b]
Bioranza	10.0±0.4	12.0±0.8	13.0±0.8	11.0±0.8	10.0	13.3	30.50	21.9	17.9[b]	15.0±0.8	16.0±0.8	18.0±0.8	13.0±0.8	10.2	12.7	30.7	21.7	17.9[b]
Chlorfluazuron	12.0± 0.8	4.0±0.4	3.0±0.4	3.0±0.4	75.0	83.3	84.20	83.7	80.8[a]	13.0±0.9	3.0±0.4	4.0±0.9	3.0±0.8	80.6	77.4	81.5	79.5	79.8[a]
Pyriproxyfen	10.0±0.4	11.0±0.9	12.0±0.8	12.0±0.8	17.5	20.0	24.20	22.1	20.6[b]	17.0±0.8	17.0±0.8	19.0±0.8	18.0±0.8	15.8	18.7	15.3	17.0	16.6[b]
Control	12.0±8	16.0±0.8	18.0±0.9	19.0±0.9	---	---	---	---	---	16.0±0.9	19.0±0.9	22.0±0.8	20.0±0.4	---	---	---	---	---
$LSD_{0.05}$									10.48									11.49
P values									000***									000***

*SE = Standard Error
**Means under each variety sharing the same letter in a column are not significantly different at P<0.05.

Table 1: Reduction percentage of the tested insecticides against *Helicoverpa armigera* larvae in cotton field during two successive seasons, 2009 and 2010 seasons.

Bioranza in 2009 and 2010, respectively. The highest residual mean of reduction percentages were 23.2 and 20.9 recorded with chlorfluazuron in 2009 and 2010 followed by 9.6, 8.6, 8.5 and 8.5% for Bioranza, Dipel DF, Pyriproxyfen and Protecto in the first season. In the second season, the tested compounds showed lowest influence reduction were 9.9, 9.6, 6.8 and 4.1% with Bioranza, pyriproxyfen, Protecto and Dipel DF compared to the highest residual mean of reduction (20.9%) with chlorfluazuron. The highest annual mean of reduction percentages were 22.4 and 21.7% recorded with chlorfluazuron compound in 2009 and 2010 seasons. The lowest annual reduction takes place with Dipel DF treatments in both seasons (2009 and 2010). It was 7.3 and 6.4%, respectively.

The statistical analysis shows that there are significant differences between chlorfluazuron and other pesticides, while no significant differences among Dipel DF, Protecto, Bioranza and pyriproxyfen. The LSD value is 5.32.

Biochemical responses

The changes in the activity of carbohydrate hydrolyzing enzymes (amylase, invertase and trehalase) and chitenase enzymes in the supernatant of the homogenated larvae of field strain of *H. armigera* were measured at two different time intervals (3 and 7 days).

Carbohydrate Hydrolyzing Enzymes

Amylase enzyme

Data tabulated in Table 3 show that, in all treatments the level of amylase activity in the supernatant of the homogenated larvae was lower than that obtained with the untreated larvae at all inspected times. The activity of amylase was decreased greatly with chlorfluazuron treatments followed by pyriproxyfen, Dipel DF, Bioranza and Protecto. It was 61.9, 73.8, 88.8, 93.9 and 95.9% compared with control.

The statistical analysis shows that a significant difference between chlorfluazuron and other pesticides after 3 and 7 days. The LSD values are 46.199 and 49.789, respectively (Table 3). The same result was take place after 7 days. Chlorfluazuron and pyriproxyfen (IGRs) caused the highest decrease at the last inspected time (59.9 and 69.9% relative to control, respectively.

Invertase enzyme

Regarding to invertase enzyme, there were decrease in the activity in *H. armigera* resulted from all treatments during all tested periods as compared to control (Table 3).

The activity of invertase enzyme tended to give the highest decrease at the first inspected time as affected by all treatments with the exception of chlorfluazuron that gave the highest reduction at last time (83.9% relative to control). Pyriproxyfen only gave the highest decrease in the activity after 3 days that recorded (75.3) comparing to other treatments. Whereas, the reduction in the invertase activity after 7 days of treatment ranged between a minimum value of 96.1% for Bioranza to a maximum value of 80.1% for pyriproxyfen.

The statistical analysis shows that a significant difference between pyriproxyfen and other treatments after 3 days, while no significant difference among all treatments after 7 days. The LDS values are 380.1 and 359.755, respectively.

Trehalase enzyme

Data presented in Table 3, indicate that the activity of trehalase

Treatments	2009 Season									2010 Season								
	Pre count ±SE*	Mean number of natural enemies ± SE			Reduction %					Pre count ± SE	Mean number of natural enemies			Reduction %				
					Initial	Residual			Annual Mean					Initial	Residual			Annual Mean
		3 days	7 days	10 days	3 days	7 days	10 days	Mean			3 days	7 days	10 days	3 days	7 days	10 days	Mean	
Dipel DF	15.0±0.8	17.0±0.8	20.0±0.9	15.0±0.9	4.6	11.1	6.1	8.6	7.3^b	17.0±0.9	16.0±0.9	22.0±0.9	15.0±0.8	11.1	4.4	3.9	4.1	6.4^b
Protecto	15.0±0.9	16.0±0.8	21.0±0.8	15.0±0.9	10.2	6.7	10.4	8.5	9.1^b	16.0±0.9	15.0±0.8	20.0±0.4	13.0±0.9	11.5	7.6	6.0	6.8	8.4^b
Bioranza	13.0±0.9	15.0±0.9	17.0±0.8	19.0±0.4	2.8	12.8	6.5	9.6	7.4^b	15.0±0.8	15.0±0.4	18.0±0.8	11.0±0.4	5.6	11.3	8.5	9.9	8.4^b
Chlorflua-zuron	17.0±0.8	16.0±0.9	19.0±0.9	21.0±0.4	20.7	25.5	20.9	23.2	22.4^a	16.0±0.9	13.0±0.9	17.0±0.9	12.0±0.9	23.3	21.5	20.3	20.9	21.7^a
Pyriproxyfen	15.0±0.8	16.0±0.8	21.0±0.8	19.0±0.8	10.2	6.7	10.4	8.5	9.1^b	16.0±0.9	15.0±0.9	19.0±0.4	14.0±0.4	11.5	12.2	7.0	9.6	10.2^b
Control	16.0±0.8	19.0±0.9	24.0±0.9	25.0±0.9	—	—	—	—	—	17.0±0.8	18.0±0.4	23.0±0.9	16.0±0.9	—	—	—	—	—
$LSD_{0.05}$									5.93									5.32
P values									0.0009***									0.0005***

*SE = Standard Error

**Means under each variety sharing the same letter in a column are not significantly different at P<0.05.

Table 2: Reduction percentage of the tested insecticides against common predators in cotton field during two successive seasons, 2009 and 2010 seasons.

enzyme in the larvae of *H. armigera* was generally decreased in Dipel DF compared to other treatments in both times. It was 73.6 and 83.7% after 3 and 7 days, respectively, followed by chlorfluazuron (84.1 and 86.4%). The activity of trehalose enzyme increased in Bioranza treatment. It was 101.1% compared to control.

The statistical analysis shows that a significant difference between Dipel DF and other treatments after 3 days, while no significant difference among all treatments after 7 days. The LDS values are 78.462 and 249.822, respectively.

Chitinase enzyme

Results obtained in Table 3 show a remarkable significant increase in the enzyme activity in *H. armigera* using pyriproxyfen (141.9 and 131.6% relative to control) and chlorfluazuron (130.0 and 122.6%) after 3 and 7 days, respectively. On the other hand, Protecto caused decrease in the activity (92.5 and 98.1%, respectively).

The statistical analysis shows that a significant difference among chlorfluazuron and pyriproxyfen, and other treatments after 3 and 7 days. The LDS values are 9.403 and 10.799, respectively.

Discussion

Generally, chlorfluazuron was the most potent insecticide against ABW, *H. armigera* in initial, residual mean and annual mean causing highly reduction percentage comparing to other treatments during the 2009 and 2010 seasons. The same results were found by other authors. Methoxyfenozide and triflumuron significantly reduced the damage caused by *H. zea* [17]. Similarly, applying lufenuron at 37 and 49 g ha effectively suppressed *H. armigera* populations and resulted significant reduction in crop damage at lower doses, while in buprofezin was not effective at any tested dose for any time of treatment [18]. Dipel DF (*B. thuringiensis*) was the least effective in the residual and annual means, whereas Bioranza (Bt) was the lowest one in the initial activity at the two successive seasons. Other researchers found that the biopesticide Agerin (Bt) which similar to Dipel DF formulation (both of them *Bacillus thuringiensis* subsp. *Kurstaki*) caused the least effective treatment against bollworms than the conventional pesticides [19]. Additionally, spinosad caused significantly less dead moths of *H. armigera, H. punctigera* and other Noctuid compared to methomyl and could be used where quick action is not needed [20].

On contrary, during 1992 – 1993 seasons, MVP (Bt) resulted in 100% mortality of larvae at 10 days after the first application. Likewise 10 days after the second application, MVP product and Karate (lambda-cyhalothrin) gave 100% mortality of *Heliothis* larvae. While, during 1993-94 cotton seasons all the strains gave better control than the pesticide Karate [21]. Also, using Dipel 2X recorded 63.04 and 44.46% against spiny and American boll worms infesting cotton bolls during 2004 and 2005 seasons, respectively; whereas, application of Dipel 2X followed by conventional insecticides gave (40.40 and 50.85%), respectively [22].

Populations of predator insects found in all treated areas with tested insecticides were reduced comparing to predator numbers registered in untreated areas during the two successive seasons. However, the highest initial, residual and annual means of reduction percentages recorded with chlorfluazuron treatment in both tested seasons did not exceed 23.3%, whereas the least initial and annual means were given by Bioranza and Dipel DF, respectively at both seasons. The same symmetry was ordered by other authors. They found that spraying with biological insecticides, chemical insecticides and Bt transgenic cotton plants

Table 3: Changes in the activity of some enzymes in the supernatant of the homogenated *H. armigera* larvae as affected by tested insecticides.

Treat-ments	Amylase: Enzyme activity (Mean± S. E**) at the indicated tested times				Invertase: Enzyme activity (Mean± S. E) at the indicated tested times				Trehalase: Enzyme activity (Mean± S. E) at the indicated tested times				Chitinase: Enzyme activity (Mean± S. E) at the indicated tested times			
	3days		7days		3days		7days		3days		7days		3days		7days	
	µg/glucose/ min./g b.wt	*%	µg glucose/ min./g b.wt	*%	µg glucose/ min./g b.wt	*%	µg glucose/ min./g b.wt	*%	µg glucose/ min./g .wt	*%	µg glucose/ min./g b.wt	*%	µg AChBr/ min./g b.wt	*%	µg AChBr/ min./g b.wt	*%
Dipel DF	578.8±11.4^{b}	88.8	784.0±15.5^{b}	92.0	1398.9±200.3^{a}	81.87	1302.1±156.4^{a}	84.9	654.4±39.1^{c}	73.6	910.4±21.0^{a}	83.7	56.3±4^{b}	108.6	65.3±2.7^{b}	92.9
Protecto	624.1±11.1^{ab}	95.9	833.2±18.7^{ab}	97.8	1435.6±178.5^{a}	84.02	1319.4±184.5^{a}	86	860.3±22.7^{a}	96.8	1058.7±82.3^{a}	97.4	47.9±3.2^{b}	92.5	68.9±3.3^{b}	98.1
Bioranza	610.5±11.6^{ab}	93.9	807.6±16.0^{ab}	94.8	1584.2±214.8^{a}	92.71	1473.8±134.6^{a}	96.1	873.3±12.9^{a}	98.3	1099.5±77.7^{a}	101.1	54.5±3.4^{b}	105.1	74.3±3.1^{b}	105.7
Chlorflua-zuron	402.3±9.8^{d}	61.9	510.1±13.4^{d}	59.9	1507.7±191.1^{a}	88.23	1286.7±150.6^{a}	83.9	747.4±27.0^{b}	84.1	939.4±19.6^{a}	86.4	67.4±2.6^{a}	130.0	86.2±4^{a}	122.6
Pyriproxy-fen	480.6±15.7^{c}	73.8	595.5±12.7^{c}	69.9	1286.3±176.4^{b}	75.28	1229.4±97.7^{a}	80.1	858.6±20.9^{a}	96.6	1037.2±55.6^{a}	95.4	73.5±2.5^{a}	141.9	92.50±4.1^{a}	131.6
Control	650.4±24.9^{a}	100	852.2±19.4^{a}	100	1708.7±163.9^{a}	100	1534.2±157.3^{a}	100	888.6±22.7^{a}	100	1087.1±62.1^{a}	100	51.8±2.3^{b}	100	70.3±3.6^{b}	100
L.S.D$_{0.05}$	46.199**		49.789**		380.100**		359.755^{ns}		78.462**		249.822^{ns}		9.403**		10.799**	
P values	0.0000		0.0000		0.0095		0.6772		0.0002		0.4863		0.0005		0.0008	

*Activity %= percentage relative to control.
**SE = Standard Error
***Means under each variety sharing the same letter in a column are not significantly different at P<0.05.

reduced predator population by 2.64-14.2% [23]. Bt cotton efficiently controls cotton bollworms, while the decrease of pesticide applications allows the buildup of high populations of predators, such as lady beetles, *Coccinella septempunctata*, lacewings, *Chrysopa sinica*, spiders and others in mid-season [24]. Buprofezin and lufenuron (IGR,s) at lower doses, appeared safe to predator populations, which did not differ significantly in IGR-treated versus untreated control plots. Population densities of coccinellids were significantly lower at high concentrations of both IGRs in treatment plots, possibly as a result of reduced prey availability [18]. Side effect of seven pesticides on beneficial arthropods was highly influence compared with Agerin (Bt) treatment which had the minimum side effect on beneficial arthropods [19].

Carbohydrates are of vital importance since they can be utilized by the insects' body for production of energy or conversion to lipids or proteins. Metabolism of carbohydrates is controlled mainly by carbohydrate hydrolyzing enzymes. The final product of carbohydrates metabolism is glucose, the increase of these enzymes during the larval stage suggested that these enzymes degrade carbohydrates to glucose for chitin build-up [6]. This clears in Table 3, chlorfluazuron decreased the activity of amylase. So, degradation of carbohydrates also decreases. This leads to disturbance in chitin building and failure of molting process.

Therefore, the inhibition of carbohydrate hydrolyzing enzymes recorded in the present study might affect the molting process and subsequently may explained the reason of mortality occurred in *H. armigera* larvae as illustrated previously in the toxicological experiments. These results are in agreement with previous research who observed pronounced decrease in the carbohydrate hydrolyzing enzymes especially amylase and invertase was observed after treated 5th instar larvae of cotton leafworm, *S. littoralis* (Lepidoptera: Noctuidae) with sub-lethal concentrations of thuringeinsin (beta-exotoxin of *B. thuringiensis*) [25]. Consult and Mimic (IGRs) decreased the invertase activity after 5 days of treatment, whereas Consult, Atabron and Cascade exhibited reduction in trehalase and invertase activities in *S. littoralis* [26]. Additionally, the activities of trehalase, invertase and amylase enzymes in *S. littoralis* larvae treated with Tracer (spinosad) and triflumuron were generally decreased than untreated larvae during different tested times [27].

Ecdysis is initiated by apolysis the process that separates epidermal cells from the old cuticle by molting fluid secretion and ecdysal membrane formation. The molting fluid contains proteases and chitinases, enzymes that digest the main constitution of the old endocuticle [28]. The insect growth regulator, diflubenzuron interferes with the development of the cuticle, to which insect skeletal muscle is attached. The effect of diflubenzuron on the ultra structure of the muscle attachment to the cuticle in larvae of Noctuid *S. littoralis* is described, and it is concluded that there is no digestion of the affected old cuticle, and no digestion of the tonofibrillae (microtubules passing through the pore canals and attached to the cuticulin layer) [29].The fluctuation in the chitinase activity in the homogenated larvae was observed by many authors. Markedly increase in chitinase activity occurred when treated 4th instar larvae of *S. littoralis* were treated with diflubenzuron [30]. Chlorfluazuron caused a significant increase in chitinase activity of *S. littoralis* [31].

These results confirmed that the insect growth regulators (chlorfluazuron and pyriproxyfen) were more effective than the biopesticides against *H. armigera*, but these pesticides have a side effect on the natural enemies compared with the biopesticides. So, the biopesticides is more suitable to integrated pest management program.

On the other hand, the activity of amylase, invertase and trehalase was clearly decreased in chlorfluazuron and pyriproxyfen (IGR,s) especially after 7 days of treatment. While, the activity of chitinase was increased in chlorfluazuron and pyriproxyfen compared to the biopesticides. This mean that chitinase play an important role in *H. armigera* resistant to insect growth regulators.

References

1. Ibrahim M.M., Metwally AG, Nazmy, NH and Ibrahim FEZ (1974) Studies on the American bollworm on cotton in Egypt. *Heliothis zea* (Boddie) *Heliothis armigera* (Lepidoptera: Noctuidae). Agric Res Rev 52: 1-8.
2. Xiulian, S, Hualin W, Xincheng S, Xinwen C, Chaomei P, Dengming P and Johannes AJ (2004) Biological activity and field efficacy of a gen- etically modified *Helicoverpa armigera* single-nucleocapsid nucleopolyhedrovirus expressing an insect-selective toxin from a chimeric promoter. Biol Control 29: 124-137.
3. Reed W, Pawar CS (1981) *Heliothis* A global problem. Proceedings of the International Workshop on *Heliothis* Management. ICRISAT Center, India.
4. Forrester NW, Cahill M, Bird LJ, Layland J K (1993) Management of pyrethroid and endosulfan resistance in *Helicoverpa armigera* (Lepidoptera: Noctuidae) in Australia. Bull of Entomol Research Supplement No.1.
5. Duffie WD, Sullivan MJ, Turnipseed SG, Dugger P, Richter D (1998) Predator mortality in cotton from different insecticide classes. Proceedings Beltwide Cotton Conferences, San Diego, California, USA.
6. Wyatt GR (1967) The biochemistry of sugars and polysaccharides in insects. Adv Insect Physiol 4: 287-360.
7. Franco OL, Rigden DJ, Melo FR, Maria F. Grossi-de-Sá (2000) Plant α-amylase inhibitors and their interaction with insect α-amylase: structure, function and potential for crop protection. Eur J Biochem 269: 397-412.
8. Franco OL, Rigden DJ, Melo FR, Grossi-de-Sa MF (2002) Activity of wheat α-amylase inhibitors towards bruchid α-amylases and structural explanation of observed specification. Eur J Biochem 267: 2166-2173.
9. Fukamizo T (2000) Chitinolyticenzymes: catalysis, substrate binding, and their application. Curr Protein Peptide Sci 1: 105-124.
10. Chamuene A, Ecole C, Sidumo A (2007) Effect of strip intercropping for management of the American bollworm, *Helicoverpa armigera* Hübner (Lepidoptera: Noctuidae) on cotton (*Gossypium hirsutum*) in Morrumbala district. African Crop Science Conference Proceedings 8: 1049-1052.
11. Henderson CF, Tilton EW (1955) Tests with acaricides against the brown wheat mite. J Econ Entomol 48: 157-161.
12. Aly MM (1990) Bioactivity of certain plant extracts of Fam. Mytraceae and other biocides on some pests attacking cotton cultivation. Ph.D. Thesis, Institute of Environmental Studies and Researchers, Ain Shams Univ, Cairo, Egypt.
13. Ishaaya I, Swirski E (1976) Trehalase, invertase and amylase activities in the black scaleSaissetia oleae, and their relation to host adaptability. J Insect Physiol 22: 1025-1029.
14. Ishaaya I, Casida JE (1974) Dietary TH 6040 alters composition and enzyme activity of housefly larval cuticle. Pestic Biochem Physiol 4: 484-490.
15. Snedecor GW, Cochran GW (1980) Statistical methods 2nd Ed. Iowa State Univ Press, Iowa, USA.
16. CoStat Statistical Software. Microcomputer program analysis version, 6.311. (2005) CoHortSoftware, Monterey, California, USA.
17. Branco MC, Pontes LA, Amaral PST, Mesquita FMV (2003) Insecticides for the controlof the South American tomato pinworm and the corn earworm and impact of those products on *Trichogramma pretiosum. Horticultura Brasileira* 21: 652-654.
18. Gogi MD, Rana M, Sarfraz LM, Dosdall MJ, Keddie AB, et al. (2006) Effectiveness of two insect growth regulators against Bemisia tabaci (Gennadius) (Homoptera: Aleyrodidae) and *Helicoverpa armigera* (Hübner) (Lepidoptera: Noctuidae) and their impact on population densities of arthropod predators in cotton in Pakistan. Pest Manag Sci 10: 982-90.
19. Abd-Elrahman M, Younis H, Sayed H, Sanaa M, Ibrahim A, Zaki A, Zeitoun M (2007) Field evaluation of certain pesticides against the cotton bollworms with special (749) reference to their negative impact on beneficial arthropods (2006 cotton season, Minia region, Egypt). 8 ACSS conference 27-31 October, Faculty of Agric., Minia University, Minia, Egypt.
20. Alice PS, Peter CG, Anthony JH (2010) Development of a synthetic plant volatile-based attracticide for female Noctuid moths. III. Insecticides for adult *Helicoverpa armigera* (Hübner) (Lepidoptera: Noctuidae). Australian J Entomol 49: 31-39.
21. Baloch AA, Korejo AK, Kalroo AM, Sanjrani MW (1996) Studies on comparativeefficacy of different strains of *Bacillus thuringiensis* against *Heliothis armigera* on cotton crop in Sindh [Pakistan]. Second International Congress of Entomological Sciences March 22-23.
22. Hegab MEM (2008) Studies on some elements of integrated control of cotton bollworms. Ph.D. Thesis, Faculty of Agric. Al-Azhar University, Cairo, Egypt.
23. Yi-zhong Y, Yi-dong S, Kun Q, Lu R, Yue-shu (2002) Effect of different control methods onnatural enemies in cotton field. Chinese J Biological Control 18: 111-114.
24. Wu K, Lin K, Miao J, Zhang Y (2005) Field abundances of insect predators and insect pests on α endotoxin-producing transgenic cotton in northern China.
25. El-Ghar GE, Radwan HS, El-Bermawy ZA, Zidan LT (1995) Inhibitory effect of thuringiensin and abamectin on digestive enzymes and non-specific esterases of *Spodoptera littoralis* (Boisd.) (Lepidoptera: Noctuidae) larvae. J Appl Entomol 119: 355-359.
26. Eid AM (2002) Esterases and phosphatases in relation to chlorpyrifos-resistance in *Spodoptera littoralis* (Boisd.) (Lepidoptera: Noctuidae). Egypt J Appl Sci 17: 275-284.
27. Mead HM, El-Sheakh AA, Soliman BA, Desuky WM, Abo-Ghalia AH (2008) Biochemical effect of some compounds on carbohydrate hydrolyzing enzymes of cotton leafworm, *Spodoptera littoralis* (Boisd.), Egypt. J Agric Res 86: 2169-2192.
28. Reynolds SE, Samuels RI (1996) Physiology and biochemistry of insect molting fluid. Adv Insect Physiol 26: 157-232.
29. Hegazy G, Degheele D (1990) The ultrastructure of muscle attachment in larvae of *Spodoptera littoralis* with special reference to diflubenzuron treatment, Mededlingen Van de Faculteit Landbouww. Rijksuniv. Gent Belgium 55: 609-620.
30. Farag AM (2001) Biochemical studies on the effect of some insect growth regulators on the cotton leafworm. MSc. thesis Fac Agric Cairo University, Egypt.
31. Abdel-Aal AE (2006) Effect of chlorfluazuron, nuclear polyhydrosis virus (SLNP) and *Bacillus thuringiensis* on some biological and enzyme activity of cotton leafworm, *Spodoptera littoralis* (Boisd.). Bull Ent Soc., Cairo University, Egypt.
32. Second International Symposium on Biological Control of Arthropods, Davos, Switzerland.

Wheat Crop Yield Losses Caused by the Aphids Infestation

Arif Muhammad Khan[1], Azhar Abbas Khan[2*], Muhammad Afzal[2] and Muhammad Shahid Iqbal[3]

[1]National Institute for Biotechnology and Genetic Engineering, Faisalabad, Pakistan

[2]Department of Entomology, University of Sargodha, Sargodha, Pakistan

[3]College of Agriculture D.G. Khan, Sub-Campus University of Agriculture, Faisalabad, Pakistan

Abstract

Present study was planned to determine the yield loss in wheat crop at different dates of observation for the population of aphids. The wheat cultivar BK-2002 was sown at Arid Zone Research Institute, Bhakkar. For two different treatments viz. sprayed and un-sprayed wheat crop was compared, sprayed crop was observed with minimum aphid's population whereas un-sprayed appeared comparatively susceptible with maximum population of aphids. Thiamethoxam (Actara®) was used as insecticide for sprayed treatment @ 25 gm/acre as recommended commercially. Last week of February was found to be very favourable for aphids in wheat fields in the study area. Third week of February was found to be the most suitable period for the bio-control agents i.e. *Coccinellids* in wheat crop. Positive correlation was found to exist between the population of aphids and *Coccinellids* among different wheat cultivars.

Keywords: Aphids; Wheat crop; Yield loss

Introduction

Wheat (*Triticum aestivum* L.) is a major crop with largest area under cultivation in Pakistan and plays a significant role in economic stability of the country [1]. Low yield of wheat per hectare in Pakistan compared to the other advanced countries is due to several abiotic and biotic factors, such as traditional methods of cultivation, varieties, lack of irrigation facilities, barani areas, soil fertility and incidence of insect pests and diseases.

Although many insect pests attack wheat (*Triticum aestivum* L.) in Pakistan, severe damage is caused by aphids. Aphids cause yield losses either directly (35-40%) by sucking the sap of the plants or indirectly (20-80%) by transmitting viral and fungal diseases [2]. Population density of aphids also depends on the abiotic factors [3-5]. During spring season (February-March) aphid population increases, at the same time biocontrol agents like *coccinellids* also increase as natural check on this pest [6].

Several control methods have been evolved for the control of aphids. These include cultural, physical, mechanical, biological, chemical and host plant resistance. Mostly, the aphid populations are maintained below the economic injury level by combination of naturally occurring population regulating factors. But sometimes, the aphids can be extremely injurious if present in large number and chemicals have to be used for control [7].

The wheat crop is generally infested with aphids during the growth stages when both the adults and nymphs take a heavy toll by sucking cell sap which reduces the vitality of the plants. The infested leaves turn pale, wilt and wear a silky appearance. Some species also have toxins in their saliva and dense infestation may kill young shoots. Honey dew excretion is often prolific and sooty moulds usually accompany aphids infestation which eventually affects the rate of photosynthesis in plants. The poor yield of wheat crop is mainly attributed to its instability to aphids attack. The aphids are considered as serious pest of wheat crop. They can multiply very rapidly under favourable conditions on leaves, stems and inflorescence. The infestation causes severe distortion of leaves and inflorescence, and can significantly decrease the yield through direct feeding. Decline in the yield in wheat crop is attributed to several abiotic factors, traditional methods of cultivation, low yielding varieties, lack of proper irrigation facilities in most of the areas, relatively low level of soil fertility and a higher incidence of insect pests and diseases. The present study was conducted to evaluate the yield losses in wheat crop due to the infestation of aphids. The investigations were projected to manage this serious pest and to boost up the wheat production keeping in view the quality and quantity of the production.

Materials and Methods

Present experiment was performed at Arid Zone Research Institute (AZRI), Bhakkar, Pakistan during 2004-05 to see yield losses in wheat crop infested by green aphid. The data of aphids and its yield losses was recorded during the whole experimental period (from January 2005 to April 2005). The wheat cultivar BK-2002 obtained from Punjab Seed Corporation Bhakkar was used as cultivar. The crop was sown in lines on 15th of November, 2004. Two treatments viz. sprayed and un-sprayed were distributed in 16 plots, 8 for each, the plot size was 2 m^2. Thiamethoxam (Actara®) was used as insecticide for sprayed treatment @ 25 gm/acre as recommended commercially. Polyethylene sheet was hanged at the height of 4 ft between treatments, the drift of aphid population from sprayed to unsprayed plots was prevented. The experiment was replicated four times by using Complete Randomized Block Design. Treatment plan is given in the Figure 1.

Data was collected after every ten days from 01-01-2005 to 15-04-2005. During each sampling date, five wheat plants from each plot were randomly selected and the number of aphids per tiller of each plant was counted. At harvest, the yield of both sprayed and unsprayed plots was compared to assess yield losses. To see the grain weight of

***Corresponding author:** Azhar Abbas Khan, Department of Entomology, University of Sargodha, Sargodha, Pakistan, E-mail: azhar512@gmail.com

sprayed and unsprayed plots, 1000 grains of every plot were counted and their weights were compared. Data regarding abiotic factors (RH% and °C) was recorded from observatory of AZRI, Bhakkar. ANOVA was made to compare the yield losses and thousand grain weight losses, after counting percentage yield losses (LSD $P \leq 0.05$). Minitab 13.3, a statistical software package was used for statistical analyses.

Results

First detection of aphids on the wheat crop was observed on 11th of January 2005 (Table 1); at this time, the population was 0.18 aphid/plant, afterwards a gradual increase in the population was observed that reached to its peak in the mid of March (51.55 aphid/plant). After mid of March, the population started declining and 1.56 aphid/plant were observed in the mid of April and gradually no aphids were seen in the wheat fields. The aphid infestation was scattered on leaves, spikes and in mid of March also observed on stem.

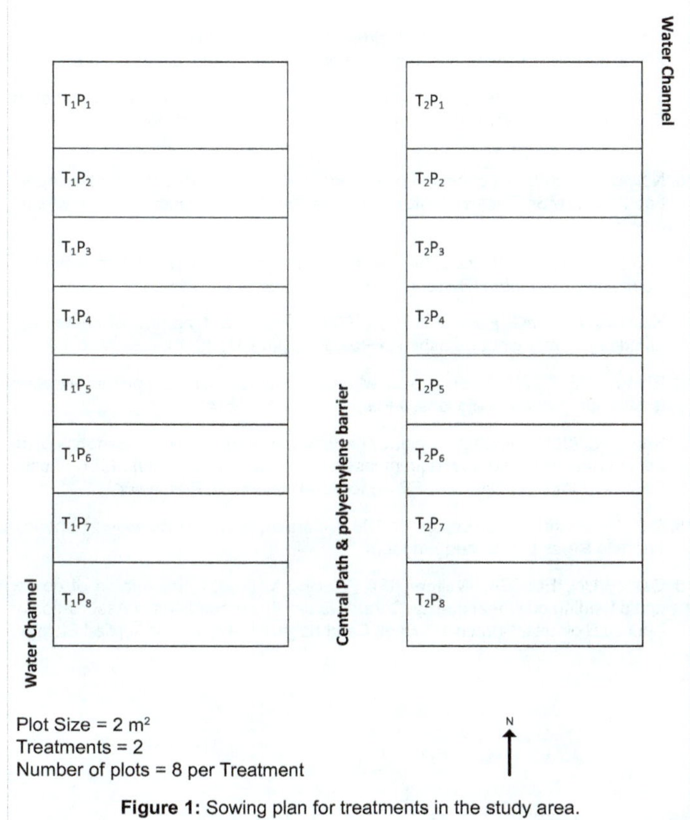

Figure 1: Sowing plan for treatments in the study area.

	Dates	Population/Plant
1.	01/01/2005	0.00
2.	11/01/2005	0.18
3.	21/01/2005	0.26
4.	31/01/2005	0.52
5.	11/02/2005	1.57
6.	21/02/2005	8.64
7.	04/03/2005	39.67
8.	15/03/2005	51.55
9.	26/03/2005	21.74
10.	05/04/2005	11.34
11.	15/04/2005	1.56

Table 1: Population Dynamics of Wheat Aphids.

Effect of temperature and relative humidity on aphid population dynamics

Aphid infestation started in 2nd week of January mostly on the leaves. At this time, the relative humidity was 72% and the temperature ranged from 17.5°C to 8.5°C, maximum and minimum respectively. Up to the end of the February, the increase in the population was gradually slow while a slight fluctuation in the relative humidity and temperature were recorded. Sharp increase of aphid population in 1st week of March was recorded which remained till mid of March (Table 1). At this time, 71% relative humidity, 26.5°C maximum temperature and 15°C minimum temperature were recorded. Afterwards, aphids population started declining and on 26th of March 2005, 21.74 aphids/plant were recorded (Figure 2). During the month of April, the RH ranged from 73% to 88% and temperature from 26.5°C to 32.0°C maximum and from 16.5°C to 13.0°C minimum respectively. RH reached to 74% and minimum temperature also increased in the mid of April and aphid population dropped down to 1.56 aphids/plant while at the end of April, no counts were observed in field.

Yield losses assessment and thousand-grain weight comparison

The results showed that the yield per plot ranged from 17.40 kg to 24.10 kg with an average of 19.79 kg in treated plots as compared to 16.00 kg to 24.00 kg with an average of 18.93 kg in untreated plots (Table 2). Further results revealed that 4.57% more yield was recorded in plots where thiamethoxam (Actara®) was sprayed. However, the statistical analysis shows that the means are not significantly different

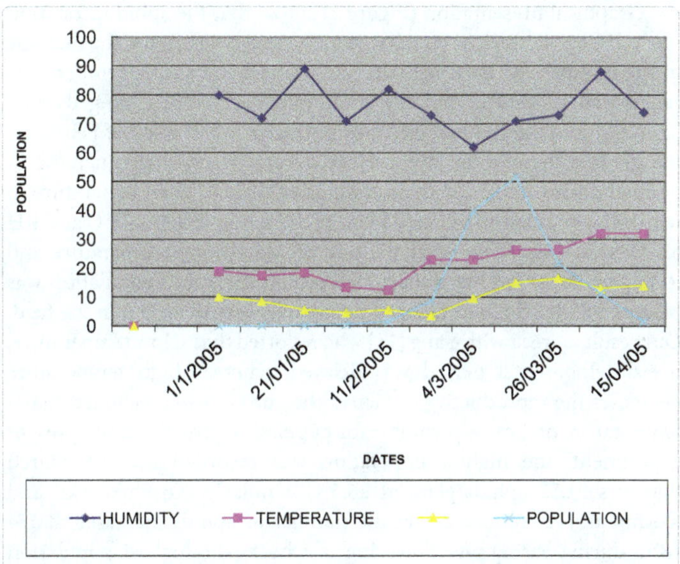

Figure 2: The Effect of Relative Humidity and Temperature on the Population Dynamics of wheat Aphids.

Source	Degree of Freedom	Mean Squares for Yield loss assessment (Kg/plot)	Mean Squares for 1000-grain weight Comparison
Replications	3	21.226	30.216
Treatments	1	1.496ns	182.787**
Error	3	0.454	106.267
LSD		1.517	23.198

$P \leq 0.05$

Table 3: Analysis of Variance for Yield Losses Assessment and 1000-Grain Weight.

	Treatments	Range	Average	% differences	LSD
Yield losses assessment (kg/plot)	T1	17.4-24.1	19.79	4.57	1.517
	T2	16.0-24.0	18.93		
Thousand Grain Weight Comparison	T1	329.02-377.90	339.52	5.92	23.198
	T2	320.01-336.32	329.96		

$P \leq 0.05$ T_1 = Sprayed T_2 = Unsprayed

Table 2: Yield Losses Assessment.

for each treatment. The 1000-grain weight per plot ranged from 329.00 to 377.90 with an average of 339.52 gm in treated plots as compared to 320.01 to 336.32 with an average of 329.96 gm in untreated plots, 5.92% increase occurred in 1000-grain weight in plot where insecticide was sprayed (Table 3).

Discussion

The aphids infestation started in the mid of January and gradually increased during the vegetative growth of wheat crop. The population reached to its peak in the mid of March during the heading stage of the wheat and gradually declined when the crop reached to maturity with 1.56 aphids/plant in the mid of April. Similar results were shown by Karimullah and Ahmad [8]; they observed that aphids infestation started in the 1st week of February and peaked in the latter half of March with fluctuations in population size thereafter up to mid of April. Our results are in conformity with Xiong [9], and Nawaz [10]; who observed that population of aphids in the field increased with the development of the wheat and peaked at the heading stage.

Graphical presentation (Figure 2) shows that the aphid infestation started in the 4th week of January up to the mid of February, the increase in the population remained very slow, where maximum temperature range was 12.5-19°C and the minimum temperature ranged from 3.5-8.5°C and the RH ranged 72-82%. In start of 3rd week of February, the gradual increase in temperature occurred and sharp increase in population was recorded up to the mid of March. The temperature in which the population increased was in the range of 9.5-26.5°C and RH 62-71% was recorded. After the mid of March, the temperature and RH again increased but a sharp decrease in the aphid population was recorded. After the mid of April, no aphids were observed in the field. Our results agreed with Yang [11] who reported that at low temperature, the developmental period was delayed, whereas high temperature decreases the reproductive capacity. The author also concluded that a temperature of 25°C is favorable for population growth. In the present experiment, the highest population was recorded on 15th March that was 51.55 aphids/plant at 26.5°C. Similarly, Kostyukovskii and Kushneuk [12] observed that the number of aphids increased at 15-18°C during earing and flowering of wheat. The highest population density was recorded during grain development and initiation of wax ripening. The decline in the aphid population could also be the result of the crop maturity as stated by Riedell [13]; that infestation of aphids on wheat crop is abundant during the heading and flowering stages and is reduced during the maturity stage of the crop. In the present study, reduction of 4.57% in the yield and 5.92% in the thousand-grain weight was recorded due to aphid infestation. Here, our results agree with Khan [14], while Gair et al. [15] and Oakley et al. [16] observed a reduction in yield and thousand-grain weight as 12% and 39% respectively which are more as in our study. Since the experimental plots were normally fertilized with NPK, the non-significant yield losses could be due to proper nutrition of the crop, as stated by Riedell [13].

References

1. Anwar J, Ali MA, Hussain M, Sabir W, Khan MA, et al. (2009) Assessment of yield criteria in bread wheat through correlation and path analysis. J Anim Plant Sci 19: 185–188.

2. Aslam M, Razaq M, Akhter W, Faheem M, Ahmad F (2005) Effect of sowing date of wheat on aphid (Schizaphis gramium RONDANI) population. Pak Entomol 27: 79–82.

3. Aheer GM, Munir M, Ali A (2007) Impact of weather factors on population of wheat aphids at Mandi Baha-ud-Din district. J Agric Res 45: 61–66.

4. Aheer GM, Ali A, Ahmad M (2008) Abiotic factors effect on population fluctuation of alate aphids in wheat. J Agric Res 46: 367–371.

5. Wains MS, Aziz-ur-Rehman, Latif M, Hussain M (2008) Aphid dynamics in wheat as affected by weather and crop planting time. J Agric Res 46: 361–366.

6. Khan AA, Khan AM, Tahir HM, Afzal M, Khaliq A et al. (2011) Effect of wheat cultivars on aphids and their predator populations. Afr J Biotechnol 10: 18399-18402.

7. Hatchett AH, Starks KJ, Webster JA (1987) Insect and mites pest of wheat: In Wheat and wheat improvement (E.G. Heyne Edition) Madison, Wisconsin, USA.

8. Karimullah, Ahmad KF (1989) Incidence of the cereal aphid *Sitobion avenae* (F) on different cultivars of wheat. Sarhad Journal of Agriculture 5: 59-61.

9. Xiong CJ (1990) Study on the relationship between the occurrence of *Rhopalosiphum padi* (L) and the growing period of wheat. Insect Knowledge 27: 5-7.

10. Nawaz H (2000) Insect pest of wheat and the effect of fertilizer (NPK) on aphid population. MSc Thesis Deptt of Entomology, NWFP Agricultural University, Peshawar.

11. Yang XW (1990) Effects of temperature and light on the population growth of *Schizaphis graminum* (Rondani). Insect Knowledge 27: 263-266.

12. Kostyukovskii MG, Kushnerik VM (1990) Population dynamics of cereal leaf aphids on winter wheat. Zashchita Rastenii (Kiev) 37: 10-13.

13. Riedell WE (1990) Tolerance of wheat to Russian Wheat aphids: Nitrogen fertilization reduces yield loss. J Plant Nutr 13: 579-584.

14. Khan SS (2000) Impact of plant phenology of various wheat genotypes on aphid population and subsequent losses in wheat due to aphids. MSc Thesis, Dept of Plant Protection, NWFP Agricultural University, Peshawar.

15. Gair R, Jenkins JEE, Lester E (1987) Cereal pests and diseases (9thedn). Farming Press Ltd, United Kingdom.

16. Oakley UN, Ellis SA, Walters KFA, Watling M (1993) The effects of cereal aphid feeding on wheat quality: Cereal Quality III, proceedings of Association of Applied Biologist, Churchill College Cambridge, UK Aspects of Applied Biology.

Toxicity and Residual Activity of a Commercial Formulation of Oil from Neem, *Azadirachta indica* A. Juss. (Meliaceae), in the Embryonic Development of *Diatraea saccharalis* F. (Lepidoptera: Crambidae)

Camila Vieira da Silva[1]*, Larissa Carla Lauer Schneider[2] and Hélio Conte[3]

[1]*Faculty of Apucarana (FAP), Biological Sciences Department, Apucarana PR, Brazil*

[2]*Veterinary Medicine Department, University of Maringa (UEM), Estrada da Paca Bairro São Cristóvão, Umuarama PR, Brazil*

[3]*Biology Cell and Genetics Department, CCB, University of Maringa (UEM), Avenida Colombo, Maringá PR, Brazil*

Abstract

The effects of a commercial formulation of oil from neem, *Azadirachta indica* A. Juss (Meliaceae), on the eggs of sugarcane borer *Diatraea saccharalis* F.,1794 (Lepidoptera: Crambidae), at different stages of its embryonic development were investigated. To evaluate the ovicidal activity of the oil, eggs were sprayed with the product at concentrations of 0.1%, 0.3%, 0.5%, 1.0% and 2.0% at 1, 2, 3, 4, 5 and 6 days after day were laid. Controls were sprayed with distilled water. The neem oil was toxic to eggs, mainly when applied to 2-, 3-, 4- and 5-days-old eggs. The percentage of viability reduction was 31-99%. Higher levels of caterpillar eclosion were obtained in eggs treated with neem oil at older ages. However, the resulting caterpillars had fatal morphological anomalies, except those exposed to 0.1% neem oil. Neem oil is highly toxic when sprayed on *D. saccharalis* eggs, as demonstrated by severely reduced hatching, increased mortality of hatched caterpillars, impaired embryonic development, and residual activity in the production of new individuals. Therefore, neem oil may be a promising agent against the sugarcane borer, during the stage in which penetrates the stalk, causing heavy damage to the sugar-cane crop.

Keywords: Sugarcane borer; Neem oil; Residual activity; Embryonic development; Toxicity

Introduction

Plant species with highly promising insecticidal characteristics belong to the Meliaceae family [1]. This is especially the case for *Azadirachta indica* A. Juss. (Meliaceae), commonly known as, the neem tree [2]. Neem extracts have been extensively studied due to their high content of insecticidal substances, including triterpenoids, azadirachtin and melantriol, which are effective against several pest species [2-6].

The inseticidal substances mentioned above have complex aromatic structures that decrease the development of resistance of insects; they are also biodegradable. Their activity is quickly reduced by light; they are only slightly toxic to mammals, and are compatible with the natural foes of several insects [7-9]. Studies on the effects on insects treated or fed with azadirachtin have reported growth inhibition; death of caterpillars during ecdysis; prolongation of the larval stage; deformation of caterpillars and adults; decrease in longevity, fecundity and fertility [10-13]. Azadirachtin in neem oil has selective effects on insects, mostly with respect, impaired feeding and interrupted growth [14,15].

Evaluations of formulations of neem oil, e.g. AZT (containing 30 mg azadirachtin ml^{-1}), NEEM-AZA (containing 3 mg azadirachtin ml^{-1}), and AZ (pure azadirachtin), on second instar *Plutella xylostella* (Lepidoptera: Plutellidae) caterpillars, revealed a 50 to 90% mortality rate. AZT had also ovicidal activity [16]. Calneem oil has been found to have insecticidal and anti-oviposition effects, as well as ovicidal and repellent properties against caterpillars of *Ephestia cautella* (Lepidoptera: Pyralidae) [12].

Studies of the effect on the embryonic development, e.g. neem oil directly applied to Lepidoptera eggs, are rare. Some studies have documented the effects on oviposition, following treatment of caterpillar adults or pupae [12,13,17,18]. Ovicide effects of neem solution were reported after application to egg masses of different ages in *Diatraea saccharalis* F., 1794 (Lepidoptera: Crambidae), but subsequent effects on the development of caterpillars have not been demonstrated [19]. Other types of extracts with ovicidal effects have been reported in connection with the following, *P. xylostella*, *Heliothis armigera* Hübner, 1805; *Spodoptera frugiperda* (Lepidoptera: Noctuidae); *Bemisia tabaci* (Hemiptera: Aleyrodidae); *Tribolium castaneum* (Coleoptera: Tenebrionidae); *Anopheles stephensi* and *Aedes (Stegomyia) aegypti* (Diptera: Culicidae); *D. saccharalis* [19-25].

The aim of this study was to identify natural insecticidal alternatives to enable a more efficient control of the sugarcane borer. Therefore, the toxicity of neem oil was evaluated with regard to embryonic development, and the residual activity in larvae that newly emerged from eggs.

Materials and Methods

Experiments were performed at the Laboratory of Morphology and Cytogenetic of Insects, State University of Maringá (Maringá, Paraná, Brazil). Bioassays were conducted at 25°C ± 1°C, a relative humidity of 70% ± 10%, and a 12-h photoperiod. For the acquisition of *D. saccharalis* eggs of 1 to 6 days of age, the adults were placed on posture chambers, and the eggs deposited were maintained in coupling chambers until hatching.

***Corresponding author:** Camila Vieira da Silva, Faculty of Apucarana (FAP), Biological Sciences Department, Apucarana PR, Brazil, E-mail: hconte@uem.br

Natuneem® (pure neem oil with an azadirachtin rate of over 1,500 ppm) is an organic product, certified by BCS OKO Garantie, Doc. Natur- 9009/09.05/7331-BR, and was used as a bio-insecticide in the experiment. Neem oil was diluted in distilled water to produce concentrations of 0.0% (control), 0.1%, 0.3%, 0.5%, 1.0% and 2.0%. Groups with approximately 50 eggs each were formed. They were maintained in polyethylene plates (6.0 cm diameter and 2.0 cm thick), which were lined with filter paper and filled with cotton; they were wetted daily with distilled water. Plates were then grouped according to concentration and age of the eggs (1, 2, 3, 4, 5 and 6 days after being laid). Four replications were user for each treatment, including the control. Approximately 100 µL of the respective neem oil solution was sprayed on the different egg groups. Data were recorded daily for up to 9 days post-treatment.

The number of hatched caterpillars of the treated eggs was counted daily, so that the possible residual effect of the agent could be analysed. Caterpillars were separated in polyethylene plates and provided with an artificial diet [26]. Plates were kept at the same experimental conditions, as described above. Mortality percentages for eggs of different ages were assessed for each treatment condition, according to Abbott [27]. The frequency (%) and type of abnormalities of hatched caterpillars and frequent changes and abnormal developments in the eggs were photographed by using a digital camera, Cannon 7.0 MG, with a stereomicroscope (Carl Zeiss, Jena, Germany). The viability rate for eggs for each treatment condition age, duration of the embryonic period were analysed using the statistical program PRISM 5.0. Differences among treatments were analysed using the ANOVA and Tukey tests ($p<0.05$).

Results and Discussion

Neem oil was toxic to *D. saccharalis* eggs in all tested concentrations, and the toxicity varied with the dose and the age of the eggs (Table 1). Neem oil has been considered to have an ovicidal action because the agent may obstruct the egg membrane and hinder the embryo´s respiratory exchanges [28]. Similar to the results obtained in previous studies, neem oil blocked the aeropyles of eggs and the body and head of 2 lice species, thus preventing the embryos of both of lice species from accessing oxygen, and from releasing carbon dioxide [29].

No significant differences ($p>0.05$) were found between treatments and control in the development time of eggs (Table 2). Treatment with neem oil at all concentrations resulted in a significant ($p<0.05$) decrease in egg viability, which varied between 38.1 and 98.9% (Table 2). The observed decrease may have occurred because of the activity of an active substance in the formulation, which interfered with the embryonic development. Azadirachtin, the main substance in neem extracts, has a selective effect on insects [30]. In fact, experiments have shown that it may impair mitochondrial functions [31]. Because mitochondria produce energy for cell processes and are required for the embryonic development, they are probably affected.

A significant decrease in the viability was obvious in all treatment groups, when compared to the control group (Table 1). This finding was observed mainly in eggs during the first 3 days of development (≈ 1, 2, and 3 days post-oviposition). The death of the insects depends on the dose and on the duration of exposure to the chemical substance/agent, which may occur some days after application [32]. The emergence of caterpillars in eggs of *Ceratothripoides claratris* (Thysanoptera: Thripidae), treated with neem formulations were significantly affected by the relationship between the age of the eggs and the concentration [33].

A lower decrease in viability has been reported in embryos at more advanced stages (eggs of 4, 5 and 6 days of age), mainly at a concentration of 0.1% (Table 1) [19]. However, the resulting caterpillars presented morphological abnormalities, with consequently high mortality rates (Figures 1E-1H). Two hypotheses could be proposed: either the substance was active in the final development stage or the neonatal caterpillars had physiological modifications, after the contact with residues, which affected the morphological changes. Studies have shown that caterpillars from neem-treated eggs may have a high mortality rate mainly because of the contact of the insects with the chorion until their hatching, or because of consumption of neem present on the chorion [34].

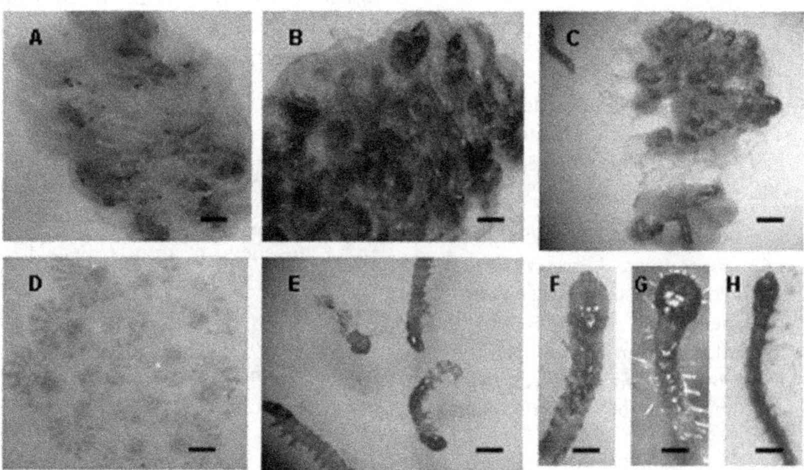

Figure 1: Morphological alterations in *Diatraea saccharalis* (Lepidoptera: Crambidae), caused by neem oil in caterpillars from neem treated eggs.

In A, 1-day-old eggs treated with neem oil (1%) after 9 days; they showed changes in their development. In B-C, 4-day-old eggs treated with neem oil (1%) after 9 days. In D-E, 2-day-old eggs treated with neem oil (2%), with abnormalities in hatched caterpillars, F shows caterpillars with slight melanisation and lack of cephalic capsule. G, shows the reduced length of body, colouring of the darkened body, and a slight increase in the cephalic region. H, shows caterpillars with greater body length and intense melanisation (bar: 3 mm).

Concentrations (%)	Age eggs (days) treated					
	1	2	3	4	5	6
0.0	83.00 ± 1.73a	85.50 ± 4.50a	79.50 ± 2.50a	78.00 ± 1.15a	67.50 ± 2.06a	75.50 ± 2.06a
0.1	27.50 ± 1.50b	36.00 ± 1.83b	54.50 ± 2.21b	56.00 ± 2.94b	63.50 ± 3.50a	56.50 ± 4.03b
0.3	1.00 ± 1.00c	3.00 ± 0.58c	9.50 ± 0.96c	37.70 ± 4.17c	32.00 ± 2.16b	35.70 ± 0.85c
0.5	1.50 ± 0.95c	1.00 ± 0.57c	3.75 ± 0.85c	38.50 ± 3.30c	30.75 ± 2.81b	27.00 ± 1.47c
1.0	2.50 ± 1.50c	0.00 ± 0.00c	6.50 ± 0.96c	36.50 ± 1.26c	20.25 ± 1.10c	26.25 ± 1.65c
2.0	0.00 ± 0.00c	0.00 ± 0.00c	5.00 ± 0.71c	0.00 ± 0.00d	0.00 ± 0.00d	7.00 ± 1.29d

Means in each column followed by the same letter are not significantly different by Tukey's test (P<0.05).

Table 1: Percentage of *Diatraea saccharalis* (Lepidoptera: Crambidae) egg hatching treated with different concentrations of neem oil and compared to control.

Concentrations (%)	Period embryonic	Hatching total (%) Mean±SE	% reduction	Larva mortality (%) Mean±SE
0.0	5.50 ± 0.64	79.17 ± 2.58a	---------	---------
0.1	5.75 ± 0.57	49.00 ± 5.70b	38.1	21.65 ± 4.04 a
0.3	5.87 ± 0.98	19.82 ± 6,91c	74.9	89.17 ± 4.90 b
0.5	6.21 ± 1.10	17.82 ± 6.82c	77.5	97.01 ± 2.29 b
1.0	5.60 ± 0.67	15.34 ± 5.97c	80.7	95.58 ± 2.36 b
2.0	5.28 ± 1.21	0.83 ± 0.83c	98.9	97.52 ± 2.56 b
F-values	0.4041 NS	28.15***		97.39***

Means in each column followed by the same letter are not significantly different by Tukey's test (P<0.05); *** P< 0.0001; NS not significant.

Table 2: Toxicity of neem oil on *Diatraea saccharalis* (Lepidoptera: Crambidae) hatching and caterpillars mortality when applied to eggs immediately after being laid.

Caterpillar formation occurred in eggs that were treated for 4 days from the start of development. However, the eggs had a darker colour, and a few caterpillars failed to break the eggs chorion (Figures 1B and 1C). These observations are similar to those obtained for neem-treated insects that completed the embryonic development, but died without rupturing the egg chorion [35]. According to these authors, the neem extract had not interfered in their embryogenesis. On the other hand, it has been reported that there are no changes in development and survival, even when neonatal caterpillars feed on neem extract-treated chorion [36]. Neem treatment of female individuals of *Rhipicephalus sanguineus* (Acari: Ixodidae), resulted in modifications of the chorion after oviposition, impairied maturation, and production of new individuals [18]. In the present study, treatment with neem altered the morphology of the caterpillars, during their formation within the egg (i.e. during the embryonic development), thus, before having contact with the chorion (Figures 1A-1D).

One- or 2-day-old eggs treated with neem oil exhibited a yellowish-orange colour, with small dark spots, which was observed after 9 days of treatment (Table 1, Figure 1D). Control eggs were either mostly hatched, or the caterpillars were fully developed and prepared for hatching (Table 1, Figure 1A). In all treatments, there was a decrease in the viability of the eggs. The strongest decrease occurred in eggs up to 3 days of age. Hatched caterpillars showed had high mortality rates when treated with neem oil, at concentrations ranging 0.3-2.0% that varied between 89.17 ± 4.9% and 97.52 ± 2.56%. No significant differences among treatments have been detected ($p<0.05$; $F=1.44$; Table 2). These results are similar to those obtained by using caterpillars of *Mamestra brassicae* (Lepidoptera: Noctuidae). Treated eggs remained at the first instar stage, whereas caterpillars of control eggs developed towards the second and third instar stages [37].

Sensitivity of neonate caterpillars to neem, i.e. aggressiveness, has been reported when the extract is sprayed immediately after the hatching of the eggs of *Acantholyda erythrocephala* (Hymenoptera: Pamphiilidae) [38]. In the case of *Cycloneda sanguinea* (Coleoptera: Coccinellidae), the authors did not report any changes in the survival and duration of caterpillars that development from neem-treated eggs [36]. After 9 days of treatment, 5 day and 9 day *D. saccharalis* eggs treated with neem oil showed few dead caterpillars, just before completion of hatching. Some died with parts of their body outside the egg. In fact, these results are very similar to those observed in eggs of *Tuta absoluta* (Lepidoptera: Gelechiidae) [39]. Ovicidal effects occurred in eggs at the start of the development (at 1, 2 and 3 days of age), which confirms that older eggs are more resistant to plant extracts due to their higher maturity, and because they hinder the entrance of external products [19,25]. Other studies report that Lepidoptera eggs have a lipid membrane in the internal part of the chorion that impedes ovicidal agents, and thus, makes them insensitive to these compounds [40].

Few *D. saccharalis* caterpillars, apparently not affected by neem, hatched and died at 1 or 2 days after completion of hatching. This occurred mainly at neem concentrations of 0.3 and 0.5%, i.e. at the concentrations suggested in the formulation (Table 3). Studies have reported the death of caterpillars, immediately after hatching from the eggs of *Neodiprion abietis* (Hymenoptera: Diprionidae), treated with Neemix 4.5 EC at a concentration of 90 ppm [41]. Since plant extracts are viscose, this effect may be due to impaired locomotion and feeding of caterpillars, especially those at the first instar stage, and thus, reduced viability in this phase [7].

The current study demonstrates that neem oil had the highest effect on hatched caterpillars of treated eggs, because of its interference during the development of the caterpillars. This has caused abnormalities, and the subsequent death of the caterpillars (Table 2, Figures 1E-1H). When eggs were treated with neem oil at a concentration of 2%, formation of caterpillars was reported on the second day of development (2 days after oviposition). However, the caterpillars had a greyish colour; their colour was lighter than that of the control. Hatched caterpillars showed a somewhat soft abdomen, light colour, and in some instances, lack of the cephalic capsule (Figures 1D and 1F). Probably, this concentration of neem oil which has a higher content of azadirachtin, caused the inhibition of ovarian ecdysteroids [42], that are produced by follicular

Age Eggs (days) treated	Treatments									
	0.1		0.3		0.5		1.0		2.0	
	LE	R*	LE	R*	LE	R*	LE	R*	LE	R*
1	27.5 ± 1.5a	67	1.0 ± 1.0a	99	1.5 ± 0.9[a]	98	2.5 ± 1.5[a]	97	0.0 ± 0.0a	----
2	36.0 ± 1.8b	58	3.0 ± 0.5b	96	1.0 ± 0.5[a]	99	0.0 ± 0.0a	----	0.0 ± 0.0ac	----
3	54.5 ± 2.2c	31	9.5 ± 0.9ab	88	3.7 ± 0.8[a]	95	6.5 ± 0.9ab	92	5.0 ± 0.7bd	94
4	56.0 ± 2.9c	28	37.7 ± 4.1c	51	38.5 ± 3.3b	50	36.5 ± 1.2c	53	0.0 ± 0.0ac	----
5	63.5 ± 3.5c	15	32 ± 2.1c	57	30.7 ± 2.8b	59	20.2 ± 1.1d	73	0.0 ± 0.0ac	----
6	56.5 ± 4.0c	25	35.7 ± 0.8c	25	27.0 ± 1.4b	64	26.2 ± 1.6e	65	7.0 ± 1.3bd	91
F-values	24.60***		70.31***		67.90***		147.3***		27.69***	

LE- Mean ± SE caterpillars hatched.

R*- % reduction over Abbott [27], mortality of caterpillars hatching.

Means in each column followed by the same letter are not significantly different by Tukey's test ($P<0.05$);); *** $P< 0.0001$.

Table 3: % Mean ± SE in the caterpillars of eggs and respective mortality in *Diatraea saccharalis* (Lepidoptera: Crambidae) after one or two days of hatching.

cells of oocytes, and was incorporated into eggs during ovarian development [43]. Embryonic ecdysone signaling is required in eggs for germ band retraction, head involution, and morphogenetic movements that shape the first instar caterpillars [44].

Conclusions

The results show that neem oil is highly toxic to eggs and first instar caterpillars when the extract is sprayed on eggs, as demonstrated by impaired embryonic development and residual activity in the production of new individuals. Thus, neem oil may be an excellent agent for use against the sugarcane borer, during the stage in which the insect penetrates the stalk, causing heavy damage to the sugar cane crop.

Acknowledgments

We are grateful to CAPES (Coordenação de Aperfeiçoamento de Pessoal de Nível Superior) for financial support.

References

1. Jacobson M (1989) Botanical pesticides: Past, present and future. In: Arnason JT, Philogene BJR, Morand P (Eds) Insecticides of plant origin. America Chemical Society, Washington, USA 1-10.
2. Biswas K, Ishita C, Ranajit KB, Uday B (2002) Biological activities and medicinal properties of neem (*Azadirachta indica*). Curr Sci 82: 1336-1345.
3. Lavie D, Jain MK, Shpan-Gabrielith SR (1967) A locust phagorepellent from two media species. Chemical Communications 13: 910-911.
4. Butterworth JH, Morgan ED (1971) Investigation of the locust feeding inhibition of the seeds of the neem tree, *Azadirachta indica*. J Insect Physiol 17: 969-977.
5. Saxena RC (1987) Neem seed derivatives for management of rice insect pest-a review of recent studies. In: Schumutterer H, Ascher KRS (Eds), Proceeding of the Third International Neem Conference, Nairobi, Kenya, Eschborn 93-98.
6. Mordue AJ, Nisbet AJ (2000) Azadirachtin from the neem tree *Azadirachta indica*: its action against insects. Anais da Sociedade Entomológica do Brasil 29: 615-632.
7. Torres AL, Barros R, Oliveira JV (2001) Effect of plant extracts aquosos the development of *Plutella xylostella* (L.) (Lepidoptera: Plutellidae). Neotrop Entomol 30: 151-156.
8. Boeke SJ, Boersma MG, Alink GM, van Loon JJ, van Huis A, et al. (2004) Safety evaluation of neem (*Azadirachta indica)* derived pesticides. J Ethnopharmacol 94: 25-41.
9. Medina P, Budia F, del Estal P, Viñuela E (2004) Influence of azadirachtin, a botanical insecticide, on *Chrysoperla carnea* (Stephens) reproduction: toxicity and ultrastructural approach. J Econ Entomol 97: 43-50.
10. Mordue AJ, Blackwell A (1993) Azadirachtin: an update. J Insect Physiol 39: 903-924.
11. Khan M, Hossain MA, Islam MS (2007) Effects of neem leaf and a commercial formulation of a neem compound on the longevity, fecundity and ovarian development of the melon fly *Bactrocera cucurbitae* (Coquilett) and the oriental fruit fly, *Bactrocera dorsalis* (Hendel) (Diptera: Tephritidae). Pak J Biol Sci 10: 3656- 3661.
12. Pineda S, Martínez AM, Figueroa JI, Schneider MI, Del Estal P, et al. (2009) Influence of azadirachtin and methoxyfenozide on life parameters of *Spodoptera littoralis* (Lepidoptera: Noctuidae). J Econ Entomol 102: 1490-1496.
13. Shehu A, Obeng-Ofori D, Eziah VY (2010) Biological efficacy of CalneemTM oil against the tropical warehouse moth *Ephestia cautella* (Lepidoptera: Pyralidae) in stored maize. Int J Trop Insect Sci 30: 207-213.
14. Martinez SS (2002) Neem: nature, multiple uses, production. In: IAPAR, Londrina, Brazil 142.
15. Pathak CS, Tiwari SK (2010) Toxicological effects of neem *Azadirachta indica* A. Juss leaf powder against the ontogeny of *Corcyra cephalonica* (Staint.) (Lepidoptera: Pyralidae). Journal of Biopesticides 3: 617-621.
16. Verkerk RHJ, Wright DJ (1993) Biological activity of neem seed kernel extracts and synthetic azadirachtin against larvae of *Plutella xylostella* L. Pestic Sci 37: 83-91.
17. El-Bokl MM, Baker RFA, El-Gammal HL, Mahmoud MZ (2010) Biological and histopathological effects of some insecticidal agents against red palm weevil *Rhynchophorus ferrugineus*. Egypt Acad J Biolog Sci 1: 7-22.
18. Denardi SE, Bechara GH, de Oliveira PR, Camargo Mathias MI (2011) Inhibitory action of neem aqueous extract (*Azadirachta indica* A. Juss) on the vitellogenesis of *Rhipicephalus sanguineus* (Latreille, 1806) (Acari: Ixodidae) ticks. Microsc Res Tech 74: 889-899.
19. Tavares WS, Cruz I, Fonseca FG, Gouveia NL, Serrão JE, et al. (2010) Deleterious activity of natural products on postures of *Spodoptera frugiperda* (Lepidoptera: Noctuidae) and *Diatraea saccharalis* (Lepidoptera: Pyralidae). Z Naturforsch C 65: 412-418.
20. Loke JH, Heng CK, Rejab A, Basirum N, Mardi HCA (1992) Studies on neem (*Azadirachta indica* A. Juss) in Malaysia. In: Ooi PAC, Lim GS, Teng PS (Eds), Proceedings Third International Conference on Plant Protection in the Tropics, Kuala Lumpur, Malaysia 103-107.
21. Jeyakumar P, Gupta GP (1999) Effect of neem seed kernel extract (NSKE) on *Helicoverpa armigera*. Pesticide Research Journal 11: 32-36.
22. Souza AP, Vendramim JD (2000) Ovicidal activity of aqueous extracts of *Meliaceae* on the whitefly *Bemisia tabaci* (Gennadius) biotype B in tomato. Sci Agric 57: 403-406.
23. Das DR, Parweem S, Faruki SI (2006) Efficacy of commercial neem-based insecticide, Nimbicidine against eggs of the red flour beetle *Tribolium castaneum* (Herbst). Univ J Zool Rajshahi Univ 25: 51-55.
24. Marimuthu G, Rajamohan S, Mohan R, Krishnamoorthy Y (2012) Larvicidal and ovicidal properties of leaf and seed extracts of *Delonix elata* (L.) Gamble (Family: Fabaceae) against malaria (*Anopheles stephensi* Liston) and dengue (*Aedes aegypti* Linn.) (Diptera: Culicidae) vector mosquitoes. Parasitol Res 111: 65-77.
25. Tavares WS, Cruz I, Petacci F, Freitas SS Serrão JE, et al. (2011) Insecticide activity of piperine: Toxicity to eggs of *Spodoptera frugiperda* (Lepidoptera: Noctuidae) and *Diatraea saccharalis* (Lepidoptera: Pyralidae) and phytotoxicity on several vegetables. J Med Plant Res 5: 5301-5306.

26. Hensley SD, Hammond AM (1968) Laboratory techniques for rearing the sugarcane borer on an artificial diet. Journal of Economical Entomology 61: 1742-1743.

27. Abbott WS (1925) A method of computing the effectiveness of an insecticide. Journal of Economical Entomology 18: 265-267.

28. Schmutterer H (1990) Properties and potential of natural pesticides from the neem tree, *Azadirachta indica*. Annu Rev Entomol 35: 271-297.

29. Mehlhorn H, Abdel-Ghaffar F, Al-Rasheid KA, Schmidt J, Semmler M (2011) Ovicidal effects of a neem seed extract preparation on eggs of body and head lice. Parasitol Res 109: 1299-1302.

30. Hummel HE, Hein DF, Schmutterer H (2012) The coming of age of azadirachtins and related tetranortriterpenoids. Journal of Biopesticides 5: 82-87.

31. Vogt H, Gonzalez M, Andan A, Smagghe G, Vinuela E (1998) Side effects of Azadirachtin *via* residual contact in young larvae of the predator *Chrysoperla carnea* (Stephens) (Neuroptera: Chrysopidae). Plant Health Bulletin: Pest 24: 67-78.

32. Schmutterer H (1988) Potential of azadirachtin-containing pesticides for integrated pest control in developing and industrialized countries. J Insect Physiol 34: 713-719.

33. Dammini WTSP, Borgemeister C, Poehing HM (2005) Effects of neem and spinosad on *Ceratothripoides claratris* (Thysanoptera: Thripidae), an important vegetable pest in Thailand, under laboratory and greenhouse conditions. Journal of Economical Entomology 98: 438-448.

34. Tanzubil PB, Mccaffery R (1990) Effects of azadirachtin and aqueous neem seed extracts on survival, growth and development of the African armyworm, *Spodoptera exempta*. Crop Prot 9: 383-386.

35. Liu TX, Stansly PA (1995) Toxicity of biorational insecticides to *Bemisia argentifolii* (Homoptera: Aleyrodidae) on tomato leaves. Journal of Economic Entomology 88: 564-568.

36. Silva FAC, Martinez SS (2004) Effect of neem seed oil aqueous solutions on survival and development of the predator *Cycloneda sanguinea* (L.) (Coleoptera: Coccinellidae). Neotrop Entomol 33: 751-757.

37. Seljasen R, Meadow R (2006) Effects of neem on oviposition and eggs and larval development of *Mamestra brassicae* L: Dose response, residual activity, repellent effect and systemic activity in cabbage plants. Crop Prot 25: 338-345.

38. Lyons DB, Helson BV, Thompson DG, Jones GC, Mcfarlane JW, et al. (2003) Efficacy of ultra-low volume aerial application of an azadiracthin-based insecticide for control of the pine false webworm, *Acantholuda erythrocephala* (L.) (Hymenoptera: Pamphilidae), in Ontario Canada. International Journal of Pest Management 49: 1-8.

39. Trindade RC, Marques JMR, Xavier HS, Oliveira JV (2000) Methanol extract of neem seed kernels and mortality of eggs and larvae of the tomato pinworm. Sci Agric 57: 407-413.

40. Smith EH, Salkeld EH (1966) The use and action of ovicides. Annu Rev Entomol 11: 331-368.

41. Li SY, Skinner AC, Rideout T, Stone DM, Crummey H, et al. (2003) Lethal and sublethal effects of a neem-based insecticide on balsam fir sawfly (Hymenoptera: Diprionidae). J Econ Entomol 96: 35-42.

42. Rembold H, Sieber KP (1981) Inhibition of oogenesis and ovarian ecdysteroid synthesis by azadirachtin in Locusta migratoria migratorioides (R. and F.). Zeitschrift für Naturforschung 36: 466-469.

43. Glass H, Emmerich H, Spindler KD (1978) Immunohistochemical localisation of ecdysteroids in the follicular epithelium of locust oocytes. Cell Tissue Res 194: 237-244.

44. Kozlova T, Thummel CS (2003) Essential roles for ecdysone signaling during Drosophila mid-embryonic development. Science 301: 1911-1914.

Ureide Content of Guar under Influence of Hexaconazole and Triazophos

Yuvraj D Kengar* **and Bhimarao J Patil**

Department of Botany, Kanya Mahavidyalaya Islampur College, Karad, Maharashtra, India

Abstract

The important part of regular human diet is vegetable, which supply minerals and nutrients for maintenance of health. India is first rank producer of guar comprise 83% of total world production but these are affected by diseases like leaf spots, leaf blotch and insect pest such as serpentine leaf miners, hairy caterpillars and jassids. The major biotic constraint in vegetable production is pest problem. Application of pesticides has now become a common practice in modern agriculture. Chemicalization of agriculture rescues the crop from pest and diseases but when applied in excessive dose affects on plant health and metabolism. The various stresses alter the plant metabolism including pesticidal stress. The ureide content is reported in this study under pesticidal stress of Hexaconazole and Triazophos in leguminous vegetables guar (*Cyamopsis tetragonoloba* (L.) Taub.) after seed treatment and foliar sprays to ensure the nitrogen fixation and metabolism. Changes in the correlation between nitrogen fixation and ureide levels accelerate disturbance in the nitrogen metabolism and nitrogen products in plant. The concentrations of both pesticides for this experiment were 0.05, 0.1, 0.15, 0.2 and 0.3%. The seeds were soaked in these concentrations of pesticides for 12 h. After 12 h seed soaking period, the treated seeds were thoroughly washed with distilled water and sown in earthen pot containing garden soil and manure. The first foliar spray of Hexaconazole and Triazophos were applied with respective concentrations on 10th day while second foliar spray was on 25th day of plant growth. Analysis of ureide was carried out on the 15th and 30th day of growth that is 5 days after each spray. The uriede content in guar was remarkable increased after seed treatment followed by first foliar spray of 0.10, 0.15 and 0.20% hexaconazole however decreased it with dose concentrations of second foliar spray. The ureide content was decreased after both the foliar sprays of Triazophos.

Keywords: Ureide; Guar; Hexaconazole; Triazophos

Introduction

Legumes are agronomical and economical important in many cropping systems because of their nitrogen assimilation ability from atmosphere through nodules, anticipated to increase with it's need in sustainable agricultural practices development. Indeed, biological N_2 fixation is the most significant natural pathway through introduction of nitrogen into the biosphere has been administered. The nodulation on legumes can be determinate or indeterminate development pattern, consisting of three major tissues involving: a central infection zone, an inner cortex that includes vascular bundles, and an outer cortex [1,2]. The carbohydrates and other metabolites are diffusion to the nodule zone via phloem in the nodule cortex and the products of N_2 fixation, either amides (mainly asparagine) or ureides (allantoin and allantoic acid), are exported to the shoot via the xylem [2,3]. Ureide represents the sum of allantoin and allantoate. The point to remobilization of nitrogen from the oldest leaves as the main source for ureide synthesis and accumulation in shoots and developing tissues in the nonnodulated plants, and for the sharp increase in ureides during early pod filling in the nodulated plants when nitrogen fixation starts to decline. Changes in the correlation between nitrogen fixation and ureide levels with transition to reproductive development have been clearly demonstrated [4]. The major accumulation of ureides occurs has completely inhibited the N_2 fixation in nodulated plants [5]. Remobilization of nitrogen from senescent tissues was suggested as the most likely alternative source of ureides and involved in the N-feedback regulation of nitrogen fixation [6]. Moreover, ureide accumulation upon drought stress has been hypothesized to be responsible for N_2 fixation inhibition [7]. The ureide composition under pesticidal stress is not been reported in leguminous vegetables plants to ensure the nitrogen fixation and metabolism altering effect.

Vegetables are essential part of our human regular diet which supply nutrients and minerals for the good health and proper functioning of human body. Guar (*Cyamopsis tetragonoloba* (L.) Taub.) is a leguminous vegetable, having importance due to gum and guran production, utilized in numerous modern manufacturing and food processesing industries. Inclusion of vegetables in daily diet is very essential [8]. However, Guar is frequently attacked by diseases and insect pests. Application of pesticides as seed treatments and foliar sprays has become a common practice in modern agriculture [9], although application of pesticides interfere biochemical process of plants [10]. Hexaconazole (fungicides) and Triazophos (insecticide) are used to control leaf spot and leaf minors respectively in guar [8]. The plant absorbs a certain amount of the pesticides applied; changes in the plant's metabolism take place. In connection with this the impact of Hexaconazole and Triazophos on several metabolism of legume vegetable, guar has not yet been studied. Hence the attempt had made to study the effect of these pesticides on ureide contents in guar after seed treatment and foliar sprays.

Materials and Methods

The seed soaking treatments of Hexaconazole 5% EC and Triazophos 40% EC were given separately to healthy seeds of guar. The seeds were soaked in 0.05, 0.1, 0.15, 0.2 and 0.3% (v/v) concentrations of these pesticides for 12 h. After 12 h seed soaking period, the treated seeds were thoroughly washed with distilled water and sown in earthen pot containing garden soil and manure (3:1). The seeds treated with distilled water for 12 hour and sown in earthen pot were used as control.

***Corresponding author:** Yuvraj D Kengar, Department of Botany, Kanya Mahavidyalaya, Karad, Maharashtra India, E-mail: yuvrajkengar@gmail.com

The first foliar spray of Hexaconazole and Triazophos were applied with respective concentrations on 10th day while second foliar spray was on 25th day of plant growth. Analysis of ureide was carried out on the 15th and 30th day of growth i.e., 5 days after each spray.

Ureide content was measured according to the method of [11]. One hundred mg of dried and powdered plant material was taken in a test tube followed by 2 ml of 0.25 N NaOH. It was stirred and heated for 10 min at 100°C. After adding 1 ml of 0.65 N HCl at 100°C again it was heated for 10 min at 100°C. 1 ml of 0.4 M sodium phosphate buffer (ph 7.0) was added and it was cooled to room temperature and centrifuged for 5 min at 400 rpm. 1.5 ml supernatant was mixed with 0.25 ml of fresh phenylhydrazine solution (0.1 g of phenylhydrazine HCl, dissolved in 30 ml distilled water, prepared few minute before use. It was kept at room temperature for 5 minutes and then cooled on an ice bath. It was mixed with 1.25 ml pre-cooled concentrated HCl followed by 0.25 ml ferricyanide solution (0.50 g potassium ferricyanide in 30 ml distilled water)). Then the sample was removed from the ice bath and absorbance of colored product (dibenzyl formazan) was read at 535 nm after 15 min. The ureide content was estimated with the help of standard curve of Allantoin (0.05 mg ml^{-1}) using different concentrations.

Results

Effect of Hexaconazole after seed treatment and foliar sprays on ureide content in guar is studied and results are recorded in Table 1. The uriede content was increased after first foliar spray of Hexaconazole however it was decreased after second foliar sprays of different concentrations. The remarkable increased values of ureide were recorded at 0.10, 0.15 and 0.20% Hexaconazole (12.39, 14.12 and 16.48 μ mols g^{-1}) in guar after first foliar spray. The higher doses of Hexaconazole were act much detrimental on ureide content after second spray. The 0.30% Hexaconazole second spray recorded 2.93 μ mols g^{-1} ureide content in guar which is much reduced as compared to control; whereas 0.05 to 0.20% Hexaconazole reported 18.75, 16.43, 14.54 and 13.38 μ mols g^{-1} ureide content which was also less than control.

The effect of Triazophos on ureide content in guar was studied and it is recorded in Table 1. The ureide content in guar was decreased after both foliar sprays of Triazophos. The ureide content recorded at 0.05 to 0.30% Triazophos were 10.32, 10.39, 9.12, 6.48 and 4.37 μ mols g^{-1} after first foliar spray whereas 14.78, 13.18, 11.56, 8.10 and 6.75 μ mols g^{-1} after second foliar sprays in guar. The ureide content was reduced after treatment of Triazophos in all doses; however 0.20 and 0.30% Triazophos recorded much reduced ureide content after seed treatment and foliar sprays.

	Ureide content (μ mols g^{-1} dry weight)	
Treatment	**First foliar spray**	**Second foliar spray**
Control	11.23	19.23
Hexaconazole (% V/V)		
0.05	10.66	18.75
0.10	12.39	16.43
0.15	14.12	14.54
0.20	16.48	13.38
0.30	11.37	2.93
Triazophos (% V/V)		
0.05	10.32	14.78
0.10	10.39	13.18
0.15	9.12	11.56
0.20	6.48	8.10
0.30	4.37	6.75

Table 1: Effect of hexaconazole and triazophos seed treatment and foliar spray on ureides content of Guar.

In general, the ureide content increased only after first foliar spray of Hexaconazole, whereas in case of Triazophos treatment, the ureide content showed a decreased value with increased concentrations after both the spray.

Discussion

Leguminous plants are categorized into two groups depending upon the transporting form of nitrogen, one includes ureide transporting plants and other consists of amide transporting plants. Ureide contents in the oldest, primary leaves, and in the youngest, leaves during the development of plants grown under nitrogen fixation conditions or of nitrate-fertilized have been studied by many workers [12]. The activity of Allantoinase reported in the primary and the uppermost leaves during the development of nitrogen-fixing plants and plants grown with nitrate as the main nitrogen source. The ureides, allantoin and allantoate are major forms of nitrogen transported from root nodules to shoots in tropical legumes. In these plants, nitrogen fixed is used for purine synthesis. Through a series of enzymatic steps, purines are oxidized to allantoin and allantoate. Ureides synthesized in the nodules are transported to the shoot where they should be degraded and their N content assimilated. De novo purine synthesis is the main route for ureide formation in nodules. However, purines involved in the biogenesis of ureides may also arise by turnover of nucleic acids [13]. However, the biosynthetic route, degradation of ureides starts with hydrolysis of the internal amide bond of allantoin which giving rise to allantoate, reaction catalysed by allantoin amidohydrolase, characterized in plants [14,15]. The pathway for degradation of allantoate into glyoxylate and ammonia, is still under debate and further work. The occurrence of several pathways for the degradation of both allantoate and ureidoglycolate has been reported [16,17]. Most recent reports suggested that plants degrade allantoate to ureidoglycolate via allantoate amidohydrolase (AAH; EC 3.5.3.9) and ureidoglycine aminohydrolase (EC 3.5.3-) [18-22]. The ureidic plants relying upon N_2 fixation as the sole nitrogen source, ureides may comprise up to 86% of the N in the xylem sap, whereas amino acids, amides and nitrate are the major forms of nitrogen translocated from the roots after fertilization with nitrate [23,24]. In these plants, it is assumed that ureides reach high concentrations only in nodulated, nitrogen-fixing plants stem or petiole. This ureide levels has been established as an easy method to determine nitrogen fixation rates [24-28]. There are several reports h showing plant development influences the level of ureides in xylem sap and in leaves [29,30]. Changes in ureide levels upon plant development have been considered an important factor for the use of the ureide as a convenient method to determination of rate of nitrogen fixation [31].

In our experiments, the uriede content was increased after first foliar spray of Hexaconazole however it was decreased after second foliar sprays of different concentrations. The remarkable increased values of ureide were recorded at 0.10 to 0.20% Hexaconazole after first foliar spray. The higher doses of Hexaconazole were act much detrimental on ureide content after second spray. The second spray of 0.30% Hexaconazole responsible for much reduction of ureide content as compared to control. The ureide content in guar was decreased after both foliar sprays of Triazophos. However the 0.20 and 0.30% Triazophos recorded much reduced ureide content after seed treatment and foliar sprays. In general, the ureide content increased only after first foliar spray of Hexaconazole, whereas in case of Triazophos treatment, the ureide content showed a decreased value with increased concentrations after both the spray. The higher doses reduced the ureide content in guar. This indicated that the pesticides impose stress on ureide metabolism at higher doses. However the ureide levels have been shown to rise under water stress conditions also, and it has been suggested that the accumulation of ureides responsible for the feedback inhibition of nitrogen fixation in adverse conditions [32,33]. Recently reported that ureide content increases considerably in nonnodulated common bean plants suffering water stress and drought-induced senescence was considered the possible source of ureides [5]. An increase in uriede content of soybean leaves has been reported by Yukimoto and Ishianj after application of some oroganophosphorus insecticides [34].

The first spray of Hexaconazole was act as inducer for enzymatic system while excess doses may act suppressor in guar. The Triazophos retards the regulation mechanism of ureide synthesis indicated the inactivation of enzymes involved in this metabolism [35]. However there is little information on the regulation of genes and enzyme activities of ureide metabolism. Moreover regulation of ureide metabolism gene expression has been investigated in a ureidic plants [36,37] will help to focus on more details about ureide metabolism in guar like plants to glance on the nitrogen fixation mechanism.

References

1. Streeter JG (1991) Transport and metabolism of carbon and nitrogen in legume nodules. Advances in Botanical Research 18: 129-187.
2. Walsh KB (1995) Physiology of the legume nodule and its response to stress. Soil Biology and Biochemistry 27: 637-655.
3. Schubert S, Serraj R, Plies BE, Mengel K (1995) Effect of drought stress on growth, sugar concentrations and amino acid accumulation in N2-fixing alfalfa. Journal of Plant Physiology 146: 541-546.
4. Herridge DF, Peoples MB (1990) Ureide assay for measuring nitrogen fixation by nodulated soybean calibrated by N methods. Plant Physiology 93: 495-503.
5. Alamillo JM, Diaz LJL, Sanchez MMV, Pineda M (2010) Molecular analysis of ureide accumulation under drought stress in Phaseolus vulgaris L. Plant, Cell and Environment 33: 1828-1837.
6. Fischinger SA, Drevon JJ, Claassen N, Schulze J (2006) Nitrogen from senescing lower leaves of common bean is re-translocated to nodules and might be involved in a feedback regulation of nitrogen fixation. Journal of Plant Physiology 163: 987-995.
7. Serraj R, Vadez VV, Denison RF, Sinclair TR (1999) Involvement of ureides in nitrogen fixation inhibition in soybean. Plant Physiology 119: 289-296.
8. Kengar YD, Kamble AB, Sabale AB (2014) Effect of hexaconazole and triazophos on seed germination and growth parameters of spinach and guar. Annals of Biological Research 5: 89-92.
9. Ahemad M, Khan MS (2010) Comparative toxicity of selected insecticides to pea plants and growth promotion in response to insecticide-tolerant and plant growth promoting Rhizobium leguminosarum. Crop Protection 29: 325-329.
10. Jerlin B (2001) Effects of atrazine on growth nodulation and nitrogen constituents of Vigna mungo. J Ecotoxicol Eniviron Monito 11: 209-214.
11. Glenister RA, Rue JL (1987) Measuring ureides in symbiotic nitrogen fixation. INC, New York.
12. Juan LDL, Gregorio GV, Javier F, Manuel P, Josefa MA (2012) Developmental effects on ureide levels are mediated by tissue-specific regulation of allantoinase in Phaseolus vulgaris L. Journal of Experimental Botany 63: 4095-4106.
13. Zrenner R, Stitt M, Sonnewald U, Boldt R (2006) Pyrimidine and purine biosynthesis and degradation in plants. Annual Review of Plant Biology 57: 805-836.
14. Webb MA, Lindell JS (1993) Purification of allantoinase from soybean seeds and production and characterization of anti- allantoinase antibodies. Plant Physiology 103: 1235-1241.
15. Yang J, Han KH (2004) Functional characterization of allantoinase genes from Arabidopsis and a nonureide-type legume black locust. Plant Physiology 134: 1039-1049.
16. Todd CD, Tipton PA, Blevins DG, Piedras P, Pineda M, et al. (2006) Update on ureide degradation in legumes. Journal of Experimental Botany 57: 5-12.
17. Munoz A, Bannenberg GL, Montero O, Cabello DJM, Piedras P, et al. (2011) An alternative pathway for ureide usage in legumes: enzymatic formation of a ureidoglycolate adduct in Cicer arietinum and Phaseolus vulgaris. Journal of Experimental Botany 62: 307-318.
18. Todd CD, Polacco JC (2006) AtAAH encodes a protein with allantoate amidohydrolase activity from Arabidopsis thaliana. Planta 223: 11080-1113.
19. Werner AK, Sparkes IA, Romeis T, Witte CP (2008) Identification, biochemical characterization, and subcellular localization of allantoate amidohydrolases from Arabidopsis and soybean. Plant Physiology 146: 418-430.
20. Werner AK, Romeis T, Witte CP (2010) Ureide catabolism in Arabidopsis thaliana and Escherichia coli. Nature Chemical Biology 6: 19-21.
21. Werner AK, Witte CP (2011) The biochemistry of nitrogen mobilization: purine ring catabolism. Trends in Plant Science 16: 381-317.
22. Serventi F, Ramazzina I, Lamberto I, Puggioni V, Gatti R, et al. (2010) Chemical basis of nitrogen recovery through the ureide pathway: formation and hydrolysis of S-ureidoglycine in plants and bacteria. ACS Chemical Biology 19: 203-214.
23. McClure PR, Israel DW (1979). Transport of nitrogen in the xylem of soybean plants. Plant Physiology 64: 411-416.
24. McClure PR, Israel DW, Volk RJ (1980) Evaluation of the relative ureide content of xylem sap as an indicator of N2 fixation in soybeans: greenhouse studies. Plant Physiology 66: 720-725.
25. Pate JS, Atkins CA, White ST, Rainbird RM, Woo KC (1980) Nitrogen nutrition and xylem transport of nitrogen in ureide-producing grain legumes. Plant Physiology 65: 961-965.
26. Herridge DF (1982) Use of the ureide technique to describe the nitrogen economy of field-grown soybeans. Plant Physiology 70: 7-11.
27. Patterson TG, LaRue TA (1983) Nitrogen fixation (C2H2) and ureide content of soybeans: Ureides as an index for fixation. Crop Science 23: 825-831.
28. Herridge DF, Bergersen FJ, Peoples MB (1990) Measurement of nitrogen fixation by soybean in the field using the ureide and natural 15N abundance methods. Plant Physiology 93: 708-716.
29. Aveline A, Crozat Y, Pinochet X, Domenach AM, Cleyet MJC (1995) Early remobilization: a possible source of error in the ureide assay method for N2 fixation measurement by early maturing soybean. Soil Science and Plant Nutrition 41: 737-751.
30. Schubert KR (1986) Products of biological nitrogen fixation in higher plants: Synthesis, transport and metabolism. Annu Rev Plant Physiol 37: 539-574.
31. Vankessel C, Roskoski JP, Keane K (1988) Ureide production by N2-Fixing and non N2-Fixing leguminous trees. Soil Biol Biochem 20: 891-897.
32. Serraj R (2003) Effects of drought stress on legume symbiotic nitrogen fixation: physiological mechanisms. Indian Journal of Experimental Biology 41: 1136-1141.
33. King CA, Purcell LC (2005) Inhibition of N2 fixation in soybean is associated with elevated ureides and amino acids. Plant Physiology 137: 1389-1396.
34. Yukimoto M, Ishianj A (1981) Phytotoxicities of organophosphorus insecticides to crops.6. Nitrogen contents in soyabean leaves applied with organophosphorus insecticides. Chem Insp Stn Tokyo 21: 50-53.
35. Tu CM (1981) Effects of pesticides on activities of enzymes and microorganisms in a clay soil. J Environ Sci Health B 16: 179-191.
36. Charlson DV, Korth KL, Purcell LC (2009) Allantoate amidohydrolase transcript expression is independent of drought tolerance in soybean. Journal of Experimental Botany 60: 847-851.
37. Yang SS, Valdes LO, Xu WW, Bucciarelli B, Gronwald JW, et al. (2010) Transcript profiling of common bean (Phaseolus vulgaris L.) using the GeneChip Soybean Genome Array: optimizing analysis by masking biased probes. BMC Plant Biology 10: 85.

36

Verbenone Plus Reduces Levels of Tree Mortality Attributed to Mountain Pine Beetle Infestations in Whitebark Pine, a Tree Species of Concern

Christopher J. Fettig[1]*, Beverly M. Bulaon[2], Christopher P. Dabney[1], Christopher J. Hayes[3] and Stephen R. McKelvey[1]

[1]*Pacific Southwest Research Station, USDA Forest Service, Davis, California 95618 USA*
[2]*Forest Health Protection, USDA Forest Service, Sonora, California 95370 USA*
[3]*Forest Health Protection, USDA Forest Service, Missoula, Montana 59802 USA*

Abstract

In western North America, recent outbreaks of the mountain pine beetle, *Dendroctonus ponderosae* Hopkins, have been severe, long-lasting and well-documented. We review previous research that led to the identification of Verbenone Plus, a novel four-component semiochemical blend [acetophenone, (*E*)-2-hexen-1-ol + (*Z*)-2-hexen-1-ol, and (–)-verbenone] that has been demonstrated to inhibit the response of a closely-related bark beetle species, western pine beetle, *D. brevicomis* LeConte, to attractant-baited traps and trees. In this study, we evaluate the efficacy of Verbenone Plus for protecting stands of whitebark pine, *Pinus albicaulis* Engelm., a species of concern being considered for listing as a threatened and endangered species, from mortality attributed to *D. ponderosae* infestations in the central Sierra Nevada, California, USA. The experimental design was completely randomized with two treatments (untreated control, Verbenone Plus) and four replicates (0.4-ha square plots) per treatment. A total of 450 trees were killed by *D. ponderosae*, 377 were *P. albicaulis* and 73 were lodgepole pine, *P. contorta* Dougl. ex Laws. Significantly fewer pines (*P. albicaulis* and *P. contorta*) and *P. albicaulis* (only) were killed by *D. ponderosae* on Verbenone Plus-treated plots compared to the untreated control. On average, there was ~78% reduction in tree mortality attributed to Verbenone Plus. We discuss the implications of these and other results to the development of Verbenone Plus as a semiochemical-based tool for tree protection.

Keywords: Acetophenone; *Dendroctonus ponderosae*; Non-host angiosperm volatiles; Pest management; *Pinus albicaulis; Pinus contorta*

Introduction

Recent outbreaks of mountain pine beetle, *Dendroctonus ponderosae* Hopkins, have been severe, long-lasting and well-documented [1]. For example, since 2001 >25 million ha of forest have been impacted by *D. ponderosae. Dendroctonus ponderosae* ranges throughout British Columbia and Alberta, Canada, most of the western USA, into northern Mexico, and colonizes several pine species, most notably, lodgepole pine, *Pinus contorta* Dougl. ex Loud., ponderosa pine, *P. ponderosa* Dougl. ex Laws., sugar pine, *P. lambertiana* Dougl., limber pine, *P. flexilis* E. James, western white pine, *P. monticola* Dougl. ex D. Don, and whitebark pine, *P. albicaulis* Engelm.[2]. Episodic outbreaks of this notable pest are a common occurrence, but the magnitude and extent of recent outbreaks have exceeded the range of historic variability, and have occurred in areas where *D. ponderosae* outbreaks were once rare (e.g., *P. albicaulis* forests) or previously unrecorded (e.g., jack pine forests, *P. banksiana* Lamb.) [1-4].

Pinus albicaulis is a wide-ranging tree species in western North America that grows at the highest elevations (Figure 1), often in association with other conifers [5]. In the last decade, extensive levels of tree mortality have occurred across much of the range of *P. albicaulis* and have been attributed to climatic changes and elevated populations of *D. ponderosae* [3,6], and white pine blister rust infections caused by a non-native invasive fungi [7]. Scientists speculate that under continued warming, the loss of *P. albicaulis* may be imminent in some areas. To that end, the U.S. Fish and Wildlife Service announced in 2011 that it determined *P. albicaulis* warranted protection under the Endangered Species Act, but that adding the species to the Federal List of Endangered and Threatened Wildlife and Plants was precluded by the need to address other listing actions of higher priority [8]. Accordingly, the U.S. Fish and Wildlife Service has added *P. albicaulis* to the list of candidate species eligible for protection, and will continue to review its status on an annual basis [8].

Pinus albicaulis plays a major ecological role in the functioning of high elevation ecosystems, surviving conditions that are often too cold, too dry and too windy for many other tree species [5]. *Pinus albicaulis* is considered a keystone species in the subalpine environment, stabilizing soils, moderating and regulating runoff, and facilitating the establishment and survival of other species [5,9]. Due to the slow growth and maturation of *P. albicaulis*, and the unique ecological services this species provides, protection of *P. albicaulis* from *D. ponderosae* is desirable, but challenging. Development of environmentally-friendly (e.g., biopesticides) and portable methods of tree protection are needed given the remote and sensitive nature of the subalpine environments where *P. albicaulis* persists.

Verbenone (4,6,6-trimethylbicyclo[3.1.1]hept-3-en-2-one) is an anti aggregation pheromone of *D. ponderosae*, western pine beetle, *D. brevicomis* LeConte, and southern pine beetle, *D. frontalis* Zimmerman [10], and is produced by auto-oxidation of the host monoterpene α-pinene via the intermediary compounds *cis*- and *trans*-verbenol [11], by the beetles themselves [12], and/or through degradation of host material typically by microorganisms associated with bark beetles [13-15]. Because of its behavioral activity, as demonstrated in

***Corresponding author:** Christopher J. Fettig, Ecosystem Function and Health Program, Pacific Southwest Research Station, Davis, California, USA,
E-mail: cfettig@fs.fed.us

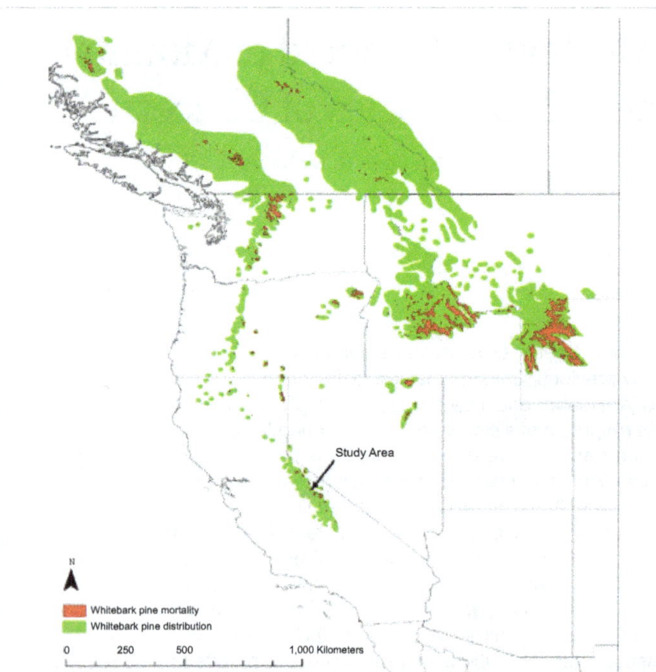

Figure 1: Distribution of *Pinus albicaulis* (green) based on Critchfield and Little [41] with areas of tree mortality (red) based on aerial survey data indicating polygons containing *P. albicaulis* killed by *Dendroctonus ponderosae* during 2007-2011.

numerous trapping bioassays, verbenone has been evaluated as a tool for mitigating coniferous tree mortality due to bark beetle infestations in western North America. Efforts have concentrated on individual tree [e.g., 16-18] or small-scale (e.g., <4 ha) stand protection, primarily from *D. ponderosae* [e.g., 19-22]. Results have been favorable, but inconsistent. Negative results have been linked to photoisomerization of verbenone to behaviorally inactive chrysanthenone [23]; inconsistent or inadequate release [24]; rapid dispersal of verbenone [25]; and/or limitations in the range of inhibition of verbenone [26], particularly when *D. ponderosae* populations were high [19-21]. A lack of efficacy may also be due to the complexity of the host selection process, which involves other visual and olfactory cues produced by hosts, non hosts and competing species [27]. Verbenone was first registered by the U.S. Environmental Protection Agency (licensed for sale and distribution) in December 1999 to control *D. frontalis* in southern forests. Since then, the label has been expanded to include *D. ponderosae* and *D. brevicomis* in forests, recreational and municipal settings, and in rights of way and other easements. Verbenone is generally deployed in individual passive release devices (pouches) by hand application to the tree bole or in bead, flake and sprayable formulations by ground or aerial application [10].

Verbenone has been found ineffective for protecting individual *P. ponderosa* [18,28] and *P. ponderosa* stands [29] from *D. brevicomis* infestations. As a result, based on the semiochemical-diversity hypothesis [30], Shepherd et al. [27] suggested that synthetic verbenone should be deployed with other beetle-derived or non host cues that more accurately reflect the complexity of the olfactory environment in forests. In the context of pest management, a diverse array of chemical cues and signals may disrupt bark beetle searching more than high doses of a single semiochemical (e.g., verbenone) or even mixtures of semiochemicals intended to mimic one type of signal (e.g., antiaggregation pheromones), because they represent heterogeneous stand conditions to foraging insects [27, 30]. Fettig et al. [31] reported that combinations of bark volatiles [benzaldehyde, benzyl alcohol, (*E*)-conophthorin, guaiacol, nonanal, and salicylaldehyde], three green leaf volatiles [(*E*)-2-hexenal, (*E*)-2-hexen-1-ol, and (*E*)-2-hexen-1-ol], or the nine compounds combined did not affect the response of *D. brevicomis* to attractant-baited traps. However, when the bark and green leaf volatiles were combined with (−)-verbenone, they reduced trap catches to levels significantly below that of verbenone alone. A nine-component blend [benzyl alcohol, benzaldehyde, guaiacol, nonanal, salicylaldehyde, (*E*)-2-hexenal, (*E*)-2-hexen-1-ol, (*Z*)-2-hexen-1-ol and (−)-verbenone] reduced trap catch by 87% compared to the attractant-baited control [31]. Based on this work, Fettig et al. [32] were first to demonstrate the successful application of a semiochemical-based tool for protecting *P. ponderosa* from mortality attributed to *D. brevicomis*. Additional research confirmed the effect [33], but initial blends were complex and likely not feasible for operational use.

Fettig et al. [34] further examined the response of *D. brevicomis* to several blends of non host angiosperm volatiles and (−)-verbenone in attractant-baited traps in hopes of improving the efficacy of their 9-component blend, and to reduce the number of components involved. Their research resulted in development of a novel four-component blend [acetophenone, (*E*)-2-hexen-1-ol + (*Z*)-2-hexen-1-ol, and (−)-verbenone; Verbenone Plus] that has been demonstrated to inhibit the response of *D. brevicomis* to attractant-baited traps and trees in several studies [28, 34]. The objective of this study was to determine the effectiveness of Verbenone Plus for protecting *P. albicaulis* from mortality attributed to *D. ponderosae* in California, USA.

Materials and Methods

This study was conducted at June Mountain Ski Area, Inyo National Forest, California, USA (37.75°N, 119.06°W; 3,012-m elevation) (Figure 1), 2010. Site selection was based on reports indicating that *D. ponderosae* infestations were causing substantial levels of tree mortality in this area (B. Bulaon, unpubl. data) and subsequent field visits. The experimental design was completely randomized with two treatments and four replicates (0.4-ha square plots) per treatment. Treatments included: (1) untreated control and (2) Verbenone Plus [acetophenone, (*E*)-2-hexen-1-ol + (*Z*)-2-hexen-1-ol and (−)-verbenone] [34] (Table 1). Plots were located in stands with a mean stand density of 48.7 m^2/ha of which ~65% was *P. albicaulis* with the remainder *P. contorta* (Table 2).

Semiochemicals were hand-applied in a ~10.6 by 10.6 m grid (50 U/plot) to the nearest tree at ~2 m in height on 10 June 2010 and remained throughout the seasonal flight period of *D. ponderosae* [35]. Treatments were removed and plots assessed for *D. ponderosae*

Semiochemical	Source*	Purity (%)	Release device	Release rate (mg/d)**
Acetophenone	Sigma-Aldrich	99	Contech 15 ml poly-ethylene bottle	18.0 (20°C)
(*E*)-2-Hexen-1-ol (*Z*)-2-Hexen-1-ol	Bedoukian	97	Contech pouch (1:1 blend)	50.0 (20°C)
Verbenone [77%-(-)]	Contech	97	7-g Contech pouch	50.0 (20°C)

*Sigma-Aldrich = Sigma-Aldrich Canada Ltd., Oakville, Ontario, Canada; Bedoukian = Bedoukian Research Inc., Danbury, Connecticut, USA; Contech = Contech Enterprises Inc., Delta, British Columbia, Canada.
**Reported by manufacturer of release device and measured in the laboratory at specified temperature.

Table 1: Description of semiochemicals, release devices and release rates used in tree protection studies in *Pinus albicaulis* stands, June Mountain Ski Area (37.75°N, 119.06°W; 3, 012-m elevation), Inyo National Forest, California, USA, 2010.

Plot	Treatment*	% Crown cover	% Slope	Mean dbh ± SE**	Basal area (m²/ha)	% *P. albicaulis****	Trees per ha	% *P. albicaulis*****
1	Untreated control	60	15	29.1 ± 3.0	76.1	79	840	94
2	Verbenone Plus	40	18	21.4 ± 2.7	17.7	100	395	100
3	Untreated control	60	25	23.5 ± 1.5	48.6	95	964	87
4	Untreated control	20	18	25.5 ± 1.6	53.1	63	914	70
5	Verbenone Plus	40	17	27.3 ± 2.4	74.7	41	988	88
6	Verbenone Plus	40	5	21.3 ± 2.1	45.5	49	939	63
7	Untreated control	40	30	24.3 ± 2.2	57.0	40	939	63
8	Verbenone Plus	20	32	21.4 ± 3.1	17.2	52	370	87

*Verbenone Plus [acetophenone, (*E*)-2-hexen-1-ol + (*Z*)-2-hexen-1-ol and (−)-verbenone] applied at 50 U/plot.
**dbh, diameter at breast height (1.37 m) in cm; SE, standard error.
***Based on basal area (cross-sectional area of trees at 1.37 m in height).
****Based on number of trees.

Table 2: Characteristics of experimental 0.4-ha plots at June Mountain Ski Area (37.75°N, 119.06°W; 3, 012-m elevation), Inyo National Forest, California, USA, 2010.

attacks 12–13 October 2010. Analyses were limited to trees successfully mass attacked by *D. ponderosae* during the treatment period. A tree was considered successfully mass attacked, and therefore killed by *D. ponderosae* if boring dust surrounded the root collar, and/or the phloem and sapwood were discolored, the bark separated readily from the sapwood, and adult (parent) galleries and larval mines were visible following bark removal. Tests of normality and equal variance were conducted to confirm data met assumptions of normality and homoscedasticity prior to analysis (SigmaStat Version 12.0, Systat Software Inc., San Jose, California, USA). The mean percentages of trees (*P. albicaulis* and *P. contorta*) and of *P. albicaulis* (only) killed by *D. ponderosae* were compared by *t*-test using α=0.05 (SigmaStat Version 12.0).

Results

Among all plots, a total of 469 trees were attacked by *D. ponderosae*. However, 19 trees exhibited strip attacks (i.e., a partial attack of the tree bole typically insufficient to cause tree mortality), and therefore were excluded from our analyses. Of the 450 trees that were killed by *D. ponderosae*, 377 were *P. albicaulis* and 73 were *P. contorta*. At the plot level, tree mortality ranged from 0 trees (plot 5, Verbenone Plus-treated) to 139 trees (plot 7, untreated control), and from 0% to 36.6% of trees, respectively. In the untreated control, levels of tree mortality exceeded 15% on all plots. For *P. albicaulis*, mortality ranged from 0 trees (plot 5, Verbenone Plus-treated) to 112 trees (plot 7, untreated control), and from 0% to 41.5% of *P. albicaulis*, respectively. A significantly lower percentage of trees (*P. albicaulis* and *P. contorta*) were killed by *D. ponderosae* on Verbenone Plus-treated compared to untreated control plots (t = -4.25, P = 0.005) (Figure 2). The effect was consistent for *P. albicaulis* as significantly fewer *P. albicaulis* died on Verbenone Plus-treated plots (t = -4.04, P = 0.007) (Figure 2). On average, there was ~78% reduction in tree mortality attributed to the application of Verbenone Plus in *P. albicaulis* stands.

Discussion

This paper is the first report on the effectiveness of Verbenone Plus for protecting *P. albicaulis* from mortality attributed to *D. ponderosae*. In 2008, we examined the effect of Verbenone Plus on the response of *D. ponderosae* to attractant-baited traps in *P. contorta* stands in Utah, USA, but the experiment failed to produce meaningful results due to adverse weather conditions that hampered *D. ponderosae* flight. In that experiment, 4.9 ± 1.5 and 0.2 ± 0.1 *D. ponderosae* (mean ± SEM, n = 42) were captured in the control and Verbenone Plus treatments, respectively. Several years of research initially resulted in the development of Verbenone Plus for protecting *P. ponderosa* from mortality attributed to *D. brevicomis* [27-29,31-34], where it serves as the only effective semiochemical-based tool for tree protection in that system [28].

Limited work has occurred regarding the development of semiochemical-based tools to protect *P. albicaulis* from *D. ponderosae* infestation [17,21,36-38]. This is likely due to its limited commercial value [5], and that until recent years levels of tree mortality attributed to *D. ponderosae* in *P. albicaulis* forests were limited throughout much of the geographic range [9] (Figure 1). In the Sierra Nevada, *P. albicaulis* has experienced significant levels of tree mortality (Figure 1), and some previous attempts to protect trees from *D. ponderosae* by application of verbenone have failed (e.g., June Mountain Ski Area in 2009; B. Bulaon, unpubl. data).

Warwell et al. [39] modeled the contemporary climate profiles of *P. albicaulis* and predicted future responses to warming. They reported rapid and large-scale declines in the area occupied by *P. albicaulis*. For example, the contemporary climate profile was predicted to decline by ~70% and move upward in elevation by ~330 m by 2030. By the end of this century, the contemporary climate profile of *P. albicaulis* was projected to decline to an area equivalent to <3% of its current distribution [39]. In 2007, the Whitebark Pine Restoration Program was initiated by the USDA Forest Service with the primary goals of protecting and enhancing *P. albicaulis* populations, providing adequate

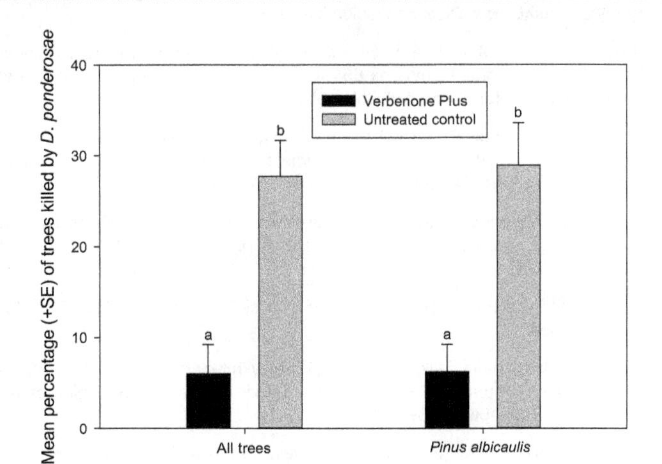

Figure 2: Mean percentage (+ SE) of trees killed by *Dendroctonus ponderosae* on 0.4-ha experimental plots treated with and without Verbenone Plus (acetophenone, (*E*)-2-hexen-1-ol + (*Z*)-2-hexen-1-ol, and (−)-verbenone) at June Mountain Ski Area (37.75°N, 119.06°W; 3,012-m elevation), Inyo National Forest, California, USA, 2010. Means followed by the same letter within groups are not significantly different (*t*-test, *P*>0.05).

regeneration, and increasing the proportion of *P. albicaulis* with natural resistance to white pine blister rust [40]. This should include maintenance and protection of mature, cone-bearing and disease-resistant trees from throughout the geographic range of *P. albicaulis* [7]. Based on our research, we suggest additional work on Verbenone Plus should concentrate on determining optimal release rates and spacing necessary to achieve adequate levels of efficacy in other areas throughout the range of *P. albicaulis* (Figure 1), and comparison of the efficacy of Verbenone to Verbenone Plus within the same *P. albicaulis* stands. Ongoing research (not presented here) indicates that Verbenone Plus is effective for protecting *P. contorta* from mortality attributed to *D. ponderosae* (C. Fettig, unpubl. data), and its efficacy compared to verbenone alone is being evaluated in that system. Such data would be useful in facilitating commercialization of Verbenone Plus (i.e., as the only effective semiochemical-based tool for *D. brevicomis*) given the recent impacts of *D. ponderosae* to forest resources.

Acknowledgements

We thank Z. Heath (Forest Health Protection, USDA Forest Service) for developing Figure 1, and M. Sprague (June Mountain Ski Area) and S. Kusumoto (Inyo National Forest, USDA Forest Service) for their contributions to this project. This work was supported, in-part, by a USDA Forest Service Forest Health Technology Enterprise Team grant (0110) to CF, BM, D. Cluck (Forest Health Protection, USDA Forest Service), SM, and CH, a U.S. President's Early Career Award for Scientists and Engineers to CF, in-kind contributions from Contech Enterprises Inc., the Pacific Southwest Research Station, and Forest Health Protection.

References

1. Bentz BJ, Allen CD, Ayres M, Berg E, Carroll A, et al. (2009) Bark beetle outbreaks in western North America: Causes and consequences. University of Utah Press, Salt Lake City, Utah, USA.
2. Gibson K, Kegley S, Bentz B (2009) Mountain pine beetle. FIDL 2. U.S. Department of Agriculture, Forest Service, Pacific Northwest Region, Portland, Oregon, USA.
3. Bentz BJ, Régnière J, Fettig CJ, Hansen EM, Hayes JL, et al. (2010) Climate change and bark beetles of the western United States and Canada: Direct and indirect effects. Bioscience 60: 602–613.
4. Krist FJ Jr, Sapio FJ, Tkacz BM (2007) Mapping risk from forest insects and diseases. FHTET Report 2007-06. U.S. Department of Agriculture, Forest Service, Washington, District of Columbia, USA.
5. Arno SF, Hoff R (1990) *Pinus albicaulis* Engelm. In: Silvics of North America. Vol. I. Conifers. Agric. Handbook 654. U.S. Department of Agriculture, Forest Service, Washington, District of Columbia, USA.
6. Jewett JT, Lawrence RL, Marshall LA, Gessler PE, Powell SL, et al. (2011) Spatiotemporal relationships between climate and whitebark pine mortality in the Greater Yellowstone Ecosystem. For. Sci. 57: 320–335.
7. Kendall K, Keane RE (2001) The decline of whitebark pine. In: Whitebark pine communities: Ecology and restoration. Island press, Washington, District of Columbia, USA.
8. Federal (U.S.) Register (2011) Proposed rules, Tuesday July 19, 2011. 76: 42631–42654.
9. Logan J, MacFarlane MW, Willcox L (2010) Whitebark pine vulnerability to climate-driven mountain pine beetle disturbance in the Greater Yellowstone Ecosystem. Ecol Appl 20: 895–902.
10. Gillette NE, Munson AS (2009) Semiochemical sabotage: Behavioral chemicals for protection of western conifers from bark beetles. In: The Western Bark Beetle Research. Group: A unique collaboration with Forest Health Protection, Proceedings of a Symposium at the 2007 Society of American Foresters Conference, October 23–28, 2007, Portland, Oregon. PNW-GTR-784. U.S. Department of Agriculture, Forest Service, Pacific Northwest Research Station, Portland, Oregon, USA.
11. Hunt DWA, Borden JH, Lindgren BS, Gries G (1989) The role of autoxidation of α-pinene in the production of pheromones of *Dendroctonus ponderosae* (Coleoptera: Scolytidae). Can J For Res 19: 1275–1282.
12. Byers JA, Wood DL, Craig J, Hendry LB (1984) Attractive and inhibitory pheromones produced in the bark beetle, *Dendroctonus brevicomis*, during host colonization: Regulation of inter- and intraspecific competition. J Chem Ecol 10: 861–877.
13. Leufven AG, Bergstrom G, Falsen E (1984) Interconversion of verbenols and verbenone by identified yeasts isolated from the spruce bark beetle *Ips typographus*. J Chem Ecol 10: 1349–1361.
14. Hunt DWA, Borden JH (1990) Conversion of verbenols to verbenone by yeasts isolated from *Dendroctonus ponderosae* (Coleoptera: Scolytidae). J Chem Ecol 16: 1385–1397.
15. Lindgren BS, Nordlander G, Birgersson G (1996) Feeding deterrence of verbenone to the pine weevil, *Hylobius abietis* (L.) (Col., Curculionidae). J Appl Ent 120: 397–403.
16. Borden JH, Pureswaran DS, Poirier LM (2004) Evaluation of two repellent semiochemicals for disruption of attack by the mountain pine beetle, *Dendroctonus ponderosae* Hopkins (Coleoptera: Scolytidae). J Entomol Soc Brit Columbia 101: 117–123.
17. Kegley S, Sandra J, Gibson K, Kenneth E (2004) Using verbenone to protect whitebark pine trees from mountain pine beetle attack. Department of Agriculture, Forest Service, Northern Region, Missoula, Montana, USA.
18. Gillette NE, Stein JD, Owen DR, Webster JN, Fiddler GO, et al. (2006) Verbenone-releasing flakes protect individual *Pinus contorta* trees from attack by *Dendroctonus ponderosae* and *Dendroctonus valens* (Coleoptera: Curculionidae, Scolytinae). Agric For Entomol 8: 243–251.
19. Progar RA (2003) Verbenone reduces mountain pine beetle attack in lodgepole pine. West J Appl For 18: 229–232.
20. Progar RA (2005) Five-year operational trial of verbenone to deter mountain pine beetle (*Dendroctonus ponderosae*; Coleoptera: Scolytidae) attack of lodgepole pine (*Pinus contorta*). Environ Entomol 34: 1402–1407.
21. Bentz BJ, Kegley S, Gibson K, Thier R (2005) A test of high-dose verbenone for stand-level protection of lodgepole and whitebark pine from mountain pine beetle (Coleoptera: Curculionidae: Scolytinae) attacks. J Econ Entomol. 98: 1614–1621.
22. Gillette NE, Erbilgin N, Webster JN, Pederson L, Mori SR, et al. (2009) Aerially applied verbenone-releasing laminated flakes protect *Pinus contorta* stands from attack by *Dendroctonus ponderosae* in California and Idaho. For Ecol Manage 257: 1405–1412.
23. Kostyk BC, Borden JH, Gries G (1993) Photoisomerization of anti aggregation pheromone verbenone: Biological and practical implications with respect to the mountain pine beetle, *Dendroctonus ponderosae* Hopkins. J Chem Ecol 19: 1749–1759.
24. Bentz BJ, Lister CK, Schmid JM, Mata SA, Rasmussen LA, et al. (1989) Does verbenone reduce mountain pine beetle attacks in susceptible stands of ponderosa pine? RN-RM-495. U.S. Department of Agriculture, Forest Service, Rocky Mountain Research Station, Ogden, Utah, USA.
25. Negron JF, Allen K, McMillin J, Burkwhat H (2006) Testing verbenone for reducing mountain pine beetle attacks in ponderosa pine in the Black Hills, South Dakota. U.S. Department of Agriculture, Forest Service, Rocky Mountain Research Station, Fort Collins, Colorado, USA.
26. Miller DR (2002) Short-range horizontal disruption by verbenone in attraction of mountain pine beetle (Coleoptera: Scolytidae) to pheromone-baited funnel traps in stands of lodgepole pine. J Entomol Soc Brit Columbia 99: 103–105.
27. Shepherd WP, Huber DPW, Seybold SJ, Fettig CJ (2007) Antennal responses of the western pine beetle, *Dendroctonus brevicomis* (Coleoptera: Curculionidae), to stem volatiles of its primary host, *Pinus ponderosa*, and nine sympatric nonhost angiosperms and conifers. Chemoecology 17: 209–221.
28. Fettig CJ, McKelvey SR, Dabney CP, Huber DPW, Lait CG, et al. (2012) Efficacy of Verbenone Plus for protecting ponderosa pine trees and stands from *Dendroctonus brevicomis* (Coleoptera: Curculionidae) attack in British Columbia and California. J Econ Entomol, in press.
29. Fettig CJ, McKelvey SR, Borys RR, Dabney CP, Hamud SM, et al. (2009) Efficacy of verbenone for protecting ponderosa pine stands from western pine

beetle (Coleoptera: Curculionidae, Scolytinae) attack in California. J. Econ. Entomol. 102: 1846–1858.

30. Zhang QH, Schlyter F (2004) Olfactory recognition and behavioural avoidance of angiosperm nonhost volatiles by conifer-inhabiting bark beetles. Agric For Entomol 6: 1–19.

31. Fettig CJ, McKelvey SR, Huber DP (2005) Nonhost angiosperm volatiles and verbenone disrupt response of western pine beetle, *Dendroctonus brevicomis* (Coleoptera: Scolytidae), to attractant-baited traps. J Econ Entomol 98: 2041–2048.

32. Fettig CJ, Dabney CP, McKelvey SR, Huber DPW (2008) Nonhost angiosperm volatiles and verbenone protect individual ponderosa pines from attack by western pine beetle and red turpentine beetle (Coleoptera: Curculionidae, Scolytinae). West J Appl For 81: 6–19.

33. Fettig CJ, McKelvey SR, Dabney CP, Borys RR, Huber DPW (2009) Response of *Dendroctonus brevicomis* to different release rates of nonhost angiosperm volatiles and verbenone in trapping and tree protection studies. J Appl Ent 133: 143–154.

34. Fettig CJ, McKelvey SR, Dabney CP, Huber DPW (2012) Responses of *Dendroctonus brevicomis* (Coleoptera: Curculionidae) in behavioral assays: Implications to development of a semiochemical-based tool for tree protection. J Econ Entomol 105: 149–160.

35. Fettig CJ, Shea PJ, Borys RR (2005) Spatial and temporal distributions of four bark beetle species (Coleoptera: Scolytidae) along two elevational transects in the Sierra Nevada. Pan Pacific Entomol 81: 6–19.

36. Kegley S, Gibson K (2009) Individual-tree tests of verbenone and green-leaf volatiles to protect lodgepole, whitebark and ponderosa pines, 2004-2007. Forest Health Protection Report 09-03. U.S. Department of Agriculture, Forest Service, Northern Region, Missoula, Montana, USA.

37. Gillette NE, Hansen EM, Mehmel CJ, Mori SR, Webster JN, et al. (2012) Area-wide application of verbenone-releasing flakes reduces mortality of whitebark pine (*Pinus albicaulis*) caused by mountain pine beetle *Dendroctonus ponderosae*. Agric For Entomol, in press.

38. Perkins DL, Jorgensen CL, Rinella M (2011) Protecting whitebark pines through a mountain pine beetle epidemic with verbenone—is it working? In: The future of high-elevation, five needle white pines in western North America: Proceedings of the High Five Symposium, 28-30 June 2010, Missoula, Montana. RMRS-P-63. U.S. Department of Agriculture, Forest Service, Rocky Mountain Research Station, Fort Collins, Colorado, USA.

39. Warwell MV, Rehfeldt GE, Crookston N (2006) Modeling contemporary climate profiles of whitebark pine (*Pinus albicaulis*) and predicting responses to global warming. In: Proceedings of the Conference on Whitebark Pine: A Pacific Coast perspective. R6-NR-FHP-2007-01. U.S. Department of Agriculture, Forest Service, Pacific Northwest Region, Ashland, Oregon, USA.

40. Schwandt J (2011) Highlights of the Forest Health Protection Whitebark Pine Restoration Program. In: The future of high-elevation, five needle white pines in western North America: Proceedings of the High Five Symposium, 28-30 June 2010, Missoula, Montana. RMRS-P-63. U.S. Department of Agriculture, Forest Service, Rocky Mountain Research Station, Fort Collins, Colorado, USA.

41. Critchfield WB, Little EL (1966) Geographic distribution of the pines of the world. Misc Publ 991.

PERMISSIONS

All chapters in this book were first published in JASFR, by OMICS International; hereby published with permission under the Creative Commons Attribution License or equivalent. Every chapter published in this book has been scrutinized by our experts. Their significance has been extensively debated. The topics covered herein carry significant findings which will fuel the growth of the discipline. They may even be implemented as practical applications or may be referred to as a beginning point for another development.

The contributors of this book come from diverse backgrounds, making this book a truly international effort. This book will bring forth new frontiers with its revolutionizing research information and detailed analysis of the nascent developments around the world.

We would like to thank all the contributing authors for lending their expertise to make the book truly unique. They have played a crucial role in the development of this book. Without their invaluable contributions this book wouldn't have been possible. They have made vital efforts to compile up to date information on the varied aspects of this subject to make this book a valuable addition to the collection of many professionals and students.

This book was conceptualized with the vision of imparting up-to-date information and advanced data in this field. To ensure the same, a matchless editorial board was set up. Every individual on the board went through rigorous rounds of assessment to prove their worth. After which they invested a large part of their time researching and compiling the most relevant data for our readers.

The editorial board has been involved in producing this book since its inception. They have spent rigorous hours researching and exploring the diverse topics which have resulted in the successful publishing of this book. They have passed on their knowledge of decades through this book. To expedite this challenging task, the publisher supported the team at every step. A small team of assistant editors was also appointed to further simplify the editing procedure and attain best results for the readers.

Apart from the editorial board, the designing team has also invested a significant amount of their time in understanding the subject and creating the most relevant covers. They scrutinized every image to scout for the most suitable representation of the subject and create an appropriate cover for the book.

The publishing team has been an ardent support to the editorial, designing and production team. Their endless efforts to recruit the best for this project, has resulted in the accomplishment of this book. They are a veteran in the field of academics and their pool of knowledge is as vast as their experience in printing. Their expertise and guidance has proved useful at every step. Their uncompromising quality standards have made this book an exceptional effort. Their encouragement from time to time has been an inspiration for everyone.

The publisher and the editorial board hope that this book will prove to be a valuable piece of knowledge for researchers, students, practitioners and scholars across the globe.

LIST OF CONTRIBUTORS

Baffi C and Trevisan M
Centro di Ricerca BIOMASS, Facoltà di Scienze Agrarie, Alimentari e Ambientali, Università Cattolica del Sacro Cuore, Piacenza, Italy

Cella F and Fumi I
Syngen, Agrosistemi, Piacenza, Italy

Nino Paul and Stephan Hafele
International Rice Research Institute, Los Baños, Laguna, Philippines

Pompe C Cruz and Edna A Aguilar
Crop Science Cluster, College of Agriculture, University of the Philippines at Los Baños, Laguna, Philippines

Rodrigo B Badayos
Agricultural Systems Cluster, College of Agriculture (CA), University of the Philippines at Los Baños, Laguna, Philippines

Wherley B, Baumann P, Senseman S and White R
Department of Soil and Crop Sciences, Texas A&M University, College Station, TX 77843, USA

Smith J
Department of Soil and Crop Sciences, Texas A&M University, College Station, TX 77843, USA
The Scotts Miracle-Gro Company, Marysville, OH 43041, USA

Falk S
The Scotts Miracle-Gro Company, Marysville, OH 43041, USA

Birhanu Messele and L.M. Pant
Department of Agro-ecology, Menschen fur Menschen, Agro-Technical and Technology College, P. O. Box 322, Harar, Ethiopia

Yao Adjrah
Laboratoire de Microbiologie et de Contrôle de qualité des Denrées Alimentaires, Ecole Supérieure des Techniques Biologiques et Alimentaires (ESTBA), Université de Lomé BP 12281, Lomé - Togo

Bakouma Laba, Koffi Koba, Komlan Sanda and Amen Y Nenonéné
Unité de recherche sur les Agroressources et la Santé Environnementale, Ecole Supérieure d'Agronomie, Université de Lomé, BP 20131, Lomé - Togo

Wiyao Poutouli
Laboratoire de Biologie Animale et de Zoologie, Faculté des Sciences, Université de Lomé, BP 1515, Lomé - Togo

Agarwal Ritu, Choudhary Anjali, Tripathi Nidhi and Bharti Deepak
Assistant Professor, Department of Biotechnology and Biochemistry, Career College, Bhopal 462 023, M.P, India

Patil Sheetal
Trainee student, Department of Biotechnology and Biochemistry, Career College, Bhopal, 462 023, M.P, India

Nigussie Lulie and Nagappan Raja
Department of Biology, Faculty of Natural and Computational Sciences, Post Box-196, University of Gondar, Gondar, Ethiopia

Tadele Shiberu and Mulugeta Negeri
Department of Plant Sciences, College of Agriculture and Veterinary Sciences, Ambo University, Ethiopia

G. Pandiarajan and B. Makesh Kumar
Department of Plant Biology and Plant Biotechnology, G.Venkataswamy Naidu College, Kovilpatti, Tamilnadu, India

N. Tenzing Balaiah
Department of Plant Biology and Plant Biotechnology, Ayya Nadar Janaki Ammal College Sivakasi, Tamilnadu, India

A. Gandhi and U. Sivagama Sundari
Department of Botany, Annamalai University, Annamalai Nagar-608 002, Tamil Nadu, India

Attia Batool Abbasi
Department of Entomology, University of Sargodha, Sargodha, Pakistan

Azhar Abbas Khan, Rehana Bibi, Muhammad Shahid Iqbal and Javairia Sherani
College of Agriculture D.G. Khan, subcampus University of Agriculture, Faisalabad, Pakistan

Arif Muhammad Khan
National Institute for Biotechnology and Genetic Engineering, Faisalabad, Pakistan

Gopal Das
Department of Entomology, Bangladesh Agricultural University, Mymensingh-2202, Bangladesh

Alemu Assefa
Fitche Soil Research Center, Oromia Agricultural Research Institute, Ethiopia

Tamado Tana and Jemal Abdulahi
Department of Plant Science, Haramaya University, Ethiopia

Mervat H Hussein and Noha I Badr El Din
Botany Department, Faculty of Science, Mansoura University, Mansoura, Egypt

Ali M Abdullah and El Sayed I Mishaqa
Reference Laboratory for Drinking Water, Holding Company for Water and Wastewater, Cairo, Egypt

Jhala YK, Shelat HN and Panpatte DG
Department of Agricultural Microbiology, Anand Agricultural University, Anand-388 110, Gujarat, India

Kefale Wagaw
Faculty of Chemical and Food Engineering, Bahir Dar University, Bahir Dar, Ethiopia

Dereje Haile
Biology Department, Hawassa University, Hawassa, Ethiopia

Firew Mekbib
Department of Plant Science, Haramaya University, Dire Dawa, Ethiopia

Fassil Assefa
Department of Microbiology, Addis Abeba University, Addis Abeba, Ethiopia

Hazwani Mohd Zaini and Normala Halimoon
Department of Environmental Sciences, Faculty of Environmental Studies, Universiti Putra Malaysia, 43400 UPM Serdang, Selangor, Malaysia

André Mancebo Mazzetto and Carlos Eduardo Pellegrino Cerri
Escola Superior de Agricultura Luiz de Queiroz, Universidade de São Paulo, Avenida Pádua Dias, 11, 13400-000, Piracicaba, São Paulo, Brazil

Arlete Simões Barneze, Brigitte Josefine Feigl and Carlos Clemente Cerri
Centro de Energia Nuclear na Agricultura, Universidade de São Paulo, Avenida Centenário, 303, 13400-970, Piracicaba, São Paulo, Brazil

Mervet H Hussein
Botany Department, Faculty of Science, Mansoura University, Mansoura, Egypt

Ali M Abdullah and Noha I Badr El-Din
Holding Water Company for Water and Wastewater, Cairo, Egypt

Eladl G Eladal
Botany Department, Faculty of Science, Mansoura University, Mansoura, Egypt

Gebrekiros Gebremedhin
Mekelle Soil Research Center, Mekelle, Ethiopia

Sofonyas Dargie, Efriem Tariku and Meresa Wslassie
Abergelle Agricultural Research Center, Abi Adi, Ethiopia

Quijano-Vicente G
Laboratorio de Biotecnología, SPR de RI. Bustamante-Parra y Asociados, Km 8 Carretera a Riito. San Luis Rio Colorado, México 83430

Ortiz-Uribe N
Departamento de Posgrado, Universidad Estatal de Sonora, Unidad Académica San Luis Rio Colorado, Carretera Sonoyta Km 6.5, San Luis Río Colorado, México 83450

Montoya-Gonzalez AH
Laboratorio de Biotecnología, SPR de RI. Bustamante-Parra y Asociados, Km 8 Carretera a Riito. San Luis Rio Colorado, México 83430
Departamento de Posgrado, Universidad Estatal de Sonora, Unidad Académica San Luis Rio Colorado, Carretera Sonoyta Km 6.5, San Luis Río Colorado, México 83450

Morales-Maza A
Departamento de Cultivos Protegidos, Instituto Nacional de Investigaciones Forestales, Agrícolas y Pecuarias. Campo Experimental Valle de Mexicali, Baja California. México, Carretera a San Felipe Km. 7.5, Colorado Dos, Mexicali, México 21700

Hernandez-Martinez R
Departamento de Microbiología, Centro de Investigación Científica y de Educación Superior de Ensenada Baja California (CICESE), Carretera Ensenada-Tijuana 3918, Ensenada, México 22860

Tamiru A
International Centre of Insect Physiology and Ecology, Ethiopia

Bayih T
Hawassa University, PO Box 05, Hawassa, Ethiopia
Haramaya University, PO Box 138, Dire Dawa, Ethiopia

Chimdessa M
Haramaya University, PO Box 138, Dire Dawa, Ethiopia

Brhane H
Mekelle Soil Research Center, Tigray Agricultural Research Institute, Mekelle, Ethiopia

Mamo T
Agricultural Commercialization Cluster (ACC), Initiative and Ethiopian Soil Information System (EthioSIS), Agricultural Transformation Agency, Addis Ababa, Ethiopia

Teka K
Department of Land Resources Management and Environmental Protection, Mekelle University, Mekelle, Ethiopia

Lakshmanan Kasi Elumalai and Ramasamy Rengasamy
Center for Advanced Studies in Botany, University of Madras, Guindy Campus, Chennai - 600025

Birhanu Gizaw, Zerihun Tsegay, Genene Tefera, Endegena Aynalem, Misganaw Wassie and Endeshaw Abatneh
Microbial Biodiversity Directorate, Ethiopian Biodiversity Institute, Addis Ababa, Ethiopia

Babak S Pakdaman
Department of Phytomedicine, Ramin Agricultural and Natural Resources University, Ahwaz, Iran

Ebrahim Mohammadi Goltapeh and Mohsen Nadepoor
Department of Plant Pathology, Agricultural Faculty, Tarbiat Modares University, Tehran, Iran

Bahram Mohammad Soltani
Department of Biological Sciences, Department of Genetics, Tarbiat Modares University, Tehran, Iran

Ali Asghar Talebi
Department of Agricultural Entomology, Agricultural Faculty, Tarbiat Modares University, Tehran, Iran

Joanna S Kruszewska and Sebastian Piłsyk
Laboratory of Fungal Glycobiology, Division of Genetics, Institute of Biochemistry and Biophysics, Warsaw, Poland

Sabrina Sarrocco and Giovanni Vannacci
Department of Tree Science, Entomology and Plant Pathology "Giovanni Scaramuzzi", Faculty of Agriculture, University of Pisa, Pisa, Italy

Francesconi W
International Center for Tropical Agriculture (CIAT), USA

Williams CO
USDA-NRCS National Soil Survey Center, Lincoln, NE, USA

Smith DR
USDA-ARS Grassland, Soil and Water Research Laboratory, USA

Williams JR and Jeong J
Blackland Research an Extension Center, Texas A&M University System, Temple, TX, USA

Natalia TG
CAPES Foundation, Ministry of Education of Brazil, Brazil

Robert MH
Sustainable Futures Institute, Michigan Technological University, Houghton, MI, USA

Ramy Hamouda, Adel Bahnasawy and Samir Ali
Agricultural Engineering Department-Faculty of Agriculture-Benha University 13736, Egypt

El-Shahat Ramadan
Microbiology Department-Faculty of Agriculture-Ain Shams University, Egypt

Karthick Raja Namasivayam, R S Arvind Bharani and Moinuddin Raqib Ansari
Department of Biotechnology, Sathyabama University, Chennai, Tamil Nadu, India

Hatem Mohamed Al-shannaf and Hala Mohamed Mead
Plant Protection Research Institute, Agriculture Research Centre, Dokki, Giza, Egypt

Al-Kazafy Hassan Sabry
Pests and Plant Protection Dep. National Research Centre, Dokki, Giza, Egypt

Azhar Abbas Khan and Muhammad Afzal
Department of Entomology, University of Sargodha, Sargodha, Pakistan

Camila Vieira da Silva
Faculty of Apucarana (FAP), Biological Sciences Department, Apucarana PR, Brazil

Larissa Carla Lauer Schneider
Veterinary Medicine Department, University of Maringa (UEM), Estrada da Paca Bairro São Cristóvão, Umuarama PR, Brazil

Hélio Conte
Biology Cell and Genetics Department, CCB, University of Maringa (UEM), Avenida Colombo, Maringá PR, Brazil

Yuvraj D Kengar and Bhimarao J Patil
Department of Botany, Kanya Mahavidyalaya Islampur College, Karad, Maharashtra, India

Christopher J. Fettig, Stephen R. McKelvey and Christopher P. Dabney
Pacific Southwest Research Station, USDA Forest Service, Davis, California 95618 USA

Beverly M. Bulaon
Forest Health Protection, USDA Forest Service, Sonora, California 95370 USA

Christopher J. Hayes
Forest Health Protection, USDA Forest Service, Missoula, Montana 59802 USA

Index